SUSTAINABLE PLANT NUTRITION

SUSTAINABLE PLANT NUTRITION

MOLECULAR INTERVENTIONS AND ADVANCEMENTS FOR CROP IMPROVEMENT

Edited by

TARIQ AFTAB

Department of Botany, Aligarh Muslim University, Aligarh, India

KHALID REHMAN HAKEEM

Department of Biological Sciences, Faculty of Science, King Abdulaziz University, Jeddah, Saudi Arabia

ELSEVIER

ACADEMIC PRESS

An imprint of Elsevier

Academic Press is an imprint of Elsevier
125 London Wall, London EC2Y 5AS, United Kingdom
525 B Street, Suite 1650, San Diego, CA 92101, United States
50 Hampshire Street, 5th Floor, Cambridge, MA 02139, United States
The Boulevard, Langford Lane, Kidlington, Oxford OX5 1GB, United Kingdom

Notices
Knowledge and best practice in this field are constantly changing. As new research and experience broaden our understanding, changes in research methods, professional practices, or medical treatment may become necessary.

Practitioners and researchers must always rely on their own experience and knowledge in evaluating and using any information, methods, compounds, or experiments described herein. In using such information or methods they should be mindful of their own safety and the safety of others, including parties for whom they have a professional responsibility.

To the fullest extent of the law, neither the Publisher nor the authors, contributors, or editors, assume any liability for any injury and/or damage to persons or property as a matter of products liability, negligence or otherwise, or from any use or operation of any methods, products, instructions, or ideas contained in the material herein.

ISBN: 978-0-443-18675-2

For information on all Academic Press publications visit our website at https://www.elsevier.com/books-and-journals

Publisher: Nikki P. Levy
Acquisitions Editor: Nancy J. Maragioglio
Editorial Project Manager: Kathrine Esten
Production Project Manager: Kumar Anbazhagan
Cover Designer: Christian J. Bilbow

Typeset by TNQ Technologies

Contents

List of contributors

Qamar Abbas
Department of Plant Science, Karakoram International University, Gilgit, Pakistan

Misbah Abbas
Institute of Soil and Environmental Sciences, University of Agriculture Faisalabad, Faisalabad, Punjab, Pakistan

Yawar Abbas
Department of Environmental Sciences, Karakoram International University, Gilgit, Pakistan

Ghulam Hassan Abbasi
Institute of Agro-Industry and Environment, Faculty of Agriculture and Environment, The Islamia University of Bahawalpur, Bahawalpur, Pakistan

Sneha Adhikari
Genetics and Plant Breeding, ICAR-Indian Institute of Wheat and Barley Research, Shimla, India

Tariq Aftab
Department of Botany, Aligarh Muslim University, Aligarh, Uttar Pradesh, India

Namrah Ahmad
Department of Plant Biotechnology, Atta-ur-Rehaman School of Applied Biosciences, National University of Sciences and Technology, Islamabad, Pakistan

Rubaz Ahmed
Institute of Soil and Environmental Sciences, University of Agriculture, Faisalabad, Pakistan

Muhammad Ahsan
Department of Horticultural Sciences, The Islamia University of Bahawalpur, Bahawalpur, Pakistan

Qurat ul Ain
Institute of Environment and Sustainable Development in Agriculture, Chinese Academy of Agricultural Sciences, Beijing, China; Division of Soil and Environmental Biotechnology, National Institute for Biotechnology and Genetic Engineering (NIBGE), Faisalabad, Pakistan

Javaid Akhtar
Institute of Soil and Environmental Sciences, University of Agriculture, Faisalabad, Pakistan

Muhammad Ali
Institute of Agro-Industry and Environment, Faculty of Agriculture and Environment, The Islamia University of Bahawalpur, Bahawalpur, Pakistan

Haibat Ali
Department of Environmental Sciences, Karakoram International University, Gilgit, Pakistan

Komal Shoukat Ali
Department of Botany, Government College Women University Sialkot, Sialkot, Pakistan

Bader Alhafi Alotaibi
Department of Agricultural Extension and Rural Society, King Saud University, Riyadh, Saudi Arabia

Muhammad Ameen
Department of Soil Science, Faculty of Agriculture and Environment, The Islamia University of Bahawalpur, Bahawalpur, Pakistan

Hirdayesh Anuragi
Genetics and Plant Breeding, ICAR-Central Agroforestry Research Institute, Jhansi, Uttar Pradesh, India

Muhammad Anwar-Ul-Haq
Institute of Soil and Environmental Sciences, University of Agriculture, Faisalabad, Pakistan

Behnam Asgari Lajayer
Department of Soil Science, Faculty of Agriculture, University of Tabriz, Tabriz, Iran

Muhammad Nadeem Ashraf
Institute of Soil and Environmental Science, University of Agriculture, Faisalabad, Pakistan

Tess Astatkie
Faculty of Agriculture, Dalhousie University, Truro, Canada

Tabinda Athar
Institute of Soil and Environmental Sciences, University of Agriculture, Faisalabad, Pakistan

Muhammad Ashar Ayub
Institute of Soil and Environmental Sciences, University of Agriculture Faisalabad, Faisalabad, Punjab, Pakistan; Horticultural Sciences Department, University of Florida, Institute of Food and Agricultural Sciences, Indian River Research and Education Centrer, Fort Pierce, FL, United States; Institute of Agro-Industry and Environment, The Islamia University of Bahawalpur, Bahawalpur, Punjab, Pakistan

Fouzieh Beigmohammadi
Department of Environment, Faculty of Natural Resource and Environment, Malayer University, Malayer, Iran

Huzaima Bibi
Department of Biological Sciences, International Islamic University, Islamabad, Pakistan

Kailash Chandra
Department of Plant Breeding and Genetics, College of Agriculture (SKN Agriculture University) Fatehpur-Shekhawati, Sikar, Rajasthan, India

Zhicong Dai
Institute of Environment and Ecology, School of The Environment and Safety Engineering, Jiangsu University, Zhenjiang, China; Jiangsu Collaborative Innovation Center of Technology and Material of Water Treatment, Suzhou University of Science and Technology Suzhou, Suzhou, China

Zahra Dehghanian
Department of Biotechnology, Faculty of Agriculture, Azarbaijan Shahid Madani University, Tabriz, Iran

Shish Ram Dhaka
Department of Entomology, College of Agriculture (SKN Agriculture University) Fatehpur-Shekhawati, Sikar, Rajasthan, India

Daolin Du
Institute of Environment and Ecology, School of The Environment and Safety Engineering, Jiangsu University, Zhenjiang, China

Mohammad Reza Gohari Kia
Department of Plant Breeding and Biotechnology, Faculty of Agriculture, University of Tabriz, Tabriz, Iran

Vahideh Gougerdchi
Department of Plant Breeding and Biotechnology, Faculty of Agriculture, University of Tabriz, Tabriz, Iran

Shahnawaz Hassan
Department of Environmental Sciences, University of Kashmir, Srinagar, India

Naveed Hayat
Department of Economics, Division of Management and Administrative Science, University of Education, Lahore, Pakistan

Kiran Hingonia
Department of Agronomy, College of Agriculture, Sumerpur (Agriculture University, Jodhpur) Rajasthan, India

Humaira
Centre of Research for Development, University of Kashmir, Srinagar, India

Babar Hussain
Department of Plant Science, Karakoram International University, Gilgit, Pakistan

Hafiz Athar Hussain
Institute of Environment and Sustainable Development in Agriculture, Chinese Academy of Agricultural Sciences, Beijing, China; Division of Soil and Environmental Biotechnology, National Institute for Biotechnology and Genetic Engineering (NIBGE), Faisalabad, Pakistan

Saddam Hussain
Department of Agronomy, University of Agriculture, Faisalabad, Pakistan

Asma Imran
Division of Soil and Environmental Biotechnology, National Institute for Biotechnology and Genetic Engineering (NIBGE), Faisalabad, Pakistan

Urfi Jahan
Plant Ecology and Environment Laboratory, Department of Botany, Aligarh Muslim University, Aligarh, Uttar Pradesh, India

Asad Jamil
Institute of Soil and Environmental Sciences, University of Agriculture, Faisalabad, Pakistan

Moazzam Jamil
Department of Soil Science, Faculty of Agriculture and Environment, The Islamia University of Bahawalpur, Bahawalpur, Pakistan

Muhammad Jafar Jaskani
Faculty of Agriculture, Institute of Horticultural Sciences, University of Agriculture, Faisalabad, Pakistan

Hanuman Singh Jatav
Department of Soil Science and Agriculture Chemistry, College of Agriculture (SKN Agriculture University) Fatehpur-Shekhawati, Sikar, Rajasthan, India

Khadija Javeed
Department of Plant Pathology, Agriculture College, Guizhou University, Guiyang, China; Faculty of Mountain Agriculture and Environmental Sciences, Kohsar University Murree, Pakistan

Jovel Varghese Jose
Amity Institute of Biotechnology, Amity University Mumbai, Mumbai, India; Blue Marble Space Institute of Science, Washington, United States

Uzma Kafeel
Plant Ecology and Environment Laboratory, Department of Botany, Aligarh Muslim University, Aligarh, Uttar Pradesh, India

Khair-Ul-Nisa
Department of Environmental Sciences, University of Kashmir, Srinagar, India

Fareed Ahmad Khan
Plant Ecology and Environment Laboratory, Department of Botany, Aligarh Muslim University, Aligarh, Uttar Pradesh, India

Sher Wali Khan
Department of Plant Science, Karakoram International University, Gilgit, Pakistan

Auqib Manzoor
Department of Biochemistry, University of Kashmir, Srinagar, India

Naveed Ul Mushtaq
Department of Bioresources, School of Biological Sciences, University of Kashmir, Srinagar, India

Ghulam Mustafa
Department of Economics, Division of Management and Administrative Science, University of Education, Lahore, Pakistan

Masarat Nabi
Department of Environmental Science, University of Kashmir, Srinagar, Jammu and Kashmir, India

Farheen Naz
Department of Biosciences, Jamia Millia Islamia, New Delhi, India

Misbah Naz
Institute of Environment and Ecology, School of The Environment and Safety Engineering, Jiangsu University, Zhenjiang, China

Qadrul Nisa
Division of Plant Pathology, SKUAST-K, Srinagar, India

Sohail Ahmed Rajper
Department of Biotechnology, Sindh Agricultural University, Tandojam, Pakistan

Ayesha Rasheed
Center of Agricultural Biochemistry and Biotechnology, University of Agriculture, Faisalabad, Pakistan

Aadil Rasool
Department of Bioresources, School of Biological Sciences, University of Kashmir, Srinagar, India

Muhammad Ammar Raza
School of Food Science and Biotechnology, Key Laboratory of Fruits and Vegetables Postharvest and Processing Technology Research, Zhejiang Gongshang University, Hangzhou, China

Reiaz Ul Rehman
Department of Bioresources, School of Biological Sciences, University of Kashmir, Srinagar, India

Luqman Riaz
College of Life Sciences, Henan Normal University, Xinxiang, China

Ali Rizwan
Institute of Soil and Environmental Sciences, University of Agriculture, Faisalabad, Pakistan

Seerat Saleem
Department of Bioresources, School of Biological Sciences, University of Kashmir, Srinagar, India

Wasifa Hafiz Shah
Department of Bioresources, School of Biological Sciences, University of Kashmir, Srinagar, India

Eisa Solgi
Department of Environment, Faculty of Natural Resource and Environment, Malayer University, Malayer, Iran

Zahara Sultan
Department of Botany, University of Kashmir, Srinagar, India

Inayatullah Tahir
Department of Botany, School of Biological Sciences, University of Kashmir, Srinagar, India

Najeebul Tarfeen
Centre of Research for Development (CORD), University of Kashmir, Srinagar, India

Muhammad Tariq
Department of Pharmacology, Lahore Pharmacy Collage, Lahore, Pakistan

Shilpa Hanmantrao Tarte
Department of Plant Biotechnology MGM College of Agricultural Biotechnology, Aurangabad, Maharashtra, India

Muhammad Umair
Institute of Soil and Environmental Sciences, University of Agriculture, Faisalabad, Pakistan

Muhammad Jawad Umer
State Key Laboratory of Cotton Biology/Institute of Cotton Research, Chinese Academy of Agricultural Science (ICR, CAAS), Anyang, China

Shafeeq ur-Rahman
School of Environment and Civil Engineering, Dongguan University of Technology, Dongguan, Guangdong, China; MOE Laboratory for Earth Surface Processes, College of Urban and Environmental Sciences, Peking University, Beijing, China

Amir Hossein Vafaei
Department of Microbial Biotechnology, Faculty of Basic Sciences, Higher Education Institute of Mizan, Tabriz, Iran

Eric D. van Hullebusch
Université Paris Cité, Institut de Physique du Globe de Paris, CNRS, Paris, France

Saba Wani
Department of Biochemistry, University of Kashmir, Srinagar, India

Mohsin Zafar
Department of Environmental Sciences, Mirpur University of Science and Technology, Mirpur-Azad Jammu and Kashmir, Pakistan

Sania Zaib
Department of Biochemistry, Biological Sciences, Quaid-i-Azam University, Islamabad, Pakistan

Qingwen Zhang
Institute of Environment and Sustainable Development in Agriculture, Chinese Academy of Agricultural Sciences, Beijing, China

Muhammad Zia ur Rehman
Institute of Soil and Environmental Sciences, University of Agriculture Faisalabad, Faisalabad, Punjab, Pakistan

About the editors

Dr. Tariq Aftab

Tariq Aftab received his PhD in the Department of Botany at Aligarh Muslim University, India, and is currently an Assistant Professor there. He is the recipient of a prestigious Leibniz-DAAD Fellowship from Germany, Raman Fellowship from the Government of India, and Young Scientist Awards from the State Government of Uttar Pradesh (India) and Government of India. After completing his doctorate, he has worked as Research Fellow at the National Bureau of Plant Genetic Resources, New Delhi, and as Postdoctorate Fellow at Jamia Hamdard, New Delhi, India. Dr. Aftab also worked as Visiting Scientist at Leibniz Institute of Plant Genetics and Crop Plant Research (IPK), Gatersleben, Germany, and in the Department of Plant Biology, Michigan State University, USA. He is a member of various scientific associations from India and abroad.

He has edited 15 books with international publishers, including Elsevier Inc., Springer Nature, and CRC Press (Taylor & Francis Group), coauthored several book chapters, and published over 80 research papers in peer-reviewed international journals. His research interests include physiological, proteomic, and molecular studies on medicinal and crop plants.

Dr. Khalid Rehman Hakeem

Khalid Rehman Hakeem, PhD, is a Professor at King Abdulaziz University, Jeddah, Saudi Arabia. After completing his doctorate (Botany; specialization in Plant Eco-physiology and Molecular Biology) from Jamia Hamdard, New Delhi, India, in 2011, he worked as a lecturer at the University of Kashmir, Srinagar, for a short period. Later, he joined Universiti Putra Malaysia, Selangor, Malaysia, and worked there as Postdoctorate Fellow in 2012 and Fellow Researcher (Associate Professor) from 2013 to 2016. Dr. Hakeem has more than 12 years of teaching and research experience in plant eco-physiology, biotechnology and molecular biology, medicinal plant research, plant—microbe—soil interactions, as well as in environmental studies. He is the recipient of several fellowships at both national and international levels; also, he has served as the Visiting Scientist at Jinan University, Guangzhou, China. Currently, he is involved with a number of international research projects with different government organizations.

So far, Dr. Hakeem has authored and edited more than 80 books with international publishers, including Springer Nature, Academic Press (Elsevier), and CRC Press. He also has to his credit more than 160 research publications in peer-reviewed international journals and 65 book chapters in edited volumes with international publishers.

At present, Dr. Hakeem serves as an editorial board member and reviewer of several high-impact international scientific journals from Elsevier, Springer Nature, Taylor and Francis, Cambridge, and John Wiley Publishers. He is included in the advisory board of Cambridge Scholars Publishing, UK. He is also a fellow of Plantae group of the American Society of Plant Biologists, member of the World Academy of Sciences, member of the International Society for Development and Sustainability, Japan, and member of Asian Federation of Biotechnology, Korea. Dr. Hakeem has been listed in Marquis *Who's Who in the World*, since 2014—2019. Currently, Dr. Hakeem is engaged in studying the plant processes at eco-physiological as well as molecular levels.

CHAPTER ONE

Physiological and molecular aspects of macronutrient uptake by higher plants

Jovel Varghese Jose[a,b]
[a]Amity Institute of Biotechnology, Amity University Mumbai, Mumbai, India
[b]Blue Marble Space Institute of Science, Washington, United States

1. Introduction

Macronutrients such as nitrogen (N), phosphorus (P), potassium (K), calcium (Ca), magnesium (Mg), and sulfur (S) are required for the survival of all living organisms and are particularly important for plant absorption in order to optimize plant growth. Soil nutrient accessibility varies greatly depending on soil type and nutrient solubility. Plants need 17 different elements for optimal growth, 14 of which are absorbed from the soil (Tilman et al., 2002). The soil derived nutrients are also classified as macro- and micronutrients, depending on the amount required. The molecular processes underlying crop responses to fertilizer depletion have been studied for decades in an effort to reduce crop nutrient provisioning. On the other hand, increasing the concentration of one or more nutrients may serve as a constraint on plant development owing to other variables such as insufficient diffusion and rhizosphere microbial activity in soil. The main constraints contributing to low agricultural yields include the prevalence of marginal soils, insufficient fertilizer efficiency, limited budgetary resources, and a lack of a suitable framework for agronomic inputs in many countries. Understanding the genetics of plant nutrition is advancing rapidly, with the potential to discover and control key regulatory minerals involved in nutrient acquisition, transport, and absorption, as well as root system design. The efficient use of nutrients improves food safety and decreases the amount of nutrients released into the environment. A proportionate amount of nourishment is required, with N being particularly important in producing crops that will proliferate the globe forthwith and in the forthcoming. As a result, the use of extra N- capable crops has the potential to play a critical

Sustainable Plant Nutrition
ISBN: 978-0-443-18675-2
https://doi.org/10.1016/B978-0-443-18675-2.00010-9

role in increasing or maintaining crop yields in the future, especially after a weather change (Spiertz & Ewert, 2009). Inquiring the physiology and genetics of nutrient use efficiency (NUE) at the single cell, molecular, and tissue scales, as well as the development and comprehension of NUE characteristics via plant breeding. Furthermore, NUE assessment and nutrition reconciliation in plants, field scales for assessing crops with improved NUE characteristics. For example, N fixation at the cell or tissue level must be verified at a larger scale that is suitable in crop culture at the farm level (Reich et al., 2014). Massive amounts of fertilizer are not absorbed by crops and instead enter natural ecosystems, resulting in eutrophication, which has negative consequences such as environmental degradation and loss of agrobiodiversity (Stoate et al., 2001). Phosphorus (P) fecundations leave a trail of heavy metals in soils, which may have negative effects on plants and have implications for human nutrition (Kratz et al., 2011). One of the key challenges for increasing NUE is Contemporizing the nutrient supplied with the nutrient requirement of plants. To overcome the limits of current soil reserves, external nutrients in the form of fertilizer are applied. The term "root system architecture" refers to a comprehensive view of the structural arrangement of roots within their growth environment, as well as their function in nutrient uptake and use by plants (Jung & McCouch, 2013). Crop yield improvement in modern agriculture is largely reliant on fertilizers. It is acceptable to use less fertilizer while still maximizing crop production in sustainable agriculture systems that are both commercially feasible and ecologically beneficial (MacDonald et al., 2011). Modern rice cultivars had greater variability in grain yield, N, P, and K absorption in controls and in the presence of NPK fecundators, and yield responsiveness to fertilizer appeal (Li et al., 2016). To guarantee more thrifty and agreeable fertilizer use, modern breeds need genetic enhancement in nutrient absorption efficiency. Numerous genes have been identified to enhance crop nutrient utilization efficiency, and these candidate genes are involved in activities such as absorption, adaptability, remission, and nutrient depots (Thomsen et al., 2014). In this chapter, we will discuss the importance of macronutrients and their role in crop production improvement. We will then discuss the physiological and molecular mechanisms by which macronutrients affect plant nutrient uptake efficiency and increase their efficiency in deficit and other external stress conditions. Finally, we will discuss how modern agriculture and new technologies can improve crop production in the current climate change scenario.

2. Importance of macronutrients for plants?

Plant development occurs as a consequence of a series of complex processes, including the utilization of solar energy, CO_2, H_2O, and soil nutrients. Sodium, phosphorus, potassium, calcium, sulfate, and magnesium macronutrients are absorbed at a rate of 0.2%–4% of total plant dry weight, while plants need a minimum of 0.01% dry weight of boron, chlorine, iron, nickel, zinc, copper, manganese, and molybdenum micronutrients. P esterases are a component of phospholipids, which are biosynthetic intermediates that control cell signaling pathways throughout development via protein-lipid phosphorylation. N is a component of chlorophyll, amino acids, and 16% of proteins 1.5%–2% of the total dry weight that plays a function in cellular metabolism, even though both P and N are essential DNA and RNA components (Fig. 1.1). K is found in plants in concentrations ranging from 1% to 3% dry weight and is involved in cell electric signaling, enzyme catalysis, water osmoregulation, and tolerance to temperature, pests, drought, and stress (Wang & Wu, 2013).

P shortage reduces yield by 10%–15%, slows tillering and fodder production, and has an effect on seed and fruit formation quality. Plants developed processes and regulatory networks that allow them to increase nutrient and mineral availability in the rhizosphere in response to unfavorable conditions, resulting in increased procurement and adequate intracellular concentrations (Heuer et al., 2017).

3. Macronutrients role for sustainability of crop system

Nutrient depletion, acidification, and erosion are some of the variables influencing agriculture, where soil fertility is determined by physical, chemical, and biological characteristics rather than just the amount and quality of nutrients, which may be insufficient for plant absorption in soil (Table 1.1). Inorganic Pi and organic or inorganic N (NO_3^- or NH_4^+) are usually inaccessible, although availability varies depending on external variables such as microbial activity, leaching, and soil physical characteristics, with Pi being most soluble between 6.5 and 7.5 pH and most soils having an alkaline or acidic pH (Kiba & Krapp, 2016). Furthermore, symbiotic or free-living bacteria may convert atmospheric nitrogen (N_2) to NH_4^+, which is taken by plants; inorganic potassium (K) accessible as K^+ in the top soil profile can be utilized in small fractions of 1%–2% from 3000 to 100,000 kg ha^{-1} K (Wang & Wu,

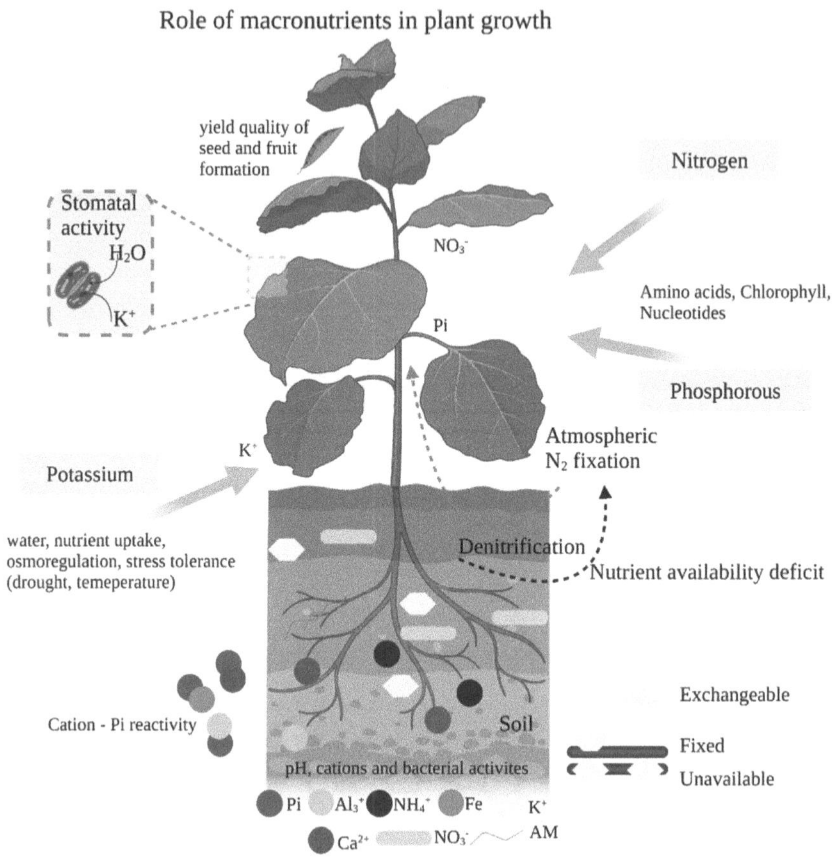

Fig. 1.1 ***Role of macronutrient in plant development and nutrient availability in soil.*** **-** While cations such as K^+, NO_3^- and Pi are required for plant development and yield, they also contribute to cellular functions such as osmoregulation. However, environmental factors such as pH and organic matter have an effect on nutrient availability in soil. Denitrification by bacteria transforms $N_{2(atm)}$ and NO_3^-, and both fixation and denitrification need K^+. AM - Arbuscular mycorrhizal, NH_4^+- Ammonium; Ca^{2+}- Calcium, Al^{3+} - Aluminum, Fe - Iron.

2013). Mineral fertilizers are critical for agricultural systems because they replenish macronutrients lost in the soil due to cultivation or erosion; therefore, fertilizer usage has steadily increased over the last 50 years. The Haber-Bosch process converts $N_{2(atm)}$ to N_2, which enables the production of limitless amounts of N fertilizers, whereas mining and refining from mineral reserves in the earth's crust is required for K^+ and Pi fertilizers 28. Pi-fertilizers are inefficient because crops utilize only 30% of the fertilizer and

Table 1.1 Macronutrient and mineralogical constituents of rock on surface land of Earth (Klein & Hurlbut, 1999).

Mineral constituent	Igneous rock (%)	Shale (%)	Sandstone (%)	Macronutrient constitutents
Amphiboles	16.5	—	11.4	Ca, Fe, Mg
Apatite	0.8	—	11.3	Ca, P
Carbonates	—	5.6	11.3	Ca, Fe, Mg
Clay	—	24.8	6.4	Ca, Fe, K, Mg, Na
Feldspars	59.4	29.9	11.8	Ca, K, Na
Iron oxides	—	5.5	1.6	Fe
Micas	3.7	—	11.2	Ca, Fe, K, Mg, Na
Titanium minerals	1.4	—	11.1	Ca, Fe
Other minerals	—	10.9	2.1	—

the remainder is lost to soil fixation and microbial conversion to nonabsorbable organic compounds. Meanwhile, emission of nitrous oxide (N_2O) depletes the ozone layer, while dry or wet NH_4^+ deposition causes acidification and eutrophication of terrestrial and aquatic ecosystems (Mannino et al., 2014).

4. Root system architecture (RSA) modulation

Phytohormones and the redox status of the root meristem are regulated by Pi shortage. For example, the growth of proteoid roots in white lupin mobilizes Pi through organic acid exudation resulting in a larger shallow root system with greater Pi acquisition efficiency. In maize, pi enrichment increases the length of the axial root without increasing the number of lateral branches. Pi absorption efficiency may reach up to 90% when root hairs are long, since longer root hairs generate more shoots and biomass efficiently scavenges Pi to produce Pi-efficient crops in which Arabidopsis ZAT lines are overexpressed, resulting in prolonged roots with higher Pi content (Devaiah et al., 2007). Epigenetic alterations in Arabidopsis mutants defective in H2A.Z deposition increase phosphate activity, density, and length of root hairs while shortening main roots (Smith et al., 2010). Overexpression of the GmEXB2 gene, encoded by β-expansin, leads in higher Pi absorption in the low or high Pi range, resulting in increased root elongation and cell division rate in Arabidopsis (Guo et al., 2011).

The patch form of NO_3^- promotes the elongation of long roots, while an excess supply hinders the development of longer roots where systemic signals cause substantial root growth in plants that utilize nitrogen from the NO_3^- precipitation site. The ANR1 TF and NRT1.1 genes encode proteins that participate in the signaling pathway that regulates root meristem activity in retort to NO_3^- (Gan et al., 2012). Phytohormones stimulate root branching through auxin, cytokinin, and abscisic acid, while auxin receptors and miR167/ARF8 module alter root system architecture via the NO_3^- response in Arabidopsis (Yong-Villalobos et al., 2015).

K^+ cation produces turgor, and osmotic pressure stimulates root cell expansion and root hair elongation, resulting in an increase in the number of long root hairs, while a lack leads in stunted root development, which is reliant on external NH_4^+ levels and phytohormones (Guo et al., 2011). The ethylene response and auxin-resistant mutants in Arabidopsis, which increase root hair extension in the trh one mutant through auxin transport, indicate that NH_4^+ toxicity slows root growth, which exacerbates K^+ shortage (Vicente-Agullo et al., 2004). It must be determined, however, if breeding for an efficient root system for one nutrient will reduce the efficiency with which other nutrients are taken up from the soil (Fig. 1.2).

5. Molecular mechanism and signaling transport of macronutrients in plants

5.1 Nitrogen uptake

Nitrogen (N) is found in amino acids, proteins, nucleic acid purine, and pyrimidine rings, chlorophyll, and enzymes. Because it cannot utilize atmospheric N_2, it is a vital nutrient that higher plants acquire, absorb, transport, and redistribute in two forms: NO_3^- and NH_4^+ under tight genetic control. It, along with phosphorus and potassium, is needed in quite significant amounts for normal development and metabolism (King et al., 1993).

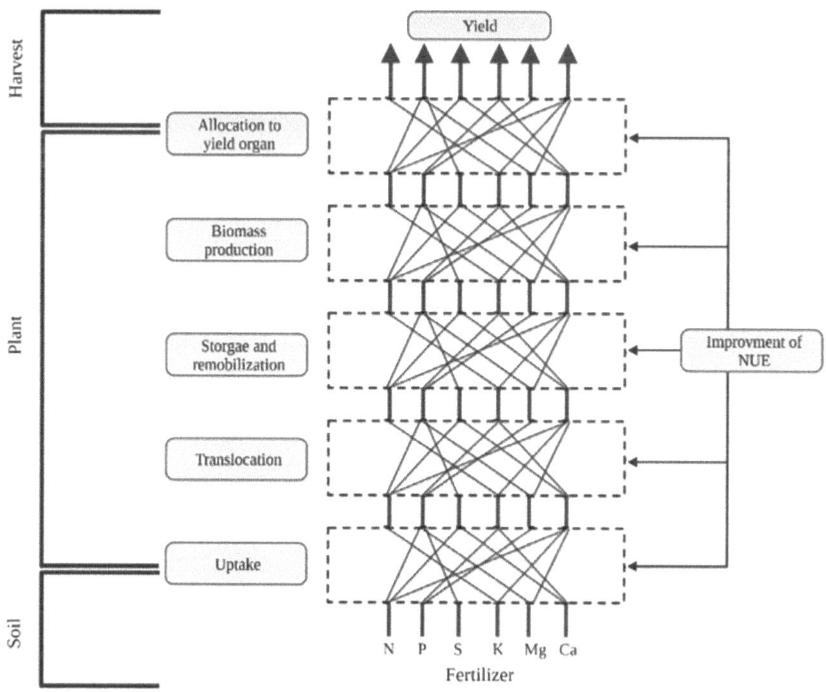

Fig. 1.2 ***NUE improvement in crops targeting in varying physiological mechanism.*** Nutrients engage with one another in all of these activities, which may result in antagonistic or synergistic property.

Nitrogen use efficiency (NUE) is a measure of the amount of biomass production per unit of N taken up from the soil and which N is converted to grain (Moll et al., 1982). The absorption of NO_3^- may be enhanced via NO_3^- transporters and control of these transport systems, however strict regulation of N uptake in roots guides to a decrease in uptake capability (Miller et al., 2007). In plants, nitrate is transported through at minimum three distinct mechanisms. Once the ions are present in small concentrations among growth medium, the transporters in the two high-affinity transport systems (HATS) induced by NO_3^- (iHATs) and constitutively expressed (cHATs) bring them athwart the plasma membrane of root cells. The transporters in the low-affinity transport system LATS carry the ions across the plasma membrane of root cells When the external NO_3^- concentration is high (1–50 mM) (Crawford & Glass, 1998). NO_3^- uptake occurs primarily across active transport through the plasma membranes of cortical and root epidermal cells. This is accomplished via the establishment of an auspicious H^+ electrochemical gradient with plasma membrane H^+-ATPases. The activity of H^+ -ATPase contends the proton motive force (Δp) and membrane potential ($\Delta\Psi$) required for ion transport. The absorption of NO_3^- occurs through the symport of $2H^+/NO_3^-$ in either low- and high-affinity transport systems (Sperandio et al., 2014).

When the NO_3^- content in the soil is elevated (>1 mM) the NRT1 genes encode low-affinity nitrate transporters (LATS). When NO_3^- concentrations are low (0.02 mM), the NRT2 genes encode high-affinity nitrate transporters (HATS). Several NRT2 genes are inductive (iHATS) whereas remnants are constitutive (cHATs). So far, these genes have been identified as being mostly expressed in the roots of higher plants (Tsay et al., 2007). The fact that root shape has a lesser of an impact on N absorption, suggesting that when modeling nutrient uptake (Burns, 1980). However, NO_3^- absorption is dependent on water movement, and root morphology plays a role in NO_3^- uptake in drying soil. In deep sands, NO_3^- is promptly leached down the soil profile besides the capability of crops to retrieve it, however plants with swiftly developing deep roots able to hinder such forfeitures and permit nutrient and water reclamation (Liao et al., 2004). Ribulose-1,5-bisphosphate carboxylase/oxygenase enzyme is essential for raising the leaf area index and optimizing radiation utilization (Imai et al., 2008). While the advancement of PsNUE is reliant on the effectiveness of Rubisco, the drawback is that it decreases the available nitrogen in the leaf. It has the potential to prolong senescence in terms of remaining green, which is good for photosynthesis nevertheless bad for NUE if it results in reduced N remobilization (Kichey et al., 2007).

5.2 Phosphorus uptake

Phosphorus (P) deficiency results in a 30% reduction in agricultural yields worldwide, and as a result of the situation, plants adopt shoot and root modification to collect phosphate (Pi) from phosphate rocks in the soil to sustain plant development. P is highly bound to weathered acid soil clay minerals near the surface, resulting in limited P availability for plant absorption. Phosphate (Pi) is mainly taken up as HPO_4^{2-} or $H_2\ PO_4^-$ ions in the rhizosphere, depending on availability and pH wherein pH 7.2, $HPO_4^{2-} \approx H_2\ PO_4^-$, below pH 7.2, $H_2\ PO_4^- > HPO_4^{2-}$but above pH 7.2 $H_2\ PO_4^- < HPO_4^{2-}$. To obtain Pi from the soil and maintain Pi equilibrium inside the plant via internal redistribution, plants undergo morphological, physiological, and metabolic alterations. Plant roots exude organic acids, enzymes, and mucilage, which determine various chemical and biological reactions around the root surface. These reactions control the availability of plant nutrients, their uptake, and use, resulting in higher productivity of crops.

Phosphorous transport, partitioning, and remobilization are all instances of adaptive processes that aim to increase the efficiency of internal P use, due to the deficiency changes in root system architecture (RSA) including aerenchyma development, postembryonic root system, adventitious rooting, elongation of the basal root, root hair density and ethylene production increases and root hair length increased, all of which was inversely related to the P concentration in plants and it is controlled the root development and changes in endogenous levels by auxin genes such as *AIR1, AIR3, AIR9, AIR12, HRGP*, and *LRP1*. When compared to plants with optimal nutrient supply allocate a greater percentage of biomass to roots with optimum (Hammond et al., 2004; Lynch, 2011; RYSER et al., 2006). Auxin signaling was regulated by many important regulators, including transport inhibitor response 1 (TIR1), TIR1-like auxin F-box (AFB) proteins, and auxin response factor (ARF) proteins 21. A shallow root architecture, three-dimensional root structure, and long roots improve Pi-uptake efficiency and yield in P-deficient soils (Lin et al., 2008). Under conditions of P shortage, the root-to-shoot dry weight ratio increases as a result of enhanced root development at the expense of shoot development (Mollier & Pellerin, 1999). Under stress, Arabidopsis' root system becomes much more branched, with decreased primary root growth and increased lateral root length and quantity (Pérez-Torres et al., 2008). Phosphorus-Starvation Tolerance 1 (PSTOL1) is a receptor-like cytoplasmic kinase that acts as a direct link between root morphology and P acquisition and is required for rice to tolerate P deficiency 19. Phosphate Deficiency Response 2 (PDR2) is needed to maintain the SCR protein's nucleus distribution in the context of low P, which limits root meristematic activity, by connecting P responses to possible root changes mediated by SCARECROW (SCR). MiRNAs serve as sensor molecules in plants, regulating P homeostasis and RSA dynamics. They may also be involved in environmentally induced adaptation responses to abiotic stresses like P deficiency. For example, miRNA399 regulates P homeostasis in Arabidopsis and is bound by the long noncoding RNA Induced by Phosphate Starvation 1 (IPS1), preventing miR399-mediated degradation of the PHO2 transcript (Franco-Zorrilla et al., 2007).

As a consequence of the deficit resulting reduced leaf quantity and delayed leaf development result in a loss in photosynthetic capability, as well as decreased auxiliary branch elongation and reduced flower development (Vance, 2010). Nevertheless, increased absorption leads to an excess of Pi buildup in older leaves, leading to toxicity and chlorosis to necrosis (Aung et al., 2006). The response by plant to the Pi deficit is through the mechanism of TCA cycle derived organic acids synthesis into the rhizosphere where an observed increase in malate dehydrogenase and phosphoenolpyruvate carboxylase (PEPC) (Gregory et al., 2009). Protons from organic acids interact with insoluble Fe_3^+, Al_3^+ and Ca_2^+ compounds in Pi-containing minerals to form chelates and release Pi for absorption, increasing Pi concentration in the soil while promoting the production of nucleases, phosphodiesterases, and other enzymes into the soil due to the deficit. Identifying crop types that are capable of obtaining Pi under inadequate circumstances without sacrificing harvest may aid in increasing phosphate use efficiency (PUE) (Plaxton & Tran, 2011).

5.3 Potassium uptake

Potassium is the fourth most prevalent mineral and nutrient in plants, making up about 2.5% of the lithosphere (K). The K^+ concentrations in soil range from 0.1 to 6 mM, accounting for 2%–10% of plant dry weight, with values less than 10 g kg^{-1} of dry weight indicating a deficiency. The processes regulated by K^+ are ion homeostasis, turgor pressure, sugar loading by phloem, and membrane polarity (Adams & Shin, 2014). All cells and organelles acquire K^+ via dual K^+ affinity uptake systems, one of which is a low-affinity K^+ uptake system that operates at K^+ concentrations around 1 mM and higher, while the other is a high-affinity K^+ uptake system that operates at K+ concentrations in the micromolar range (Epstein et al., 1963).

K^+ channels from three types of plant voltage-gated channels engage in persistent K^+ absorption from the soil solution, long-distance K^+ transport in the xylem and phloem, and K^+ fluxes in guard cells throughout stomatal motions (Chérel, 2004). While AtCNGC1 and AtCNGC4 have comparable K^+ and Na^+ permeability, AtCNGC2 is directly involved in K^+ absorption in Arabidopsis due to its high K^+ permeability and extensive-expression in roots (Talke et al., 2003). In Arabidopsis, AtKC1 colocalizes with certain other inward channels such as AKT1, AtKAT1, AtKAT2, and AKT2/3 to react to a wide spectrum of K+ conductivity concentrations 42–44. When K^+ is depleted, AtRCI3 interacts with AtRHD2 to generate ROS, implying an increase in AtHAK5 expression 135, 137, 138. A high-affinity K^+ transporter (Trk/HKT) is essentially a Na^+/K^+ symporter in which Trk superfamily cation transporters are topologically linked to K^+ channels that mediate Na^+/K^+ transport and control Na^+ homeostasis (Munns & Tester, 2008). Plant Cation/H^+ Exchangers are Na^+ transporters that control plant cell homeostasis and osmolality. A single NADPH oxidase, AtRHD2, AtrbohC, performs important functions in K^+ shortage signaling by modulating gene expression and root hair growth through activation of the high-affinity K^+ absorption pathway (Shin et al., 2005). All plant genomes contain high-affinity potassium transporters (HAK) and potassium uptake permeases (KUP) constitute a major high-affinity potassium acquisition system. K+ shortage decreases the quantity of free indole-3-acetic acid in Arabidopsis and results in the downregulation of the r2r3-type myb transcription factor. By decreasing the quantity and duration of lateral root development, certain transcription factors, such as an AtHAK5 K+ deficiency-induced promoter, enhance K^+ absorption efficiency (Hong et al., 2013). The greater the root surface area, the greater the potential for K^+ absorption by plants, whereas overexpression of genes encoding MYB-like transcription factors AtCPC1, a positive regulator of root hair differentiation, improves root hair growth in Arabidopsis plants, and overexpression of K^+ channel transporter did not increase K+ uptake in plants due to their control (Song et al., 2014). Tetra ploidy may account for the fact that Arabidopsis plants containing the *AtKUP2, ATK2, AtKAT2,* and *AtTPK3* genes on a single chromosome accumulate more K+ than haploid or diploid plants (Hammond et al., 2004). To maintain cytosolic K^+ homeostasis and photosynthetic activity, the K^+ concentration in leaf tissues fluctuates, and the K^+ concentration in epidermal cells decreases. Reduced leaf osmotic potential, enhanced water binding, water intake, and turgor efficacy improve crop drought resistance by regulating SOD activity and reducing reactive oxygen damage to the plasma membrane caused by drought stress (Li et al., 2006). It regulates the activity of active oxygen generated in response to drought stress and protects the plasma membrane from damage, as well as regulating the levels of ABA, ethylene, and CTK internal hormones. The nonexchangeable K that cucumber, pepper, and other crops may release by inoculating them with K-solubilizing bacteria like Bacillus mucilaginous has a beneficial impact on boosting K availability. Thus, increasing the quantity of soil-borne K^+ accessible to plants by cultivating a big population of helpful microflora or engineering for metabolic specialties. Improved K^+ use efficiency would aid in lowering agricultural production costs, protecting the environment, and increasing crop yields (Saiyad et al., 2015).

5.4 Calcium uptake

Plants get a calcium (Ca) content of up to 5% of shoot dry weight and delayed uneven distribution of calcium. Increased Ca^{2+} concentrations inside the cytoplasm are cytotoxic, and the depositing of insoluble oxalic and phytic acid salts during vacuolar migration is tightly regulated (Borchert, 1990). Calcium channels, Ca-ATPases, and Ca^{2+}/H^+ antiporters all contribute to the regulation of Ca^{2+} flow and therefore to the maintenance of calcium homeostasis. The concentration of Ca^{2+} in the cytoplasm controls hyperpolarization-activated Ca^{2+} channels (HACCs), which are only present in the root apex and Ca^{2+} uptake demanded regions for cell division (White, 1998). In response to a range of environmental stimuli, Depolarization-Activated Cation Channels (DACCs) cause a fast and transient influx of Ca^{2+} (Kudla et al., 2010). Cyclic Nucleotide-Gated Channels (CNGCs) become inactive when they bind to cAMP or cGMP in the plasma membrane, resulting in Ca^{2+} influx (Kudla et al., 2010). Glutamate receptors (GLRs) have a role in Ca^{2+} influx as well as Ca^{2+}-mediated responses to cold stress and aluminum toxicity (Sivaguru et al., 2003). Ca^{2+} homeostasis is controlled by calcium proton antiporters, which create a pH gradient athwart the tonoplast via H^+- ATPase, which transports cations from the cytoplasm toward the vacuole in place of H^+. The plasma membrane, nucleus, endoplasmic reticulum, and cytoskeleton all contain Calcium-Dependent Protein Kinases (CDPKs) that convert cytosolic Ca^{2+} signals into biochemical and genetic adaptations through phosphorylating ion and water channels, membrane solute transporters, and DNA-binding proteins (Sanders et al., 2002). The physiological response of plants to water stress conditions includes increasing their ability to absorb water from the soil, storing water inside cells, and reducing transpiration loss by regulating stomatal closure induced by cytosolic Ca^{2+} and inhibiting stomatal reopening. The identification of pathogen-associated molecular pattern (PAMP) molecules causes increments in Ca^{2+} influx throughout the plasma membrane, raising cytosolic Ca^{2+} levels. Calcium usage efficiency (CaUE) is defined as grain yield or shoots per unit of deposited calcium. In today's environment, genotypically adaptable crops with a high tolerance for abiotic and biotic constraints have a high CaUE (Fageria, 2016). Acid soils with high leaching capacity, weathering, and low base saturation are shown, as well as metal toxicity of Fe, Mn, and Al. Ca shortage limits plant growth development. After gypsum application and liming, most industrial wastes used as liming materials provide good yields in crop efficiency and soil structure (McLaughlin, 1999).

5.5 Magnesium uptake

Magnesium (Mg), a free divalent cation with a concentration of 0.1%—0.4% in plant cytosol and 2—10 mM in leaf chloroplast, regulates photosynthesis and membrane currents. Mg^{2+}equilibrium in the cytosol and chloroplast is regulated by the organelle vacuole (Leigh & Wyn Jones, 1986). Mg^{2+} fluctuation and deficiency affect photosynthetic enzymes and root development, which in turn affects nutrient and water absorption (Shaul, 2002). Mg^{2+} transport is aided by transporter proteins athwart biological membranes, whither the bacterial membrane transport protein cobalt-resistant phenotype of bacterial mutants (CorA) binds the cation to the extracellular binding loop linking the C-terminal transmembrane domains (TMs) carrying exceedingly high concentrations of positive and negative charges controlling Mg^{2+} flux where at the C-terminus of TMs, CorA homologs exhibit a conserved GMN (Gly-Met-Asn) motif, and mutation of GMN abolishes Mg^{2+} transport (Moomaw & Maguire, 2008). In acidic soils, Al^{3+} toxicity slows root growth and impairs nutrient and water absorption, while overexpression of AtMGT family genes enhanced Al^{3+} tolerance in plants (Delhaize et al., 2012). Mg^{2+} retranslocation from older to younger leaves may be seen as reddish-brown tissues between green veins or as chlorosis (Tanoi & Kobayashi, 2015), which causes Mg^{2+} dechelation from chlorophyll in favor of young tissue growth (White & Broadley, 2009).

However, due to the deficit, transpiration is hindered before electron transport imbalance, and sucrose loading by phloem in leaves is also disturbed, which may indicate an earlier response to Mg deficiency (Cakmak & Yazici, 2010). Weathering of ultramafic rocks leads to serpentine soils containing ferromagnesian minerals with a low Ca: Mg ratio and a decreased Ca^{2+} concentration (Brady et al., 2005), indicating that an excess of Mg^{2+} may act as a replacement for Ca^{2+} (Guo et al., 2014). Agronomic biofortification improves soil and foliar Mg^{2+} fertilization, extending plant growth and maintaining a high magnesium content in crops by utilizing magnesium ammonium phosphate (struvite) as a sustainable source of P and Mg and therefore relieving nutrient deficit. In genetic biofortification, a gene implicated in Mg^{2+} homeostasis has been identified in a variety of crop species through forward or reverse genetics; furthermore, producing Al^{3+}- resistant plants may improve Mg^{2+} transit and accumulation (Bose et al., 2013).

5.6 Sulfur uptake

Sulfur (S) constitutes 0.06%—0.10% of the earth's crust but also 0.5%—1.5% of the dry weight of plant tissues, and it is the most abundant sulfate ion in saltwater, constituting 7.7% of total weight. It is a plant defense molecule that functions as an antimicrobial peptide and is a constituent of critical metabolites and amino acids cysteine and methionine, as well as other cofactors/vitamins that assist in photosynthetic respiration and electron transfer (Bloem et al., 2015; Forieri et al., 2013). The inorganic-S form of sulfate (SO_4^{2-}) is efficiently transported from the soil to the xylem through membrane-localized transporter proteins that transport S with H^+ against the concentration gradient and are ultimately retained in the vacuole. Before converting SO_4^{2-} to sulfide (S^{2-}), it produces adenosine-5′-phosphosulphate (APS) with ATP-sulphurylase (ATPS), which is subsequently reduced to sulfite by APS reductase and finally introduced to O-acetyl serine (OAS) to form cysteine amino acid (Falk et al., 2007; Davidian & Kopriva, 2010). Sulfate transporters SULTR1—SULTR4 induced S starvation, leading to low- and high-affinity sulfate absorption capacity in the plasma membrane through roots (Kataoka et al., 2004). The Sulfur Deficiency Induced 1 (SDI1) gene in Arabidopsis is activated in response to S deficit and controls gene expression in reaction to S nutritional status. Glucosinolates, or mustard oil bombs, are produced as a defensive mechanism in plants, where glutathione-rich S-containing molecules needed for root meristem cell division simultaneously signal for S starvation retort, and Cys—Gly dipeptide degradation by 5-oxoproline a peptidase results in Cys synthesis (Wittstock & Burow, 2010). Genetic modification of SULTR genes such as SULTR1; 1 and SULTR1; 2 increases S compound absorption by 60% 50. In Arabidopsis, the PaAPR bacterial sulfate assimilation gene increases the absorption of sulfite, Cys, and S-containing substances while miRNA395 overexpression may increase sulfate buildup in shoots twice by knocking down the APS4 gene (Tsakraklides et al., 2002).

6. Nutrient interaction with plants to enhance NUE and decrease ecological footprints

Without reducing production, it is critical to improving the efficiency of agronomic goods, which is a major issue given the world's growing population. Following Liebscher's rule of the optimum, nutrients have positive synergistic effects on increasing NUE provided by nutrient-plant interaction to optimal fertilization (Mengel & Kirkby, 2012). Nutrient absorption, storage, and transit are all affected by macronutrient availability (Fig. 1.3) in both antagonistic and synergistic ways. Plants control macronutrient absorption by reducing K uptake and boosting Ca, Mg, and Na uptake, even though K is a countercation for nutrients absorbed as an anion through the plasma membrane via the roots, such as S (Forster & Mengel, 1969).

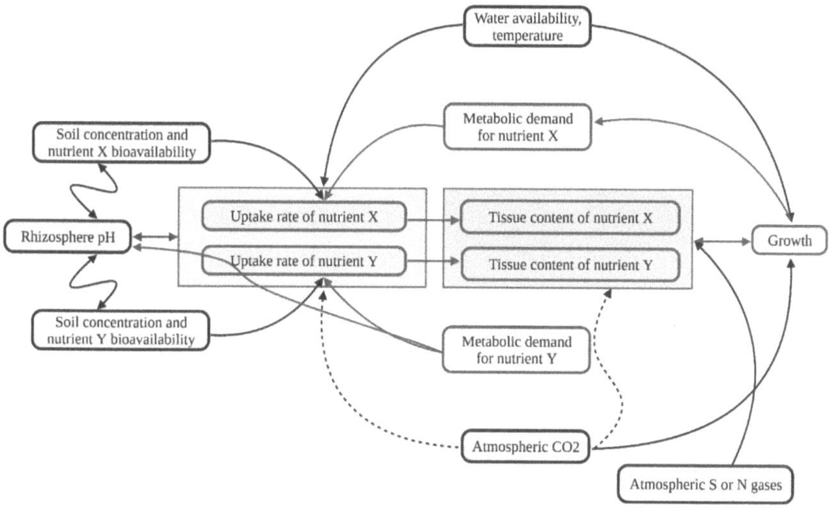

Fig. 1.3 ***Environmental and metabolic regulation of crop nutrition.*** Environmental factors (blue [dark gray in print version]) and metabolic demand (red [light gray in print version]) regulate nutrient absorption, which leads to the identification of tissue nutrient growth and content.

In soil, two phyto mineral forms of nitrogen NO_3^- and NH_4^+, resist cation absorption by generating H+ and anion and fighting for uptake with OH^- anion (Kirkby & Mengel, 1967). Additionally, an overabundance of NO_3^- reduces Pi and S, while NH_4^+ decreases the other cations. Additionally, increased Mg, Ca and K supply decreases root vascular storage absorption (Lu et al., 2005). Increased Na levels result in cation balance and the resorption of K and Ca from cell walls and membranes, as well as their expulsion from the cytosol. Precipitation reduces mineral nutrient storage, decreasing NUE efficiency and absorption mostly owing to ionic charge interaction (Reich et al., 2017).

7. Nutrient use efficiency dependent crop production

Even if more fertilizer is applied, only a small amount is used; for example, nitrogen immobilized by microbes is lost through denitrification and is not absorbed by crops, resulting in environmental issues. reducing the increase in nutrient uptake by plants (Canfield et al., 2010). NUE attempts a mass balance between total plant nutrient content at harvest and crop yield in agronomy, implying a long period of development in nutrient utilization efficiency, absorption, and harvest product concentration (Weih et al., 2011). Overall, NUE indicates the ultimate accumulation of needed nutrients for yield in proportion to plant-internal reserve storage for the input in situations while external factors such as agronomic, physiological, and crop recovery influence the designation of better crops for production. Plant nutrition analysis focuses on the developmental phases between seed grain and harvest to estimate soil nutrient content, where utilization efficiency, concentration, and grain specific effectiveness are comparable and indicate the crop's capacity to generate a yield from nutrients taken up (Moll et al., 1982). Furthermore, NUE links ultimate yield output to seed N intake, showing the potential for N assemblage from seed to harvested grains. The complete integrated study of crop improvement through selection and breeding is more appropriate for assessing the farm-level effects of agricultural innovations.

8. Future prospects and conclusion

Macronutrients are essential for plant development, and their excess or shortage is harmful to plants. To complete their life cycles, plants need macronutrients such as phosphorus, nitrogen, potassium, calcium, magnesium, and sulfur. Plant scientists must optimize a technique for conserving vital nutrients by matching nutrient supply to crop requirements or for minimizing hazardous elements and crop growth's negative environmental effects. Improved water, nutrient, and soil management techniques are critical for addressing existing environmental problems and increasing agricultural systems' nutrient absorption efficiency. Technological advancements in the monitoring of nutrient flow in the rhizosphere soil ecology, root absorption mechanisms, and plant-nutrient distribution aid improve crop quality and yield. Plants developed sophisticated biochemical mechanisms to respond to species-specific deficiency responses that differ in metabolic activity. Sensing is one of these mechanisms, in which plants react to shortages by expanding their root surface and decreasing branch development in the early stages. Local and systemic downstream signaling cascades transport nutrients from roots to xylem to phloem, including macronutrients, hormones, enzymes, miRNAs, and transcription factors. Advanced theoretical modeling in combination with genomics and ionomics will be an excellent concept for agricultural systems-related contemporary scientific study. The phenotyping of plants is critical for identifying genes and evaluating genome editing methods in the laboratory and the field. CRISPR/Cas9 results in commercially viable crops and is particularly well-suited for agricultural development in species with a high ploidy genome and long-life cycles.

References

Adams, E., & Shin, R. (2014). Transport, signaling, and homeostasis of potassium and sodium in plants. *Journal of Integrative Plant Biology, 56*(3), 231–249.

Aung, K., Lin, S.-I., Wu, C.-C., Huang, Y.-T., Su, C., & Chiou, T.-J. (2006). Pho2, a phosphate overaccumulator, is caused by a nonsense mutation in a microRNA399 target gene. *Plant Physiology, 141*(3), 1000–1011.

Bloem, E., Haneklaus, S., & Schnug, E. (2015). Milestones in plant sulfur research on sulfur-induced-resistance (SIR) in Europe. *Frontiers in Plant Science, 5*, 779.

Borchert, R. (1990). Ca 2+ as developmental signal in the formation of Ca-oxalate crystal spacing patterns during leaf development in Carya ovata. *Planta, 182*(3), 339–347.

Bose, J., Babourina, O., Shabala, S., & Rengel, Z. (2013). Low-pH and aluminum resistance in Arabidopsis correlates with high cytosolic magnesium content and increased magnesium uptake by plant roots. *Plant and Cell Physiology, 54*(7), 1093–1104.

Brady, K. U., Kruckeberg, A. R., & Bradshaw, H. D., Jr. (2005). Evolutionary ecology of plant adaptation to serpentine soils. *Annual Review of Ecology, Evolution and Systematics, 36*, 243–266.

Burns, I. G. (1980). Influence of the spatial distribution of nitrate on the uptake of N by plants: A review and a model for rooting depth. *Journal of Soil Science, 31*(2), 155–173.

Cakmak, I., & Yazici, A. M. (2010). Magnesium: A forgotten element in crop production. *Better Crops, 94*(2), 23–25.

Canfield, D. E., Glazer, A. N., & Falkowski, P. G. (2010). The evolution and future of Earth's nitrogen cycle. *Science, 330*(6001), 192–196.

Chérel, I. (2004). Regulation of K+ channel activities in plants: From physiological to molecular aspects. *Journal of Experimental Botany, 55*(396), 337–351.

Crawford, N. M., & Glass, A. D. (1998). Molecular and physiological aspects of nitrate uptake in plants. *Trends in Plant Science, 3*(10), 389–395.

Davidian, J.-C., & Kopriva, S. (2010). Regulation of sulfate uptake and assimilation—the same or not the same? *Molecular Plant, 3*(2), 314–325.

Delhaize, E., Ma, J. F., & Ryan, P. R. (2012). Transcriptional regulation of aluminium tolerance genes. *Trends in Plant Science, 17*(6), 341–348.

Devaiah, B. N., Karthikeyan, A. S., & Raghothama, K. G. (2007). WRKY75 transcription factor is a modulator of phosphate acquisition and root development in Arabidopsis. *Plant Physiology, 143*(4), 1789–1801.

Epstein, E., Rains, D. W., & Elzam, O. E. (1963). Resolution of dual mechanisms of potassium absorption by barley roots. *Proceedings of the National Academy of Sciences of the United States of America, 49*(5), 684.

Fageria, N. K. (2016). *The use of nutrients in crop plants*. CRC press.

Falk, K. L., Tokuhisa, J. G., & Gershenzon, J. (2007). The effect of sulfur nutrition on plant glucosinolate content: Physiology and molecular mechanisms. *Plant Biology, 9*(05), 573–581.

Forieri, I., Wirtz, M., & Hell, R. (2013). Toward new perspectives on the interaction of iron and sulfur metabolism in plants. *Frontiers in Plant Science, 4*, 357.

Forster, H., & Mengel, K. (1969). The effect of a short term interruption in the K supply during the early stage on yield formation, mineral content and soluble amino acid content. *Z. Acker-Pflanzenbau, 130*, 203–213.

Franco-Zorrilla, J. M., Valli, A., Todesco, M., Mateos, I., Puga, M. I., Rubio-Somoza, I., Leyva, A., Weigel, D., García, J. A., & Paz-Ares, J. (2007). Target mimicry provides a new mechanism for regulation of microRNA activity. *Nature Genetics, 39*(8), 1033–1037.

Gan, Y., Bernreiter, A., Filleur, S., Abram, B., & Forde, B. G. (2012). Overexpressing the ANR1 MADS-box gene in transgenic plants provides new insights into its role in the nitrate regulation of root development. *Plant and Cell Physiology, 53*(6), 1003–1016.

Gregory, A. L., Hurley, B. A., Tran, H. T., Valentine, A. J., She, Y.-M., Knowles, V. L., & Plaxton, W. C. (2009). In vivo regulatory phosphorylation of the phosphoenolpyruvate carboxylase AtPPC1 in phosphate-starved *Arabidopsis thaliana. Biochemical Journal, 420*(1), 57–65.

Guo, W., Cong, Y., Hussain, N., Wang, Y., Liu, Z., Jiang, L., Liang, Z., & Chen, K. (2014). The remodeling of seedling development in response to long-term magnesium toxicity and regulation by ABA–DELLA signaling in Arabidopsis. *Plant and Cell Physiology, 55*(10), 1713–1726.

Guo, W., Zhao, J., Li, X., Qin, L., Yan, X., & Liao, H. (2011). A soybean β-expansin gene GmEXPB2 intrinsically involved in root system architecture responses to abiotic stresses. *The Plant Journal, 66*(3), 541–552.

Hammond, J. P., Broadley, M. R., & White, P. J. (2004). Genetic responses to phosphorus deficiency. *Annals of Botany, 94*(3), 323–332.

Heuer, S., Gaxiola, R., Schilling, R., Herrera-Estrella, L., López-Arredondo, D., Wissuwa, M., Delhaize, E., & Rouached, H. (2017). Improving phosphorus use efficiency: A complex trait with emerging opportunities. *The Plant Journal, 90*(5), 868–885.

Hong, J.-P., Takeshi, Y., Kondou, Y., Schachtman, D. P., Matsui, M., & Shin, R. (2013). Identification and characterization of transcription factors regulating Arabidopsis HAK5. *Plant and Cell Physiology, 54*(9), 1478–1490.

Imai, K., Suzuki, Y., Mae, T., & Makino, A. (2008). Changes in the synthesis of Rubisco in rice leaves in relation to senescence and N influx. *Annals of Botany, 101*(1), 135–144.

Jung, J. K. H. M., & McCouch, S. R. M. (2013). Getting to the roots of it: Genetic and hormonal control of root architecture. *Frontiers in Plant Science, 4*, 186.

Kataoka, T., Hayashi, N., Yamaya, T., & Takahashi, H. (2004). Root-to-shoot transport of sulfate in Arabidopsis. Evidence for the role of SULTR3; 5 as a component of low-affinity sulfate transport system in the root vasculature. *Plant Physiology, 136*(4), 4198–4204.

Kiba, T., & Krapp, A. (2016). Plant nitrogen acquisition under low availability: Regulation of uptake and root architecture. *Plant and Cell Physiology, 57*(4), 707–714.

Kichey, T., Hirel, B., Heumez, E., Dubois, F., & Le Gouis, J. (2007). In winter wheat (*Triticum aestivum* L.), post-anthesis nitrogen uptake and remobilisation to the grain correlates with agronomic traits and nitrogen physiological markers. *Field Crops Research, 102*(1), 22–32.

King, B. J., Siddiqi, M. Y., Ruth, T. J., Warner, R. L., & Glass, A. D. (1993). Feedback regulation of nitrate influx in barley roots by nitrate, nitrite, and ammonium. *Plant Physiology, 102*(4), 1279–1286.

Kirkby, E. A., & Mengel, K. (1967). Ionic balance in different tissues of the tomato plant in relation to nitrate, urea, or ammonium nutrition. *Plant Physiology, 42*(1), 6–14.

Klein, C., & Hurlbut, C. S. (1999). Mineral assemblages: Introduction to rock types and vein mineralization. In *Manual of mineralogy*. John Wiley & Sons. Inc.

Kratz, S., Godlinski, F., & Schnug, E. (2011). Heavy metal loads to agricultural soils in Germany from the application of commercial phosphorus fertilizers and their contribution to background concentration in soils. In *The new uranium mining boom* (pp. 755–762). Springer.

Kudla, J., Batistič, O., & Hashimoto, K. (2010). Calcium signals: The lead currency of plant information processing. *The Plant Cell, 22*(3), 541–563.

Leigh, R. A., & Wyn Jones, R. G. (1986). Cellular compartmentation in plant nutrition: The selective cytoplasm and the promiscuous vacuole. In *Advances in plant nutrition (USA)*.

Liao, M., Fillery, I. R., & Palta, J. A. (2004). Early vigorous growth is a major factor influencing nitrogen uptake in wheat. *Functional Plant Biology, 31*(2), 121–129.

Li, X. K., Hill, P. W., Li, Y. C., Wang, Y., Ren, T., Cong, R. H., & Lu, J. W. (2016). Physiological processes associated with high yield traits in modern rice varieties. *International Journal of Plant Production, 10*(2), 221–231.

Lin, S.-I., Chiang, S.-F., Lin, W.-Y., Chen, J.-W., Tseng, C.-Y., Wu, P.-C., & Chiou, T.-J. (2008). Regulatory network of microRNA399 and PHO2 by systemic signaling. *Plant Physiology, 147*(2), 732–746.

LI, W.-Y. F., WONG, F.-L., TSAI, S.-N., PHANG, T.-H., Shao, G., & LAM, H.-M. (2006). Tonoplast-located GmCLC1 and GmNHX1 from soybean enhance NaCl tolerance in transgenic bright yellow (BY)-2 cells. *Plant, Cell & Environment, 29*(6), 1122–1137.

Lu, Y. X., Li, C. J., & Zhang, F. S. (2005). Transpiration, potassium uptake and flow in tobacco as affected by nitrogen forms and nutrient levels. *Annals of Botany, 95*(6), 991–998.

Lynch, J. P. (2011). Root phenes for enhanced soil exploration and phosphorus acquisition: Tools for future crops. *Plant Physiology, 156*(3), 1041–1049.

MacDonald, G. K., Bennett, E. M., Potter, P. A., & Ramankutty, N. (2011). Agronomic phosphorus imbalances across the world's croplands. *Proceedings of the National Academy of Sciences, 108*(7), 3086–3091.

Mannino, A., Novak, M. G., Hooker, S. B., Hyde, K., & Aurin, D. (2014). Algorithm development and validation of CDOM properties for estuarine and continental shelf waters along the northeastern US coast. *Remote Sensing of Environment, 152*, 576–602.

McLaughlin, S. B. (1999). Calcium physiology and terrestrial ecosystem processes. *New Phytology, 142*, 373–417.

Mengel, K., & Kirkby, E. A. (2012). *Principles of plant nutrition.* Springer Science & Business Media.

Miller, A. J., Fan, X., Orsel, M., Smith, S. J., & Wells, D. M. (2007). Nitrate transport and signalling. *Journal of Experimental Botany, 58*(9), 2297–2306.

Mollier, A., & Pellerin, S. (1999). Maize root system growth and development as influenced by phosphorus deficiency. *Journal of Experimental Botany, 50*(333), 487–497.

Moll, R. H., Kamprath, E. J., & Jackson, W. A. (1982). Analysis and interpretation of factors which contribute to efficiency of nitrogen utilization 1. *Agronomy Journal, 74*(3), 562–564.

Moomaw, A. S., & Maguire, M. E. (2008). The unique nature of Mg^{2+} channels. *Physiology, 23*(5), 275–285.

Munns, R., & Tester, M. (2008). Mechanisms of salinity tolerance. *Annual Review of Plant Biology, 59*, 651–681.

Pérez-Torres, C.-A., López-Bucio, J., Cruz-Ramírez, A., Ibarra-Laclette, E., Dharmasiri, S., Estelle, M., & Herrera-Estrella, L. (2008). Phosphate availability alters lateral root development in Arabidopsis by modulating auxin sensitivity via a mechanism involving the TIR1 auxin receptor. *The Plant Cell, 20*(12), 3258–3272.

Plaxton, W. C., & Tran, H. T. (2011). Metabolic adaptations of phosphate-starved plants. *Plant Physiology, 156*(3), 1006–1015.

Reich, M., Aghajanzadeh, T., & De Kok, L. J. (2014). Physiological basis of plant nutrient use efficiency—concepts, opportunities and challenges for its improvement. In *Nutrient use efficiency in plants* (pp. 1–27). Springer.

Reich, M., Aghajanzadeh, T., Helm, J., Parmar, S., Hawkesford, M. J., & De Kok, L. J. (2017). Chloride and sulfate salinity differently affect biomass, mineral nutrient composition and expression of sulfate transport and assimilation genes in Brassica rapa. *Plant and Soil, 411*(1–2), 319–332.

Ryser, P., Hill, J., Simpson, R., Moore, A., & Chapman, D. (2006). Morphology and response of roots of pasture species to phosphorus and nitrogen nutrition. Commentary. *Plant and Soil, 286*(1–2).

Saiyad, S. A., Jhala, Y. K., & Vyas, R. V. (2015). Comparative efficiency of five potash and phosphate solubilizing bacteria and their key enzymes useful for enhancing and improvement of soil fertility. *International Journal of Scientific and Research Publications, 5*(2), 1–6.

Sanders, D., Pelloux, J., Brownlee, C., & Harper, J. F. (2002). Calcium at the crossroads of signaling. *The Plant Cell, 14*(Suppl. l_1), S401–S417.

Shaul, O. (2002). Magnesium transport and function in plants: The tip of the iceberg. *Biometals, 15*(3), 307–321.

Shin, R., Berg, R. H., & Schachtman, D. P. (2005). Reactive oxygen species and root hairs in Arabidopsis root response to nitrogen, phosphorus and potassium deficiency. *Plant and Cell Physiology, 46*(8), 1350–1357.

Sivaguru, M., Pike, S., Gassmann, W., & Baskin, T. I. (2003). Aluminum rapidly depolymerizes cortical microtubules and depolarizes the plasma membrane: Evidence that these responses are mediated by a glutamate receptor. *Plant and Cell Physiology, 44*(7), 667–675.

Smith, A. P., Jain, A., Deal, R. B., Nagarajan, V. K., Poling, M. D., Raghothama, K. G., & Meagher, R. B. (2010). Histone H2A. Z regulates the expression of several classes of phosphate starvation response genes but not as a transcriptional activator. *Plant Physiology, 152*(1), 217–225.

Song, Z., Yang, S., Zhu, H., Jin, M., & Su, Y. (2014). Heterologous expression of an alligatorweed high-affinity potassium transporter gene enhances salinity tolerance in *Arabidopsis thaliana. American Journal of Botany, 101*(5), 840–850.

Sperandio, M. V. L., Santos, L. A., de Araújo, O. J. L., Braga, R. P., Coelho, C. P., de Matos Nogueira, E., Fernandes, M. S., & de Souza, S. R. (2014). Response of nitrate transporters and PM H+-ATPase expression to nitrogen flush on two upland rice varieties contrasting in nitrate uptake kinetics. *Australian Journal of Crop Science, 8*(4), 568–576.

Spiertz, J. H. J., & Ewert, F. (2009). Crop production and resource use to meet the growing demand for food, feed and fuel: Opportunities and constraints. *NJAS: Wageningen Journal of Life Sciences, 56*(4), 281–300.

Stoate, C., Boatman, N. D., Borralho, R. J., Carvalho, C. R., De Snoo, G. R., & Eden, P. (2001). Ecological impacts of arable intensification in Europe. *Journal of Environmental Management, 63*(4), 337–365.

Talke, I. N., Blaudez, D., Maathuis, F. J., & Sanders, D. (2003). CNGCs: Prime targets of plant cyclic nucleotide signalling? *Trends in Plant Science, 8*(6), 286–293.

Tanoi, K., & Kobayashi, N. I. (2015). Leaf senescence by magnesium deficiency. *Plants, 4*(4), 756–772.

Thomsen, H. C., Eriksson, D., Møller, I. S., & Schjoerring, J. K. (2014). Cytosolic glutamine synthetase: A target for improvement of crop nitrogen use efficiency? *Trends in Plant Science, 19*(10), 656–663.

Tilman, D., Cassman, K. G., Matson, P. A., Naylor, R., & Polasky, S. (2002). Agricultural sustainability and intensive production practices. *Nature, 418*(6898), 671–677.

Tsakraklides, G., Martin, M., Chalam, R., Tarczynski, M. C., Schmidt, A., & Leustek, T. (2002). Sulfate reduction is increased in transgenic *Arabidopsis thaliana* expressing 5′-adenylylsulfate reductase from *Pseudomonas aeruginosa*. *The Plant Journal, 32*(6), 879–889.

Tsay, Y.-F., Chiu, C.-C., Tsai, C.-B., Ho, C.-H., & Hsu, P.-K. (2007). Nitrate transporters and peptide transporters. *FEBS Letters, 581*(12), 2290–2300.

Vance, C. P. (2010). Quantitative trait loci, epigenetics, sugars, and microRNAs: Quaternaries in phosphate acquisition and use. *Plant Physiology, 154*(2), 582–588.

Vicente-Agullo, F., Rigas, S., Desbrosses, G., Dolan, L., Hatzopoulos, P., & Grabov, A. (2004). Potassium carrier TRH1 is required for auxin transport in Arabidopsis roots. *The Plant Journal, 40*(4), 523–535.

Wang, Y., & Wu, W.-H. (2013). Potassium transport and signaling in higher plants. *Annual Review of Plant Biology, 64*, 451–476.

Weih, M., Asplund, L., & Bergkvist, G. (2011). Assessment of nutrient use in annual and perennial crops: A functional concept for analyzing nitrogen use efficiency. *Plant and Soil, 339*(1), 513–520.

White, P. J. (1998). Calcium channels in the plasma membrane of root cells. *Annals of Botany, 81*(2), 173–183.

White, P. J., & Broadley, M. R. (2009). Biofortification of crops with seven mineral elements often lacking in human diets—iron, zinc, copper, calcium, magnesium, selenium and iodine. *New Phytologist, 182*(1), 49–84.

Wittstock, U., & Burow, M. (2010). Glucosinolate breakdown in Arabidopsis: Mechanism, regulation and biological significance. *The Arabidopsis Book/American Society of Plant Biologists, 8*.

Yong-Villalobos, L., González-Morales, S. I., Wrobel, K., Gutiérrez-Alanis, D., Cervantes-Peréz, S. A., Hayano-Kanashiro, C., Oropeza-Aburto, A., Cruz-Ramírez, A., Martínez, O., & Herrera-Estrella, L. (2015). Methylome analysis reveals an important role for epigenetic changes in the regulation of the Arabidopsis response to phosphate starvation. *Proceedings of the National Academy of Sciences, 112*(52), E7293–E7302.

CHAPTER TWO

Plant nutrition and soil fertility: physiological and molecular avenues for crop improvement

Seerat Saleem[a,1], Naveed Ul Mushtaq[a,1], Aadil Rasool[a], Wasifa Hafiz Shah[a], Inayatullah Tahir[b], Reiaz Ul Rehman[a]
[a]Department of Bioresources, School of Biological Sciences, University of Kashmir, Srinagar, India
[b]Department of Botany, School of Biological Sciences, University of Kashmir, Srinagar, India

1. Introduction

Food security can only be achieved if agricultural output is expanded on existing land, and thus crop production per unit of agricultural area must be raised. It is minerals that are crucial to the production of crops, the sustainable use of soil, and the prevention of soil degradation. The application of fertilizers and the preservation of soil fertility have been essential to boosting food production since the introduction of the 'Green Revolution' (Loneragan, 1997). During the past 40 years, nitrogen fertilizer use increase 6.9 fold and phosphorus raise by 3.5 fold (Tillman, 1999).

Plant nutrition is critical for commercial crop production. Nutritional programmes frequently employ commercially produced inorganic or organic fertilisers delivered to the soil by broadcast, irrigation, or foliar sprays.

Controlling key nutrients for plant growth is a vital stage of soil fertility and plant nutrition. These requirements are normally provided by 17 elements, which are classified as macro- and micronutrients. Macronutrients are often found in greater than 0.2% concentrations in plant tissue, whereas micronutrients are present in less than 0.01% concentrations (dry weight basis). The necessary mineral macronutrients are classified as basic macronutrients (N, K, and P) and secondary macronutrients (Ca, Mg, and S). Cu, Mn, Fe, B, Ni, Mo, Cl, and Zn are the eight micronutrients. Other mineral elements may be important for particular plants or boost crop quality or development while not being required for metabolic activities or the

[1] Equal Contribution.

Sustainable Plant Nutrition
ISBN: 978-0-443-18675-2
https://doi.org/10.1016/B978-0-443-18675-2.00009-2

conclusion of the plant's life cycle; these elements are frequently referred to as beneficial elements. Co, Na, Si, Se, and V are all beneficial elements (Behboudian et al., 2016).

2. Soil fertility and plant nutrition

Regulating soil nutrients to promote optimal plant nutrition has been an important aspect of crop production throughout history. For many centuries, recycling was an important method in agricultural output. Industrialization and population growth in the nineteenth century needed increased agricultural production, which coincided with the introduction of commercially made P fertilisers. Fertilizers including N, P, and K are expected to provide 40%–60% of yield and 40%–48% of global protein supply (Erisman et al., 2008; Stewart et al., 2005). The notion that nutrients required to be provided outside the body in the form of fertilisers in order to increase yield led to the development of the ideas of soil and plant nutrition. The soil nutrient and plant nourishment program's purpose is to address the factors influencing crop nutrient utilization sustainably in order to improve plant performance. As a consequence, the goal is to have most of each important nutrient accessible in plant tissue to support the metabolic functions of the plant. There is a critical concentration of each essential element inside plant tissue above which more nutrients will not improve plant performance, and below which plant performance will likely worsen due to nutrient deficiency. The presence of excessive amounts of several essential plant components may be dangerous and impair crop development. This is sometimes referred to as the toxic range. Fertilization also includes the analysis of soil nutrient content prior to fertiliser application. Soil fertility may be measured in a variety of methods. The analysis of data to provide fertiliser recommendations is an essential component of every soil research. The nutrient content of the soil is frequently studied and evaluated in order to propose which fertilisers to apply and which not to apply. Excess nutrients must also be avoided in order to avoid negative repercussions.

The most crop yield limiting factor especially in developing nation's world wide is soil infertility. The basic principles of a good farming practice focus on the maintenance of soil quality that can reduce land degradation, soil infertility and declining production levels. When the same piece of land is cultivated continuously, it results in a decline in soil fertility that is not easy to overcome even with the use of chemical inorganic fertilizers. The application of chemical fertilizers and pesticides has increased with increasing population. Chemical fertilizers accelerate soil acidification, cause air and ground water pollution weakens the roots of plants (Itelima et al., 2018). Remediation of degraded soils by sustainable methods is the need of the hour. Reportedly biochar is one such alternative. It not only improves the physical and chemical properties of soil but also microbial properties of soil (Ding et al., 2016). Biochar is composed of carbon, nitrogen, hydrogen and nutrients such as Ca, Na, K and Mg. Ash content in biochar ranges from 0.35% to 59.05% which is rich in cationic elements such as K (0−560 mmol kg-1), Mg (0.325 mmol kg-1), Na (0−413 mmol kg-1) and Ca (3−1210 mmol kg-1). Biochar affects soil properties by enhancing organic/mineral complexes, enhance nutrient cycles, decrease nutrient leaching and hence improve plant growth (Rajkovich et al., 2012). A collaborative strategy for enhancing soil fertility and crop productivity is the use of intercropping systems. Intercropping is an agricultural practice where two or more crops are cultivated at the same time. Intercropping is advantageous as it may lead to greater yield by making an efficient use of a piece of land and available resources. This is achieved by using a mixing of crops with different rooting ability, height, canopy structure and nutritional requirements. Intercropping improves soil fertility by fixing atmospheric nitrogen (150 tons/year) with the use of legumes and enhances soil conservation by increased ground cover (Dwivedi et al., 2015). Microbes colonize the rhizosphere, phyllosphere and some are endophytes i.e., they live inside the plant tissues. Microbes associated with plants participate in biological nitrogen fixation, produce plant growth hormonessuch as indoleacetic acid and gibberellic acid, solubilization of nutrients i.e., phosphorus, potassium and zinc, enhance plant defense reactions against pathogens and thus plant growth under challenging environments (Verma et al., 2017).

3. Micronutrients

Elements that are essential for the plant in much smaller quantities are known as micronutrients. Micronutrients are required in smaller amounts than the primary nutrients i.e., nitrogen, phosphorus, sulfur, and potassium. The various essential micronutrients for all higher plants are: boron (B), chloride (Cl), copper (Cu), iron (Fe), manganese (Mn), molybdenum (Mo), nickel (Ni), and zinc (Zn). Virtually all metabolic and cellular functions, like primary and secondary metabolism, energy metabolism, cell protection, gene regulation, hormone perception, signal transduction, and reproduction involve micronutrients (Hänsch & Mendel, 2009). Micronutrients play important roles in human body tissues. Their deficiencies result in serious health and nutritional consequences in human beings causing dietary micronutrient malnutrition. Micronutrients are the constituents of various coenzymes, macro-molecules, amino acids. In plants micronutrients play help in maintaining water potential, provide protection against environmental extremities and pathogen resistance (Nakandalage & Seneweera, 2018).

Micronutrients are mined from the earth's crust and are processed for various industrial applications. Soils contain sufficient amounts of micronutrients but with variable availability to the plants. Micronutrient availability is regulated by various edaphic and biological factors such as pH, competing cations, anions, organic matter, soil geomorphology, soil parent materials, and soil microbiology. Rapid reaction of micronutrients with soil components such as phosphates and carbonates lead to the formation of chemical precipitates and other mineral complexes leave the micronutrients unavailable to the crop. The various adaptation strategies of plants for micronutrient acquirement include elaboration of extensive root systems, excretion of root exudates such as organic acids and phytosiderophores that dissolve fixed minerals (Dimkpa & Bindraban, 2016). One of the best practices in order to provide the plants with needed micronutrients is foliar application. Foliar application of micronutrients such as Fe, Mn or Zn has reportedly promoted vegetative growth and root parameters in various medicinal and aromatic plants such *Cymbopogon flexuosus, Matricaria chamomilla, Trachyspermum ammi, Mentha piperita, Ocimum basilicum.* Foliar application of Fe, Mn or Zn also increases pigment content, carbohydrates, total free amino acids, protein, macronutrients and micronutrients (Hanafy et al., 2018). To mention the exact number of micronutrients is challenging as some elements are still not clearly categorized as essential or beneficial such as silicon (Si), which is reported as a micronutrient by some authors while as some categorize it as a beneficial element (Vatansever et al., 2017).

Micronutrient application not only enhances food grain production but also helps in sustaining soil health and strengthening the country's nutritional security. Feeding and nourishing a burgeoning human population with limited land resources for productive agriculture is the biggest challenge that we are facing. Soils with multi micronutrient deficiencies lead to poor quality crop produce (Shukla et al., 2018). Micronutrients in soil are present in ample quantities but the plants obtain these in trace amounts. Hence, these are regarded as trace elements. Micronutrients are required in low amounts but these play vital roles in plant growth and development such as fruit and seed development, production of carbohydrates, chlorophyll synthesis, reproductive growth and nutrient regulation. Micronutrient deficiency has become more prevalent in recent years and it results in abnormal growthin plants and degradation of agricultural crops. This deficiency is further increased by high soil erosion, modern crop cultivars. On the other hand, accelerated concentrations of micronutrients have also proven toxic and threaten plant growth (Tripathi et al., 2015). Adequate supply of balanced mineral nutrients increases plant production and tolerance capacity to various abiotic stresses. Among the 92 natural elements present in plants, 17 are recognized as essential nutrients. Among these 17 elements, eight are micronutrients i.e., iron (Fe), zinc (Zn), copper (Cu), manganese (Mn), molybdenum (Mo), chlorine (Cl), boron (B), and nickel (Ni). These are found in less than 0.01% of the total dry weight. Micronutrient deficiency in plants is known as "hidden hunger" and it leads to chlorosis, interveinal necrosis, and defoliation of leaves. There is a narrow range between deficiency and toxicity for micronutrients. Micronutrient deficiency in plants hampers photosynthesis, seed germination and chlorophyll biosynthesis. Micronutrients in plant system can be mobile or immobile. Mo and Mg are mobile micronutrients and their deficiency appears on the older or lower leaves, while, Fe, Cu, Mn, Ni, and B are immobile, and their deficiency appears in younger upper leaves (Noreen et al., 2018).

4. Role of Mn in plants

Manganese (Mn) is vital to plant metabolic functions; it exists in several oxidation states such as II, III, and IV, although the +2 oxidation state is more common in biological systems (Hebbern et al., 2009). Mn is an essential component for growth and development of plants, as well as for maintaining metabolic functions within various plant cell compartments, including photosynthesis, ROS scavenging, respiration, pathogen defense and sigalling. The metal is a necessary cofactor for the photosynthetic machinery's oxygen-evolving complex (OEC), which catalyses the water-splitting mechanism in photosystem II (PSII) (Bricker et al., 2012). ROS are primarily generated in chloroplasts, peroxisomes, mitochondria or in cytoplasm of plant cells. Mn acts as a cofactor for MnSODs, which are found in peroxisomes and mitochondria (Corpas et al., 2017). As a result, it aids in detoxifying reactive species. Mn is also a key cofactor for key enzymes in isoprenoid production (Kollner et al., 2008). Mn participates in lignin production via acting as a cofactor of PAL enzyme or even when lignin is produced using monolignols (Onnerud et al., 2002). Mn plays a crucial role in abscisic acid and auxin signaling (Schweighofer et al., 2004) Purple acid phosphatases (PAPs), NAD malic enzyme, and phosphoenolpyruvate carboxylase are other enzymes that need Mn as a cofactor (Gregory et al., 2009; Venkidasamy et al., 2019).

Mn insufficiency can be a significant plant nutritional problem if Mn bioavailability falls well below the amount necessary for optimal plant development (Broadley et al., 2012). In average, the significant value for Mn insufficiency is less than 10–20 mg kg^{-1} dry weight (Broadley et al., 2012). One of several effects of Mn shortage in plants is reduced development, which results in a reduction in biomass (Hebbern et al., 2005). This can all be attributed to fewer chloroplasts, lower net photosynthetic efficiency, and a decrease in chlorophyll (Alejandro et al., 2017), as well as increased susceptibility to pathogen infections (Heine et al., 2011), an imbalance in plant water relations, and decreased tolerance to cold temperatures (Ihnatowicz et al., 2014). Poorly drained and acidic soils lead to Mn toxicity. Hence, plants securely regulate Mn uptake, trafficking, and storage (Ye et al., 2020). Mn enhances the carbohydrate flow from leaves to roots, increases the sugar concentration in roots of sugar beets, protein content of the grain and is involved in Vitamin C synthesis. Mn deficiency leads to Chlorosis between veins, necrotic spots, reduced photosynthesis and chlorophyll, reduced plant development and yield (Rakhmatjonovna, 2021). Mn transport involves proteins that are not specific for the metal. Various divalent cations, such as $Fe^{2+,}$ Cu^{2+}, Zn^{2+}, Ca^{2+}, Cd^{2+}, Co^{2+}, and Ni^{2+} are transported via plant Mn^{2+} transporters (Chu et al., 2017). Mn homeostatic network involves transport proteins that are classified into importers and exporters. Mn from the extracellular space or from internal compartments is transported into the cytosol by the importers while exit of Mn out of the cytosol into intracellular compartments or into the apoplast is aided by the exporters. NRAMP family, YSL family, ZIP family are the various importers. While the exporters include, Cation Diffusion Facilitator/Metal Transport Protein (CDF/MTP) family, the Ca2+/Cation Antiporter (CaCA) superfamily, the Vacuolar Iron Transporter (VIT) family, the Bivalent Cation Transporter (BICAT) family, and the P2A-type ATPase family (Alejandro et al., 2020).

5. Role of Cu in plant development

Copper (Cu) is a crucial element for plants. Because Cu may emerge in many oxidation states, it participates in a wide range of physiological activities. Cu occurs as Cu^{2+} and Cu+ in physiological circumstances. Cu^{2+} is frequently associated with nitrogen in histidine side chains; however $Cu+$ tends to bind with sulfur in methionine or cysteine side chains. Cu serves as a structural component in many metalloproteins which are involved in electron transport reactions in chloroplasts and mitochondria. Similarly this element has a role in plant oxidative stress reaction. In addition, it is required for ethylene sensing, cell wall metabolism, and biosynthesis of molybdenum cofactors (Yruela, 2009). Apart from this a variety of enzymes use copper ions as cofactors, including the Cu/ZnSOD complex, ascorbate oxidase, cytochrome-c oxidase, amino oxidase, laccase, polyphenol oxidase and plastocyanin. Among Cu's many functions, Cu is critical for cell wall metabolism, signaling, iron mobilization, oxidative phosphorylation, as well as biogenesis (Puig et al., 2007). The various other metabolic processes involving Cu are hormone signaling, mitochondrial respiration, cell wall metabolism, photosynthetic electron transport and superoxide scavenging. Cu also acts as a structural component of a variety of enzymes such as cytochrome *c* oxidase, amino oxidase, plastocyanin and polyphenol oxidases, laccase, and Cu/Zn superoxide dismutase (Ng et al., 2014).

Superoxide dismutase extracted from chloroplasts contains 2 g of Cu atoms. Plastocyanin is a basic blue protein present in the chloroplasts of green photosynthetic tissues of higher plants. Plastocyanin is essential in this process because it transports electrons from photosystem II to photosystem I via cytochrome f. Plantacyanin, also known as cusacyanin that contains one Cu atom per molecule. It was extracted from cucumber initially, and subsequently proved that plantacyanin is found in chloroplasts (Droppa & Horváth, 1990). In sugar beet, a 60% fall in leaf Cu was associated with a considerable decrease in pigments and plastoquinone, while Cu replenishment triggered a concurrent increase in each of these components. This highlights the importance of Cu in the efficient working of pigments (Horvath et al., 1983).

Several enzymes are inhibited by excess Cu, affecting key biochemical functions in plants, such as photosynthesis, pigment production, and membrane integrity. By inhibiting photosynthetic electron transport, it initiates peroxidative chain reactions involving membrane lipids which is perhaps its most significant negative role of excessive Cu. Excess copper has a wide range of impacts on plant physiology, including interference with fatty acid and protein metabolism as well as suppression of respiration and N_2 fixation. Cu is an efficient opponent of vegetative development and generates widespread senescence symptoms at the plant level (Fernandes & Henriques, 1991). Raised Cu concentrations are toxic to plants and hence plants control their intracellular levels by controlling its uptake and minimizing free Cu concentrations via metallochaperones.

Cu deficiency results in stunted plant growth, chlorosis of leaves with necrotic spots on leaf margins, male sterility, decreased stem length, light colored leaves, stem lodging, reduced resistance to diseases, leaf wilting. On the other hand, Cu toxicity caused by increased concentrations of Cu cause nutrient imbalances, stunted growth with yellow or brown colored leaves (Rehman et al., 2019). In *Arabidopsis thaliana* Cu transport into the cytosol is mediated by six high affinity transporters COPT1−6and its exit is aided by P-type ATPases such as HMA1, HMA5, PAA1, PAA2, and RAN1. Metallochaperones such as CCS1, ATX1, and CCH help with the intracellular distribution of Cu (Vatansever et al., 2017). OsATX1, antioxidant protein1 from *Oryza sativa* plays an important role in translocation of Cu from root-to-shoot and from older leaves to developing seeds and tissues (Zhang et al., 2018).

6. Role of iron in plants

To attain optimal development, plants require Fe concentrations of 10^{-9}−10^{-4} M (Guerinot & Yi, 1994). Iron is mostly present in soil in the Fe^{+3} form (Samaranayke et al., 2012). As a micronutrient, iron is necessary for every living creature because it contributes to metabolic processes (Kim & Guerinot, 2007). Iron also activates several metabolic processes and is a prosthetic group component of several enzymatic reactions involved in plant hormone production, including like ethylene, lipoxygenase, and abscisic acid (Rout & Sahoo, 2015; Siedow, 1991). Plant metabolism requires this element. In addition to serving as a structural element for proteins and enzymes, iron plays a vital role in fundamental biological processes such as photosynthesis, chlorophyll synthesis, respiration, nitrogen fixation, uptake mechanisms, and DNA generation (Pushnik et al., 1984). Iron is required for the production of chlorophyll and for maintaining chloroplast structure and function. When iron levels reach dangerously high levels, it becomes poisonous. It can catalyze the Fenton reaction, generating hydroxyl damage by free radicals. Thus it can damage lipids, proteins, and DNA. The result of this is that plants need to be able to deal with iron shortages as well as excesses of iron (Connolly & Guerinot, 2002). Because of the limited solubility of the oxidized ferric form in aerobic settings, iron is the third most limiting nutrient for plant growth and metabolism. In soils, iron, which is the fourth most common metal in the lithosphere, is often found in high concentrations; however, it has limited bioavailability (Guerinot & Yi, 1994). Thus, increases the chances of deficiency in plants. Plants that are iron deficient have lower yields and lower nutritional quality, leading to decreased nutritional quality.

Iron uptake, mobilization, utilization and partitioning in plants require a tight regulation at cellular and molecular levels as large quantities of Fe in free state can lead to the excessive production of reactive oxygen species (ROS) (Kobayashi & Nishizawa, 2012; Vatansever et al., 2017). Plants have divided iron uptake into Strategy I and Strategy II i.e., reducing and chelating strategies respectively. The uptake involves ferrous Fe^{2+} for Strategy I and ferric Fe^{3+} for Strategy II. Iron-Regulated Transporter 1 (IRT1) mediates the uptake and transport of reduced iron in *Arabidopsis*. Rice and barley comprise of a functional homologue of IRT1 for the uptake and transport of iron (Connorton et al., 2017). NRAMP (The Natural Resistance Associated Macrophage Protein) and YSL (Yellow Stripe-Like) gene families mediate translocation of Fe under various stress conditions. Fe acts as regulatory element and protects the plants against various stresses (Tripathi et al., 2018). 16 bHLH (basic helix-loop-helix) TFs (Transcription factors) have been characterized in *Arabidopsis* that maintain iron homeostasis (Gao et al., 2019).

7. Role of boron in plants

Boron (B) is required by plant at low concentrations, so classified as micronutrient. Micronutrients are required at modest quantities are required for appropriate growth and development (Brown et al., 2002). Boron is required by crops, its availability in soil and irrigation water is a powerful determinant of agricultural yield (Reid, 2014). B has been claimed to play a crucial role in higher plants due to its ability to synthesize borate esters with rhamnogalacturonan II apiose residues (RG-II). The creation of this complex is necessary for the structure, function, tensile strength, and porosity of the cell walls (O'Neill et al., 2004). The quantity of this element in the wall varies according to type of plant, tissue, and boron availability. Plants can absorb boron as tiny uncharged boric acid molecules (approximately 96%) and to a much smaller amount, borate anions, which enter the cell via the phospholipid bilayer (Bolaos et al., 2004). It generates cis diol complex structures with glycoproteins and glycolipids in the plasma membrane, allowing it to retain its shape. It is engaged in enzymatic processes, transfer of ions, metabolites and hormones via playing a role in membrane physiology and functionality (Goldbach & Wimmer, 2007). Apart from this, boron is involved in transfer of sugar in membrane, production of secondary metabolites (flavonoids), harmones (auxin), nitrogen containing compounds and phenol metabolism (Seth & Aery, 2017). In case of legumes and mycorrhiza, this element is associated with cell mitosis, tissue development and differentiation, N_2 fixing and in transcription.

Boron has an extremely narrow range between deficiency and toxicity and is required by the plants in as low as 1μ/g—1mg/g quantities (Brdar-Jokanović, 2020). Besides its helpful character, the shortage of this nutrient causes a variety of developmental disorders. The suppression of DNA synthesis in root is reported under situations of low boron availability (Brdar, 2020). Its shortage influences photosynthetic activity and transport of photosynthetic products, as well as stomatal conductance and carbon dioxide assimilation. Boron insufficiency reduces male fertility by lowering microsporogenesis, germination and pollen tube elongation (Shah et al., 2017). Crops exhibit a range of deficient indications such as reduced root development, limited apical growth, tough leaves, decreased photosynthetic pigments and activity, disturbance in transport of ions, elevated phenolic and lignin concentrations, lower crop production etc (Shireen et al., 2018). In order to maintain optimal B content in their cells, plants regulate B homeostasis. Two classes of B transporters are involved in B uptake and transport i.e., BA channels from major intrinsic protein (MIP) family and borate transporters from BOR family (Matthes et al., 2020).

8. Role of selenium in plants

Soil salinity has a negative impact on plant development and crop production (Mushtaq et al., 2021) Salt stress plants exhibit nutritional imbalances, ion toxicity, severe osmotic and oxidative stress, reduced photosynthetic rates, and decreased output. Se, in particular, plays critical functions in mitigating the detrimental effects of abiotic stress, if used at low concentrations. It can boost plant growth and development (Hasanuzzaman et al., 2014; Pilon-Smits, 2015). Even though its involvement in plants has been thought to be helpful, se has not been categorized as an essential element. It inhibited lipid per oxidation by acting as an antioxidant (Rasool et al., 2020; Shanker, 2006). Although Se is toxic when applied large quantities, it can be helpful in moderate amounts. Senescence and plant apoptosis like processes are slowed owing to its increased antioxidation, which is linked to an elevation in enzyme activity. Se has already been demonstrated to increase plant development in the presence of ultraviolet light induced oxidative damage and salt stress (Rasool et al., 2020;

Xue and Hartikainen, 2000). Se can also be used to postpone senescence and boost the development (Xue et al., 2001). It helps in homeostasis of essential nutrients, maintain cell structure and function, thus enhancing plant growth and improve photosynthesis. It also inhibits the uptake and accumulation of toxic metals in plants (Chauhan et al., 2019). Selenium toxicity caused to the accumulation of excess selenium leads to a competition between Se and S due to their chemical structural similarity. Se phytotoxicity symptoms include foliar chlorosis, shoot stunting, necrosis, leaf withering, drying, reduced shoot and root fresh weight (FW) and dry weight (DW) (Hasanuzzaman et al., 2020). Selenate is more common in well-drained/oxidized and alkaline soils while wetlands and anaerobic soils having neutral to acidic pH mostly contain selenite. A study on model plant *A. thaliana* concluded that, SULTR1; 2, i.e., group 1 root high affinity sulfate transporter conducting active sulfate uptake is responsible for selenate uptake into the plants. H_2SeO_3 form of selenite in rice is absorbed via aquaporins and silicon (Si) influx transporter OsNIP2; 1. While as $HSeO_3$ uptake involves phosphate transporters (Lima et al., 2018).

9. Zinc (Zn)

Zinc plays an important role in the growth and development of plants and animals. The various roles of zinc in plants involve membrane function, gene expression, photosynthesis, protection against drought and pathogens and in the synthesis of hormones which are required for plant growth and development. Zinc deficiency in plants mat result due to poor soils. In crop plants zinc is transported directly from the soil eitherin Zn^{2+} form or bound to an organic acid. Hence, Zn can get accumulated in the roots of plants before being translocated to the shoots and leaves via xylem. Zn deficiency in plants cause stunted growth and chlorosis on leaves. High accumulation of zinc in seeds suggests its importance in plant fertility. Most of the world's agricultural lands are deficient in zinc. While alkaline and calcareous soils are rich in zinc, this Zn is not available to the plants because it is tightly bound to the $CaCO_3$ present in the soil (Hefferon, 2019). Zn is a catalytic element that functions in more than 300 enzymes, such as alkaline phosphatase, carbonic anhydrase, alcohol dehydrogenase, and Cu–Zn superoxide dismutase. It also has a structural role in the stabilization of proteins such as Zn cluster, Zn fingerand RING finger domains/motifs. In plants Zn homeostasis is maintained by ZIP (Zn, iron-permease family/ZRT, IRT proteins) family uptake transporters in a coordinated regulation mechanism (Sadeghzadeh & Rengel, 2011, pp. 335–375). Other proteins involved in the translocation of Zn are the heavy metalATPase (HMA) family and the metal tolerant proteins (MTP) family. ZIP family participates in the Zn influx into cell cytosol, while HMA mediate Zn efflux into the apoplast. Zn sequestration into the vacuoles and endoplasmic reticulum is facilitated by MTP family (Hassan et al., 2020).

10. Molybdenum (Mo)

Mo availability in the soil depends on various factors such as soil pH, organic compounds present in soil, concentration of absorbing oxides. Molybdenum in soil is present in various complexes like molybdenite, wulfenite and ferrimolybdenite. In alkaline soils Mo is much more soluble and hence easily accessible by plants in the form of MoO4. Plants need Mo in a small amount in the range of 12–32 g ha^{-1} for proper physiological function. Depending on plant species the critical deficiency and toxicity levels vary from 0.1 to 1 mg kg^{-1}. Mo deficiency in plants results in mottling lesions on the leaves, altered leaf morphology, leaf cupping, gray tinting and flaccid leaves (Naqib & Jahan, 2017). Mo is involved in vital metabolic functions such as synthesis of molybdenum cofactor (Moco) that acts as active site of molybdoenzymes participating in phytohormone biosynthesis, purine metabolism, sulfite detoxification, and nitrate assimilation. These enzymes include aldehyde oxidase (AO), sulfite oxidase (SOX), xanthine dehydrogenase (XDH), and nitrate reductase (NR) (Kruse et al., 2010). Plant proteins belonging to the MOT1 and MOT2 families are known as high affinity and specific $MoO4^{2-}$ transporters. MOT1 (Molybdate transporters type (1)) family are the first identified proteins facilitating $MoO4^{2-}$ transport in *Arabidopsis thaliana*. MOT1 family transporters have been reported from algae, plants, fungi, and bacteria. AtMOT1 absence led Mo deficiency in *Arabidopsis* plants characterized by reduced growth and curved and yellow leaves. AtMOT2 is the second member of MOT1 family that is identified in *A. thaliana* and it is suggested to participate in the intracellular transport of Mo by vacuolar $MoO4^{2-}$ transport (Manuel et al., 2018). MOT2 is the second $MoO4^{2-}$ transport family and CrMOT2 has been identified in Chlamydomonas. CrMOT2 is characterized as a specific and high affinity transporter of$MoO4^{2-}$ (Tejada-Jiménez et al., 2011).

11. Nickel (Ni)

Ni is an essential micronutrient that is required for the normal plant growth and development in addition to various biological functions. It is an essential part of various enzymes such as glyoxalase-I, urease that are required for nitrogen metabolism. It also helps plants to cope with various biotic and abiotic stresses. It plays important roles in maintaining the cellular redox state, ureolysis, acitogenesis, methane biogenesis, and hydrogen metabolism. Ni deficiency in plants results in stunted growth reduced N metabolism and decreased Fe uptake. Excessive Ni leads to Ni toxicity that causes leaf chlorosis, growth inhibition, reduced photosynthesis; mineral nutrition disorders (Shahzad et al., 2018). Ni uptake and photo availability in plants depends upon the Ni content in the soil solution, organic matter, and soil Ph, and iron-manganese oxide concentration. Acidic soils (low pH) increase the mobility and solubility of Ni. Ni is mostly accumulated in roots while in some species it translocated to upper plant parts (Hassan et al., 2019). Plant root system carries out Ni^{2+} uptake via both passive diffusion and active transport. The uptake mechanism used depends upon the plant species, oxidation state, soil acidity, concentration of Ni^{2+} in the soil. Ni^{2+} uptake may involve basipetal transport that takes place primarily via epidermal and cortical cell layers in phloem. It is then translocated to expanding leaves, growing tips and root parts (Sachan & Lal, 2017). Various cation transport systems such as Mg^{2+}, Fe^{2+}, Cu^{2+}, and Zn^{2+}; chelators such as citric acid, nicotianamine (NA), histidine (His) and different proteins such as permeases, metallothionein (MT), YSLs and metallochaperones can transport solubilized nickel compounds (Vatansever et al., 2017).

12. Chlorine (Cl)

Cl is an essential micronutrient that is beneficial for higher plants. Cl concentration in soil depends on bedrock material, depositions of airborne Cl. Cl is highly water soluble and mobile in soil solution (Geilfus, 2018). Cl plays an integral role in the opening and closing of stomata. It is also involved in root growth, osmoregulation, regulation of pH, stabilization of membrane potential (Hassan, 2019). Cl deficiency hampers leaf growth and causes stunted roots, reduced fruit size, wilting with chlorosis and necrosis (White & Broadley, 2001). On exposure to high levels of Cl, lower leaves undergo premature yellowing with burning of leaf margins and tips; wilting and leaf abscission in woody plants (Kathpalia & Bhatla, 2018). Net Cl^{-} uptake in plant cells involves combined activities of both influx and efflux transporters where electrochemical gradient acts as a main driving force. Cl^- influx occurs by active transport that transportsCl^- against the electrochemical gradient via high- and low-affinity Cl^-/H^+ symport mechanisms. Cell Cl^- exit takes place down the electrochemical gradient and is mediated by anion channels. Nitrate transporter 1/Peptide transporter Family (NPF) members are involved in Cl^-influx. Recent identification and characterization of genes involved in encoding transporters mediating Cl^- influx, Cl^- efflux and Cl^- compartmentalization include (ZmNPF6.4 and ZmNPF6.6); (AtSLAH3 and AtSLAH1); (AtDTX33, AtDTX35, AtALMT4, and GsCLC2) respectively (Colmenero-Flores et al., 2019).

13. Macronutrients

Plants require 14 essential inorganic elements to complete full life cycle and these essential elements on the basis of their concentration in plant dry matter are grouped into macronutrients and micronutrients. The macronutrients comprise of nitrogen (N), phosphorus (P), sulfur (S), potassium (K), calcium (Ca) and magnesium (Mg). Nutrition deficiency can be mild, transient to severe and chronic. Nutrient deficiencies negatively impact crop production, resulting in reduced yields and poor quality of food and feed (de Bang et al., 2021). Relatively large quantities of macronutrients are taken up by plants. Among these P, N and K are essential for optimal plant development. P and N are vital components of DNA and RNA. K provides the cells with electrical charges, plays an important role in osmoregulation and stress tolerance. In order to access the macronutrients under harsh conditions highly specialized mechanisms and regulatory networks are evolved by plants in order to increase nutrient availability in the rhizosphere, and thus, enhancing their acquisition and intracellular concentrations. P-, N-, and K deficiencies in forages, fruits and vegetables affect the crop quality adversely. P, N, and K fertilization is crucial for crop plants for improving acquisition, yields and internal utilization (López-Arredondo et al., 2017).

14. Nitrogen (N)

N is a vital constituent of purines and pyrimidines that form the nucleic acids (DNA and RNA). N is also an essential component of amine group of amino acids present in all peptides, proteins and enzymes. Plants have the majority of shoot N present as RuBisCo in the chloroplasts and also in photosystem proteins and chlorophyll molecules containing 4 N atoms in the tetrapyrrole (chlorin) ring. Important signaling molecules such as nitrate (NO_3^-), ammonium (NH_4^+) and nitric oxide (NO) contain N. Plants take up N mainly in the form of NO_3^- and NH_4^+ (de Bang et al., 2021). Atmospheric N_2 is the indirect source of N for plants, that is either fixed through symbiotic root legume bacteria or taken up by nonsymbiotic N fixating soil microorganisms. Plants contain 1.5%–8% total N per dry matter, 16% of which is introduced into proteins. In sufficient N supply, the total N concentration in shoots and roots is about 100 mM, that is mostly stored in vacuoles. Uptake, transportation, and redistribution of N is controlled genetically. Plants facing N deprivation respond by root growth, and proliferation in the direction of N-rich regions. Application of nitrogenous fertilizers is required for high crop yield because N availability is a limiting factor for agriculture (Weissert & Kehr, 2017). Optimum N concentration enhances photosynthetic processes, leaf area production, leaf area duration as well as net assimilation rate. Maximum use of N globally has resulted in increased yield of various crops. Application of 180 kg N yielded 3.848 tons ha^{-1} of wheat. Similarly, 120 kg ha^{-1} N enhanced the height of rice, dry weight, number of tillers, number of filled grains/panicle, length of panicle, straw yield, biological yield, harvest index, benefit cost ratio and grain yield 4.66 tons ha^{-1}. Increasing rate of N at 195.5 kg ha^{-1} significantly increased theyield and quality of sugarcane (Leghari et al., 2016). N imparts dark-green color to the plants, promotes root growth, leaves, stem and other vegetative part's growth, and enhances fruit quality, protein content, and growth of leafy vegetables. It also increases the uptake and utilization various nutrients such as potassium, phosphorous (Hemerly, 2016). Nitrogen deficiency hampers growth, results in chlorosis, appearance of red and purple spots on the leaves, restricted lateral bud growth and leaf senescence (Bianco et al., 2015). Excess N also causes negative effects on the plant such as excessive vegetative growth, extradark-green colored leaves, reduced fruit quantity and quality (Leghari et al., 2016). The two basic metabolic pathways in plants are carbon and metabolism pathways. Carbon source and energy for the nitrogen metabolism are provided by the carbon metabolism while the enzymes and photosynthetic pigments for the carbon metabolism are provided by the nitrogen metabolism. These two metabolisms are closely coordinated. Carbon and nitrogen metabolism is primarily impacted by nitrogen availability. In rice high N by regulates the carbon metabolism pathways i.e., the tricarboxylic acid (TCA) cycle and the pentose phosphate pathway

(PPP) that inhibit nitrogen assimilation and aromatic metabolism pathways. Under low nitrogen, the TCA cycle is enhanced in order to produce more energy and α-ketoglutarate, thereby enhancing nitrogen transport and assimilation. Low N also inhibits PPP, which may be consistent with the lower NADPH demand under low nitrogen (Xin et al., 2019). Most crops absorb Nitrate (NO_3^-) which is a major form of N and also a vital signaling molecule. Nitrate transporters (NRT1s and NRT2s) absorb the nitrate from the external environment into the roots. Various enzymes such as nitrate reductase (NR), nitrite reductase (NiR), glutamine synthase (GS), and glutamate synthase (GOGAT) are involved in the reduction and assimilation of nitrate into amino acids (Zhao et al., 2018).

15. Phosphorus (P)

Animals, plants and microorganisms require Phosphorus (P) for their growth, metabolism, development, and regulatory processes. Vital bimolecules such as ATP, NADPH, phospholipids, nucleic acids, and sugar-phosphates are composed of P. It also participates in the metabolism of proteins and lipids, transpiration, photosynthesis, and secondary metabolism. P is the second most important plant nutrient after N and constitutes of about 0.5% of cell dry weight. Optimum P concentration is necessary for sustainable and profitable agriculture as it impacts the chemical and biological properties of the soil. Total soil P comprises of organic and inorganic phosphorus (Pi), whose percentage varies with soil type. Pi deficiency in plantsleads to reduction of vegetative growth, chlorosis and necrosis, purple-green leaves (Ova et al., 2015).30%—40% of the world's arable land is affected by P deficiency. P enhances seed germination; stalk and stem strength; development of roots, flower and seed formation; crop yield; and quality. It also increases the N-fixing capacity of leguminous plants (Malhotra et al., 2018). Low P conditions lead to stunted growth, dark green foliage with reduced leaf surface area, upward tilting and curling of leaves (Malhotra et al., 2018). Nonutilization of P present in the soil by plants is due to its complexation with inorganic metal ions or organic compounds. To overcome this, chemical fertilizers are used by farmers which prove quite ineffective due to the precipitation of a larger proportion of soluble P and also can threaten the environment. An eco-friendly way to tackle this situation is the use of phosphate solubilizing bacteria (PSB) that can convert lockedP into plant assimilable form. Some of the dominated PSB are from genus *Bacillus, Enterobacter, Pantoea, Pseudomonas* (Kumar, 2016).

Low Pi stress in plants has resulted in the evolution of various strategies such as morphological and root anatomy alteration, induction of Pi transporters, biochemical transformation of metabolites, proteins and genes and Pi remobilization (Maharajan et al., 2018). Pi deficient plants undergo various morphological, physiological, and biochemical adaptations, collectively known as Pi-starvation response (PSR). PSR includes the induction of hundreds of Pi-starvation inducible (PSI) genes that code enzymes responsible for the reprioritization of internal Pi use and maximization of external Pi acquisition. Important responses include *Phosphate Transporter1 (PHT1)* induction that is a family of genes localized in the plasma membrane. These are responsible for increasing the Pi-influx capacity by several fold (Plaxton & Tran, 2011). The various other genes including transcription factors and miRNA, that are induced by phosphate starvation and aid in the regulation of P homeostasis include phosphate responsive (PHO), E3 ligase (SIZ1), phosphate starvation response (PHR), phosphate transporter traffic facilitator (PHF), WRKY, ZAT6 and miR399 (Malhotra et al., 2018).

16. Sulfur (S)

S is one of the essential macronutrients in plants that performs numerous plant functions and is important for the metabolic processes. S is present in several important amino acids and metabolites. S in plants plays important roles in normal conditions as well as in the abiotic stress tolerance. Abiotic stresses hamper the plant growth, reproductive development, nutrient uptake and mitochondrial reaction. S compound either act as antioxidants directly or regulate antioxidant defense system e.g., Glutathione (GSH). After N, P, and K, S is the fourth most important macronutrient required by the plants. Plants take up S via the roots in the form of sulfate (SO_4^{2-}) integrated into cysteine (Cys) which acts as a precursor of important S compounds such as methionine (Met), S-adenosylmethionine, glutathione (GSH), homo-GSH (h-GSH), phytochelatins (PCs), sulfolipids (Hasanuzzaman, Hossain, et al., 2018). S interacts with various biomolecules such as polyamines, phytohormones, nitric oxide (NO), and someother plant nutrients, that produce some derivatives essential for abiotic stress tolerance (Hasanuzzaman, Bhuyan, et al., 2018). S protects the plants against the accumulation of toxic heavy metals by phytochelatins. Phytochelatins are synthesized from glutathione. In addition to this sulfur thioredoxins participate in the regulation of enzyme activities. Iron-sulphur clusters (Fe—S), form a vital part of several proteins; reduced sulfur is a part of various vitamins and coenzymes such as biotin, coenzyme A and thiamine pyrophosphate (Pandey, 2018). S scarcity leads to insufficient use of nutrients like carbon and N affecting protein biosynthesis, chlorophyll concentration and ultimately crop yield (Capaldi et al., 2015). Due to the deficiency of S in the arable soil around the world and need for high productivity, the demand for S fertilizers is continuously increasing. The global S fertilizer market worth US $4.52 billion in 2017 is expected to reach US $7.03 billion by 2026. In plant tissuesthe total S content ranges from 0.3% to 7.6%. S in soil exists as organic sulfur compounds, sulfide (S^{2-}); elemental sulfur (S^0); and sulfate (SO_4^{2-}). It is transformed from one form to another via mobilization, mineralization, immobilization, oxidation and reduction. Total soil S is composed of 98% organic sulfur compounds, that undergoes mineralization by microorganisms to release $SO_4{}^{2-}$ for plant uptake. Increased sulfate uptake capacity is a classical response to sulfate scarcity. Increased transcriptional regulation of root localized sulfate transporters, *SULTR1* and *SULTR1* are responsible for this response (Zenda et al., 2021). The plant physiological responses to S deficiency include growth retardation, leaf chlorosis, and accumulation of anthocyanins, severe reduction in levels of nonpolar amino acids in S-deprived roots. Transcriptomic changes include induction of 'S-deficiency marker genes' i.e., genes responsible for transport and assimilation of S. S deficiency in wheat lead to the induction of *SDI1* gene that plays a role in saving the available sulfate resources for primary rather than secondary metabolism (Aarabi et al., 2020).

17. Potassium (K)

Being an essential macronutrient K is involved in various physiological processes such as enzyme activation, osmoregulation, membrane potential maintenance and electrical neutralization. K+ regulates the cell turgor pressure, stomata movement, pollen tube growth, carbohydrate translocation and metabolism. Additionally, optimum K concentration provides tolerance toward various biotic and abiotic stresses (Wang and Wu 2017). K is also involved in inducing the growth in young tissues and cell enlargement and also in improving the plant quality and oil content in plants and enhancing the shelf life of crops. Use of imbalanced fertilizer has led to the depletion of available K level in soils. It has been reported that use of biofertilizers in the form of beneficial microorganisms mobilizes the availability of nutrients, especially N—P—K and helps in the formation of positive microflora by their biological activity, hence enhancing the soil health

(Rawat et al., 2016). Reportedly *Bacillus mucilaginosus* which is a potassium-solubilizing bacteria increased macronutrient uptake in eggplant, pepper and cucumber, hence enhancing the yields (Han et al., 2006). Sensing and signaling systems in plants in response to K deficiency include specific signal molecules, ROS, Ca^{2+}, phytohormones and microRNAs. Hyperpolarization of the plasma membrane is one of the initial responses of K deficiency. Visual symptoms include chlorosis at leaf tips that turn into necrosis (de Bang et al., 2021). K deficiency in the arable lands hampers sustainable agricultural production. This calls for the genetic improvement of crop K^+ utilization efficiency (KUE). In plant cells K transport and ion homeostasis is regulated by K^+ transporters. These transporters are from the HAK/KUP/KT family e.g., In Arabidopsis root, AtAKT1 channel and AtHAK5 transporter that are two major K^+ uptake components. Similarly in rice, three HAK/KUP/KT K^+ transporters, *OsHAK1/5/21*, were reported (Wang and Wu 2017).

18. Calcium (Ca)

Ca in plants acts as structural element and participates in cellular signaling. It performs dual function in plants i.e., as a structural component and secondary messenger. Cellular signaling by Ca is accomplished by transducing, integrating and multiplying incoming signals and hence inducing physiological responses to various environmental stimuli. Majority of Ca is found in the cell walls, cell organelles and outside the cell membrane. It provides the cell wall with rigidity (Hocking et al., 2016). Cytosolic Ca^{2+} levels initiate signaling cascade in response to various factors such as light, gravity, temperature, pathogen infection and touch. Ca^{2+} has an integral participation in nutrient signaling and subsequent adaptation to nutritional status. External nutrient availability and transport activities comprises of Ca^{2+} dependent CBL-CIPK pathway where CIPK23 acts as a master regulator of K^+, NO_3^-, Fe, Mg and NH_4^+ equilibrium (Kudla et al., 2018). Ca from the soil solution is taken up by the plant via plasma membrane channels expressed in roots. These channels are classified into depolarization activated channels (DACCs) and hyperpolarization-activated channels (HACCs). Reportedly, CNGC2 (cyclic nucleotide-gated channel 2) from *Arabidopsis* is responsible for influx of Ca^{2+} into the leaf cells after unloading from the vasculature. Cytosolic Ca^{2+} concentration serves as signal and thus free Ca^{2+} levels in the cytosol are kept at low levels i.e., 0.1 μM. This level is maintained by chemical buffers and various H^+/Ca^{2+} antiporters and

Ca^{2+}-ATPases (Thor, 2019). Approximately 70%—90% of Ca in plants is present in the leaves while 10%—30% is found in the root. Ample amount of Ca is present in cell walls, where it cross links with carboxyl groups of pectin chains that enhances cell wall integrity. High acid content in soils may lead to calcium deficiency in addition to severe weathering and leaching. Ca deficiency results in reduced yield and quality of crops and hence economic losses. Developing tissues and young leaves are the first to show the deficiency symptoms as compared to the older leaves that may contain sufficient Ca concentration. Ca deficiency affects cell wall development, causes necrosis of roots and young leaves and lower shoot-root ratio (González-Fontes et al., 2017). In meristematic tissues, Ca plays a vital role in cell enlargement and cell division (spindle formation). Plant roots take up Ca as divalent cations (Ca^{2+}). Acidic soils are enriched with lime fertilizer to raise soil pH (Kathpalia & Bhatla, 2018).

19. Magnesium (Mg)

Mg is vital to plants as 75% of leaf Mg participates in protein synthesis and 15%—20% of Mg is linked to chlorophyll pigments and also acting as a cofactor to various enzymes that participate in carbon fixation and metabolism. K^{+} is an antagonist for Mg^{2+} absorption in plants. Optimal plant growth requires 1.5—3.5 g per kg Mg in vegetative parts. Approximately 125 μmol L^{-1}—8.5 mmol L^{-1} Mg content in soil sufficient for optimal plant growth although reduced Mg is due to high rainfall leading to Mg leaching; almunium toxicity, long-term unbalanced crop fertilization (NPK); acidic soils resulting in low cation exchange capacity. Mg deficiency in plants results in smaller roots and shoots, necrosis of leaves, reduced chlorophyll and carbon fixation (Guo et al., 2016). Magnesium transporter (MGT) genes are expressed in a variety of plant tissues and the proteins are involved in Mg transport in and out of the cell. Reportedly *AtMGT1* gene from *Arabidopsis thaliana* that is localized in the plasma membrane was highly induced in roots during Mg deficiency and hence suggesting its role toward Mg uptake and growth during Mg scarcity. In rice, *OsMRS2-1, -3, -6* and *-9* showed Mg transport activity when expressed ectopically in yeast CM66 mutant. More recently, *PtrMGT5* from *Poncirus trifoliata* was characterized and it was concluded that this gene was over expressed in Mg deficiency and played an important role in Mg absorption, transport and maintaining Mg homeostasis (Liu et al., 2019).

20. Conclusion

Plant nutrition has a critical role in maintaining all vital processes in plants. Various nutrient elements are present in association with enzymes and help in proper working of metabolic processes. Similarly such elements have role in aiding plant to with stand stress full environment, diseases or internal toxic compounds. A proper amount of each nutrient is cardinal, below the threshold level deficiency symptoms arise and result in plant death. Same is the case with overdose of nutrients. Thus proper soil management steps should be taken to optimize the nutritional profile for crops that in turn will result in crop improvement and alleviated yield. Further, mutual attempts among plant biologists, ecologists, molecular biologist and agronomists are required to improve crop production and sustainable soil fertility.

References

Aarabi, F., Naake, T., Fernie, A. R., & Hoefgen, R. (2020). Coordinating sulfur pools under sulfate deprivation. *Trends in Plant Science, 25*, 1227–1239.

Alejandro, S., Cailliatte, R., Alcon, C., Dirick, L., Domergue, F., Correia, D., & Curie, C. (2017). Intracellular distribution of manganese by the trans-Golgi network transporter NRAMP2 is critical for photosynthesis and cellular redox homeostasis. *The Plant Cell, 29*(12), 3068–3084.

Alejandro, S., Höller, S., Meier, B., & Peiter, E. (2020). Manganese in plants: From acquisition to subcellular allocation. *Frontiers in Plant Science, 11*, 300.

Behboudian, M. H., Pickering, A. H., Dayan, E., & Hanegev, I. (2016). Deficiency diseases, principles. *Encyclopedia of Applied Plant Sciences, 1*, 219–224.

Bianco, M. S., CecílioFilho, A. B., & de Carvalho, L. B. (2015). Nutritional status of the cauliflower cultivar Verona grown with omission of out added macronutrients. *Plos One, 10*(4), e0123500.

Bolaños, L., Lukaszewski, K., Bonilla, I., & Blevins, D. (2004). Why boron? *Plant Physiology and Biochemistry, 42*(11), 907–912.

Brdar-Jokanović, M. (2020). Boron toxicity and deficiency in agricultural plants. *International Journal of Molecular Sciences, 21*(4), 1424.

Bricker, T. M., Roose, J. L., Fagerlund, R. D., Frankel, L. K., & Eaton-Rye, J. J. (2012). The extrinsic proteins of Photosystem II. *Biochimica et Biophysica Acta (BBA)-Bioenergetics, 1817*(1), 121–142.

Broadley, M., Brown, P., Cakmak, I., Rengel, Z., & Zhao, F. (2012). Function of nutrients: Micronutrients. In *Marschner's mineral nutrition of higher plants* (pp. 191–248). Academic Press.

Brown, P. H., Bellaloui, N., Wimmer, M. A., Bassil, E. S., Ruiz, J., Hu, H., … Römheld, V. (2002). Boron in plant biology. *Plant Biology, 4*(02), 205–223.

Capaldi, F. R., Gratão, P. L., Reis, A. R., Lima, L. W., & Azevedo, R. A. (2015). Sulfur metabolism and stress defense responses in plants. *Tropical Plant Biology, 8*(3), 60–73.

Chauhan, R., Awasthi, S., Srivastava, S., Dwivedi, S., Pilon-Smits, E. A., Dhankher, O. P., & Tripathi, R. D. (2019). Understanding selenium metabolism in plants and its role as a beneficial element. *Critical Reviews in Environmental Science and Technology, 49*(21), 1937–1958.

Chu, H. H., Car, S., Socha, A. L., Hindt, M. N., Punshon, T., & Guerinot, M. L. (2017). The Arabidopsis MTP8 transporter determines the localization of manganese and iron in seeds. *Scientific Reports,* 7(1), 1–10.

Colmenero-Flores, J. M., Franco-Navarro, J. D., Cubero-Font, P., Peinado-Torrubia, P., & Rosales, M. A. (2019). Chloride as a beneficial macronutrient in higher plants: New roles and regulation. *International Journal of Molecular Sciences, 20*(19), 4686.

Connolly, E. L., & Guerinot, M. L. (2002). Iron stress in plants. *Genome Biology, 3*(8), 1–4.

Connorton, J. M., Balk, J., & Rodríguez-Celma, J. (2017). Iron homeostasis in plants—a brief overview. *Metallomics, 9*(7), 813–823.

Corpas, F. J., Barroso, J. B., Palma, J. M., & Rodriguez-Ruiz, M. (2017). Plant peroxisomes: A nitro-oxidative cocktail. *Redox Biology, 11,* 535–542.

de Bang, T. C., Husted, S., Laursen, K. H., Persson, D. P., & Schjoerring, J. K. (2021). The molecular–physiological functions of mineral macronutrients and their consequences for deficiency symptoms in plants. *New Phytologist, 229*(5), 2446–2469.

Dimkpa, C. O., & Bindraban, P. S. (2016). Fortification of micronutrients for efficient agronomic production: A review. *Agronomy for Sustainable Development, 36*(1), 7.

Ding, Y., Liu, Y., Liu, S., Li, Z., Tan, X., Huang, X., … Zheng, B. (2016). Biochar to improve soil fertility. A review. *Agronomy for Sustainable Development, 36*(2), 1–18.

Droppa, M., & Horváth, G. (1990). The role of copper in photosynthesis. *Critical Reviews in Plant Sciences, 9*(2), 111–123.

Dwivedi, A., Dev, I., Kumar, V., Yadav, R. S., Yadav, M., Gupta, D., … Tomar, S. S. (2015). Potential role of maize-legume intercropping systems to improve soil fertility status under smallholder farming systems for sustainable agriculture in India. *International Journal of Life Sciences Biotechnology and Pharma Research, 4*(3), 145.

Erisman, J. W., Sutton, M. A., Galloway, J., Klimont, Z., & Winiwarter, W. (2008). How a century of ammonia synthesis changed the world. *Nature Geoscience, 1*(10), 636–639.

Fernandes, J. C., & Henriques, F. S. (1991). Biochemical, physiological, and structural effects of excess copper in plants. *The Botanical Review, 57*(3), 246–273.

Gao, F., Robe, K., Gaymard, F., Izquierdo, E., & Dubos, C. (2019). The transcriptional control of iron homeostasis in plants: A tale of bHLH transcription factors? *Frontiers in Plant Science, 10,* 6.

Geilfus, C. M. (2018). Review on the significance of chlorine for crop yield and quality. *Plant Science, 270,* 114–122.

Goldbach, H. E., & Wimmer, M. A. (2007). Boron in plants and animals: Is there a role beyond cell-wall structure? *Journal of Plant Nutrition and Soil Science, 170*(1), 39–48.

González-Fontes, A., Navarro-Gochicoa, M. T., Ceacero, C. J., Herrera-Rodríguez, M. B., Camacho-Cristóbal, J. J., & Rexach, J. (2017). Understanding calcium transport and signalling, and its use efficiency in vascular plants. In *Plant macronutrient use efficiency* (pp. 165–180). Academic Press.

Gregory, A. L., Hurley, B. A., Tran, H. T., Valentine, A. J., She, Y. M., Knowles, V. L., & Plaxton, W. C. (2009). In vivo regulatory phosphorylation of the phosphoenolpyruvate carboxylase AtPPC1 in phosphate-starved *Arabidopsis thaliana. Biochemical Journal, 420*(1), 57–65.

Guerinot, M. L., & Yi, Y. (1994). Iron: Nutritious, noxious, and not readily available. *Plant Physiology, 104*(3), 815.

Guo, W., Nazim, H., Liang, Z., & Yang, D. (2016). Magnesium deficiency in plants: An urgent problem. *The Crop Journal, 4*(2), 83–91.

Hanafy, M. S., Ashour, H. A., & Sedek, F. M. (2018). Effect of some bio-stimulants and micronutrients on growth, yield and essential oil production of *Majorana hortensis* plants. *International Journal of Environment,* 7(1), 37–52.

Hänsch, R., & Mendel, R. R. (2009). Physiological functions of mineral micronutrients (cu, Zn, Mn, Fe, Ni, Mo, B, cl). *Current Opinion in Plant Biology, 12*(3), 259–266.

Hartikainen, H., Xue, T., & Piironen, V. (2000). Selenium as an anti-oxidant and pro-oxidant in ryegrass. *Plant and Soil, 225*(1), 193–200.

Hasanuzzaman, M., Nahar, K., Alam, M. M., & Fujita, M. (2014). Modulation of antioxidant machinery and the methylglyoxal detoxification system in selenium-supplemented Brassica napus seedlings confers tolerance to high temperature stress. *Biological Trace Element Research, 161*(3), 297–307.

Hasanuzzaman, M., Bhuyan, M. H. M. B., Mahmud, J. A., Nahar, K., Mohsin, S. M., Parvin, K., & Fujita, M. (2018). Interaction of sulfur with phytohormones and signaling molecules in conferring abiotic stress tolerance to plants. *Plant Signaling & Behavior, 13*(5), Article e1477905.

Hasanuzzaman, M., Hossain, M. S., Bhuyan, M. B., Al Mahmud, J., Nahar, K., & Fujita, M. (2018). The role of sulfur in plant abiotic stress tolerance: Molecular interactions and defense mechanisms. In *Plant nutrients and abiotic stress tolerance* (pp. 221–252). Singapore: Springer.

Hasanuzzaman, M., Bhuyan, M. B., Raza, A., Hawrylak-Nowak, B., Matraszek-Gawron, R., Al Mahmud, J., ... Fujita, M. (2020). Selenium in plants: Boon or bane? *Environmental and Experimental Botany*, 104170.

Hassan, M. U., Chattha, M. U., Khan, I., Chattha, M. B., Aamer, M., Nawaz, M., ... Khan, T. A. (2019). Nickel toxicity in plants: Reasons, toxic effects, tolerance mechanisms, and remediation possibilities—a review. *Environmental Science and Pollution Research, 26*(13), 12673–12688.

Hassan, S. M. (2019). Important role of micronutrients: An updated review. *LGUJLS, 3*(1), 31–39.

Hebbern, C. A., Pedas, P., Schjoerring, J. K., Knudsen, L., & Husted, S. (2005). Genotypic differences in manganese efficiency: Field experiments with winter barley (*Hordeum vulgare* L.). *Plant and Soil, 272*(1), 233–244.

Hebbern, C. A., Laursen, K. H., Ladegaard, A. H., Schmidt, S. B., Pedas, P., Bruhn, D., ... Husted, S. (2009). Latent manganese deficiency increases transpiration in barley (Hordeum vulgare). *Physiologia Plantarum, 135*(3), 307–316.

Hefferon, K. (2019). Biotechnological approaches for generating zinc-enriched crops to combat malnutrition. *Nutrients, 11*(2), 253.

Heine, G., Max, J. F., Führs, H., Moran-Puente, D. W., Heintz, D., & Horst, W. J. (2011). Effect of manganese on the resistance of tomato to Pseudocercospora fuligena. *Journal of Plant Nutrition and Soil Science, 174*(5), 827–836.

Hemerly, A. (2016). Genetic controls of biomass increase in sugarcane by association with beneficial nitrogen-fixing bacteria. In *Plant and animal genome XXIV conference. Plant and animal genome, during month of january*.

Hocking, B., Tyerman, S. D., Burton, R. A., & Gilliham, M. (2016). Fruit calcium: Transport and physiology. *Frontiers in Plant Science, 7*, 569.

Horvath, G., Droppa, M., & Terry, N. (1983). Changes in photosynthetic attributes in response to copper depletion in sugar beets (*Beta vulgaris* L). *Journal of Plant Nutrition, 6*(11), 971–981.

Ihnatowicz, A., Siwinska, J., Meharg, A. A., Carey, M., Koornneef, M., & Reymond, M. (2014). Conserved histidine of metal transporter at NRAMP 1 is crucial for optimal plant growth under manganese deficiency at chilling temperatures. *New Phytologist, 202*(4), 1173–1183.

Itelima, J. U., Bang, W. J., Onyimba, I. A., & Oj, E. (2018). A review: Biofertilizer; a key player in enhancing soil fertility and crop productivity. *Journal of Microbiology and Biotechnology Reports, 2*, 22–28.

Kathpalia, R., & Bhatla, S. C. (2018). Plant mineral nutrition. In *Plant physiology, development and metabolism* (pp. 37–81). Singapore: Springer.

Kim, S. A., & Guerinot, M. L. (2007). Mining iron: Iron uptake and transport in plants. *FEBS Letters, 581*(12), 2273–2280.

Kobayashi, T., & Nishizawa, N. K. (2012). Iron uptake, translocation, and regulation in higher plants. *Annual Review of Plant Biology, 63*, 131–152.
Kollner, T. G., Held, M., Lenk, C., Hiltpold, I., Turlings, T. C., Gershenzon, J., & Degenhardt, J. (2008). A maize (E)-β-caryophyllene synthase implicated in indirect defense responses against herbivores is not expressed in most American maize varieties. *The Plant Cell, 20*(2), 482–494.
Kruse, T., Gehl, C., Geisler, M., Lehrke, M., Ringel, P., Hallier, S., et al. (2010). Identification and biochemical characterization of molybdenum cofactor-binding proteins from *Arabidopsis thaliana*. *Journal of Biological Chemistry, 285*, 6623–6635.
Kudla, J., Becker, D., Grill, E., Hedrich, R., Hippler, M., Kummer, U., Parniske, M., Romeis, T., & Schumacher, K. (2018). Advances and current challenges in calcium signaling. *New Phytologist, 218*, 414–431.
Kumar, A. (2016). Phosphate solubilizing bacteria in agriculture biotechnology: Diversity, mechanism and their role in plant growth and crop yield. *International Journal of Advanced Research, 4*(4), 116–124.
Leghari, S. J., Wahocho, N. A., Laghari, G. M., HafeezLaghari, A., MustafaBhabhan, G., HussainTalpur, K., … Lashari, A. A. (2016). Role of nitrogen for plant growth and development: A review. *Advances in Environmental Biology, 10*(9), 209–219.
Lima, L. W., Pilon-Smits, E. A., & Schiavon, M. (2018). Mechanisms of selenium hyperaccumulation in plants: A survey of molecular, biochemical and ecological cues. *Biochimica et Biophysica Acta (BBA)-General Subjects, 1862*(11), 2343–2353.
Liu, X., Guo, L. X., Luo, L. J., Liu, Y. Z., & Peng, S. A. (2019). Identification of the magnesium transport (MGT) family in *Poncirus trifoliata* and functional characterization of *PtrMGT5* in magnesium deficiency stress. *Plant Molecular Biology, 101*(6), 551–560.
Loneragan, J. F. (1997). Plant nutrition in the 20th and perspectives for the 21st century. In T. Ando, K. Fujita, T. Mae, H. Matsumoto, S. Mori, & J. Sekiya (Eds.), *Plant nutrition for sustainable food production and environment. Developments in plant and soil Sciences* (pp. 3–14). Dordrecht: Springer. Vol. 78.
Maharajan, T., Ceasar, S. A., Ajeesh krishna, T. P., Ramakrishnan, M., Duraipandiyan, V., Naif Abdulla, A. D., & Ignacimuthu, S. (2018). Utilization of molecular markers for improving the phosphorus efficiency in crop plants. *Plant Breeding, 137*(1), 10–26.
Malhotra, H., Sharma, S., & Pandey, R. (2018). Phosphorus nutrition: Plant growth in response to deficiency and excess. In *Plant nutrients and abiotic stress tolerance* (pp. 171–190). Singapore: Springer.
Manuel, T. J., Alejandro, C. A., Angel, L., Aurora, G., & Emilio, F. (2018). Roles of molybdenum in plants and improvement of its acquisition and use efficiency. In *Plant micronutrient use efficiency* (pp. 137–159). Academic Press.
Matthes, M. S., Robil, J. M., & McSteen, P. (2020). From element to development: The power of the essential micronutrient boron to shape morphological processes in plants. *Journal of Experimental Botany, 71*(5), 1681–1693.
Mushtaq, N. U., Saleem, S., Rasool, A., Shah, W. H., Hakeem, K. R., & Rehman, R. U. (2021). Salt stress threshold in millets: Perspective on cultivation on marginal lands for biomass. *Phyton, 90*(1), 51.
Nakandalage, N., & Seneweera, S. (2018). Micronutrients use efficiency of crop-plants under changing climate. In *Plant micronutrient use efficiency* (pp. 209–224). Academic Press.
Naqib, S. A., & Jahan, M. S. (2017). The function of molybdenum and boron on the plants. *Journal of Agricultural Research, 2*(3), 136–141.
Ng, I. S., Zheng, X., Wang, N., Chen, B. Y., Zhang, X., & Lu, Y. (2014). Copper response of Proteus hauseri based on proteomic and genetic expression and cell morphology analyses. *Applied Biochemistry and Biotechnology, 173*(5), 1057–1072.
Noreen, S., Fatima, Z., Ahmad, S., & Ashraf, M. (2018). Foliar application of micronutrients in mitigating abiotic stress in crop plants. In *Plant nutrients and abiotic stress tolerance* (pp. 95–117). Singapore: Springer.

O'Neill, M. A., Ishii, T., Albersheim, P., & Darvill, A. G. (2004). Rhamnogalacturonan II: Structure and function of a borate cross-linked cell wall pectic polysaccharide. *Annual Review of Plant Biology, 55*, 109–139.

Onnerud, H., Zhang, L., Gellerstedt, G., & Henriksson, G. (2002). Polymerization of monolignols by redox shuttle–mediated enzymatic oxidation: A new model in lignin biosynthesis I. *The Plant Cell, 14*(8), 1953–1962.

Ova, E. A., Kutman, U. B., Ozturk, L., & Cakmak, I. (2015). High phosphorus supply reduced zinc concentration of wheat in native soil but not in autoclaved soil or nutrient solution. *Plant and Soil, 393*(1), 147–162.

Pandey, N. (2018). Role of plant nutrients in plant growth and physiology. In *Plant nutrients and abiotic stress tolerance* (pp. 51–93). Singapore: Springer.

Pilon-Smits, E. A. (2015). Selenium in plants. In *Progress in botany* (pp. 93–107). Springer International Publishing.

Plaxton, W. C., & Tran, H. T. (2011). Metabolic adaptations of phosphate-starved plants. *Plant Physiology, 156*, 1006–1015.

Puig, S., Andres-colas, N. U. R. I. A., Garcia-Molina, A. N. T. O. N. I., & Penarrubia, L. (2007). Copper and iron homeostasis in Arabidopsis: Responses to metal deficiencies, interactions and biotechnological applications. *Plant, Cell & Environment, 30*(3), 271–290.

Pushnik, J. C., Miller, G. W., & Manwaring, J. H. (1984). The role of iron in higher plant chlorophyll biosynthesis, maintenance and chloroplast biogenesis. *Journal of Plant Nutrition, 7*(1–5), 733–758.

Rajkovich, S., Enders, A., Hanley, K., Hyland, C., Zimmerman, A. R., & Lehmann, J. (2012). Corn growth and nitrogen nutrition after additions of biochars with varying properties to a temperate soil. *Biology and Fertility of Soils, 48*(3), 271–284.

Rakhmatjonovna, K. S. (2021). The importanceof micronutrientsin plant life.(IN the example OF the elements boron and manganese). *World Bulletin of Public Health, 1*(1), 4–6.

Rasool, A., Shah, W. H., Tahir, I., Alharby, H. F., Hakeem, K. R., & Rehman, R. (2020). Exogenous application of selenium (Se) mitigates NaCl stress in proso and foxtail millets by improving their growth, physiology and biochemical parameters. *Acta Physiologiae Plantarum, 42*(7), 1–13.

Rawat, J., Sanwal, P., & Saxena, J. (2016). Potassium and its role in sustainable agriculture. In *Potassium solubilizing microorganisms for sustainable agriculture* (pp. 235–253). New Delhi: Springer.

Rehman, M., Liu, L., Wang, Q., Saleem, M. H., Bashir, S., Ullah, S., & Peng, D. (2019). Copper environmental toxicology, recent advances, and future outlook: A review. *Environmental Science and Pollution Research, 26*(18), 18003–18016.

Reid, R. (2014). Understanding the boron transport network in plants. *Plant and Soil, 385*(1), 1–13.

Rout, G. R., & Sahoo, S. (2015). Role of iron in plant growth and metabolism. *Reviews in Agricultural Science, 3*, 1–24.

Sachan, P., & Lal, N. (2017). An overview of nickel (Ni2+) essentiality, toxicity and tolerance strategies in plants. *Asian Journal of Biology*, 1–15.

Sadeghzadeh, B., & Rengel, Z. (2011). *Zinc in soils and crop nutrition. The molecular and physiological basis of nutrient use efficiency in crops* (pp. 335–375). John Wiley & Sons, Inc..

Samaranayake, P., Peiris, B. D., & Dssanayake, S. (2012). Effect of excessive ferrous ($Fe^{2+)}$ on growth and iron content in rice (Oryza sativa). *International Journal of Agriculture & Biology, 14*(2).

Schweighofer, A., Hirt, H., & Meskiene, I. (2004). Plant PP2C phosphatases: Emerging functions in stress signaling. *Trends in Plant Science, 9*(5), 236–243.

Seth, K., & Aery, N. C. (2017). Boron induced changes in biochemical constituents, enzymatic activities, and growth performance of wheat. *Acta Physiologiae Plantarum, 39*(11), 1–10.

Shah, A., Wu, X., Ullah, A., Fahad, S., Muhammad, R., Yan, L., & Jiang, C. (2017). Deficiency and toxicity of boron: Alterations in growth, oxidative damage and uptake by citrange orange plants. *Ecotoxicology and Environmental Safety, 145*, 575–582.

Shahzad, B., Tanveer, M., Rehman, A., Cheema, S. A., Fahad, S., Rehman, S., & Sharma, A. (2018). Nickel; whether toxic or essential for plants and environment-A review. *Plant Physiology and Biochemistry, 132*, 641–651.

Shanker, A. K. (2006). Countering UV-B stress in plants: Does selenium have a role? *Plant and Soil, 282*(1), 21–26.

Shireen, F., Nawaz, M. A., Chen, C., Zhang, Q., Zheng, Z., Sohail, H., ... Bie, Z. (2018). Boron: Functions and approaches to enhance its availability in plants for sustainable agriculture. *International Journal of Molecular Sciences, 19*(7), 1856.

Shukla, A. K., Behera, S. K., Pakhre, A., & Chaudhari, S. K. (2018). Micronutrients in soils, plants, animals and humans. *Indian Journal of Fertilisers, 14*(3), 30–54.

Siedow, J. N. (1991). Plant lipoxygenase: Structure and function. *Annual Review of Plant Biology, 42*(1), 145–188.

Stewart, W. M., Dibb, D. W., Johnston, A. E., & Smyth, T. J. (2005). The contribution of commercial fertilizer nutrients to food production. *Agronomy Journal, 97*(1), 1–6.

Tejada-Jiménez, M., Galván, A., & Fernández, E. (2011). Algae and humans share a molybdate transporter. *Proceedings of the National Academy of Sciences, 108*(16), 6420–6425.

Thor, K. (2019). Calcium—nutrient and messenger. *Frontiers in Plant Science, 10*, 440.

Tilman, D. (1999). Global environmental impacts of agricultural expansion: The need for sustainable and efficient practices. *Proceedings of the National Academy of Sciences, 96*(11), 5995–6000.

Tripathi, D. K., Singh, S., Singh, S., Mishra, S., Chauhan, D. K., & Dubey, N. K. (2015). Micronutrients and their diverse role in agricultural crops: Advances and future prospective. *Acta Physiologiae Plantarum, 37*(7), 1–14.

Tripathi, D. K., Singh, S., Gaur, S., Singh, S., Yadav, V., Liu, S., ... Sahi, S. (2018). Acquisition and homeostasis of iron in higher plants and their probable role in abiotic stress tolerance. *Frontiers in Environmental Science, 5*, 86.

Vatansever, R., Ozyigit, I. I., & Filiz, E. (2017). Essential and beneficial trace elements in plants, and their transport in roots: A review. *Applied Biochemistry and Biotechnology, 181*(1), 464–482.

Venkidasamy, B., Selvaraj, D., & Ramalingam, S. (2019). Genome-wide analysis of purple acid phosphatase (PAP) family proteins in Jatropha curcas L. *International Journal of Biological Macromolecules, 123*, 648–656.

Verma, P., Yadav, A. N., Kumar, V., Singh, D. P., & Saxena, A. K. (2017). Beneficial plant-microbes interactions: Biodiversity of microbes from diverse extreme environments and its impact for crop improvement. In *Plant-microbe interactions in agro-ecological perspectives* (pp. 543–580). Singapore: Springer.

Wang, Y., & Wu, W. H. (2017). Regulation of potassium transport and signaling in plants. *Current Opinion in Plant Biology, 39*, 123–128.

Weissert, C., & Kehr, J. (2017). Macronutrient sensing and signaling in plants. In *Plant macronutrient use efficiency* (pp. 45–64). Academic Press.

White, P. J., & Broadley, M. R. (2001). Chloride in soils and its uptake and movement within the plant: A review. *Annals of Botany, 88*, 967–988.

Xin, W., Zhang, L., Zhang, W., Gao, J., Yi, J., Zhen, X., ... Zhao, C. (2019). An integrated analysis of the rice transcriptome and metabolome reveals differential regulation of carbon and nitrogen metabolism in response to nitrogen availability. *International Journal of Molecular Sciences, 20*(9), 2349.

Ye, Y., Cota-Ruiz, K., Hernández-Viezcas, J. A., Valdés, C., Medina-Velo, I. A., Turley, R. S., ... Gardea-Torresdey, J. L. (2020). Manganese nanoparticles control salinity-modulated molecular responses in Capsicum annuum L. through priming: A sustainable approach for agriculture. *ACS Sustainable Chemistry & Engineering, 8*(3), 1427–1436.

Yruela, I. (2009). Copper in plants: Acquisition, transport and interactions. *Functional Plant Biology, 36*(5), 409–430.

Zenda, T., Liu, S., Dong, A., & Duan, H. (2021). Revisiting sulphur—the once neglected nutrient: It's roles in plant growth, metabolism, stress tolerance and crop production. *Agriculture, 11*, 626.

Zhang, Y., Chen, K., Zhao, F. J., Sun, C., Jin, C., Shi, Y., ... Lian, X. (2018). OsATX1 interacts with heavy metal P1B-type ATPases and affects copper transport and distribution. *Plant Physiology, 178*(1), 329–344.

Zhao, L., Liu, F., Crawford, N. M., & Wang, Y. (2018). Molecular regulation of nitrate responses in plants. *International Journal of Molecular Sciences, 19*(7), 2039.

CHAPTER THREE

Nutrient acquisition, transport and metabolism within the plant cells

Muhammad Ameen[a], Javaid Akhtar[b], Muhammad Anwar-Ul-Haq[b], Ghulam Hassan Abbasi[c], Moazzam Jamil[a], Muhammad Ali[c], Tabinda Athar[b]

[a]Department of Soil Science, Faculty of Agriculture and Environment, The Islamia University of Bahawalpur, Bahawalpur, Pakistan
[b]Institute of Soil and Environmental Sciences, University of Agriculture, Faisalabad, Pakistan
[c]Institute of Agro-Industry and Environment, Faculty of Agriculture and Environment, The Islamia University of Bahawalpur, Bahawalpur, Pakistan

1. Introduction

Plants absorb the different forms of 16 different nutrients. These nutrients reach in different cells of the plant via different absorption structures, which fulfills the different nutritional requirements. Carbon, hydrogen and oxygen are obtained from air and water. No complicated pathway is attached with their uptake. Plant reserves the transport system for all nutrients from soil or water (Geisler et al., 2000). The transport systems are not enough, however plant select the nutrient uptake mechanism according to electro-chemical gradient of the membranes, like as H^+ ATPases in plasma membranes. Micronutrient uptake mechanism is not well understood as compared to macronutrients. About 70% of the earth surface is covered with water. Its distribution is not uniform. Worldwide dry climates restrict the water availability for overall production. For nitrogen acquisition arbuscular mycorrhizal symbioses is not much involved. However inorganic nitrogen uptake is involved by arbuscular mycorrhizal symbioses (Jin et al., 2005; Govindarajulu et al., 2005). Ectomycorrhiza and ericoid mycorrhizae are important (Read et al., 1991) between 61% and 86% of the plant nitrogen in Tundra and Alaska respectively by these symbioses (Hobbie & Hobbie, 2006). Nutrient availability is influenced by water availability, temperature, and all soil parameters.

Sustainable Plant Nutrition
ISBN: 978-0-443-18675-2
https://doi.org/10.1016/B978-0-443-18675-2.00005-5

When fluctuations occur from optimal conditions then plant faces the stress (Henriet et al., 2021). Plants uses the multiple omics to eliminate these stresses, that can be monitored by proteomic analysis (Henriet et al., 2019). Multiomic technique is good to regulate the interactions, which quantify the surrounding stages of translations (Wang & Chu, 2020). Novel system wide translation methods in conjuction with mRNA and protein quantification are starting to uncover the complex interplay across the omic scale (Kage et al., 2020). Puromycine-associated nascent chain proteomics (PUNCH−P) is one of these latest techniques (Aviner et al., 2013). It leads to a snapshot of the translatome by globally labeling newly synthesize protein that can be quantified using MS analysis, versus the more familiar ribosomes profiling techniques, ribose, which arrest translation and sequesnces protected mRNA fragments (Choudhary et al., 2020). By accounting for post-translational modifications, using both mRNA and translation abundance, this provides powerfull data for statistical prediction of protein abundance. Ribosome profiling technologies will revolutionize our ability to understand translational control of gene express. The outcome of this translational rates have significantly higher predictive power for protein abundance than transcript level alone (Parkes & Niranjan, 2019).

2. Nutrient acquisition

Plant growth is highly affected due to nitrogen and phosphorus being a essential nutrients in different ecosystems (Gruntman et al., 2017; Jackson & Caldwell, 1993). Potassium is required by all plants on the earth for improving the quality as well as disease resistance (Ahmad et al., 2018). The plants usually have internal traits for the adoptability and eliminate the limitations, also includes the mechanisms that enhance the resorption, as well as recycling and nitrogen (Umar et al., 2020) and phosphorus usage for growth (McNickle & Cahill, 2009). Most of the acquisition strategies are relevant to N and P (Jackson & Caldwell, 1989). The strategies change the chemical form of nutrients for their acquisition (Fransen, 1999), speed of the chemical from conversion (Cahill & McNickle, 2011), their absorption area and releasing them from the soil (Wolfsberg et al., 2018). The nutrient availability is not relevant to its chemical form but it also deals with the redistribution and reabsorption of these nutrients. The ecosystem of the plants determines the distribution and absorption strategies of these nutrients. The new work has revealed that the arbuscular mycorrhizal symbiotic pathway can be intricate in inorganic N commitment (Klauck et al.,

2018), but the measureable influence to plant food is now unknown. Two types of mycorrhizae (Ecto and ericoid) can absorb the similar dissolvable inorganic sources, entered to added foundation containing nitrogen called as inorganic insoluble forms (Table 3.1).

Table 3.1 Plant preferred forms of macro and micronutrients present in soil.

Sr.#	Nutrient	Preferred form in soil	Availability for plants	References
Macro nutrients				
1	Nitrogen (N)	$NO_3^- NH_4^+$	Available	Miller et al. (2007)
2	Phosphorus (P)	HPO_4^{-2} $H_2PO_4^-$	Available	Pierzynski et al. (2005)
3	Potassium (K)	K^+	Available	Hasanuzzaman et al. (2018)
4	Calcium (Ca)	Ca^{+2}	Available	White (2000)
5	Magnesium (Mg)	Mg^{+2} MgO	Available	Tränkner et al. (2018)
6	Sulfur (S)	SO_4^{-2}	Available	Davidian and Kopriva (2010)
Micro nutrients				
7	Iron (Fe)	Fe^{+2} Fe^{+3} Fe_2O_3	Available	Morrissey and Guerinot (2009)
8	Zinc (Zn)	Zn^{+2}	Available	Ishimaru et al. (2011)
9	Manganese (Mn)	Mn^{+2} Mn^{+3} Mn^{+4}	Available Unavailable Unavailable	Millaleo et al. (2010) Rengel (2000)
10	Copper (Cu)	Cu^{+2} Cu (II) Cu (I)	Available	Yusefi-Tanha et al. (2020)
11	Molybdenum (Mb)	MoO_4^-	Available	Kaiser et al. (2005)
12	Nickel (Ni)	Ni^{2+}	Available	Antonkiewicz et al. (2016); Ma et al. (2009); Rathor et al. (2014)
13	Cobalt (Co)	Co^{+2}	Available	International Plant Nutrition Institute (2015); Kosiorek and Wyszkowski (2019)
14	Aluminum (Al)	Al^{+3}	Available	Zhao and Shen (2018)
15	Silicon (Si)	SiO_2 H_4SiO_4	Available	Farmer et al. (2005)
16	Boron (B)	H_3BO_3 $B(OH)_4^-$	Available	Klochko et al. (2006)

The arbuscular mycorrhizae association is formed with vast majority of plants (Bowsher & Swain, 2014). When nutrients are scavenged from a long distance due to increased soil volume in root symbiotic relationship. The concept of inorganic nitrogen is under construction while the main phosphorus role is already defined (van Gestel et al., 2015). In western Australia Banksia and dryandra species are well known for the production of compound clusters (Dal et al., 2019) which is more old and poor part of the world. Missing of P and its removal from soil and plant can not be attributed to the addition of phosphorus through mycorrhizae (Mast et al., 2005). At high pH phosphorus become unavailable due to the formation of calcium phosphate which diffuses very slowly in the soil solution(Ameen et al., 2019). Similarly at low pH, it forms the complexes with Al and Fe (Smith et al., 2004). About 10 mM phosphorus is rarely achieved in soil solution, which is because of low solubility (Parfitt, 1979). Nitrogen is mobile and water soluble, so nitrate is highly mobile. The organic form of nitrogen is necessary for whole plant life cycle (Comerford, 1998). Cluster roots are formed in many plants like as lupin species which exudes a carboxylates, that helps in mobilization of sorbed P from highly P deficient soils (Lambers et al., 2006; Shane & Lambers, 2005).

Table 3.1 shows the different forms of the nutrients preferred by the plant for acquisition, uptake and transportation in the soil and plant. These forms depend upon the oxidation, reduction reactions in the medium, release of root exudates, chelation with organic material, modification of the root system and microbial symbiosis.

3. Nutrient acquisition strategies of plants

Plant alters its growth and physiology of shoot and root (Lambers et al., 2008a,b), inducing several different mechanisms for obtaining soil nutrients (Richardson et al., 2004). Sometimes root axis influences a lateral root development according to environmental conditions (Birks & Birks, 2004). The nutrients are distributed in patches naturally. It is assumed that more productive soils have more root volume as compared to low productive soils (Vitousek et al., 1995). Talking about the soils with patchy nutrients, these soils can be occupied in increased rate, which can capture the nutrients easily (Hedin, 2004). The foraging matters are influenced highly in term of quality as well as the resource supply with environment (Sahrawat, 2009). Forceful actions of conspecifics can also amend scavenging plans as in territoriality even the occurrence of peaceful conspecifics can decline the worth allocated to a environment of a particular reserve supply by persons. As macro nutrients are required in large amount for proper plant growth, so the uptake mechanism depends upon their availability within the soil. In many cases ion channels/transporter mostly on membranes improve the nutrient transportation.

Although plants are non-motile and often face nutrient shortages in their environment, they utilize a plethora of sophisticated mechanisms in an attempt to acquire sufficient amounts of the macro and micronutrients required for proper growth, development and reproduction. These mechanisms include changes in the developmental program and root structure to better "mine" the soil for limiting nutrients, induction of high affinity transport systems and the establishment of symbioses and associations that facilitate nutrient uptake. Together, these mechanisms allow plants to maximize their nutrient acquisition abilities while protecting against the accumulation of excess nutrients, which can be toxic to the plant. It is clear that the ability of plants to utilize such mechanisms exerts significant influence over crop yields as well as plant community structure, soil ecology, ecosystem health, and biodiversity.

How plant roots react to high-value soil spots may be particularly significant in antagonism for nutrients with additional florae, as root production can provide plants a reasonable improvement by permitting plants to apprehension assets more rapidly than their contestants. When the outcome of plant—plant relations on nutrient seeking has been examined, there is indication that plants change root location in reaction to the communal atmosphere (Comerford, 1998). In revisions linking more classes, there is sign that approximately plants rise the settlement of roots within truncated-superiority soil in the existence of neighbors, so decreasing struggle in greater eminence zones of the soil that are extremely argued (McNeill & Unkovich, 2007). Nitrogen is the furthermost lavish inorganic nutrient essential by plants. Soil N accessibility prominently disturbs the progress and enlargement of crops internationally (Gutiérrez, 2012). Nitrogen shortage is a well-known difficulty for plants cultivated in native ecologies (Vitousek & Howarth, 1991), and it is a main issue restraining plant progression in low pH soils (Fageria & Baligar, 2001).

Huge quantities of N nourishments are recycled in farming to produce plants that nourish growing worldwide populace each year. Erisman et al. (2008) assessed that nitrogen stimulant has sustained about four billion public born from the time when 1908, accounting for nearly 27% of the world's populace over the earlier era. At the similar period, additional nitrogen fertilization is producing ecological difficulties such as aquatic eutrophication, loss of nitrate (NO_3^-), low pH rain due to reduced nitrogen-usage (Ju et al., 2009). High revenues and extraordinary efficient nutrient utilization are crucial for existing agriculture. Thus, there is an serious necessity to proliferation plant nitrogen-use productivity by considering the replies to nitrogen (Kant et al., 2011). Aluminum (Al) is the best ample metallic compound in the Earth's outside. It is not a crucial component for florae, and surplus Al is poisonous to many plants. The chief indication of Al- plant poisonousness is the reticence of root length, which can happen afterward contact to Al^{3+} at meditations as little as μM ranks within 1 h (Matsumoto, 2000).

This spontaneity can be triggered by diminutions in cell progress and cell partition, which are credited to Al intervention with the membrane, cell wall, the cytoskeleton, cytoplasm calcium flux, magnesium endorsement, and auxin polar transportation (Ma, 2007). Nitrogen is accessible as nitrate and ammonium while K is accessible to plants, as amino acid, provisional on the chemical class and soil pH. At high pH higher amount of nitrogen, ammonia, is expected to quickly come in root cells from the soil (Coskun et al., 2013). As the leading methods of N existing to plants in maximum soils are the mineral ions. Another discrepancy is that potassium ions, when absorbed by plants, persist in humble ionic form, whereas N-comprising mixtures undertake abundant chemical responses and develop bonded interactions within carbon-based particles within the plant. This important alteration is mirrored in the chief biological functions. N is a vital component of a massive collection of chemical compounds and administrative composites, together with chlorophyll, proteins, nucleic acids, co-enzymes, plant hormones and minor metabolites, while the foremost roles of K^+ are as a most important osmolyte and foundation of +ve charge for enzyme activation and electrical ionic fluxes (Marschner, 2011).

K is of supreme status in cell functioning of plant. It is an necessary macro-nutrient that achieves acute tasks connected to enzyme stimulation, ionic modification, internal pressure generation, cell extension, ruling of membrane ionic potential, and pH regulation (Hawkesford et al., 2012). Furthermore, plants store huge quantities of potassium in vacuoles, exceeding indecently dietary supplies. About 2% cell growth and 10% plant dry weight is due to potassium (White & Karley, 2010). There is a sharp wavy correlation among the tissue K quantity and plant progress, since which a serious absorption of K^+ associate 90% of extreme produce, can be recorded. Overhead this dose, progression has no relationship with the amplified K^+ dose, but at inferior K^+ meditations, progress drops quickly. Therefore, K^+ supplementation is collective exercise in recent agriculture and around 40%–60% of production of crop are characterize to marketable use of fertilizer (Stewart et al., 2005).

4. Cellular mechanism of plant for nutrient acquisition and uptake

The assimilation of ammonium and nitrate in roots or from roots to shoots after their absorption depends upon the environmental conditions and species of the plant. ATP, cytoskeleton and reducing equivalents are required for energy production to assimilate the inorganic nitrogen, which is difficult to attain without these analytes. The above fact is true when nitrate is accessible for nitrogen. Nitrogen is assimilated directly or indirectly in photosynthesis. N assimilation is correlated with photosynthesis, respiration and photorespiration at variable intera and inter cellular levels of different tissues and organs. The transcription and translation occur as nitrogen availability affect the assimilated carbon as organic acids, starch and sucrose through their partioning (Foyer et al., 2002).

Present information specifies that plants retain a convoluted supervisory equipment that organizes the ability of N integration by carbon absorption, nutrient accessibility and additional ecological elements, and the burdens located by plant evolution and progress. Gene expression and activation of enzymes are controlled by available metabolites, substrate and other products which allow the interaction of carbon and nitrogen, assimilated at different places. The regulation of carbon and nitrogen collaboration founds tremendous compound system, linking warning sign originating from NO_3^-, NH_4^+, and N-comprising chemical analyte such as aspartate, glutamate, and glutamine (Coruzzi & Zhou, 2001; Miller et al., 2008). In accumulation, indication is developing that the NADH states take part a serious part in assimilation of nitrate and NH_3 (Dutilleul et al., 2005). These chemical indications intermingle with chemical substances, like cytokinin, that react to the nitrogen stream, and control breakdown and expansion (Coruzzi & Zhou, 2001).

Plant biomass and growth needs different pre-requisites like as H_2O, CO_2, water and mineral elements, all are very important. N is serious, meanwhile being a amino acid constituent and construction material for protein, chlorophyll, neocletide and other cellular part. The essential part of this food is emphasized, which spectacles the chief ways of CN absorption in the brightened leaf. In detail, nitrogen is mandatory by florae in larger amounts than other inorganic elements. This principally imitates the great quantity of nitroegn participated in the food forming tool, predominantly light harvesting compounds and Rubisco (Zhu et al., 2008). This participated nitrogen maintains the light-involved practice of CO_2, H_2O, and mineral N to yield sweetener, carbon-based acids, and amino acids, the elementary construction chunks of biomass buildup. Like, nitrogen accessibility is obviously a very substantial element of photosynthetic capability and plant yield (Foyer et al., 2002).

Reversely, it is also extended recognized that throughout the absorption of mineral nitrogen, noteworthy quantities of immovable carbon are compulsory to deliver the carbon frameworks that doing as acceptors during nitrogen incorporation and usage, and substantial quantities of NAD(P)H and ATP are necessary to motivate these progressions. Mineral nitrogen adjustment, amino acid creation, and the subsequent practice of protein production also signify a key constituent of plant inhalation (Hachiya et al., 2007; Piques et al., 2009). Consequently, a complete thought of the genomic and breakdown foundation of carbon to nitrogen collaborations is vital for development of crop produce as well as trauma revision and confrontation. In foliage, NO_3^- decline comprises the action of NO_3^- and NO reductase to produce NH_4^+, which is integrated through the GS cycle or GOGAT pathway. Glutamate is formed just only due to GS/GOGAT cycle, aids the biosynthesis of other amino acids act as source of carbon and nitrogen (Forde & Lea, 2007). The presence of nitrate, an oxidized form aids to the reduction reactions of nitrogen. After reduction of six ferredoxins and one NADH are utilized to produce one ammonium from one nitrate. This process is accelerated in light as compared to dark conditions. This is the outcome of fluctuations in gene copying and the extent of NO_3^- reductase enzyme, similar as a light-focused stimulation of NO_3^- reductase, concerning de-phosphorylation and discharge of an inhibitory protein (Lillo, 2008). After cloud over, failure of NO_3^- reductase happens with a half-time of about 2 min. Reduced ferredoxin is less available from nitrite deduction as in dark condition, so the above said conditions decrease the accumulation of nitrite. Photorespiration, assimilatory enzymes and ATP are the main required factors which affect the nitrogen assimilation.

In accumulation, it has been established, that the management of carbon and nitrogen digestion is attained by detecting the cell base carbon and nitrogen equilibrium and controlling the genomic process convoluted in numerous procedures such as respiration, photosynthesis, and nitrogen assimilation (Gutie'rrez et al., 2007). Plants have distinct receptive systems of carbon and nitrogen which maintain the different metabolic products up to optimum levels (Foyer et al., 2002). In principle, NO_3^-, NH_4^+, and other nitrogen metabolic products encourage genomics that are obligatory for the nitrogen usage, while feedback processes suppress these replies. This interplay is well demonstrated in the occasion of NO_3^- reductase, where manifestation is encouraged by NO_3^- and inhibited by amino acids (Vincentz et al., 1993), and the inductive achievement of NO_3^- is wholly eradicated through C malnourishment (Klein et al., 2000). The P-II protein is first proved in *E. Coli* and then in every bacterial and plant cell which perform the carbon to nitrogen reactions (Ninfa & Jiang, 2005; Templeton & Moorhead, 2004).

In plant, the nucleus related genome (GLB1), which encrypts a plastid-directed similar fragments of the multicellular P-II protein, was primary isolated in Arabidopsis (Hsieh et al., 1998) and later on several other plant classes, with oil crop castor bean (*Ricinus communis*), cereal rice (*Oryza sativa*), fodder as alfalfa (*Medicago sativa*), as well as tomato (*Solanum lycopersicum*) (Moorhead & Smith, 2003). Remarkably, enzyme dynamic examination displayed that P-II rises NAGK action, but the key purpose of interaction is to dismiss hang-up of catalytic action by the cycle of production, arginine (Chen et al., 2006). Arginin and glutamate are responsible to diminish the P-II-NAGK compound formed with 2OG, however negatively correlate the P-II-NAGK complex, showed by the retailed analysis (Feria-Bourrellier et al., 2009). Many of the chloroplast chemical cycles are undergoes due to P-II protein as well as involved in nitrogen breakdown mechanism (Feria-Bourrellier et al., 2010). In spite of these latest developments, it is uncertain how PII protein join in discrete signs and interrelates with other supervisory schemes, which might have reputed parts in the detection of nitrogen and carbon position, such as TOR/RSK indicating cycle (Deprost et al., 2007).

5. Nutrient transport within the plant

Nitrogen breakdown and reconstructive system during transport are highly sensced by highly sophisticated systems demanded by carbon and nitrogen on environmental and development basis. Although, revisions at the molecular basis have exposed an even advanced difficulty, with numerous positions of controlling, with post-transcriptional switch by microRNAs (Gutie'rrez et al., 2007; Pant et al., 2009). The diverse metabolites such as NO_3^-, NH_4, 2OG, glutamate and glutamine are regulated to monitor alterations with the help of different sensory systems of carbon and nitrogen in plants (Foyer et al., 2002). In crux, NH_4^+, NO_3^- and other nitrogen metabolites encourage genes that are essential for the consumption of nitrogen, whereas disparagement contrivances down these reactions. The optimum difficulty essential to this situation is demonstrated by the information that this proteinious enzyme is also undergoes to a multifarious guideline at translational level which is more than skillful of amendable NO_3^- acclimatization in altered form in which the NIA gene is inducevely communicated, or where the number of useful NIA gene counterfeits has been reduced four-fold (Scheible et al., 1997). The nitrogen cycles, which signals of other entities, not explained yet to plants (Vidal & Gutiérrez, 2008). Arabidopsis gave the first evidence of the nuclear gene (GLB1) in plants, which encrypts a plastid-directed similar type of multicellular PII protein (Hsieh et al., 1998) and afterward numerous other plant types (Moorhead & Smith, 2003). Xylem vessels are responsible for the transportation of nitrate and ammonium out of the roots to the leaves, where they assimilate.

Due to progressive variations and conservational loads on carbon and nitrogen assets, plants have developed a extremely stylish and multifarious sensual system that intermingles to control carbon and nitrogen integration, breakdown, and passage. Nevertheless, revisions at the genomic level have discovered an even advanced intricacy, with numerous levels of instruction, together with post-gene constructional regulator by microRNAs (Gutie'rrez et al., 2007). In calculation, it has been confirmed that the management of carbon and nitrogen breakdown is attained by detecting the cellular carbon and nitrogen stability and controlling the gene copying responsible in some procedures such as respiration, photosynthesis, and nitrogen assimilation (Palenchar et al., 2004). The P-II protein, which organizes carbon to nitrogen exchanges was first exposed in *E. coli* and recently in almost each bacterial and historic genome inspected, as well as in floras (Ninfa & Jiang, 2005; Templeton & Moorhead, 2004). Remarkably, enzyme dynamic examination exhibited that P-II rises NAGK action, nevertheless the prime occupation of tie is to discharge reserve of enzyme action by the different cycle products, as arginine (Chen et al., 2006). This outcome proposes that P-II is not solitary convoluted in the control the nitrogen breakdown, but correspondingly pays to control a number of other chemical cycles working in the food forming cells the chloroplast (Feria-Bourrellier et al., 2010).

The uptake of N, P and K triggers the mechanism for Aluminum uptake which is due to the stimulation of root growth. This phenomenon in tea plant is attributed to increases micronutrient uptake. Similarly, the formation of Al—P complex enhances the alkaline phosphatase activity, which provide more phosphorus in organic form to the diatoms in marine environment. Nutrients are obtained by the plants mostly as diffusion and mass flow, especially those nutrients that are mobile in the soil. While the immobile plant nutrients are reached in the plants for assimilation with the root interception, microbial interaction and chelation process.

6. Mechanisms and molecular characters involved for nutrient transport

Inorganic elements are chief factors of plant progress and productiveness in environment and farming. Various diseases during growth and development of the plant are due to deficiency in 17 essential plant nutrients, are also required by the plants for the optimum completion of life cycle. Some of them like about 14 nutrients including N, P, K, S, Ca, Mg, Fe, Mn, Zn, Cu, Mo, B, Cl, and Ni are absorbed from the soil (Marschner, 2011). The soluble forms of these minerals as ions are absorbed by the plants from soil solution. The organic matter, pH, microorganism, redox potential soil moisture and environmental factors are involved, affect the chemical form and solubility of all the nutrients. Meanwhile native soil atmospheres are diverse and modify freely, soil nutrient accessibility is extremely flexible and often partial. In accumulation, nutrient motion in soils is subjective by their chemistry and ecological elements. The approachability of inorganic ions with narrow agility, as well as phosphorus and iron, in soil is regulated to roots in front of the position of the nutrient. The motion of those essentials, which are comparatively movable in soil, hang on the water situations. These alterations, with soil nutrient motion, outcome in varied three-dimensional allocations (Jobbágy & Jackson, 2001). Consequently, plants must manage with the irregular three-dimensional supplies and sequential variations of nutrients to improve nutrient absorption through their whole life. One valuable policy for effective nutrient endorsement is variation of the root system construction with respect to the nutrient circumstances. Nutrient amounts disturb the size, quantity, direction and diameters of the chief roots and adjacent roots, and root hair advancement. These progressive deviations result in nutrient-reliant on root system outlines. For instance, under phosphorus malnourishment, primary root elongation is restricted, while sided roots development is enriched concurrently (Lopez-Bucio et al., 2002; Williamson et al., 2001), ensuing in the establishment of a thin root physical structures and amplified root structural area. Phosphate inclines to gather in the surface layer, where it is restrained (Jobbágy & Jackson, 2001). Hence, underground structural responses of plants to phosphorus food shortage undoubtedly donate to effective uptake of phosphate by revealing a big root surface area to phosphorus. In divergence, for water-resolvable nutrients like as NO_3^- and SO_4^-, which are cheerfully percolated to depth soil layers, root progress toward inferior layers is favored. Alternative key policy is intonation of transference movement within and between plant structures. Nutrient carriers controls nutrient acceptance into root cells and succeeding transportation within the plant. Increasing indication specifies that florae have progressed miscellaneous carriers with discrete substrate qualities, love to transport, cell type manifestation and sub-cellular localization to confirm suitable fluctuation and compartmentalization of every uptake procedure within the plant cell. Numerous instances of the supplement subordinate guideline of carrier articulation are accessible. Nutrients are transported within the plant system and within the cells mainly due to the apoplastic and symplastic movements. These movements are the result of transpiration pull. The movement of the nutrients is encouraged by the xylem and phloem vessels, their loading and unloading results the transfer of nutrients form source to ultimate sink. Fig. 3.1 showed the different molecular structures responsible for the nutrients uptake and transfer, designated as gated ion channels, carriers, transporters, transcriptional factors undergoes cellular ionic homeostasis.

7. Nutrient metabolism within the cell

Ecological conditions are continually varying on Earth, and a focal essential for any lifeform is the capacity to adapt to these vacillations. Microorganisms living in, for example the dirt or the creature stomach, experience solid variances in the amount and nature of supplements in their current circumstance. To manage these vacillations, microorganisms have developed a huge assortment of components for adjusting their digestion to new supplement conditions (Black, 1993a,b). Ongoing examinations have shown that hereditarily indistinguishable cells can fluctuate in the time they need to adjust to a similar supplement switch and frequently a negligible part of cells can't adjust at all. Respiration of sucrose, which is imported from roots, provides the energy and cytoskeleton for the assimilation of inorganic nitrogen. Carbon and nitrogen after breakdown improve the growth and development of the plant. Due to optimum supply of nitrogen food forming

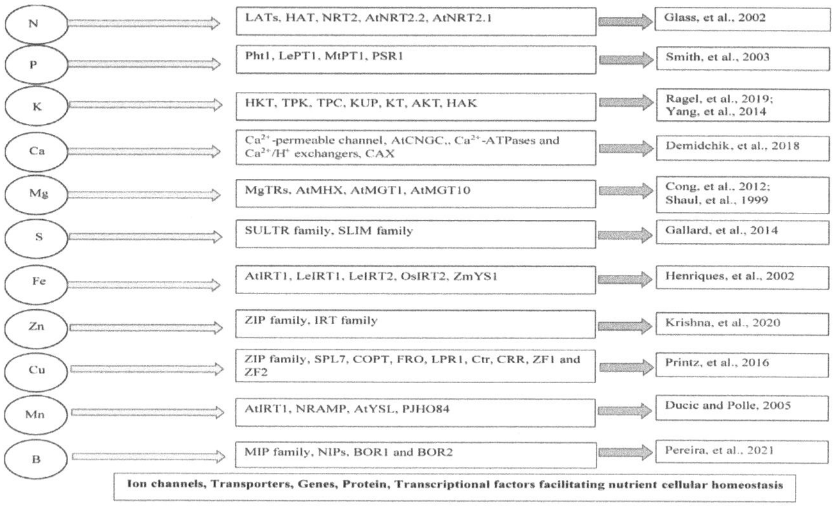

Fig. 3.1 Nutrient uptake and transport facilitation by channels, transporters and related cellular structures.

apparatus use the invested nitrogen and CO_2 assimilation occurs which ultimately improves the photosynthesis.

This heterogeneity accordingly is essentially because of phenotypic contrasts among cells at the hour of the switch, for example, in their protein levels or metabolic motions. For instance, the capacity of *Escherichia coli* cells to change from development on glucose to development on lactose relies upon the number of proteins of the lactose pathway the cells have at the hour of the switch (Lambert & Kussell, 2014) Likewise, their capacity to change from development on glucose to development on a gluconeogenic carbon source (for example acetic acid derivation) relies upon their gluconeogenic motion at the hour of the switch. At whatever point cells share a homogeneous climate, contrasts in protein levels and metabolic motions are fundamentally because of stochastic vacillations in their quality articulation (Elowitz et al., 2002).

Homogeneous conditions are probable a unique case in nature on the grounds that numerous microscopic organisms live in thick spatially organized gatherings, for example, biofilms and miniature states (Flemming & Wuertz, 2019) Thus, cells at various areas adjust to various microenvironments and differ in their quality articulation and metabolic motions (Lima et al., 1987). In spatially organized populaces, cells in this manner contrast in aggregate both because of stochastic changes in quality articulation (as in homogeneous societies) and because of physiological variation to various nearby microenvironments. Thus, organized populaces have more phenotypic variety than homogeneous populaces. Phenotypic variety can have significant ramifications for the capacity of a populace to adjust to natural changes (McNickle & Brown, 2014) Adjusting to new natural conditions can be tedious or even outlandish for cells because of energy or asset limits (Gersani et al., 1998). Actual properties, for example, water accessibility, temperature, and soil boundaries all impact supplement accessibility. Imperfect conditions lead to biotic or abiotic stresses, may happen in blends, and are frequently divergent in overall setting. Increasingly more multi-omics investigations of higher plants are being performed. Duruflé et al. (2020) consolidated phenomics, metabolomics, cell divider proteomics, and transcriptomics to investigate regular varieties inside a populace of Arabidopsis adjusted to different conditions along height slopes in on cell divider highlights. The creators guarantee that just through the blend of the multi-omics information could they recognize novel qualities associated with transformation to imperfect development conditions. A metabolomics, proteomics, and transcriptomics study on Arabidopsis presented to SO_4^{-2}, NO_3^-, and PO_4^- starvation (Luo et al., 2020) substantiated past investigations (Bielecka et al., 2015; Morcuende et al., 2007; Scheible et al., 1997; Scheible et al., 2004) yet in addition gave new bits of knowledge of the administrative systems included. Considering the inclusion of glutathione (GSH) in ROS detoxification pathways, it isn't shocking that utilization of sulfate compost builds the exercises of the particular chemicals (Usmani et al., 2020). Moreover, sulfate application keeps up with the photosynthetic limit under dry spell pressure and advances the photosynthetic digestion of sulfate to create Cys, which thus might be utilized to combine Met or to shape GSH.

8. Conclusion

Nutrients are necessary for the optimum growth of plant. Macro nutrients are required in larger quantities as compared to the micronutrients, which are required relatively in minute quantities. Macronutrients are involved in the plant structural components while micronutrients aid the plant growth and development indirectly. Plants take up the nutrients by mass flow, diffusion and root interception from soil to the plant. When plants need a nutrient, it activates some carriers or channel protein which in reaction passon the nutrients from outside of the cell to inside the cell through ionic homeostasis. The response of plant under normal and stress condition for nutrient uptake and transport is entirely different. The nutrients availability deals with absorption, redistribution and reabsorption of these nutrients. These distributions of nutrients is dependent on the ecosystem of the plant. Major roles in inorganic P uptake are well established, and evidence for involvement in inorganic N nutrition is building. Absence of a positive effect of mycorrhizas on plant growth cannot be interpreted because of a lack of significant P uptake through the mycorrhiza hyphae. The plant induces internal metabolism system, which break the nutrient complexes into their individual components, which are again utilized by the plant.

References

Ahmad, Z., Anjum, S., Waraich, E. A., Ayub, M. A., Ahmad, T., Tariq, R. M. S., Ahmad, R., & Iqbal, M. A. (2018). Growth, physiology, and biochemical activities of plant responses with foliar potassium application under drought stress—a review. *Journal of Plant Nutrition, 41*(13), 1734—1743.

Ameen, M., Akhtar, J., & Sabir, M. (2019). Effect of phosphoric acid and potassium humate on growth and yield of maize in saline-sodic soil. *Pakistan Journal of Agricultural Sciences, 56*(4), 781—790.

Antonkiewicz, J., Jasiewicz, C., Koncewicz-Baran, M., & Sendor, R. (2016). Nickel bioaccumulation by the chosen plant species. *Acta Physiol Plant, 38*, 40.

Aviner, R., Geiger, T., & Elroy-Stein, O. (2013). PUNCH-P for global translatome profiling. *Translation, 1*, e27516.

Bielecka, M., Watanabe, M., Morcuende, R., Scheible, W.-R., Hawkesford, M. J., Hesse, H., & Hoefgen, R. (2015). Transcriptome and metabolome analysis of plant sulfate starvation and resupply provides novel information on transcriptional regulation of metabolism associated with sulfur, nitrogen and phosphorus nutritional responses in Arabidopsis. *Frontiers in Plant Science, 5*, 805.

Birks, H. J. B., & Birks, H. H. (2004). Paleoecology: The rise and fall of forests. *Science, 305*, 484—485.

Black, C. A. (1993a). *Soil fertility evaluation and control.* Boca Raton, FL, USA: Lewis Publishers.

Black, C. A. (1993b). *Soil fertility evaluation and control* (Vols. 155—269, pp. 271—452). Boca Raton, FL: Lewis Publishers.

Bowsher, C. G., & Swain, P. S. (2014). Environmental sensing, information transfer, and cellular decision-making. *Current Opinion in Biotechnology, 28*, 149—155.

Cahill, J. F., & McNickle, G. G. (2011). The behavioral ecology of nutrient foraging by plants. *Annual Reviews of Ecology, Evolution and Systematics, 42*, 289—311.

Chen, Y. M., Ferrar, T. S., Lohmeier-Vogel, E. M., Morrice, N., Mizuno, Y., Berenger, B., Ng, K. K., Muench, D. G., & Moorhead, G. B. (2006). The PII signal transduction protein of *Arabidopsis thaliana* forms an arginine-regulated complex with plastid N-acetylglutamate kinase. *Journal of Biological Chemistry, 281*, 5726—5733.

Choudhary, S., Li, W., & Smith, A. D. (2020). Accurate detection of short and long active ORFs using Ribo-seq data. *Bioinformatics, 36*, 2053—2059.

Comerford, N. B. (1998). Soil phosphorus bioavailabilty. In J. P. Lynch, & J. Deikman (Eds.), *136—147Soil phosphorus bioavailability (*. American Society of Plant Physiologists.

Coruzzi, G. M., & Zhou, L. (2001). Carbon and nitrogen sensing and signaling in plants: Emerging 'matrix effects. *Current Opinion Plant Biology, 4*, 247–253.
Coskun, D., Britto, D. T., Li, M., Becker, A., & Kronzucker, H. J. (2013). Rapid ammonia gas transport accounts for futile transmembrane cycling under NH3/NH4+ toxicity in plant roots. *Plant Physiology, 163*, 1859–1867.
Dal Co, A., van Vliet, S., Kiviet, D. J., Schlegel, S., & Ackermann, M. (2019). Short-range interactions govern cellular dynamics in microbial multi-genotype systems. *bioRxiv*, 530584.
Davidian, J.-C., & Kopriva, S. (2010). Regulation of sulfate uptake and assimilation the same or not the same? *Molecular Plant, 3*(2), 314–325.
Deprost, D., Yao, L., Sormani, R., Moreau, M., Leterreux, G., Nicolai, M., Bedu, M., Robaglia, C., & Meyer, C. (2007). The Ara- bidopsis TOR kinase links plant growth, yield, stress resistance and mRNA translation. *EMBO Rep, 8*, 864–870.
Dutilleul, C., Lelarge, C., Prioul, J. L., De Paepe, R., Foyer, C. H., & Noctor, G. (2005). Mitochondria-driven changes in leaf NAD sta- tus exert a crucial influence on the control of nitrate assimilation and the integration of carbon and nitrogen metabolism. *Plant Physiology, 139*, 64–78.
Elowitz, M. B., Levine, A. J., Siggia, E. D., & Swain, P. S. (2002). Stochastic gene expression in a single cell. *Science, 297*, 1183–1186.
Erisman, J.W., Bleeker, A., Hensen, A., Vermeulen, A. 2008. Agricultural air quality in Europe and the future perspectives. Atmospheric Environment (2008), 42, 3209–3217.
Fageria, N. K., & Baligar, V. C. (2001). Improving nutrient use efficiency ofannual crops in Brazilian acid soils for sustainable crop production. *Communications in Soil Science and Plant Analaysis, 32*, 1303–1319.
Farmer, V., Delbos, E., & Miller, J. D. (2005). The role of phytolith formation and dissolution in controlling concentrations of silica in soil solutions and streams. *Geoderma, 127*, 71–79.
Feria-Bourrellier, A. B., Ferrario-Mery, S., Vidal, J., & Hodges, M. (2009). Metabolite regulation of the interaction between Arabi- dopsis thaliana PII and N-acetyl-l-glutamate kinase. *Biochemistry, Biophysics and Research Communication, 387*, 700–704.
Feria-Bourrellier, A. B., Valot, B., Guillot, A., Ambard-Bretteville, F., Vidal, J., & Hodges, M. (2010). Chloroplast acetyl-CoA carbox- ylase activity is 2-oxoglutarate-regulated by interaction of PII with the biotin carboxyl carrier subunit. *Proccedings of the National Academy of Science United States of America, 107*, 502–507.
Flemming, H.-C., & Wuertz, S. (2019). Bacteria and archaea on Earth and their abundance in biofilms. *Natural Review of Microbiology, 17*, 247.
Forde, B. G., & Lea, P. J. (2007). Glutamate in plants: Metabolism, regulation, and signaling. *Journal of Experimnetal Botany, 58*, 2339–2358.
Foyer, C. H., Vanacker, H., Gomez, L. D., & Harbinson, J. (2002). Regulation of photosynthesis and antioxidant metabolism in maize leaves at optimal and chilling temperatures. Review. *Plant Physiology and Biochemistry, 40*, 659–668.
Fransen, B. (1999). Disentangling the effects of root foraging and inherent growth rate on plant biomass accumulation in heterogeneous environments: A modelling study. *Annals of Botany, 84*, 305–311.
Geisler, M., Nadeau, J., & Sack, F. D. (2000). Oriented asymmetric divisions that generate the stomatal spacing pattern in *Arabidopsis* are disrupted by the *too many mouths* mutation. *Plant Cell, 12*, 2075–2086.
Gersani, M., Abramsky, Z., & Falik, O. (1998). Density- dependent habitat selection in plants. *Evolution in Ecology, 12*, 223–234.
Govindarajulu, M., et al. (2005). Nitrogen transfer in the arbuscular mycorrhizal symbiosis. *Nature, 435*, 819–823.
Gruntman, M., Groß, D., Májeková, M., & Tielbörger, K. (2017). Decision-making in plants under competition. *Natural Communication, 8*, 2235.

Gutiérrez, R. A. (2012). Systems biology for enhanced plant nitrogen nutrition. *Science, 336*, 1673–1675.
Gutiérrez, R. A., Lejay, L., Dean, A., Chiaromonte, F., Shasha, D. E., & Coruzzi, G. M. (2007). Qualitative network models and genome- wide expression data define carbon/nitrogen-responsive molec- ular machines in Arabidopsis. *Genomes in Biology, 8*, R7.
Hachiya, T., Terashima, I., & Noguchi, K. (2007). Increase in respiratory cost at high growth temperature is attributed to high protein turnover cost in Petunia3hybrida petals. *Plant Cell Environment, 30*, 1269–1283.
Hasanuzzaman, Mirza, Bhuyan, M. H. M. Borhannuddin, Nahar, Kamrun, Shahadat Hossain, Md, Mahmud, Jubayer Al, Shahadat Hossen, Md, Awal, Abdul, Moumita, Chowdhury Masud, & Fujita, Masayuki (2018). Potassium: A vital regulator of plant responses and tolerance to abiotic stresses. *Agronomy, 8*, 31, 2018.
Hawkesford, M., Horst, W., Kichey, T., Lambers, H., Schjoerring, J., Møller, I. S., et al. (2012). In P. Marschner (Ed.), *Chapter 6—functions of macronutrients" in Marschner's mineral nutrition of higher plants* (3rd ed., pp. 135–189). San Diego: Academic Press.
Hedin, L. O. (2004). Global organization of terrestrial plant-nutrient interactions. *Proccedings of the National Academy of Sciences United States of America, 101*, 10849–10850.
Henriet, C., Aimé, D., Térézol, M., et al. (2019). Water stress combined with sulfur deficiency in pea affects yield components but mitigates the effect of deficiency on seed globulin composition. *Journal of Experimental Botany, 70*, 4287–4304.
Henriet, C., Balliau, T., Aime, D., Le Signor, C., Kreplak, J., Zivy, M., Gallardo, K., & Vernoud, V. (2021). Proteomics of developing pea seeds reveals a complex antioxidant network underlying the response to sulfur deficiency and water stress. *Journal of Experimental Botany, 72*, 2611–2626.
Hobbie, J. E., & Hobbie, E. A. (2006). 15N in symbiotic fungi and plants estimates nitrogen and carbon flux rates in arctic tundra. *Ecology, 87*, 816–822.
Hsieh, M., Lam, H., Van de loo, F. J., & Coruzzi, G. (1998). A PII-like protein in Arabidopsis: Putative role in nitrogen sensing. *Proccedings of the National Academy of Sciences United States of America, 95*, 13965–13970.
International Plant Nutrition Institute. (2015). *Cobalt. Nutri-Facts. Agronomic fact sheets on crop nutrients.* North American Edition. http://www.ipni.net/publication/nutrifacts-na.nsf/0/5D2097137F73C07F85257EA8006297B0/$FILE/NutriFacts-NA- 15.pdf.
Ishimaru, Y., Bashir, K., & Nishizawa, N. K. (2011). Zn uptake and translocation in rice plants. *Rice, 4*, 21–27.
Jackson, R. B., & Caldwell, M. M. (1989). The timing and degree of root proliferation in fertile-soil microsites for three cold-desert perennials. *Oecologia, 81*, 149–153.
Jackson, R. B., & Caldwell, M. M. (1993). The scale of nutrient heterogeneity around individual plants and its quantification with geostatistics. *Ecology, 74*, 612–614.
Jin, H., et al. (2005). The uptake, metabolism, transport and transfer of nitrogen in an arbuscular mycorrhizal symbiosis. *New Phytology, 168*, 687–696.
Jobbágy, E. J., & Jackson, R. B. (2001). The distribution of soil nutrients with depth: Global patterns and the imprint of plants. *Biogeochemistry, 53*, 51–77.
Ju, X. T., Xing, G. X., Chen, X. P., Zhang, S. L., Zhang, L. J., Liu, X. J., et al. (2009). Reducing environmental risk by improving N management in intensive Chinese agricultural systems. *Proccedings of the National Academy of Sciences United States of America, 106*, 3041–3046.
Kage, U., Powell, J. J., Gardiner, D. M., & Kazan, K. (2020). Ribosome profiling in plants: What is not lost in translation? *Journal of Experimental Botany, 71*, 5323–5332.
Kaiser, B. N., Gridley, K. L., Brady, J. N., Phillips, T., & Tyerman, S. D. (2005). The role of molybdenum in agricultural plant production. *Annals of Botany, 96*, 745–754.

Kant, S., Bi, Y. M., & Rothstein, S. J. (2011). Understanding plant response to nitrogen limitation for the improvement of crop nitrogen use efficiency. *Journal of Experimental Botany, 62*, 1499–1509.

Klauck, G., Serra, D. O., Possling, A., & Hengge, R. (2018). Spatial organization of different sigma factor activities and c-di-GMP signalling within the three- dimensional landscape of a bacterial biofilm. *Open Biology, 8*, 180066.

Klein, D., Morcuende, R., Stitt, M., & Krapp, A. (2000). Regulation of nitrate reductase expression in leaves by nitrate and nitrogen metabolism is completely overidden when sugars fall below a critical level. *Plant Cell Environment, 23*, 863–871.

Klochko, K., Kaufman, A. J., Yao, W., Byrne, R. H., & Tossell, J. A. (2006). Experimental measurement of boron isotope fractionation in seawater. *Earth Planet Science Letters, 248*, 261–270.

Kosiorek, M., & Wyszkowski, M. (2019). Remediation of cobalt-polluted soil after application of selected substances and using oat (*Avena sativa* L.). *Environmental Science and Pollution Research, 26*, 16762–16780.

Lambers, H., et al. (2006). Root structure and functioning for efficient acquisition of phosphorus: Matching morphological and physiological traits. *Annals of Botany, 98*, 693–713.

Lambers, H., et al. (2008a). *Plant physiological ecology*. Springer Verlag.

Lambers, H., Raven, J. A., Shaver, G. R., & Smith, S. E. (2008b). Plant nutrient-acquisition strategies change with soil age. *Trends in Ecology and Evolution, 23*(2), 95–103.

Lambert, G., & Kussell, E. (2014). Memory and fitness optimization of bacteria under fluctuating environments. *PLoS Genetics, 10*, e1004556.

Lillo, C. (2008). Signaling cascades integrating light-enhanced nitrate metabolism. *BiochemistyJournal, 415*, 11–19.

Lima, E., Boddey, R. M., & Döbereiner, J. (1987). Quantification of biological nitrogen fixation associated with sugar cane using 15N aided nitrogen balance. *Soil Biology and Biochemistry, 19*, 165–170.

Lopez-Bucio, J., Hernandez-Abreu, E., Sanchez-Calderon, L., Nieto-Jacobo, M. F., Simpson, J., & Herrera-Estrella, L. (2002). Phosphate avail- ability alters architecture and causes changes in hormone sensitivity in the Arabidopsis root system. *Plant Physiology, 129*, 244–256.

Ma, J. F. (2007). Syndrome of aluminum toxicity and diversity of aluminum resistance in higher plants. *International Review of Cytology, 264*, 225–252.

Ma, Y., Rajkumar, M., & Freitas, H. (2009). Improvement of plant growth and nickel uptake by nickel resistant-plant-growth promoting bacteria. *Journal of Hazardous Material, 166*(2–3), 1154–1161.

Marschner, H. (2011). *Marschner's mineral nutrition ofHigher plants*. Cambridge, MA: Academic press.

Mast, A. R., et al. (2005). An assessment of old and new DNA sequence evidence for the paraphyly of Banksia with respect to Dryandra (Proteaceae). *Australian Systematic Botany, 18*, 75–88.

Matsumoto, H. (2000). Cell biology of aluminum toxicity and tolerance in higher plants. *International Review of Cytology*, 1–46, 2000.

McNeill, A., & Unkovich, M. (2007). The nitrogen cycle in terrestrial ecosystems. In P. Marschner, & Z. Rengel (Eds.), *Soil biology Series: Vol. 10. Nutrient cycling in terrestrial ecosystems* (pp. 37–64). Springer.

McNickle, G. G., & Brown, J. S. (2014). An ideal free distribution explains the root production of plants that do not engage in a tragedy of the commons game. *Journal of Ecology, 102*, 963–971.

McNickle, G. G., & Cahill, J. F. (2009). Plant root growth and the marginal value theorem. *Proccedings of the National Academy of Sciences United States of America, 106*, 4747–4751.

Millaleo, R., Reyes-Díaz, M., Ivanov, A. G., Mora, M. L., & Alberdi, M. (2010). Manganese as essential and toxic element for plants: Transport, accumulation and resistance mechanisms. *Journal of Soil Science and Plant Nutrition, 10*(4), 476–494.

Miller, A. J., Fan, X., Orsel, M., Smith, S. J., & Wells, D. M. (2007). Nitrate transport and signalling. *Journal of Experimental Botany, 58*, 2297–2306.

Millerm, G., Shulaev, V., & Mittler, R. (2008). Reactive oxygen signaling and abioticstress. *Physiological Plant, 133*, 481–489.

Moorhead, G. B., & Smith, C. S. (2003). Interpreting the plastid car- bon, nitrogen, and energy status: A role for PII? *Plant Physiology, 133*, 492–498.

Morcuende, R., Bari, R., Gibon, Y., Zheng, W., Pant, B. D., Blasing, O., Usadel, B., Czechowski, T., Udvardi, M. K., Stitt, M., & Scheible, W. R. (2007). Genome-wide reprogramming of metabolism and regulatory networks of Arabidopsis in response to phosphorus. *Plant, Cell and Environment, 30*, 85–112.

Morrissey, J., & Guerinot, M. L. (2009). Iron uptake and transport in plants: The good, the bad, and the ionome. *Chemical Reviews, 109*, 4553–4567.

Ninfa, A. J., & Jiang, P. (2005). PII signal transduction proteins: Sen- sors of alpha-ketoglutarate that regulate nitrogen metabolism. *Current Opinion in Microbiology, 8*, 168–173.

Palenchar, P. M., Kouranov, A., Lejay, L. V., & Coruzzi, G. M. (2004). Genome-wide patterns of carbon and nitrogen regulation of gene expression validate the combined carbon and nitrogen (CN)-signaling hypothesis in plants. *Genome Biology, 5*, R91.

Pant, B. D., Musialak-Lange, M., Nuc, P., May, P., Buhtz, A., Kehr, J., Walther, D., & Scheible, W. R. (2009). Identification of nutrient- responsive Arabidopsis and rapeseed microRNAs by comprehen- sive real-time PCR profiling and small RNA sequencing. *Plant Physiology, 150*, 1541–1555.

Parfitt, R. (1979). The availability of P from phosphate-goethite bridging complexes. Desorption and uptake by ryegrass. *Plant Soil, 53*, 55–65.

Parkes, G. M., & Niranjan, M. (2019). Uncovering extensive post-translation regulation during human cell cycle progression by integrative multi-'omics analysis. *BMC Bioinformatics, 20*, 536.

Pierzynski, G. M., McDowell, R. W., & Sims, J. T. (2005). Chemistry, cycling, and potential moment of inorganic phosphorus in soils. In *Sims JT, Sharpley an , phosphorus: Agriculture and the environment* (pp. 53–86). Madison, WI: American Society of Agronomy, Crop Science Society of America, Soil Science Society of America, Inc.

Piques, M., Schulze, W. X., Hohne, M., Usadel, B., Gibon, Y., Rohwer, J., & Stitt, M. (2009). Ribosome and transcript copy numbers, polysome occupancy and enzyme dynamics in Arabi- dopsis. *Molecular Systematic Biology, 5*, 314.

Rathor, G., Chopra, N., & Adhikari, T. (2014). Effect of variation in nickel concentration on growth of maize plant: A comparative overview for pot and hoagland culture. *Reserach Journal of Chemical Science, 4*(10), 30–32.

Read, D. J. (1991). Mycorrhizas in ecosystems. *Experientia, 47*, 376–391.

Rengel, Z. (2000). Uptake and transport of manganese in plants. In A. Sigel, & H. Sigel (Eds.), *Metal ions in biological systems* (pp. 57–87). New York: Marcel Dekker.

Richardson, S. J., et al. (2004). Rapid development of phosphorus limitation in temperate rainforest along the Franz Josef soil chronosequence. *Oecologia, 139*, 267–276.

Sahrawat, K. L. (2009). The role of tolerant genotypes and plant nutrients in reducing acid-soil infertility in upland rice ecosystem: An appraisal. *Arch Agron Soil Science, 55*, 597–607.

Scheible, W. R., GonzalesFontes, A., Morcuende, R., Lauerer, M., Geiger, M., Glaab, J., Gojon, A., Schulze, E. D., & Stitt, M. (1997). Tobacco mutants with decreased number of functional nia genes compensate by modifying the diurnal regulation of transcription, post-translational modification and turnover of nitrate reductase. *Planta, 203*, 304–319.

Scheible, W.R., Morcuende, R., Czechowski, T., Fritz, C., Osuna, D., Palacios-Rojas, Schindelasch, D., Thimm, O., Udvardi, M.K. and Stitt, M. 2004. Genome-wide reprogramming of primary and secondary metabolism, protein synthesis, cellular growth processes, and the regulatory infrastructure of Arabidopsis in response to nitrogen. *Plant Physiology 136*, 2483–2499.

Shane, M., & Lambers, H. (2005). Cluster roots: A curiosity in context. *Plant Soil, 274*, 101–125.

Smith, S. E., et al. (2004). Functional diversity in arbuscular mycorrhizal (AM) symbioses: The contribution ofthe mycorrhizal P uptake pathway is not correlated with mycorrhizal responses in growth or total P uptake. *New Phytology, 162*, 511–524.

Stewart, M. W., Dibb, W. D., Johnston, E. A., & Smyth, J. T. (2005). The contribution of commercial fertilizer nutrients to food production. *Agron Journal, 97*, 1–6.

Templeton, G. W., & Moorhead, G. B. G. (2004). A renaissance of metabolite sensing and signaling: From modular domains to riboswitches. *Plant Cell, 16*, 2252–2257.

Tränkner, Merle, Tavakol, Ershad, & Jákli, Bálint (2018). Functioning of potassium and magnesium in photosynthesis, photosynthate translocation and photoprotection. *Physiologia Plantarum, 163*, 414–431.

Umar, W., Ayub, M. A., Ahmad, H. R., Farooqi, Z. U. R., Shahzad, A., Rehman, U., Mustafa, A., & Nadeem, M. (2020). Nitrogen and phosphorus use efficiency in agroecosystems. In *Resources Use Efficiency in Agriculture* (pp. 213–257). Singapore: Springer.

Vidal, E. A., & Gutiérrez, R. A. (2008). A systems view of nitrogen nutrient and metabolite responses in Arabidopsis. *Current Opinion in Plant Biology, 11*, 521–529.

Vincentz, M., Moureaux, T., Leydecker, M. T., Vaucheret, H., & Caboche, M. (1993). Regulation of nitrate and nitrite reductase expression in Nicotiana plumbaginifolia leaves by nitrogen and carbon metabolites. *Plant Journal, 3*, 313–324.

Vitousek, P. M., et al. (1995). Foliar nutrients during long-term soil development in Hawaiian montane rain forest. *Ecology, 76*, 712–720.

Vitousek, P. M., & Howarth, R. W. (1991). Nitrogen limitation on land and in the sea: How can it occur? *Biogeochemistry, 13*, 87–115.

Wang, Y., & Chu, C. (2020). S-nitrosylation control of ROS and RNS homeo- stasis in plants: The switching function of catalase. *Molecular Plant, 13*, 946–948.

White, P. J. (2000). Calcium channels in higher plants. *Biochimica et Biophysica Acta, 1465*, 171–189.

White, P., & Karley, A. (2010). In R. Hell, & R.-R. Mendel (Eds.), *Potassium" in Cell biology of metals and nutrients* (pp. 199–224). Berlin/Heidelberg: Springer).

Williamson, L. C., Ribrioux, S., Fitter, A. H., & Leyser, O. (2001). Phosphate availability regulates root system architecture in Arabidopsis. *Plant Physiol, 126*, 875–882.

Wolfsberg, E., Long, C. P., & Antoniewicz, M. R. (2018). Metabolism in dense microbial colonies: 13C metabolic flux analysis of *E. coli* grown on agar identifies two distinct cell populations with acetate cross-feeding. *Metabology in Engineering, 49*, 242–247.

Yusefi-Tanha, E., Fallah, S., Rostamnejadi, A., & Pokhrel, L. R. (2020). Root system Architecture, copper uptake and tissue distribution in soybean (*Glycine max* (L.) Merr.) grown in copper oxide nanoparticle (CuONP)-Amended soil and implications for human nutrition. *Plants, 9*, 1326.

Zhao, X. Q., & Shen, R. F. (2018). Aluminum–nitrogen interactions in the soil–plant system. *Frontiers in Plant Science, 9*, 807.

Zhu, Y. G., Sun, G. X., Lei, M., Teng, M., Liu, Y. X., Chen, N. C., Wang, L. H., Carey, A. M., Deacon, C., Raab, A., Meharg, A. A., & Williams, P. N. (2008). High percentage inorganic arsenic content of mining impacted and non-impacted Chinese rice. *Environmental Science and Technology, 42*, 5008–5013.

van Gestel, J., Vlamakis, H., & Kolter, R. (2015). Division of labor in biofilms: The ecology of cell differentiation. *Microbiol. Spectr., 3*. MB-0002-2014.

CHAPTER FOUR

Climate change and plant nutrient availability: challenges and assessment strategies

Khair Ul Nisa[a], Najeebul Tarfeen[b], Qadrul Nisa[c], Saba Wani[d]

[a]Department of Environmental Science, University of Kashmir, Srinagar, India
[b]Centre of Research for Development (CORD), University of Kashmir, Srinagar, India
[c]Division of Plant Pathology, SKUAST-K, Srinagar, India
[d]Department of Biochemistry, University of Kashmir, Srinagar, India

1. Introduction

The global human population continues to grow at the tremendous rate and is expected to reach 9.6 billion by 2050 (Godfray et al., 2010), convincingly so the global food demands. Talking about the global cultivated land area, there has not been any significant increase in the last 25 years (O'Mara, 2012). Agricultural intensification practices as well as plant breeding approaches have undoubtedly managed to increase crop yields. The most important factor determining the crop production and nutrient availability revolves around the optimal climatic condition. The global threat to food security particularly in the lower income countries can be closely linked to changing climatic conditions (Vos & Bellù, 2019). Precisely the major climatic variables i.e., temperature, precipitation and atmospheric carbon dioxide (CO_2) can have direct effect on agricultural productivity. Furthermore, fluctuation in the normal climatic conditions may change the delicate relation between crops, weeds, pests and pathogens (Soares et al., 2019). In addition, it can pose serious impact on natural pollinators and water availability (Brouder & Volenec, 2008). Regardless, the threat climate change poses to agricultural productivity, plant nutrient availability may also get diminished (Soares et al., 2019). Overall the global climate change can directly (i) effect the plants physiological makeover to efficiently use nutrients, (ii) can alter the availability of nutrients in soil and (iii) can disrupt the nutrient transportation through soil and across root system (Prieto & Querejeta, 2020; Brouder & Volenec, 2008). It is therefore need of the hour to understand the implications and possible impact of

Sustainable Plant Nutrition
ISBN: 978-0-443-18675-2
https://doi.org/10.1016/B978-0-443-18675-2.00015-8

changing climate on plant nutrition and to work on the possible way outs to optimize this unwanted change. In this review, we will try to evaluate the impact of global climate change and its impact on plant nutrient availability and uptake mechanisms. We will emphasize on the key environmental parameters that may directly limit the nutrient cycling and availability to a great extend. We will also try to provide important assessment strategies to overcome the possible constrain with regard to plant nutrient availability in the current changing environmental conditions.

2. Climate change as a global challenge

Climate change has been regarded as a multifactorial stress (Gray & Brady, 2016). Talking about its impact on our agro systems, significant shifting in the major pathways governing the developmental processes, molecular functioning, morphological traits, and basic physiological responses has been affected. Intensive release of green house gases, unending population growth, unplanned economical growth, insatiable appetite for fossil fuel, industrial bloom, deforestation are the major risking factors associated with the global climate change (Stern & Kaufmann, 2014). These has been a constant release of green house gases (CO_2, CO, CH_4, N_2O) leading to global warming imparting more severe and permanent effect on human health and agriculture (Myers et al., 2017). It has been reported that the expected concentration of CO_2 by 2050 can reach to 550 μmol mol $^{-1}$ by 2050 (Raven & Karley, 2006). Furthermore, as per IPCC report 2018 the average increase in the global temperature is expected to be 1°C by 2050. More than two billion people is expected to face complete water shortage and 65% of human population to face partial water insufficiencies by the end of this century (Nezhadahmadi et al., 2013). The fluctuations in the global climatic conditions have negatively affected the bioavailability of plant nutrients because of change in the soil moisture, which is considered as a key medium for nutrient allocation (Fischer et al., 2019). The relative increase in the crop production have been reported (probably due to increased CO_2 and elevated temperatures) in some parts of globe. For e.g., wheat production across Europe (*Triticum sativum*) in early 1960s, however at the same time it has decreased in Southern countries like Portugal and Spain (Schär et al., 2004) indicating regional differential impact of global warming. To conclude this, the global impact of climate change on agricultural systems is expected to be negative owing to the diminishing food and nutritional quality of food crops and soil (Nelson et al., 2009).

3. Impact of excessive CO_2 and plant nutrient availability

There has been a progressive increase in the concentration of CO_2 from the past few decades (280–400 ppm currently) IPCC 2018. Studies have proposed that the increased concentration of CO_2 might improve the crop performance as it increases the photosynthetic rates and water efficiencies chiefly in C3 plants and has been regarded as "CO_2 fertilization effect" (Ainsworth & Long, 2005). Numerous agricultural crops like wheat, maize, rice, barley, beans have been reported to behave positively with the increased CO_2 concentrations. However, examining the other side of the coin, the increasing the CO_2 treatments might result in the increased concentration of soluble sugars consequently leading to photosynthetic acclimation, leaf senesce and may affect the overall growth rate (Ainsworth & Long, 2005). Cook et al. (1998) reported 25% decrease in the photosynthetic rates in *Nardus strictus* growing in eCO_2 environment for more than 100 years (>790 μM) compared to same plant sp. when grown in natural environment (<360 μM). They further communicated that the decrease in photosynthetic rates may be due to decrease in the chlorophyll content along with other photosynthetic proteins (Cook et al., 1998). Likewise, Loladze (2014) meta-analyzed FACE and NON-FACE studies and concluded that eCO_2 concentrations have negative effect on the overall mineral concentration (~8%) in the foliar and edible tissues of C3 plants. Furthermore, the overall decrease in mineral concentration in various plant groups namely crop plants, wild sp. herbaceous and woody plants was about 7.2%, 9.7%, 7.5% and 9.2% respectively (Loladze, 2014). Myers et al. (2014) reported that a significant decrease in the protein concentration in C3 grasses (6.3% in wheat and 7.8% in rice) under excessive CO_2 levels. Such data signify how global climate change might draw a significant effect on the availability of important micronutrients chiefly in the food crops thus risking the world population. Similar results were generated in case of Zinc mineral availability by Mayer et al. (2015), which showed eCO_2 decreased the availability of Zn in major food crops thus exacerbating the condition of almost 138 million people already Zn deficient residing mainly in African and South Asian countries (Myers et al., 2015). Studies have shown variable results with respect to the nutrient bioavailability with in soil and concentrations of CO_2. Kimball et al. (2002) reported positive effect of excessive CO_2 concentrations and soil nutrient availability. Similarly, in another study it was observed that increased concentration of CO_2 increased the bioavailability of P, K, Fe, Mn and Zn in soil (Kumeleh et al., 2009). Moreover, elevated CO_2 levels have been associated with reduced leaf level transpiration in C3 plants (Ainsworth et al., 2002) consequently decreasing the availability of two important nutrients; Ca and Si as their transportation is dependent on total transpiration factor (Marschner, 1995). In addition, excessive CO_2 concentration is suggested to impart positive effect on arbuscular mycorrhiza thus increasing the nutrient allocation and mobilization (Lynch & Clair, 2004). Similarly, nitrogen-fixing plants are reported to show positive response to the elevated CO_2 levels as they can directly utilize excessive photosynthetase to alleviate N limitation (Hartwig et al., 2002).

4. Impact of eCO_2 on protein accumulation

Studies have revealed that eCO_2 levels can directly decrease the protein concentration in many crop plants (Dong et al., 2018; Loladze, 2002; Myers et al., 2017). For example study conducted by Medek et al., 2017 (Medek et al., 2017) reported that the protein content of many C3 grains; chiefly rice, wheat and barley decreased 7.8%, 7.6% and 14.1% respectively (CO_2 concentration up to 500–700 ppm). The study further confirmed some of the legume species showed net decrease of 3.5% in the overall protein content, however not significant effect was seen on C4 oil plants. Furthermore, it is estimated if the current CO_2 concentration will elevate at the same rate it would result in more than 1.4 billion people globally to face severe protein deficiencies by 2050 (Soares et al., 2019). The result will be more drastic in third world countries where nutrient deficiencies have already caused havoc.

5. Impact of excessive temperature and plant nutrient availability

Global temperatures are changing at a very fast pace and consequently affect normal nutrient demands and functional biology of plants. It seems to be very difficult for the plants to adapt or to keep pace with the changing climatic conditions (Zandalinas et al., 2018). Very little is known about the influence of elevated temperature on the nutrient acquisition potential of plants, as it is difficult to simplify the possible soil and especially air temperature on nutrient availability compared to CO_2. Small scale studies suggest that the nutrient cycles that totally are associated with biological systems (C and N) show different reaction when compared to other type of nutrient cycles such as P, S and K (biological and geological processes) or predominantly geological (such as K, Ca, Mg, S or micronutrients), e.g (Watanabe et al., 2007; Wood et al., 2006). It is quiet known fact that the Rubisco activity is optimal at 25°C, however temperatures greater than this may result in the photosynthesis acclimation as more O_2 is dissolved at higher temperatures thus replacement of CO_2 at Rubisco sites in the chloroplasts (Bernacchi et al., 2002). The nutrient availability in relation to the increasing temperatures might present contrasting results for arctic and equatorial type of climatic conditions as in case of former the higher temperatures are accompanied by the increased soil nutrient availability. When compared to arid and semi-arid regions increased temperatures might favor drought like conditions. Furthermore, the nutrient uptake potential of plants is enzyme dependent process (Rennenberg et al., 2006). The direct effect of increased temperature on nutrient uptake process cannot be quantified in field conditions as it is often accompanied by drought like conditions. Moreover, studies have proposed variable results while defining the impact of elevated temperatures on soil nutrient availability and suggested that the enzymatic processes might get enhanced only up to threshold level, beyond that the excessive temperatures might seize nutrient uptake processes (Gessler et al., 1998). Numerous studied supports the fact that increase in soil temperature might enhance the bioavailability of soil nutrients to the plants. It is important to mention here that the studies carried out by BassiriRad et al. (1996) confirmed that the combined effect of the excessive temperature and excessive CO_2 in the mobilization of phosphorus and nitrogen. Higher temperatures might also favor the good growth which may ultimately accelerates the process of nutrient uptake and allocation as confirmed by Repo et al. (2005) and Fotelli et al. (2005). It is noteworthy to mention here that the nature and the total span of the elevated temperatures are the main governing factors in defining the overall impact. A period of strong, high annual maximum might alter the normal nutrient uptake patterns (Grassi et al., 2003). In another study, it was observed that the elevated temperature and drought could directly reduce the above ground biomass accumulation of plants (Sardans, Peñuelas, & Prieto, et al., 2008). Furthermore, Poorter and Perez (2001) concluded that the elevated temperatures might reduce the growth stimulating effect of CO_2. Root characteristics like morphology, architecture, respiration and membrane absorbance are markedly affected by soil temperatures, which ultimately influence the plant nutrient availability and accumulation processes (Bassirirad, 2000).

Increased temperatures can enhance the nutrient cycling processes as it favors litter decomposition and mineralization of soil nutrients (Lukac et al., 2010). Talking about the nitrogen cycling process and excessive temperature (eT) relation variable results can be witnessed depending upon the type of ecosystem and the particular season (Beier et al., 2008; Rustad et al., 2001). Elevated temperatures enhance the N cycling chiefly during the cold conditions when temperature is limiting or in the spring when the plant-microbial activity is at peak (Beier et al., 2008, Sardans, Peñuelas, & Estiarte, et al., 2008). Pendall et al. (2004) reported that the eT might accelerate the soil nitrogen mineralization mainly due to nitrate leaching. Rustad et al. (2001) also concluded that the eT result in the increased N mineralization with positive impact on the growth of plants (Rustad et al., 2001). However, Beier et al. (2008) communicated contrasting results by confirming that eT have no relation with N mineralization and highlighted the role of soil moisture as the main active parameter. In addition, the litter decomposition rate is predicted to be accelerated by 4%—7% with the increase in temperature and precipitation from a double atmospheric CO_2 scenario (Moore et al., 1999). Summing up it can be concluded that an increase in N mineralization in soil with increased temperatures can be predicted in such ecosystems where temperature is limiting factor following the adequate quantity of soil moisture and substrate availability (Hyvönen et al., 2007).

6. Impact of excessive light and plant nutrient availability

Light intensity plays a significant role in determining the plant growth and yield in agro ecosystems (from light). Studies have conveyed that increased UV and visible radiation may directly affect the foliar phenolic content (Caldwell et al., 2007; Lavola et al., 2000). As a result, this change in secondary may directly decrease nutrient cycling, decomposition process and soil chemistry (Lynch & Clair, 2004). Several studies reported that the excessive UV and visible intensities can lead to photoinhibition and ultimately to photo-oxidative stress and the result can be more vulnerable in presence of other stresses (Foyer et al., 1994; Xu et al., 1998). Additionally, the interaction between the high light intensities along with the nutrient deficient conditions and metal toxicity can accelerate photo-oxidative stress (González et al., 1998) consequently reducing the overall productivity (Fig. 4.1).

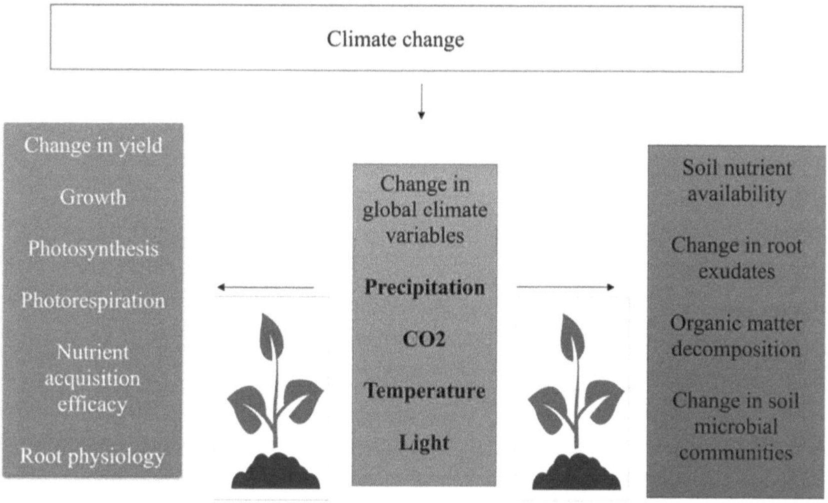

Fig. 4.1 Potential impact of climate change on overall on plant physiology and nutrient bioavailability.

7. Molecular adaptations in plants in response to the changing climate

The extend to which these changing climatic conditions effect plant growth, development and other physiological processes, posses a great threat to sustainable agriculture. As discussed earlier, environmental factors like drought, elevated temperatures,-both individually, or more commonly, in combination may alter the nutrient availability and plant's ability to accumulate the same (Myers et al., Ahuja et al., 2010). Climate change catastrophes disrupt the overall plant architecture and other cellular activities consequently hamper plant growth and development. Plants possess great reprogramming abilities (transcriptome, metabolome and proteome) to adapt in stressful environmental conditions (Ahuja et al., 2010; Soares et al., 2019; Zandalinas et al., 2018). For e.g., the in response to the drought conditions plants reduces transpiration, photosynthetic activities or may alter cell wall elasticity. ABA is the key player governing the essential functions related to water deficit state like regulating expression of drought resistant genes, stomatal opening/closing (Li et al., 2008; Ramírez et al., 2009; Seo et al., 2009). Nuclear factor Y (NFYA5) is an important transcriptional factor mediated via abscisic acid (ABA)-dependent manner and is important for drought resistance; induction at both the transcriptional and posttranscriptional levels was strongly induced by drought stress in *Arabidopsis thalian* (Li et al., 2008). Like wise OCP3 was found to play important role in

ABA-mediated stomatal closure during drought stress (Ramírez et al., 2009). Ubiquitination is also plays active role in mediating drought stress responses. Transcriptional factor DREB2A and ubiquitination in *Arabidopsis* is reported to be actively involved in the modulation of drought response through DREB2A-regulated gene expression (Lee et al., 2009; Sakuma et al., 2006). Furthermore, plants are exposed to excessive temperatures throughout their multi seasonal life (Sarkar et al., 2009). Plants specifically cope up with excessive temperatures by important transcriptional factors regulating the expression of heat shock proteins (HSP) (Liu et al., 2008; Schramm et al., 2008). Gallas and Waters reported two important marker genes- Ascorbate peroxidase 2 (APX2) and Heat shock factor A2 (HSFA2) in response to combined stress of excessive temperature and light (Gallas & Waters, 2015). Excessive CO_2 may have variable effect on the over all physiology of plant and is greatly determined by the availability of other factors like water. Plants can accumulate higher biomass provided enough water is available (Wu et al., 2004). Also increased CO_2 may improve plant water relation under drought conditions as the former favor the closure of stomata and decreases transpiration (Kumar et al., 2019; Yuhui et al., 2017). Vicente et al. (2018) studied the impact of elevated CO_2, temperature and nitrogen in dandrum wheat. They communicated that their occurred co-expression of the genes associated with synthesis of organic N compounds and C metabolism (N fixation and respiratory pathways) (Vicente et al., 2018). Wang, Lam, et al., 2018 reported the elevation in CO_2 has no effect on maximal photochemical efficiency of photosystem II (PSII) in rice however; it significantly affected the quantum yield of PSII photochemistry in soybean (Wang, Hu, et al., 2018). Rizhsky et al. (2002) concluded that the combined effect of drought and elevated temperature resulted in the suppressed photosynthesis and increased respiration rate along with increased leaf temperature in *Nicotiana tobacum* (Rizhsky et al., 2002). Furthermore, *Cicer arietinum* showed increased Rubisco activity when exposed to excessive heat exposures and significantly decreased when the plant is exposed too combined effect of excessive temperatures and drought (Awasthi et al., 2017).

8. Assessment strategies to overcome nutrient loss in changing climate

The changing climatic conditions will have a detrimental impact on nutrient availability and other ecosystem functioning ultimately hampering normal processes governing the plant nutrient uptake and allocation. Clearly, as discussed above changing climatic conditions appears to have a direct effect on plant growth and other physiological efficiencies (Brouder & Volenec, 2008) (Fig. 4.2). Progress in the understanding of plant responses to edaphic stress and global climatic variables, numerous strategies have been put forward to preserve the nutrient loss in the future climatic conditions. The main target in order to mitigate the climate change effects and the nutritional loss focuses mainly on the agricultural crops. Proper agricultural planning and farmers' sensitization to protect the crop production and to promote food security in the extreme weather events can be viewed as the major highlights of this adaptational framework (Prior et al., 2018). Changing climate has markedly affected the soil microbial communities, their degradation potential consequently leading to the nutrient deprived soil (Chen et al., 2016). Therefore, in order to compensate the nutrient loss optimal field operations like smart fertilization holds a key role. The precision agriculture technologies has revolutionized the agricultural practices and has markedly (i) reduced the amount of inputs (responsible for green house gas emission), (ii) maintained the efficient crop yield and (iii) optimized the use of fertilizers. For e.g., the online data based on soil nutrient values can provide timely and accurate information regarding the fertilization and nutritional requirements specific to external environmental conditions (Elia & Conversa, 2015; He et al., 2011). Similarly, nutrient monitoring technologies like hyperspectral imaging have allowed the researchers to successfully monitor N requirements in tea plants (Wang, Lam, et al., 2018). Likewise, N deficiencies were detected in rice and wheat plants via multispectral imagery systems (Zhang et al., 2018; Zhu et al.,

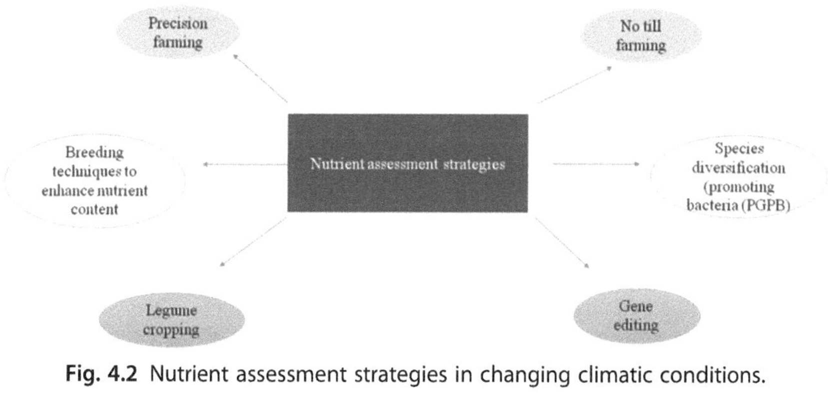

Fig. 4.2 Nutrient assessment strategies in changing climatic conditions.

2018). Another sustainable alternative that can be utilized in current changing environments to suffice nutritional loss includes plant growth-promoting bacteria (PGPB) inoculants. Olanrewaju et al. (2017) communicated PGPB as biological fertilizers as they can improve not only plant health but also can enhance soil nutrient conditions (Olanrewaju et al., 2017). Nie et al. (2015) reported that when grass sp. was inoculated with *Pseudomonas fluorescens* increased the overall performance along with the growth rate in excessive CO_2 concentrations (Nie et al., 2015). In another study, it was observed that PGPB markedly regulate the soil nutrient conditions like nitrogen fixation, siderophore and phytohormone production (Lobo et al., 2019). In another study conducted on *Medicago scutellata* it was observed that the application of N fixing bacteria chiefly under drought condition not only compensate the requirements of chemical fertilizers but also regulated soil macronutrients and micronutrients under stressful conditions (Shabani et al., 2015). Undertaking this strategy we will be able to tackle specific environmental constrains by pre-selecting the type of PGPB to optimize plant growth in field conditions (da Silva et al., 2018). However, the economic, large-scale applicability and field stability challenges cannot be overlooked (Lobo et al., 2019).

Legume farming has also been reported as an effective strategy to mitigate climate change negatives. Ma et al. (2018) reported that the legume farming in crop rotation system can effectively reduce about 25% of green house gases (Ma et al., 2018). Legume crops have number of add-ons besides improving N fixing properties. Legumes can reduce CO_2 emissions, maintains soils physical and chemical nature, decreases pest susceptibility and so on (Karkanis et al., 2018). Xue et al. (2016) reported that the legume cropping could improve the availability of other essential nutrients like P, Fe and Zn. No tilling or zero tilling practice can reduce the nutrient loss and thus can been opted as a promising sustainable strategy to mitigate the same. No till farming can improve soil structure, enhance nutrient cycling and water infiltration rates (Barão et al., 2019). Repots have confirmed that such practices not only reduces less efforts but can equally prove fruitful in reducing methane emissions (Zhao et al., 2016), reduced P loss (Daryanto et al., 2017). Apart from above mentioned assessment strategies genetic engineering technologies can open new ways by understanding regulatory mechanisms of differential gene expression under changing climatic conditions and pursue new technologies that will certainly mitigate negative impacts of climate change and enhance advantages for future plant growth through transgenic manipulation (Brouder & Volenec, 2008). It is note worthy to mention here that the focus of the plants breeders was mainly to improve yield and to enhance resistance. The progressive decreases in the nutrient content and adverse climatic events have lead researches to shift their attention. Promising technologies like Quantitative Trait Loci (QTL) for the enhancement of nutrient content under excessive temperatures and CO_2 concentrations (Pilbeam, 2015), studying the expressional analysis of various genes under eCO_2 (Zhang et al., 2018) have marked the new beginnings. Furthermore, CRISPR/Cas9 system has also enabled researchers to develop successful rice lines with increased amylase content and resistant starch (Mishra et al., 2018).

9. Conclusion

The global climate change casts a drastic shadow on the plant growth and development. The abiotic stresses associated with it, can directly hamper the nutrient availability in soil and reduce the overall nutrient allocation efficiencies. Excessive CO_2, drought, temperature, and light are crucial indicators of environmental stress. Both positive and negative interactions can be found between individual climatic variable and soil nutrient availability. Therefore, it is very important to work on optimizing the plant growth and development under such abiotic stress conditions. Breeding practices will help to develop climate resilient crops. Apart from this, genetic engineering approach like application of transgenic plants, advanced tools (CRISPR/Cas9 system) have emerged with a new hope and possibilities to negate the stress related to climate change in agro systems.

References

Ahuja, I., de Vos, R. C., Bones, A. M., & Hall, R. D. (2010). Plant molecular stress responses face climate change. *Trends in Plant Science, 15*(12), 664–674.

Ainsworth, E. A., Davey, P. A., Bernacchi, C. J., Dermody, O. C., Heaton, E. A., Moore, D. J., … Zhu, X.g. (2002). A meta-analysis of elevated [CO_2] effects on soybean (Glycine max) physiology, growth and yield. *Global Change Biology, 8*(8), 695–709.

Ainsworth, E. A., & Long, S. P. (2005). What have we learned from 15 years of free-air CO_2 enrichment (face)? A meta-analytic review of the responses of photosynthesis, canopy properties and plant production to rising CO_2. *New Phytologist, 165*(2), 351–372.

Awasthi, R., Gaur, P., Turner, N. C., Vadez, V., Siddique, K. H., & Nayyar, H. (2017). Effects of individual and combined heat and drought stress during seed filling on the oxidative metabolism and yield of chickpea (Cicer arietinum) genotypes differing in heat and drought tolerance. *Crop and Pasture Science, 68*(9), 823–841.

Barão, L., Alaoui, A., Ferreira, C., Basch, G., Schwilch, G., Geissen, V., … Morugán-Coronado, A. (2019). Assessment of promising agricultural management practices. *Science of the Total Environment, 649*, 610–619.

Bassirirad. (2000). Kinetics of nutrient uptake by roots: responses to global change. *The New Phytologist.*

BassiriRad, H., Tissue, D., Reynolds, J., & CHAPIN, F., III (1996). Response of eriophorum vaginatum to CO2 enrichment at different soil temperatures: Effects on growth, root respiration and PO43− uptake kinetics. *New Phytologist, 133*(3), 423−430.

Beier, C., Emmett, B., Peñuelas, J., Schmidt, I., Tietema, A., Estiarte, M., ... Sowerby, A. (2008). Carbon and nitrogen cycles in European ecosystems respond differently to global warming. *Science of the Total Environment, 407*(1), 692−697.

Bernacchi, C. J., Portis, A. R., Nakano, H., Von Caemmerer, S., & Long, S. P. (2002). Temperature response of mesophyll conductance. Implications for the determination of Rubisco enzyme kinetics and for limitations to photosynthesis in vivo. *Plant Physiology, 130*(4), 1992−1998.

Brouder, S. M., & Volenec, J. J. (2008). Impact of climate change on crop nutrient and water use efficiencies. *Physiologia Plantarum, 133*(4), 705−724.

Caldwell, M. M., Bornman, J., Ballaré, C., Flint, S. D., & Kulandaivelu, G. (2007). Terrestrial ecosystems, increased solar ultraviolet radiation, and interactions with other climate change factors. *Photochemical & Photobiological Sciences, 6*(3), 252−266.

Chen, Z., Wang, B., Wang, J., Pan, G., & Xiong, Z. (2016). Contrasting effects of elevated CO 2 and warming on temperature sensitivity of soil organic matter decomposition in a Chinese paddy field. *Environmental Monitoring and Assessment, 188*(10), 1−8.

Cook, A., Tissue, D., Roberts, S., & Oechel, W. (1998). Effects of long-term elevated [CO_2] from natural CO_2 springs on Nardus stricta: Photosynthesis, biochemistry, growth and phenology. *Plant, Cell and Environment, 21*(4), 417−425.

Daryanto, S., Wang, L., & Jacinthe, P. A. (2017). Meta-analysis of phosphorus loss from no-till soils. *Journal of Environmental Quality, 46*(5), 1028−1037.

Dong, J., Gruda, N., Lam, S. K., Li, X., & Duan, Z. (2018). Effects of elevated CO2 on nutritional quality of vegetables: A review. *Frontiers in Plant Science, 9*, 924.

Elia, A., & Conversa, G. (2015). A decision support system (GesCoN) for managing fertigation in open field vegetable crops. Part I—methodological approach and description of the software. *Frontiers in Plant Science, 6*, 319.

Fischer, S., Hilger, T., Piepho, H.-P., Jordan, I., & Cadisch, G. (2019). Do we need more drought for better nutrition? The effect of precipitation on nutrient concentration in east african food crops. *Science of the Total Environment, 658*, 405−415.

Fotelli, M. N., Rudolph, P., Rennenberg, H., & Geßler, A. (2005). Irradiance and temperature affect the competitive interference of blackberry on the physiology of European beech seedlings. *New Phytologist, 165*(2), 453−462.

Foyer, C. H., Lelandais, M., & Kunert, K. J. (1994). Photooxidative stress in plants. *Physiologia Plantarum, 92*(4), 696−717.

Gallas, G., & Waters, E. R. (2015). Boechera species exhibit species-specific responses to combined heat and high light stress. *PLoS One, 10*(6), e0129041.

Gessler, A., Schneider, S., Von Sengbusch, D., Weber, P., Hanemann, U., Huber, C., ... Rennenberg, H. (1998). Field and laboratory experiments on net uptake of nitrate and ammonium by the roots of spruce (Picea abies) and beech (Fagus sylvatica) trees. *The New Phytologist, 138*(2), 275−285.

Godfray, H. C. J., Beddington, J. R., Crute, I. R., Haddad, L., Lawrence, D., Muir, J. F., ... Toulmin, C. (2010). Food security: The challenge of feeding 9 billion people. *Science, 327*(5967), 812−818.

González, A., Steffen, K. L., & Lynch, J. P. (1998). Light and excess manganese: Implications for oxidative stress in common bean. *Plant Physiology, 118*(2), 493−504.

Grassi, G., Millard, P., Gioacchini, P., & Tagliavini, M. (2003). Recycling of nitrogen in the xylem of Prunus avium trees starts when spring remobilization of internal reserves declines. *Tree Physiology, 23*(15), 1061−1068.

Gray, S. B., & Brady, S. M. (2016). Plant developmental responses to climate change. *Developmental Biology, 419*(1), 64–77.

Hartwig, U. A., Lüscher, A., Nösberger, J., & Van Kessel, C. (2002). Nitrogen-15 budget in model ecosystems of white clover and perennial ryegrass exposed for four years at elevated atmospheric pCO2. *Global Change Biology, 8*(2), 194–202.

He, J., Wang, J., He, D., Dong, J., & Wang, Y. (2011). The design and implementation of an integrated optimal fertilization decision support system. *Mathematical and Computer Modelling, 54*(3–4), 1167–1174.

Hyvönen, R., Ågren, G. I., Linder, S., Persson, T., Cotrufo, M. F., Ekblad, A., ... Jarvis, P. G. (2007). The likely impact of elevated [CO_2], nitrogen deposition, increased temperature and management on carbon sequestration in temperate and boreal forest ecosystems: A literature review. *New Phytologist, 173*(3), 463–480.

Karkanis, A., Ntatsi, G., Lepse, L., Fernández, J. A., Vågen, I. M., Rewald, B., ... Olle, M. (2018). Faba bean cultivation—revealing novel managing practices for more sustainable and competitive European cropping systems. *Frontiers in Plant Science, 9*, 1115.

Kimball, B., Kobayashi, K., & Bindi, M. (2002). Responses of agricultural crops to free-air CO_2 enrichment. *Advances in Agronomy, 77*, 293–368.

Kumar, A., Nayak, A., Das, B., Panigrahi, N., Dasgupta, P., Mohanty, S., ... Pathak, H. (2019). Effects of water deficit stress on agronomic and physiological responses of rice and greenhouse gas emission from rice soil under elevated atmospheric CO_2. *Science of the Total Environment, 650*, 2032–2050.

Kumeleh, A. S., Sharmila, P., Uprety, D., & Saradhi, P. (2009). Effects of elevated CO_2 on soil physicochemical characteristics under Free Air CO_2 Enrichment (FACE) technology. Paper presented at the IOP Conference Series. *Earth and Environmental Science, 6*.

Lavola, A., Julkunen-Tiitto, R., de la Rosa, T. M., Lehto, T., & Aphalo, P. J. (2000). Allocation of carbon to growth and secondary metabolites in birch seedlings under UV-B radiation and CO_2 exposure. *Physiologia Plantarum, 109*(3), 260–267.

Lee, H. K., Cho, S. K., Son, O., Xu, Z., Hwang, I., & Kim, W. T. (2009). Drought stress-induced Rma1H1, a RING membrane-anchor E3 ubiquitin ligase homolog, regulates aquaporin levels via ubiquitination in transgenic Arabidopsis plants. *The Plant Cell, 21*(2), 622–641.

Li, W.-X., Oono, Y., Zhu, J., He, X.-J., Wu, J.-M., Iida, K., ... Zhu, J.-K. (2008). The Arabidopsis NFYA5 transcription factor is regulated transcriptionally and posttranscriptionally to promote drought resistance. *The Plant Cell, 20*(8), 2238–2251.

Liu, H. T., Gao, F., Li, G. L., Han, J. L., Liu, D. L., Sun, D. Y., & Zhou, R. G. (2008). The calmodulin-binding protein kinase 3 is part of heat-shock signal transduction in *Arabidopsis thaliana*. *The Plant Journal, 55*(5), 760–773.

Lobo, C. B., Tomás, M. S. J., Viruel, E., Ferrero, M. A., & Lucca, M. E. (2019). Development of low-cost formulations of plant growth-promoting bacteria to be used as inoculants in beneficial agricultural technologies. *Microbiological Research, 219*, 12–25.

Loladze, I. (2002). Rising atmospheric CO2 and human nutrition: Toward globally imbalanced plant stoichiometry? *Trends in Ecology & Evolution, 17*(10), 457–461.

Loladze, I. (2014). Hidden shift of the ionome of plants exposed to elevated CO_2 depletes minerals at the base of human nutrition. *Elife, 3*, e02245.

Lukac, M., Calfapietra, C., Lagomarsino, A., & Loreto, F. (2010). Global climate change and tree nutrition: Effects of elevated CO2 and temperature. *Tree Physiology, 30*(9), 1209–1220.

Lynch, J. P., & Clair, S. B. S. (2004). Mineral stress: The missing link in understanding how global climate change will affect plants in real world soils. *Field Crops Research, 90*(1), 101–115.

Marschner, H. (1995). *Mineral nutrition of higher plants* (2nd ed.). Great Britain: Academic.

Ma, Y., Schwenke, G., Sun, L., Li Liu, D., Wang, B., & Yang, B. (2018). Modeling the impact of crop rotation with legume on nitrous oxide emissions from rain-fed agricultural systems in Australia under alternative future climate scenarios. *Science of the Total Environment, 630*, 1544–1552.

Medek, D. E., Schwartz, J., & Myers, S. S. (2017). Estimated effects of future atmospheric CO_2 concentrations on protein intake and the risk of protein deficiency by country and region. *Environmental Health Perspectives, 125*(8), 087002.

Mishra, R., Joshi, R. K., & Zhao, K. (2018). Genome editing in rice: Recent advances, challenges, and future implications. *Frontiers in Plant Science, 9*, 1361.

Moore, T., Trofymow, J., Taylor, B., Prescott, C., Camire, C., Duschene, L., … Morrison, I. (1999). Litter decomposition rates in Canadian forests. *Global Change Biology, 5*(1), 75–82.

Myers, S. S., Smith, M. R., Guth, S., Golden, C. D., Vaitla, B., Mueller, N. D., … Huybers, P. (2017). Climate change and global food systems: Potential impacts on food security and undernutrition. *Annual Review of Public Health, 38*(1), 259–277.

Myers, S. S., Wessells, K. R., Kloog, I., Zanobetti, A., & Schwartz, J. (2015). Rising atmospheric CO2 increases global threat of zinc deficiency. *The Lancet Global Health, 3*(10), e639.

Myers, S. S., Zanobetti, A., Kloog, I., Huybers, P., Leakey, A. D. B., Bloom, A. J., Carlisle, E., Dietterich, L. H., Fitzgerald, G., Hasegawa, T., Holbrook, N. M., Nelson, R. L., Ottman, M. J., Raboy, V., Sakai, H., Sartor, K. A., Schwartz, J., Seneweera, S., Tausz, M., & Usui, Y. (2014). Increasing CO_2 threatens human nutrition. *Nature, 510*, 139–142.

Nelson, G. C., Rosegrant, M. W., Koo, J., Robertson, R., Sulser, T., Zhu, T., … Batka, M. (2009). *Climate change: Impact on agriculture and costs of adaptation* (Vol. 21). Intl Food Policy Res Inst.

Nezhadahmadi, A., Prodhan, Z., & Faruq, G. (2013). *Drought tolerance in wheat* (p. 610721). Science World J.

Nie, M., Bell, C., Wallenstein, M. D., & Pendall, E. (2015). Increased plant productivity and decreased microbial respiratory C loss by plant growth-promoting rhizobacteria under elevated CO_2. *Scientific Reports, 5*(1), 1–6.

O'Mara, F. P. (2012). The role of grasslands in food security and climate change. *Annals of Botany, 110*(6), 1263–1270.

Olanrewaju, O. S., Glick, B. R., & Babalola, O. O. (2017). Mechanisms of action of plant growth promoting bacteria. *World Journal of Microbiology and Biotechnology, 33*(11), 1–16.

Pendall, E., Bridgham, S., Hanson, P. J., Hungate, B., Kicklighter, D. W., Johnson, D. W., … Olsrud, M. (2004). Below-ground process responses to elevated CO_2 and temperature: A discussion of observations, measurement methods, and models. *New Phytologist, 162*(2), 311–322.

Pilbeam, D. J. (2015). Breeding crops for improved mineral nutrition under climate change conditions. *Journal of Experimental Botany, 66*(12), 3511–3521.

Poorter, H., & Pérez-Soba, M. (2001). The growth response of plants to elevated CO 2 under non-optimal environmental conditions. *Oecologia, 129*(1), 1–20.

Prieto, I., & Querejeta, J. I. (2020). Simulated climate change decreases nutrient resorption from senescing leaves. *Global Change Biology, 26*(3), 1795–1807.

Prior, J. H., Connon, I. L., McIntyre, E., Adams, J., Capon, A., Kent, J., … Westcott, H. (2018). Built environment interventions for human and planetary health: integrating health in climate change adaptation and mitigation. *Public Health Research and Practice, 28*(4).

Ramírez, V., Coego, A., López, A., Agorio, A., Flors, V., & Vera, P. (2009). Drought tolerance in Arabidopsis is controlled by the OCP3 disease resistance regulator. *The Plant Journal, 58*(4), 578–591.

Raven, J. A., & Karley, A. J. (2006). Carbon sequestration: Photosynthesis and subsequent processes. *Current Biology, 16*(5), R165–R167.
Rennenberg, H., Loreto, F., Polle, A., Brilli, F., Fares, S., Beniwal, R., & Gessler, A. (2006). Physiological responses of forest trees to heat and drought. *Plant Biology, 8*(05), 556–571.
Repo, T., Kalliokoski, T., Domisch, T., Lehto, T., Mannerkoski, H., Sutinen, S., & Finér, L. (2005). Effects of timing of soil frost thawing on Scots pine. *Tree Physiology, 25*(8), 1053–1062.
Rizhsky, L., Liang, H., & Mittler, R. (2002). The combined effect of drought stress and heat shock on gene expression in tobacco. *Plant Physiology, 130*(3), 1143–1151.
Rustad, L., Campbell, J., Marion, G., Norby, R., Mitchell, M., Hartley, A., ... Gurevitch, J. (2001). A meta-analysis of the response of soil respiration, net nitrogen mineralization, and aboveground plant growth to experimental ecosystem warming. *Oecologia, 126*(4), 543–562.
Sakuma, Y., Maruyama, K., Osakabe, Y., Qin, F., Seki, M., Shinozaki, K., & Yamaguchi-Shinozaki, K. (2006). Functional analysis of an Arabidopsis transcription factor, DREB2A, involved in drought-responsive gene expression. *The Plant Cell, 18*(5), 1292–1309.
Sardans, J., Peñuelas, J., & Estiarte, M. (2008). Changes in soil enzymes related to C and N cycle and in soil C and N content under prolonged warming and drought in a Mediterranean shrubland. *Applied Soil Ecology, 39*(2), 223–235.
Sardans, J., Peñuelas, J., Prieto, P., & Estiarte, M. (2008). Drought and warming induced changes in P and K concentration and accumulation in plant biomass and soil in a Mediterranean shrubland. *Plant and Soil, 306*(1), 261–271.
Sarkar, N. K., Kim, Y.-K., & Grover, A. (2009). Rice sHsp genes: Genomic organization and expression profiling under stress and development. *BMC Genomics, 10*(1), 1–18.
Schär, C., Vidale, P. L., Lüthi, D., Frei, C., Häberli, C., Liniger, M. A., & Appenzeller, C. (2004). The role of increasing temperature variability in European summer heatwaves. *Nature, 427*(6972), 332–336.
Schramm, F., Larkindale, J., Kiehlmann, E., Ganguli, A., Englich, G., Vierling, E., & Von Koskull-Döring, P. (2008). A cascade of transcription factor DREB2A and heat stress transcription factor HsfA3 regulates the heat stress response of Arabidopsis. *The Plant Journal, 53*(2), 264–274.
Seo, P. J., Xiang, F., Qiao, M., Park, J.-Y., Lee, Y. N., Kim, S.-G., ... Park, C.-M. (2009). The MYB96 transcription factor mediates abscisic acid signaling during drought stress response in Arabidopsis. *Plant Physiology, 151*(1), 275–289.
Shabani, G., Ardakani, M., Chaichi, M., Friedel, J., & Khavazi, K. (2015). Effect of different fertilizing treatments on nutrient uptake in annual medic (Medicago scutellata cv. Robinson) under irrigated and dry farming systems. *Journal of Agricultural Science and Technology, 17*(2), 299–310.
da Silva, J. F., da Silva, T. R., Escobar, I. E. C., Fraiz, A. C. R., Dos Santos, J. W. M., do Nascimento, T. R., ... Signor, D. (2018). Screening of plant growth promotion ability among bacteria isolated from field-grown sorghum under different managements in Brazilian drylands. *World Journal of Microbiology and Biotechnology, 34*(12), 1–10.
Soares, J. C., Santos, C. S., Carvalho, S. M., Pintado, M. M., & Vasconcelos, M. W. (2019). Preserving the nutritional quality of crop plants under a changing climate: Importance and strategies. *Plant and Soil, 443*(1), 1–26.
Stern, D. I., & Kaufmann, R. K. (2014). Anthropogenic and natural causes of climate change. *Climatic Change, 122*(1), 257–269.
Vicente, R., Martínez-Carrasco, R., Pérez, P., & Morcuende, R. (2018). New insights into the impacts of elevated CO_2, nitrogen, and temperature levels on the regulation of C and N metabolism in durum wheat using network analysis. *New Biotechnology, 40*, 192–199.

Vos, R., & Bellù, L. G. (2019). Global trends and challenges to food and agriculture into the 21st century. *Sustainable Food and Agriculture*, 11–30.
Wang, Y., Hu, X., Hou, Z., Ning, J., & Zhang, Z. (2018). Discrimination of nitrogen fertilizer levels of tea plant (Camellia sinensis) based on hyperspectral imaging. *Journal of the Science of Food and Agriculture, 98*(12), 4659–4664.
Wang, A., Lam, S. K., Hao, X., Li, F. Y., Zong, Y., Wang, H., & Li, P. (2018). Elevated CO_2 reduces the adverse effects of drought stress on a high-yielding soybean (Glycine max (L.) Merr.) cultivar by increasing water use efficiency. *Plant Physiology and Biochemistry, 132*, 660–665.
Watanabe, T., Broadley, M. R., Jansen, S., White, P. J., Takada, J., Satake, K., ... Osaki, M. (2007). Evolutionary control of leaf element composition in plants. *New Phytologist, 174*(3), 516–523.
Wood, T. E., Lawrence, D., & Clark, D. A. (2006). Determinants of leaf litter nutrient cycling in a tropical rain forest: Soil fertility versus topography. *Ecosystems, 9*(5), 700–710.
Wu, D.-X., Wang, G.-X., Bai, Y.-F., & Liao, J.-X. (2004). Effects of elevated CO_2 concentration on growth, water use, yield and grain quality of wheat under two soil water levels. *Agriculture, Ecosystems & Environment, 104*(3), 493–507.
Xue, Y., Xia, H., Christie, P., Zhang, Z., Li, L., & Tang, C. (2016). Crop acquisition of phosphorus, iron and zinc from soil in cereal/legume intercropping systems: A critical review. *Annals of Botany, 117*(3), 363–377.
Xu, Z.-F., Luo, G.-H., Wang, A.-G., & Cheng, Y. (1998). *Effects of strong light and exogenous active oxygen species on chlorophyll fluorescence in lettuce.*
Yuhui, W., Denghua, Y., Junfeng, W., Yi, D., & Xinshan, S. (2017). Effects of elevated CO_2 and drought on plant physiology, soil carbon and soil enzyme activities. *Pedosphere, 27*(5), 846–855.
Zandalinas, S. I., Mittler, R., Balfagón, D., Arbona, V., & Gómez-Cadenas, A. (2018). Plant adaptations to the combination of drought and high temperatures. *Physiologia Plantarum, 162*(1), 2–12.
Zhang, G., Zhang, T., Liu, J., Zhang, J., & He, C. (2018). Comprehensive analysis of differentially expressed genes reveals the molecular response to elevated CO_2 levels in two sea buckthorn cultivars. *Gene, 660*, 120–127.
Zhao, X., Liu, S. L., Pu, C., Zhang, X. Q., Xue, J. F., Zhang, R., ... Chen, F. (2016). Methane and nitrous oxide emissions under no-till farming in China: A meta-analysis. *Global Change Biology, 22*(4), 1372–1384.
Zhu, H., Liu, H., Xu, Y., & Guijun, Y. (2018). UAV-based hyperspectral analysis and spectral indices constructing for quantitatively monitoring leaf nitrogen content of winter wheat. *Applied Optics, 57*(27), 7722–7732.

CHAPTER FIVE

Role of mineral nutrients in biological nitrogen fixation

Uzma Kafeel, Urfi Jahan and Fareed Ahmad Khan
Plant Ecology and Environment Laboratory, Department of Botany, Aligarh Muslim University, Aligarh, Uttar Pradesh, India

1. Introduction

Nitrogen (N) is an important component of a cell and present in several biomolecules such as amino acids, nucleotide phosphates, nucleic acids and proteins. N enters natural ecosystems mainly because of the action of living entities like algae, cyanobacteria and land plants. The process of reduction of N_2 to NH_3 consequently conformed into amino acids aided by prokaryotes, is termed as biological nitrogen fixation (BNF). BNF responsible for the retrieval of N in environment, is crucial to fulfill the nutritive needs of the organisms dwelling in the environment. In the biogeochemical N cycle, N changes to many forms, from NH_3 (or NH_4 β, highly reduced), to nitrate ion (NO_3^-, highly oxidized), the steps of this cycle are influenced by both abiotic and biotic events. Major processes that take place in N cycle (Fig. 5.1) include the following:

- Ammonification: N present in organic residues (dead plants/animals) and litter is released into the soil by microorganisms present in the soil. These decomposers remove NH_3 from organic nitrogenous complexes, transform it to ammonium (NH_4) after removal and discharge this reduced form of inert N in the surroundings. NH_3 produced during decomposition is sequentially used for other natural processes.
- Nitrification- In this process NH_3 is transformed into NO_3^- by chemosynthetic bacteria present in soil. Oxidation of NH_3 with the help of Nitrosomonas (bacterium sp.) leads to the formation of Nitrite (NO_2^-). NO_2^- is further converted into NO_3^- by *Nitrobacter* (bacterium sp.). The bacteria gain energy from oxidation of NH_3 to NO_2^- (*Nitrosomonas*)

Sustainable Plant Nutrition
ISBN: 978-0-443-18675-2
https://doi.org/10.1016/B978-0-443-18675-2.00004-3

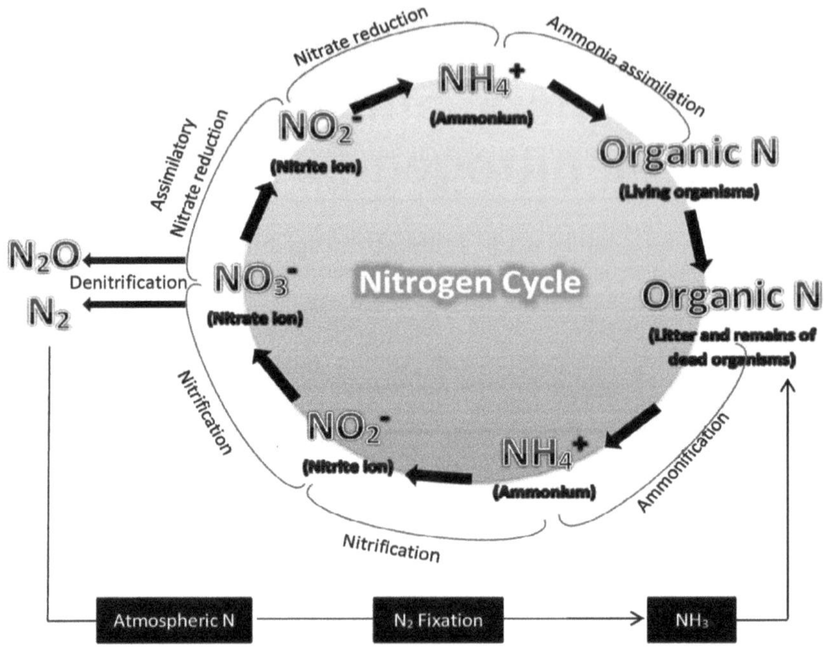

Fig. 5.1 Schematic representation of the processes occurring during nitrogen cycle.

as well as NO_2^- to NO_3^- (*Nitrobacter*). NH_3 gas is harmful for plants so this transformation is essential. Reactions taking place during Nitrification are given below:

$$2NH_4^+ + 3O_2 \rightarrow 2NO_2^- + 4H^+ + 2H_2O$$

$$NO_2^- + O_2 \rightarrow 2NO_3^-$$

Denitrification- Denitrification or assimilatory reduction of NO_3^-, the N compounds reenter the atmosphere by transforming NO_3^- into gaseous N. This phase takes place in the absence of oxygen (O) and is the last step of N cycle. *Clostridium* and *Pseudomonas* (denitrifying bacteria) act on NO_3^- to gain O resulting in release of free N gas, leading to the retrieval of biological N, therefore completing the cycle.

This is a highly significant phenomenon, making room for the reclamation of N that gets vanished in surroundings owing to bacterial actions. Nitrogen fixation provides the biosphere with approximately 200 million tons N/year. 85%–90% of N accessible to the plants in natural ecosystems, has been assessed to originate from BNF (Rascio & Rocca, 2008).

2. Mechanism of nitrogen fixation

Plants and microorganisms utilize inert nitrogen complexes like NH_4 or NO_3^- present in water and soil. Fixation of N_2 (Fig. 5.2) in biologically available complexes for cell uptake is aided by specific bacteria. These are defined as diazotrophs, because they obtain all of their N by fixing N_2. Throughout BNF, N_2 gets reduced in various electron transport mechanisms, causing production of NH_3 and hydrogen liberation (Kim & Rees, 1992). Later on, NH_4 is then utilized for the production of biomolecules. The nitrogenase enzyme complex catalyzes this reduction of N_2 to NH_4 in every nitrogen-fixing organism, through an ATP-dependent and tremendous energy taking chemical reaction. The complex of nitrogenase (Fig. 5.3) is made up of two different enzymes-dinitrogenase reductase (azoferredoxin) and dinitrogenase (molybdo-ferredoxin), both of them show no enzymatic function individually (Hageman & Burris, 1978). These two enzymes show high sensitivity to O_2, as it quickly deactivates them. At the time of nitrogen reduction, dinitrogenase reductase (oxidized) receives

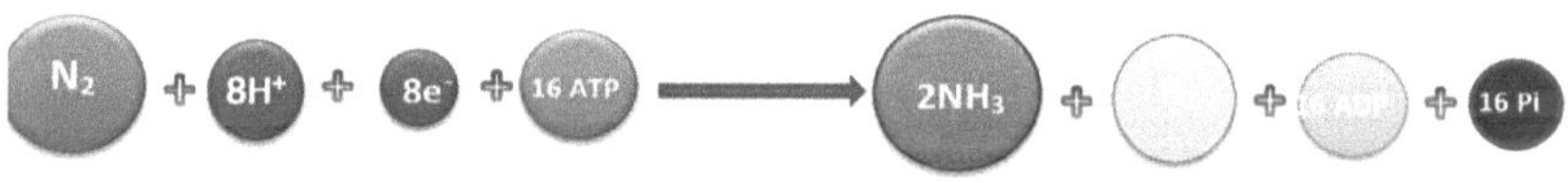

Fig. 5.2 Overall reaction of molecular nitrogen fixation.

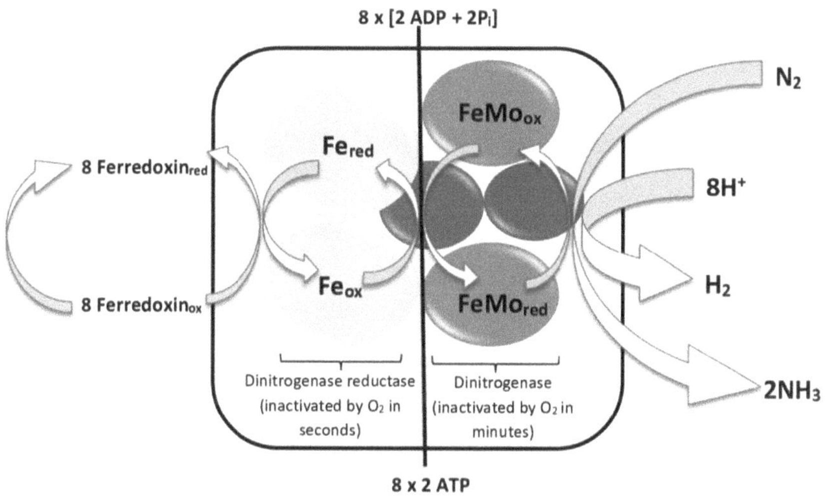

Fig. 5.3 Functioning of nitrogenase enzyme complex and nitrogen fixation.

an e^- from a benefactor species (i.e., reduced ferredoxin or flavodoxin) and joins two ATP molecules. This induces a conformational shift in redox potential of the Fe protein, lowering it from −300 to −400 mV. Hydrolysis of both ATP molecules takes place simultaneously with the transport of e^- from reduced Fe protein to oxidized dinitroreductase. Lastly, the FeMo protein performs the e^- (plus proton) transmission to the N_2 coupled to the MoFe cofactor. Because it takes six e^- to reduce N_2 to $2NH_3$, the hydrolysis of two ATP molecules results in six successive reduction events for every single electron going over the nitrogenase. Nevertheless, the nitrogenase moreover identifies protons (H) inside the cell, as a result, for every N_2 reduction (to $2NH_3$), 2 H ions are reduced to H_2, requiring two additional electrons and another 4ATP hydrolysis. Strong inhibition is also caused by the reaction by-products NH_3 and H_2. As a result, N_2 fixation is an inferential mechanism triggered by diazotrophic species only when alternative more cost-effective nitrogen sources are unavailable. The nitrogenase enzyme complex has a significant feature: its remarkable sensitivity to even low oxygen concentrations. Nitrogenase activity must be protected in aerobic conditions and photoautotrophic cyanobacteria, where O_2 is generated through light reactions during photosynthesis (Fay, 1992). Single celled cyanobacteria integrate N_2 fixation and photosynthesis in same cell, however these two routes are temporarily separated, with BNF limited to the dark time when O_2 intensity is stumpy (Falkowski et al., 2003). In addition to these defences, metabolic processes such as the Mehler reaction or specific O_2 foraging molecules like cyanoglobin and leghemoglobin, the second of which is important in rhizobia-plant interactions, can reduce oxygen concentrations (Appleby, 1984; Hill et al., 1996).

3. Factors affecting biological nitrogen fixation

Diffusion, mass transfer, and nutrient concentration are all elements that drive soil microorganism development. Because soils with bigger pores and pore spaces store less moisture, soil pore space is related to soil water retention. As a result, soil combinations with smaller internal pore spaces support the development of rhizobia and the majority of soil microorganisms (Turco & Sadowsky, 1995). Soil-water constituency has an indirect and direct influence on the proliferation of rhizosphere bacteria such as rhizobia by reducing water activity under acute tolerance parameters and fluctuating plant development, root modeling, and seepage. Reduced legume nodulation in desert soil is most likely attributable to reduced rhizobia population densities all through the arid conditions. The effect of soil and water interactions on plant development and vitality, in addition to nodulation, should not be underestimated. Rhizobia have developed a multitude of osmotic stress-response techniques, the most prevalent of which is the amassing of mineral and biological solutes within the cell. *R. meliloti*, for example, is resistant to osmotic stress-prompted growth suppression by storing appropriate solutes such as K^+, glutamate, proline, glycine betaine, proline betaine, trehalose, and the dipeptide N-acetyl glutaminyl glutamine amide (Boscari, 2002). Because some suitable solutes may be utilized as both N and C sources for development, their dissimilation can be synchronized to avoid degradation in the course of osmotic stress.

3.1 Soil pH stress

Because of the wide range of acid soils on the globe, the effect of soil pH on the nodulation method has received a great deal of attention. Low soil pH is often regarded as a predictor of conditions in which further soil qualities might hinder crop progress instead of a fundamental reason of reduced crop growth. Aside from the direct effects of soil acidity on plants, nodulation and nitrogen fixation may be hindered, indirectly decreasing legume growth. In this sense, Rhizobia might be relatively forbearing of soil acidity factors than the host plant. Rhizobial nodules could or could not fix N, or the rate at which they fix nitrogen may be scanty. Over 1.5 g ha^{-1} of acid soils impede agricultural productivity worldwide (Graham et al., 2000), and soil acidity concerns affect up to 25% of the world's croplands. According to Brockwell et al. (1991), the population of *S. meliloti* was roughly 10-3 lower in soils with a pH of six than in soils with a pH greater than 7. Soil pH, like soil moisture, has a range of impacts on rhizobia, although only a few develop and endure well under pH range of 4.5—5.0. Even though microsymbiont seems to be extrapH sensitive than the host plant (Hungria & Vargas, 2000), soil acidity impacts both the development and infection processes of the legume plant. This is presumably attributable to a breakdown in signal interchange among macrosymbionts and microsymbionts, as well as rhizobial inhibition of nodulation genes and Nod factor excretion. Interestingly, nodulated legumes tend to be more sensitive to Al and Mn metal damage than their N-fed competitors (Hungria & Vargas, 2000).

3.2 Soil temperature

Soil temperature has a considerable effect on the persistence and vitality of rhizobial strains in soils. Cowpea rhizobia, for instance, flourish at 37°C in West Africa's hot, dry Sahel-Savannah area, and more than 90% of the strains isolated from this site propagated finely in 40°C (Werner & Newton, 2005). Temperature appears to have an effect on rhizobia that is strain and soil reliant. *Bradyrhizobium* sp., for instance, was not as heat tolerant as *R. leguminosarum bv. trifolii*, but adding montmorillonite and illite to sandy soils rectified the problem. The temperature of the soil has a significant influence on nodulation competition (Triplett & Sadowsky, 1992). This result might be ascribed to temperature-prompted nodulation or nodule confinement to the subsurface. Aranjuelo et al. (2007) explored the influence of high temperature and water accessibility on CO_2 altercation and N-fixation in nodulated alfalfa plants. They discovered temprature had an impact on plant and nodule dry weight. Infection, N_2-fixation ability, and legume development have all been found to be influenced by relatively high root temperature (Hungria & Vargas, 2000), as well as certain cultivar and strain interchanges (Arayankoon et al., 1990). All legume and Rhizobium combinations seem to favor a temperature range of roughly 25–30°C for common bean (*Phaseolus vulgaris*), 32°C for clover (*Trifolium*) and pea and 36–40°C for soybean (*Glycine max*), peanut and cowpea *(Vigna unguiculata)* (Long, 2001).

3.3 Soil nutrients

As one might expect, soil nutritional quality has a considerable effect on the relationship as well as the independent development and existence of both the partners. As a result of the parallel growth in soil nitrogen, fixation diminishes with legume age. When N was applied to 0–20 cm of top soil or the soil surface, a negative exponential relationship between N fertiliser rate and N_2 fixation was seen. The variance in this connection is explained by a variability in N supply acquired from indigenous sources (net soil N mineralization, irrigation, and atmospheric deposition), as well as perhaps other variables impacting growth and N_2 fixation (for example, soil pH and drought). In this sense, an unfertilized grass component in a short-term mixed soil layer may play a beneficial function by competing for available soil N, causing the legume to get a larger proportion of N via N_2 fixation (Kafeel et al., 2022; Ledgard & Steele, 1992). It is important to note that nutritional stress can be caused indirectly by changes in soil matric potential or acidity, which restrict nutrient bioavailability, rather than directly by a lack of nutrients. While studying nutritional constraints to symbiotic nitrogen fixation, it's critical to distinguish between variables that effect host development and those that affect the microorganism or the symbiotic link. Acid and water stress, for example, change root development, which can affect nodulation and nitrogen fixation indirectly. This activity is thought to be mediated by abscisic acid. Mohammadi et al. (2011) investigated the effects of several soil fertility approaches on chickpea absorption, metabolism, and N_2 fixation. Simultaneous applying carbon-based manure and bio-fertilizer treatment resulted in the greatest number of nodules, nodule biomass, and N_2 fixation. The mixture of compost, farmyard manure, and biofertilizers increased the dry mass of the roots. Farmyard manure (FYM) and compost are biological sources of useful nutrients that increase soil organic matter and so enhance grain quality (Al Tawaha et al., 2020; Mohammadi & Rokhzadi, 2012). Unfortunately, several of the studies that were described did not measure the original soil nitrate content. To account for differences in total supply of mineral N from soil and fertilizer sources, a margin for 0.99 quartile was built to indicate the greatest practicable quantity of N_2 fixation as a function of fertilizer-N rate. When no N fertilizer was added, the maximum amount of N_2 fixation reached 336 kg ha^{-1}, if 100 and 300 kg ha^{-1} of fertilizer-N were applied to the upper soil layer, and as a result of the exponential decline, a maximum N_2 fixation of 130 and 20 kg N ha^{-1} would be expected if 100 and 300 kg ha^{-1} of fertilizer-N were applied to the upper soil layer, respectively (Salvagiotti et al., 2008). According to Isaac et al. (2011) the number of nodules and the N obtained from the atmosphere determined using the 15 N natural abundance technique did not correspond with the growing p gradient. Phosphorus boosted growth and enhanced mineral N intake from solution under nonlimiting N circumstances without influencing the quantity of N absorbed from the atmosphere. As P availability grew, *Acacia senegal's* N acquisition techniques shifted, with decreased dependence on N_2 fixation until the rhizosphere reached a significant N absorption zone. Both plants and microsymbionts appear to have higher requirements for critical components like Ca^{2+}, P, and N when they are stressed. In some cases, Ca^{2+} may counteract the negative effects of low pH on root development and ion accumulation, as well as boost nod-gene expression and induction (Richardson et al., 1988). Rhizobial attachment to root hairs, nodule formation as well as nodulation, are all affected by calcium unavailability, with or without the confusing impact of low pH (Alva et al., 1990). Finally, nodulation factors and rhizobia cause a calcium-spiking phenomena in legume root-hair cells, implying Ca^{2+} partakes a significant role in symbiotic relationships on molecular level. An Al-induced Ca deficit has been blamed for reduced nodulation of soybeans in acidic soil (Biswas et al., 2003). Soil improvement has been demonstrated to be beneficial in peanut production in numerous studies. Several other nutritional variables influence the growth and survival of rhizobia in soils, in addition to macronutrients (Brockwell et al., 1995; Kafeel et al., 2022). Rhizobia have a broad metabolic profile and have been proven to grow on a number of soil and plant derived chemicals. Some growth-promoting substances have also been shown to trigger nod genes and be hemotactic. In addition, adding organic matter glycerol, and glutamate to soil and inoculants has been shown to increase rhizobia survival and populations in soils, as well as timely nodulation besides N_2 fixation (Rynne et al., 1994). The finding demonstrates that, while rhizobia can exist in soils, their efficiency may be increased by adding C, suggesting that these are C restricted in their usual state. Basu et al. (2008) discovered a codependent relationship between N fixation and fertilization.

4. Function of mineral nutrients in BNF

In world agriculture, a number of significant abiotic variables limit legume output and nitrogen fixation. Extreme temperatures, nutrition availability, water availability as well as harmful soil characteristics such as acidity and pH, can be summarized. Mineral nutrients may have an impact on N_2 fixation in legumes and nonlegumes at various stages of the association, including nodule development and infection nodule function, and growth of host plant (O'Hara, 2001). As a way of measuring symbiotic sensitivity to supply or concentration, Robson (1978) summarized the effects of nutrient supply combined with N on legume development. He concluded that larger quantities of Mo and Co are essential for associated N fixation in host plants than Ca, Cu and P. Even though investigational proof exists for 11 nutrients (B, Ca, Co, Cu, Fe, K, Mo, Ni, P, Se and Zn) for mutual development in certain legume classes, just four of them (Ca, Fe, Mo and P) seem to cause substantial restrictions on symbiotic legume efficiency in certain cultivated soils (O'Hara, 2001). Each important nutrient performs a specific biochemical, physiological and metabolical function. Nominal nutrient levels are essential to maintain metabolic activity in both legumes and rhizobia.

4.1 Phosphorus

Phosphorus (P) is essential for a variety of biochemical and molecular plant functions, including energy procurement, storing and usage (Epstein & Bloom, 2005). P scarcity continues to be a major stumbling block for N fixation and interdependent relationships. The demand could be larger than that for the host plant's root or shoot growth. Plant and rhizobial P supplies varied significantly (Beck & Munns, 1985; Pereira & Bliss, 1989), with slowly growing rhizobia showing more tolerance to low P concentrations than wildly growing rhizobia (Beck & Munns, 1985). P availability has a substantial influence on N_2 fixation and nodulation because nodules are robust sinks for P (Hart, 1989; Leung, 1987; Saxena, 1991; Singleton et al., 1985). Legumes that rely on symbiotic N may experience nitrogen-deficit if their P supply is insufficient. P concentration per unit dry mass is frequently much greater in nodules than in underground and upperground plant parts, especially when P supply is limited externally (Adu-Gyamfi & Fujika, 1989). Plants that fix N have a higher P need than those that get direct nitrogen fertilization, possibly due to the necessity for nodule formation and signal transduction, as well as the presence of P-lipids in a huge population of bacterioids (Al-Tawaha, 2020; Graham & Vance. 2000). In addition, the capability of growing nodules to contend for P with other somatic sinks (root and leaf meristems) with restricted peripheral streaming could differ among legume species (Jakobsen, 1985; Robson & Bottomley, 1991).

4.2 Potassium

Although potassium (K) is not a part of any metabolite, it is used as a counter ion, in activation of a variety of enzymes and as the primary mineral cation in cellular osmosis (Epstein & Bloom, 2005). Internode growth is slowed (Benlloch, 1980), and certain dicotyledonous species form rosettes (Bergmann, 1993). Old leaves display the first signs of K insufficiency as K is translocated from old to young leaves and growing tips through the phloem in such situations. Some rhizobia have been shown to have a qualitative requirement for K (Sherwood, 1970; Vincent, 1977). When K is removed from a specified media, Vincent (1977) claims *R. meliloti* and *R. trifolii* display limited progress and a lined retort in cell-produce up to 0.005 mM were synthesized in batch cultures.

4.3 Calcium

Despite the fact that calcium (Ca) is a major nutrient in plants, it is aggressively rejected from their cytoplasm. Ca serves a variety of purposes in higher plants. Many legume species planted in inorganic acid soils have nodulation inhibition as a primary limiting factor in N_2 fixation. Liming increases soil pH, which is particularly efficient in boosting nodule number in plants like common bean (Buerkert, 1990) and Alfalfa (Pijnenberg, 1990). Excessive monomers and protonic Al concentrations, as well as low Ca concentrations, all contribute to reduced nodulation in acidic soils (Alva et al., 1991). Subclover plant chlorosis caused by low N_2 fixation has been documented in bean plants with N_2-fixing symbiosis (Banath et al., 1966). Reduced N fixation in *T. subterraneum* (Banath et al., 1966), *Glycine* max (Blevins et al., 1977), and *Madia sativa* (Miller & Sirois. 1983) nodules also impacts rhizobia adhesion to root hairs (Smit et al., 1992), nodule development and nodulation (Alva et al., 1991). Finally, nodulation agents and rhizobia cause a calcium-spiking phenomena in bean root-hair cells (Wais et al., 2002). Numerous studies on the function of Ca reallocation during infection have been published since then (Downie & Walker, 1999; Niebel et al., 1999), together with the evidence that Nod factors may produce such changes even in tissue culture cells (Cardenas et al., 1999; Downie & Walker, 1999; Niebel et al., 1999; Vokoyama et al., 2000). Recognition of Nod factor receptor and unraveling the downstream motion transduction conduit would likely be a significant trial for biologists in the pursuit to better identify the growth route leading to development of the root nodule.

4.4 Sulfur

Sulfur (S) is a necessary component for plant development and biological function. In vegetative plant tissue, the S-containing amino acids cysteine (cyst) and methionine (meth) play an important part in the function construction, and conformation of enzymes and proteins (Tabatabai, 1986). Although S is often found in synthetic media for rhizobia development (Bergersen, 1961; Sherwood, 1970), not much effort has been made to quantify S requirements until recently (Munns & Mosse, 1980). Using batch and chemostat cultures, O'Hara et al. (1987a,b,c) investigated the S nutrition of two strains of *Brady japonicum* plus two strains of *Brady rhizobium* species. Before S-limiting could occur in batch culture, high quantities of infecting S in media constituents had to be displaced. When S concentrations in the inflowing media were lower than 20 mM, the growth of the four *Brady* rhizobia strains turned out to be restricted in chemostat culture. A dynamic S-uptake system in addition to the alkaline enzyme (sulfatase) was unblocked in S-deficient cells (Graham et al., 1988).

4.5 Boron

Boron (B) is one of the eight important trace nutrients, that most plants require for normal growth. In nitrogen-fixing pea plants, the distribution of boron is shown. Yamagishi and Yamamoto (1994) found that inadequate B supply caused significant changes in N_2 fixation in soybeans. Bolanos et al. (1996) investigated the influence of B on Rhizobium-legume cell-surface exchange and pea nodule growth. The amount of Rhizobia contaminating host cells and the no. of infection threads were lowered in plants with under-supplied boron, and the contamination threads acquired morphological abnormalities. Bonilla et al. (1997) found the cell walls of plants deficit in boron-with organizational abnormalities lacking in covalently bonded hydroxyproline/proline rich proteins that back an O_2 obstruction restraining nitrogenase deactivation and a reduction in N fixing. Studies of B scarce beans and peas revealed that nodulation was completely impeded under severe deficit (Mulder, 1948), although nodule formation was impacted under less severe B shortage.

4.6 Copper

Separate from its function in respirational proteins in rhizobia, essential for N_2 fixation (Delgado et al., 1998), copper (Cu) also has a part in a protein viz. produced in tandem by means of the nif-genes and may impact the effectiveness of bacteroid activity. Numerous rhizobial strains, notably *R. leguminosarum bv. phaseoli,* generate a pigment, melanin. Melanin-producing genes are present on the same huge Sym plasmid that contains nod plus nif-genes (Johnston et al., 1978). Because of the regulating *R. leguminosarum* nifA gene, the melA gene, which codes for the Cu-containing enzyme tyrosinase, is highly expressed in bacteroids (Hawkins & Johnston, 1988). Finally, there is a growing interest in the phenomena of bacteria being "viable but nonculturable." A recent study found that addition of Cu to *Agrobacterium* or *R. leguminosarum* cells causes them to enter this condition for unknown reasons (Alexander et al., 1999). In subterranean clover, a lack of Cu reduced nitrogen fixation (Snowball et al., 1980).

4.7 Zinc

Zinc (Zn) is a micronutrient that crop plants require in tiny levels, yet its significance in crop productivity has grown in current years. Zn administration on plants exposed to salinity stress improved photosynthesis, mesophyll efficiency, water consumption efficiency and quantum revenue significantly in comparison to plants bared to salinity stress only (Weisany et al., 2011). Weisany et al. (2012) also found Zn application greatly reduced lipid peroxidation and hydrogen peroxide concentrations under salt treatments. A description of protein engineered by Chauhan and O'Brian (1995) that pertains to Zn and *B. japonicum*, in addition to the putative relevance of Zn in the activity of the Ros/MucR transcriptional regulators. S-aminolaevulinic acid dehydratase-an enzyme (the product of hem-B) that generally requires Mg^{2+} as a cofactor in these bacteria. Plants, on the other hand, have a Zn^{2+}containing enzyme. They replaced the N-terminal amino acids of *B. japonicum* enzyme, causing the altered protein to bind Zn^{2+} rather than Mg^{2+}, using site-focused mutagenesis of *B. japonicum* hem-B. Despite the fact that a functional hemB is required for N_2 fixation (Chauhan & O'Brian, 1995), this had no effect on symbiotic N_2 fixation. As a result, the plant can provide the additional Zn^{2+} necessary by this unique inoculant strain.

4.8 Iron

Several major enzymes require iron (Fe) in the nitrogenase complex and the e^- carrier ferredoxin as well as certain hydrogenases. For the heme part of hemoglobin, beans have a very high iron demand. As a result, in legumes such as lupins (Tang et al., 1990) and peanuts, Fe is vital in high quantities for nodule development than for host plant development. In Fe-deficient peanut nodules, specific rates of nitrogenase activity were reduced (O'Hara et al., 1988), showing that Fe shortage may be a direct limitation on nodule function. Leghaemoglobin is a protein that binds oxygen. Leghaemoglobin, an iron protein, is the most copious protein produced by the host plant in the nodule. In bacteria, FeS clusters are found in nitrogenase and nitrogenase reductase, with the former having the cofactor FeMoCo in the dynamic spot for N_2 reduction. Additionally, bacteroids have a great respiratory mandate, necessitating a large number of cytochromes and other e^- suppliers, each with its personal Fe center (Delgado et al., 1998). Although iron deficiency had no effect on branch growth, it had a substantial impact on nodule mass, specifically leghemoglobin concentration, the quantity of bacteroids and nitrogenase activity, parallel to plants that had received an iron foliar spray

5 days prior. Fe is not transferred into the nodules following a foliar spray in lupin (*Lupinus angustifolius*), unlike peanut, and Fe supply directed at the contamination spots on the roots is compulsory for successful nodulation (Tang et al., 1990). The lack of Fe has a profound influence on nodule formation in laboratory circumstances. Nodule growth is substantially more vulnerable to a lack of Fe in lupin and peanut than other characteristics like root and shoot biomass (O' Hara et al., 1988).

4.9 Manganese

Manganese (Mn) helps in the creation of polyamines, which are crucial for plant development and growth as well as the reclamation of reactive O_2 species (Evans & Malmberg, 1989). When *Rhizobium leguminosarum* is deprived of Mn, attachment of rhizobia to juvenile root hairs is boosted, which is one of the first phases in the contamination procedure (Kijne et al., 1988). It's unclear if this is due to Mn influences the amount and kind of rhizobial exopolysaccharide (Appanna & Preston, 1987).

4.10 Molybdenum

Molybdenum (Mo) is a trace element used by plants that create root nodules using N-fixing bacteria, while it is also used in minute amounts by plants that do not form nodules in a protein entailed in N metabolism and uptake (Wiedenhoeft, 2006). Mo in the 'FeMoCo' cofactor is fundamental to the N-reduction phenomenon - in any case for many nitrogenases, its applicability to N_2 fixation is obvious. Mo—Fe protein features two molybdenum atoms and two types of oxidation—reduction centers: two iron—molybdenum cofactors (FeMoco) plus 4 Fe—S (4Fe—4S) centers. The activation site of the molybdenum-consisting nitrogenase protein is FeMoco of nitrogenase in N_2-fixing organisms (Allen et al., 1999). Even though Mo is favorably transported inside nodules when there is a limited supply (Brodrick & Giller, 1991), N-deficit caused by Mo scarcity is prevalent in leguminous plants that rely on N_2fixation, particularly in inorganic acid soils of humid as well as sub-humid regions. Several studies have found that spraying Mo to grain legumes in the field boosts nodule weight and, N_2 fixation leading to increased total N concentration and seed production (Vieira et al., 1998; Yanni, 1992). When inoculated into soybean roots, a *B. japonicum* strain lacking in Mo transport demonstrated reduced N-fixation association (Delgado et al., 2006). Numerous other leguminous plants that were rigorously deficient in molybednum exhibited more vivid signs of insufficiency in laboratory circumstances (Dilworth & Loneragan, 1991). In this recent study, experts from the University of Tokyo and Kyushu University successfully produced and isolated the intermediate material that holds the key to the new ammonia synthesis process, as well as clarified the reaction mechanism. e^{-} transfer took place between two molybdenum atoms coupled by a nitrogen molecule during the ammonia production process.

4.11 Nickel

Bertrand and de Wolf (1967) stated the soil-nickel treatment to field-grown soybean (*Glycine* max Merr.) ensued in a substantial rise in nodule mass and seed output, indicating that nickel (Ni) application helped the establishment of N-fixing plant species. In certain leguminous plants, insignificant amounts of Ni are necessary for root nodule development and hydrogenase activity. It is now widely accepted, thanks to the finding of a nickel-dependent hydrogenase in numerous rhizobial bacteria (Buerkert, 1990; Naikoo et al., 2020) that the efficiency of N fixation is closely related to hydrogenase action, as the latter's oxidation of H_2 supplies the ATP needed for N reeduction to NH_3. When *Glycine* max matured in soil culture were moistened once in 2 weeks' time with a nutrient solution comprising 1 mM $NiCl_2$, the hydrogenase action of *Rhizobium japonicum* nodules was approximately 45% higher at day 52 than that of control plants, though the promoting effect faded by day 100, apparently due to increased Ni toxicity (Dalton, 1985).

4.12 Cobalt

Nitrogen-fixing microorganisms such as cyanobacteria require cobalt (Co) for mutual N_2 fixation by leguminous (Ahmed & Evans, 1960; Dilworth et al., 1979) and nonleguminous plants (Hewitt & Bond, 1966; Johnson et al., 1966). For example, *Glycine* max grown with no inorganic N_2 but just atmospheric N_2, exhibit fast N-fixation and development when Co is introduced at 1/0.1 g Co ml 1, but show insignificant development when Co is not added (Ahmed & Evans, 1960). Cobalt's importance in N_2 fixation stems mostly from its function as a cofactor of cobalamine (Vit-B_6), which works as a coenzyme involved in N_2 fixation plus nodule development (Dilworth et al., 1979; Jordan & Reichard, 1998; Licht et al., 1996). Wilson and Nicholas (1967) revealed that Co is required for nodule forming in wheat and leguminous plant. According to Cowles et al. (1969) Co is crucial for rhizobial development. In addition, Co is a requisite in the bacterial enzyme complex. It is one of the essential nutrients for N-fixing bacteria because they play significant structural roles in proteins. Cobalt deficiency has diverse impacts on nodule growth and functions at different levels and to varying degrees.

5. Conclusion

Mineral nutrients serve a variety of purposes including protein synthesis, nucleic acids synthesis, cell wall synthesis, osmotic content of cell sap maintenance, e^- transportation systems, contents of chlorophyll, enzyme action, and are chief components of coenzymes macromolecules, and nitrogen fixation. These are just a few of the many metabolic processes they participate in the plant. In comparison to alternative methods, such as sewage sludge and N-chemical fertilizer, that are heavy metal contaminated, this chapter highlights the significance of certain mineral elements in BNF as an eco-friendly and less expensive strategy to boost soil fertility. Mineral nutrients can affect N_2 fixation in leguminous plants and help in overcoming the limitations of inorganic N present in soil (it suppresses nodule formation plus nitrogenase activity) (Sprent et al., 1988). N_2 fixation as well as mutual interactions is severely hampered by a lack of phosphorus supply and availability. At the molecular level, calcium plays an essential part in symbiotic connections. Boron impacts pea nodule growth and surface contact between cells of Rhizobium and legumes (Bolanos et al., 1996). In subterranean clover, copper shortage reduced nitrogen fixation (Snowball et al., 1980). Nitrogen fixation is impossible without sufficient amounts of molybdenum and iron, which are found in the nitrogen-fixing enzyme nitrogenase. Substantial quantities of Nickel are required for hydrogenase action and root nodule forming in some leguminous plants. Cobalt is important for the growth of N-fixing bacteria, as well as cyanobacteria. Evidently, BNF is a very vital natural process and continuous progress in understanding the role of mineral nutrients in BNF is required to ensure the global food supply.

References

Adu-Gyamfi, J. J., Fujika, K., & Ogata, S. (1989). Phosphorus absorption and utilization efficiency of pigeonpea (*Cajanus cajan* (L.) Millsp.) in relation to dry matter production, and dinitrogen fixation. *Plant Soil, 119*, 315–324.

Ahmed, S., & Evans, H. J. (1960). Cobalt: A micronutrient for the growth of soybean plants under symbiotic conditions. *Soil Science, 90*, 205–210.

Al-Tawaha, A. R. M. S., Singh, S., Kafeel, U., Singh, V., Naikoo, M. I., Kumari, A., Imran, Amanullah, Al-Tawaha, A. R., Qaisi, A. M., Khanum, S., Thangadurai, D., Sangeetha, J., Islam, S., Etesami, H., Kerkoub, N., Amrani, A., Labidi, Z., Maaref, H., Nasri, H., Sanmukh, S. G., & Serra, E. T. (2020). Improving water use efficiency and nitrogen use efficiency in rice through breeding and genomics approaches. In A. Roychoudhury (Ed.), *Vol. 15. Rice research for quality improvement: Genomics and genetic engineering* (pp. 978–981). Springer.

Alexander, N. J., McCormick, S. P., & Hohn, T. M. (1999). TRI12, a trichothecene efflux pump from *Fusarium sporotrichioides*: Gene isolation and expression in yeast. *Molecular and General Genetics, 261*, 977–984. PUBMED: 10485289.
Allen, R. M., Roll, J. T., Rangaraj, P., Shah, V. K., Roberts, G. P., & Ludden, P. W. (1999). Incorporation of molybdenum into the iron-molybdenum cofactor of nitrogenase. *Journal of Biology and Chemistry, 274*, 15869–15874.
Alva, A. K., Assher, C. J., & Edwards, D. G. (1990). Effect of solution pH, external calcium concentration, and aluminum activity on nodulation and early growth of cowpea. *Australian Journal of Agricultural Research, 41*, 359–365.
Alva, A. K., Kerven, G. L., Edwards, D. G., & Asher, C. J. (1991). Reduction in toxic aluminum to plants by sulfate complexation. *Soil Science, 152*, 351–359.
Appanna, V. D., & Preston, C. M. (1987). Manganese elicits the synthesis of a novel exopolysaccharide in an arctic Rhizobium. *FEBS Letters, 215*, 79–82.
Appleby, C. A. (1984). Leghemoglobin and rhizobium respiration. *Annual Reviews of Plant Physiology, 35*, 443–478.
Aranjuelo, I., Irigoyen, J., & Sánchez-Díaz, M. (2007). Effect of elevated temperature and water availability on CO_2 exchange and nitrogen fixation of nodulated alfalfa plants. *Environment and Experimental Botany, 59*, 99–108.
Arayankoon, T., Schomberg, H. H., & Weaver, R. W. (1990). Nodulation and N_2 fixation of guar at high root temperature. *Plant Soil, 126*, 209–213.
Banath, C. L., Greenwood, E. A. N., & Loneragan, J. F. (1966). Effects of calcium deficiency on symbiotic nitrogen fixation. *Plant Physiology, 4l*, 760–763.
Basu, M., Bhadoria, P. B. S., & Mahapatra, S. C. (2008). Growth, nitrogen fixation, yield and kernel quality of peanut in response to lime, organic and inorganic fertilizer levels. *Bioresearch and Technology, 99*, 4675–4683.
Beck, D. P., & Munns, D. N. (1985). Effect of calcium on the phosphorus nutrition of Rhizobium meliloti. *Soil Science Society of America Journal, 49*, 334–337.
Benlloch, M. (1980). Effects of potassium and gibberellic acid on stem growth of whole sunflower plants. *Physiologia Plantarum, 49*(4), 443–448.
Bergersen, F. J. (1961). The growth of rhizobia in synthetic media. *Australian Journal of Biological Sciences, 14*, 349–360.
Bergmann, W. (1993). Ernährungsstörungen bei Kulturpflanzen—Entstehung und Diagnose Jena. *VEB Gustav Fischer, 10*, 384–394.
Bertrand, D., & de Wolf, A. (1967). Nickel, a dynamic trace element for higher plants. *CR Academic Science, 265*, 1053–1055.
Biswas, P., Hosain, D., Ullah, M., Akter, N., & Bhuiya, M. A. A. (2003). Performance of groundnut (*Arachis hypogaea* L.) under different levels of bradyrhizobial inoculums and nitrogen fertilizer. *Journal of Agriculture, 1*, 61–68.
Blevins, D. G., Barnett, N. M., & Bottino, P. J. (1977). The effects of calcium deficiency and the ionophore A23187 on nodulation, nitrogen fixation and growth of soybeans. *Physiology of Plant, 41*, 235–258.
Bolanos, L., Brewin, N. J., & Bonilla, I. (1996). Effects of boron on Rhizobium-legume cell-surface interactions and nodule development. *Plant Physiologist, 110*, 1249–1256.
Bonilla, I., Mergold-Villasefisor, C., Campos, M. E., Sanches, N., Perez, H., Lopez, L., Castrejon, L., Sanchez, F., & Cassab, G. I. (1997). The aberrant cell walls of boron-deficient bean root nodules have no covalantly bound hydroxyproline- proline-rich proteins. *Plant Physiology, 115*, 1329–1340.
Boscari, A., Mandon, K., Dupont, L., Poggi, M. C., & Le Rudulier, D. (2002). BetS is a major glycine betaine/proline betaine transporter required for early osmotic adjustment in *Sinorhizobium meliloti*. *Journal of Bacteriology, 184*, 2654–2663.

Brockwell, J., Pilka, A., & Holliday, R. A. (1991). Soil pH is a major determinant of the numbers of naturally occurring Rhizobium meliloti in non-cultivated soils in central New South Wales. *Australian Journal Of Experimental Agriculture, 31*(2), 211–219.

Brockwell, J., Bottomley, P. J., & Thies, J. E. (1995). Manipulation of rhizobia microflora for improving legume productivity and soil fertility, a critical assessment. *Plant Soil, 174*, 143–180.

Brodrick, S. J., & Giller, K. E. (1991). Genotypic difference in molybdenum accumulation affects N fixation in tropical *Pharsalus vulgaris*. *Journal of Experimental Botany, 42*, 1339–1343.

Buerkert, A. (1990). Soil acidity and liming effects on stand, nodulation, and yield of common bean. *Journal of Agronomy, 82*, 749–754.

Cardenas, L., Feijo, J. A., Kunkel, J. G., Sanchez, F., Holdaway-Clarke, T., Hepler, P. K., & Quinto, C. (1999). Rhizobium Nod factor induces increases in intracellular free calcium and extracellular calcium in fluxess in bean root hairs. *Journal of Plant, 19*, 347–352.

Chauhan, S., & O'Brian, M. R. (1995). A mutant Brady *Rhizobium japonicum* delta-amino levulinic acid dehydratase with an altered metal requirement functions in situ for tetrapyrrole synthesis in soybean root nodules. *Journal of Biological Chemistry, 270*, 19823–19827.

Cowles, J. R., Evans, H. J., & Russell, S. A. (1969). B12 coenzyme-dependent ribonucleotide reductase in *Rhizobium* species and the effects of cobalt deficiency on the activity of the enzyme. *Journal of Bacteriology, 97*(3), 1460–1465.

Dalton, D. A., Evans, H. J., & Hanus, F. J. (1985). Stimulation by nickel of soil microbial urease activity and urease and hydrogenase activities in soybeans grown in a low-nickel. *Soil and Plant Soil, 88*, 245–258.

Delgado, M. J., Bedmar, E. J., & Downie, J. A. (1998). Genes involved in the formation and assembly of rhizobial cytochromes and their role in symbiotic nitro-gen fixation. *Advance Microbiology and Physiology, 40*, 191–231.

Delgado, J. A., Shaffer, M., Hu, C., Lavado, R. S., Wong, J. C., Joosse, P., ... Sotomayor, D. (2006). A decade of change in nutrient management: a new nitrogen index. *Journal of Soil and Water Conservation, 61*(2), 62a–71a.

Dilworth, M. J., & Loneragan, J. E. (1991). An alternative nitrogenase is not expressed in molybdenum-deficient legume root nodules. *New Phytology, 118*, 303–308.

Dilworth, M. J., Robson, A. D., & Chatel, D. L. (1979). Cobalt and nitrogen fixation in *Lupinus angustifolius* L. 11.Nodule formation and function. *New Phytologist, 83*, 63–79.

Downie, J. A., & Walker, S. A. (1999). Plant responses to nodulation factors. Current Opinion on. *Plant Biology, 2*, 483–489.

Epstein, E., & Bloom, A. J. (2005). *Mineral nutrition of plants: Principles and perspectives* (2nd ed., Vol. 12, pp. 321–345). Sunderland, Massachusetts: Sinauer Associates.

Evans, P., & Malmberg, T. (1989). Do polyamines have roles in plant development. *Annual Reviews of Plant Molecular Biology, 40*, 239–269.

Falkowski, P., Berman-Frank, I., & Lundgren, P. (2003). Nitrogen fixation and photosynthetic oxygen evolution in cyanobacteria. *Research in Microbiology, 154*(3), 157–164.

Fay, P. (1992). Oxygen relations of nitrogen fixation in cyanobacteria. *Microbiology Reviews, 56*, 340–373.

Graham, W., O'hara, I., Nantakorn, B., Michael, J., & Dilworth, I. (1988). Mineral constraints to nitrogen fixation. *Plant and Soil, 108*, 93–110.

Graham, P. H., & Vance, C. P. (2000). Nitrogen fixation in perspective, an overview of research and extension needs. *Field Crops Research, 65*, 93–106.

Hageman, R. V., & Burris, R. H. (1978). Dinitrogenase and nitrogenase reductase associate and dissociate with each catalytic cycle. *Procedings of the National Academy of Sciences, 75*(8), 2699–2702.

Hart, A. L. (1989). Nodule phosphorus and nodule activity in white clover. *New zealand Journal of Agriculture Research, 32*, 145–149.

Hawkins, F. K. L., & Johnston, A. W. B. (1988). Transcription of a *Rhizobium* leguminous *arum biovar phaseoli* gene needed for melanin synthesis is activated by nifA of Rhizobium and *Kiebsiella pneumoniae*. *Molecular Microbiology, 2*, 331–337.

Hewitt, E. J., & Bond, G. (1966). The cobalt requirement of non-legume root nodule plants. *Journal of Experimental Botany, 17*, 480–491.

Hill, D. R., Belbin, T. J., Thorsteinsson, M. V., Bassam, D., Brass, S., Ernst, A., Boger, P., Paerl, H., Mulligan, M. E., & Potts, M. (1996). GlbN (cyanoglobin) is a peripheral membrane protein that is restricted to certain *Nostoc* spp. *Journal of Bacteriology, 178*, 6587–6598.

Hungria, M., & Vargas, M. A. T. (2000). Environmental factors affecting N_2 fixation in grain legumes in the tropics, with emphasis on Brazil. *Field Crops Research, 65*, 151–164.

Isaac, M. E., Harmand, J. M., & Drevon, J. J. (2011). Growth and nitrogen acquisition strategies of *Acacia senegal* seedlings under exponential phosphorus additions. *Journal of Plant Physiology, 168*(8), 776–781.

Jakobsen, I. (1985). The role of phosphorus in nitrogen fixation by young pea plants (*Pisum sativum*). *Physiology of Plants, 64*, 190–196.

Johnson, G. V., Mayeux, P. A., & Evans, H. J. (1966). A cobalt requirement for symbiotic growth of *Azolla filiculoides* in the absence of combined nitrogen. *Plant Physiology, 41*, 852–855.

Johnston, A. W. B., Beynon, J. L., Buchanan-Wollaston, A. V., Setchell, S. M., Hirsch, P. R., & Beringer, J. E. (1978). High frequency transfer of nodulating ability between strains and species of Rhizobium. *Nature (London), 276*, 634–636.

Jordan, A., & Reichard, P. (1998). Ribonucleotide reductases. *Annual Review Biochemistry, 67*, 71–98.

Kafeel, U., Jahan, U., Raghib, F., & Khan, F. A. (2022). Global importance and cycling of nanoparticles. In V. D. Rajput, K. K. Verma, N. Sharma, & T. Minkina (Eds.), *The role of nanoparticles in plant nutrients under soil pollution*. Cham: Springer Nature.

Kijne, J. W., Smit, G., Diaz, C. L., & Lugtenberg, B. J. J. (1988). Lectin-enhanced accumulation of manganese-limited *Rhizobium leguminosarum* cells on pea root hair tips. *Journal of Bacteriology, 170*, 2994–3000.

Kim, J., & Rees, D. C. (1992). Structural models for the metal centers in the nitrogenase molybdenum-iron protein. *Science, 257*, 1677–1682.

Ledgard, S. F., & Steele, K. W. (1992). Biological nitrogen fixation in mixed legume grass pastures. *Plant Soil, 141*, 137–153.

Leung, K., & Bottomley, P. J. (1987). Influence of phosphate on the growth and nodulation characteristics of *Rhizobium trifolii*. *Applied and Environmental Microbiology, 53*, 2098–2105.

Licht, S., Gerfen, G. J., & Stubbe, J. (1996). Thiel radicals in ribonucleotide reductases. *Science, 271*, 477–481.

Long, S. R. (2001). Genes and signals in the Rhizobium-legume symbiosis. *Plant Physiology, 125*, 69–72.

Miller, R. W., & Sirois, J. C. (1983). Calcium and magnesium effects on symbiotic nitrogen fixation in the alfalfa (*M. sativa) Rhizobium meliloti* system. *Physiology of Plant, 58*, 464–470.

Mohammadi, K., Ghalavand, A., & Aghaalikhani, M. (2011). Effect of different soil fertility strategies on absorption metabolism and molecular nitrogen fixation in chickpea (*Cicer arietinum*). *Iran Journal, 91*, 78–89.

Mohammadi, K., & Rokhzadi, A. (2012). An integrated fertilization system of canola (*Brassica napus* L.) production under different crop rotations. *Indian Crop Production, 37*, 264–269.

Mulder, E. G. (1948). Investigation on the nitrogen nutrition of pea plants. *Plants and Soil, 1*, 179–212.

Munns, D. N., & Mosse, B. (1980). Mineral nutrition of legume crops. In R. J. Summerfield, & A. H. Butting (Eds.), *H.M. Stationary office, London* (pp. 115–125). Advances in legume science.

Naikoo, M. I., Kafeel, U., Naushin, F., & Khan, F. A. (2020). Halophytes in India and their role in phytoremediation. In M. N. Grigore (Ed.), *Handbook of halophytes*. Cham: Springer. https://doi.org/10.1007/978-3-030-17854-3_115-1

Niebel, A., Gressent, F., Bono, J. J., Ranjeva, R., & Cullimore, J. (1999). Recent advances in the study of Nod factor perception and signal transduction. *Biochimie, 81*, 669–674.

O'Hara, G. W. (2001). Nutritional constraints on root nodule bacteria affecting symbiotic nitrogen fixation: A review. *Australian Journal of Experimental Agriculture, 41*, 417–433.

O'Hara, G. W., Franklin, M., & Dilworth, M. J. (1987b). Sulfur nutrition of free living and symbiotic Bradyrhizobium japoni- cure and Bradyrhizobium sp. (*Arachis*). In , *Vol. 18. Proc. Int. Symposium on food legume improvement for asian farming systems 1-5 sept. 1986* (p. 275). Thailand: Khon Kaen.

O'Hara, G. W., Franklin, M., & Dilworth, M. J. (1987c). Effect of sulfur supply on sulfate uptake, and alkaline sulfatase activity in free-living and symbiotic bradyrhizobia. *Archives Microbiology, 149*, 163–167.

O'Hara, G. W., Boonkerd, N., & Dilworth, M. J. (1988). Mineral constraints to nitrogen fixation. *Plant Soil, 108*, 93–110.

O'Hara, G. W., Franklin, M., & Dilworth, M. J. (1987a). Sulphur nutrition of bradyrhizobia. In , *150. UNESCO regional symposium and workshop on biotechnology of nitrogen fixation in the tropics* (pp. 25–29). Serdang, Malaysia: UPM.

Pereira, P. A. A., & Bliss, F. A. (1989). Selection of common bean (*Phaseolus vulgaris L.*) for N2 fixation at different levels of available phosphorus under field and environmentally-controlled conditions. *Plant Soil, 115*, 75–82.

Pijnenberg, J. W., & Lie, M. (1990). Effect of lime-pelleting on the nodulation of lucerne (*Medicago sativa L.*) in acid soil: A comparative study carried out in the field, in pots and in rhizotrons. *Plant Soil, 121*, 225–234.

Rascio, N., & Rocca, N La (2008). Biological nitrogen fixation. *Ecological Processes Elsevier, 255*, 1677–1682.

Richardson, A. E., Simpson, R. J., Djordjevic, M. A., & Rolfe, B. G. (1988). Expression of nodulation genes in *Rhizobium leguminosarum bvtrifolii* is affected by low pH and by Ca^{2+} and Al ions. *Applied and Environmental Microbiology, 54*, 2541–2548.

Robson, A. D. (1978). Mineral nutrients limiting nitrogen fixation in legumes. In C. S. Andrew, & E. J. C. Kamprath (Eds.), *Vol. 10. Mineral nutrition of legumes in tropical and subtropical soils* (pp. 277–293).

Robson, A. D., & Bottomley, P. J. (1991). Limitations in the use of legumes in agriculture and forestry. In M. J. Dilworth, & A. R. Glenn (Eds.), *Vol. 10. The biology and biochemistry of nitrogen fixation* (pp. 320–349).

Rynne, F. G., Glenn, A. R., & Dilworth, M. J. (1994). Effect of mutations in aromatic catabolism on the persistence and competitiveness of. *Rhizobium Leguminosarum Bv Trifolii Soil Biology and Biochemistry, 26*, 703–710.

Salvagiotti, F., Cassman, K. G., Specht, J. E., Walters, D. T., Weiss, A., & Dobermann, A. (2008). Nitrogen uptake, fixation and response to fertilizer N in soybeans: A review. *Field Crops Research, 108*, 1–13.

Saxena, A. K., & Rewari, R. B. (1991). The influence of phosphate and zinc on growth, nodulation and mineral composition of chickpea (*Cicer arietinum L.)* under salt stress. *World Journal of Microbiology Biotechnology, 7*, 202–205.

Sherwood, M. T. (1970). Improved synthetic medium for the growth of Rhizobium. *Journal of Applied Bacteriology, 33*, 708–713.

Singleton, P. W., Abel Magid, H. M., & Tavares, J. W. (1985). Effect of phosphorus on the effectiveness of strains of *Rhizobium japonicum*. *Soil Science Society of America Journal, 49*, 613–616.

Smit, G., Swart, S., Lugtenberg, B. J. J., & Kijne, J. W. (1992). Molecular mechanisms of attachment of Rhizobium bacteria to plant roots. *Molecular Microbiology, 6*, 2897–2903.

Snowball, K., Robson, A. D., & Loneragan, J. F. (1980). The effect of copper on nitrogen fixation in subterranean clover (*Trifolium subterraneum*). *New Phytologist, 85*, 63–72.

Sprent, J. I., Stephens, J. H., & Rupela, O. P. (1988). Environmental effects on nitrogen fixation. In R. J. Sumerfield (Ed.), *Vol. 10. World crops: Cool season food legumes* (pp. 801–810).

Tabatabai, M. A. (1986). *Sulfur in agriculture* (Vol. 17, pp. 198–221). Madison, Wisconsin: American Society of Agronomy.

Tang, C., Robson, A. D., & Dilworth, M. J. (1990). A split-root experiment shows that iron is required for nodule initiation in *Lupinus angustifolius* L. *New Phytology, 115*, 61–67.

Triplett, E. W., & Sadowsky, M. J. (1992). Genetics of competition for nodulation. *Annual Reviews of Microbiology, 46*, 399–428.

Turco, R. F., & Sadowsky, M. J. (1995). Understanding the microflora of bioremediation. In H. D. Skipper, & R. F. Turco (Eds.), *Vol. 43. Bioremediation: Science and applications. Soil science (special publication)* (pp. 87–103). Soil science society of America journal.

Vieira, R. F., Cardoso, E. J. B. N., Vieira, C., & Cassini, S. T. A. (1998). Foliar application of molybdenum in common beans I. Nitrogenase and reductase activities in a soil of high fertility. *Journal of Plant Nutrition, 21*, 169–180.

Vincent, J. M. (1977). Rhizobium: General microbiology. In R. W. F. Hardy, & W. S. Silver (Eds.), *A treatise on dinitrogen fixation III* (pp. 277–366). New York: Wiley.

Vokoyama, T., Kobayashi, N., Kouchi, H., Minamisawa, K., Kaku, H., & Tsuchiya, K. (2000). A lipochito oligosaccharide, Nod factor, induces transient calcium influx in soybean suspension-cultured cells. *Plant Journal, 22*, 71–78.

Wais, R. J., Keating, D. H., & Long, S. R. (2002). Structure-function analysis of nod factor-induced root hair calcium spiking in Rhizobium-legume symbiosis. *Plant Physiology, 129*, 211–224.

Weisany, W., Sohrabi, Y., Heidari, G., Siosemardeh, A., & Ghassemi Golezani, K. (2011). Physiological responses of soybean (*Glycine max L.*) to zinc application under salinity stress. *Australian Journal of Crop Science, 5*(11), 1441 1447.

Weisany, W., Sohrabi, Y., Heidari, G., Siosemardeh, A., & Ghassemi Golezani, K. (2012). Changes in antioxidant enzymes activity and plant performance by salinity stress and zinc application in soybean (*Glycine max* L). *Journal of Plant Omics, 5*(2), 60–67.

Werner, D., & Newton, W. E. (2005). *Nitrogen fixation in agriculture, forestry, ecology, and the environment.* Springer Publication.

Wiedenhoeft, A. C. (2006). *Plant nutrition (the green world)* (p. 10001). 132 West 31st Street New York NY: By Infobase Publishing..

Wilson, S. B., & Nicholas, D. J. D. (1967). A cobalt requirement for non-nodulating legumes and for wheat. *Phytochemistry, 6*, 1057–1066.

Yamagishi, M., & Yamamoto, Y. (1994). Effects of boron on nodule development and symbiotic nitrogen fixation in soybean plants. *Soil Science and Plant Nutrition, 40*, 265–274.

Yanni, Y. G. (1992). Performance of chickpea, lentil and lupin nodulated with indigenous or inoculated rhizobia micropartners under nitrogen, boron, cobalt and molybdenum fertilization schedules. *World Journal Microbiology and Biotechnology, 8*, 607–613.

CHAPTER SIX

Molecular basis of plant nutrient use efficiency - concepts and challenges for its improvement

Sneha Adhikari[a], Hirdayesh Anuragi[b], Kailash Chandra[c], Shilpa Hanmantrao Tarte[d], Shish Ram Dhaka[e], Hanuman Singh Jatav[f], Kiran Hingonia[g]

[a]Genetics and Plant Breeding, ICAR-Indian Institute of Wheat and Barley Research, Shimla, India
[b]Genetics and Plant Breeding, ICAR-Central Agroforestry Research Institute, Jhansi, Uttar Pradesh, India
[c]Department of Plant Breeding and Genetics, College of Agriculture (SKN Agriculture University) Fatehpur-Shekhawati, Sikar, Rajasthan, India
[d]Department of Plant Biotechnology MGM College of Agricultural Biotechnology, Aurangabad, Maharashtra, India
[e]Department of Entomology, College of Agriculture (SKN Agriculture University) Fatehpur-Shekhawati, Sikar, Rajasthan, India
[f]Department of Soil Science and Agriculture Chemistry, College of Agriculture (SKN Agriculture University) Fatehpur-Shekhawati, Sikar, Rajasthan, India
[g]Department of Agronomy, College of Agriculture, Sumerpur (Agriculture University, Jodhpur) Rajasthan, India

1. Introduction

The word "efficiency" refers to "the level of output obtained per unit of input". Concerning plant system output could be growth, yield and physiological activities in relation to various inputs such as land, nutrient, water, or energy. When considering nutrients as input, the term nutrient use efficiency is used and is defined in terms of plant growth, physiological activity and harvestable yield acquired per unit of nutrient. Nutrient use efficiency is typically divided into two interactive components: the amount of nutrients taken up by plants concerning nutrient supply (nutrient acquisition efficiency, NAcE) and biomass produced by the unit of nutrient incorporated by plants (nutrient utilization efficiency, NUtE). The most commonly used definition of NUE, i.e., "yield obtained per unit of nutrient supplied," was originally used for nitrogen (Moll et al., 1982) and afterward used for other nutrients too (Ortiz-Monasterio et al., 2001). Another frequently used definition is determined by comparing the yield obtained under low nutrient supply with yield obtained under sufficient nutrient supply conditions, which is defined by Graham (1984) as the potential of a plant

Sustainable Plant Nutrition
ISBN: 978-0-443-18675-2
https://doi.org/10.1016/B978-0-443-18675-2.00001-8

genotype to generate higher grain yield when grown in the soil in which certain nutrient is limiting. Later on, another definition of NUE, i.e., the ability of genotype to obtain nutrients from the growth medium and further efficient utilization in root and shoot biomass production, and economical plant material (grain, seed, forage, and fruit) was proposed by Blair (1993).

The estimates of nutrient use efficiency varied among nutrients such as for N 30%–50%, for P 15%–20%, for K 70%–80%, for S 8%–10%, and for micronutrient (Fe, Zn, Cu, Mn and B) it is 1%–2%. The most probable cause of the low efficiency of these nutrients is depicted in Fig. 6.1. To date, for measuring NUE, several terms have been given and are described as the question they tackle (Table 6.1). To differentiate nutrient efficient genotype from inefficient genotype, Gerloff (1983) has used the term nutrient efficiency ratio (NER). Genotypes that can produce high yield under limited nutrient availability are classified as nutrient use efficient genotypes. The NUE of a genotype is influenced by three factors (Baligar et al., 2001) in a sequential manner of which uptake efficiency is the first one which is influenced by influx rate, the kinetics of nutrient influx as well as radial transportation in roots, and the proportion of the particular nutrient present or

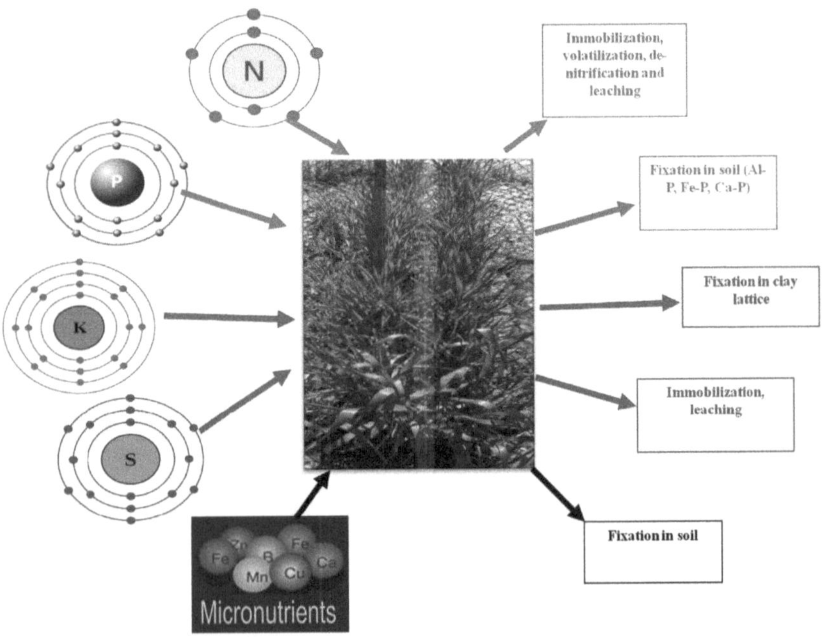

Fig. 6.1 Probable causes of low efficiency of nutrient.

Table 6.1 Commonly used NUE terms.

NUE term	Formula	Question tackle
Partial factor productivity (PFP)	Yield/Nutrient applied *100	How productive is this cropping system in comparison to its nutrient input?
Agronomic use efficiency (AE)	(Yield in the fertilized field- yield in the unfertilized field)/ Nutrient applied *100	How much productivity improvement was gained by use of nutrient input?
Physiological use efficiency (PE)	(Yield in fertilized treatment - yield in unfertilized treatment)/(nutrient uptake in fertilized treatment- nutrient uptake in fertilized treatment	What is the ability of the plant to transform nutrients acquired from the source applied into economic yield?
Partial nutrient balance (PNB)	Nutrient content of the harvested portion of the crop/Amount of nutrient applied	How much nutrient is being taken out of the system in relation to how much is applied?
Internal utilization efficiency (IF)	The yield of the harvested portion of the crop with nutrients applied /	What is the ability of the plant to transform nutrients acquired from all sources into

(*Continued*)

Table 6.1 Commonly used NUE terms.—cont'd

NUE term	Formula	Question tackle
	Total nutrient uptake in aboveground crop biomass with nutrient applied	economic yield (grain, etc.)?
Apparent recovery efficiency (ARE)	(Nutrient uptake in fertilized treatment- nutrient uptake in fertilized treatment)/ Nutrient applied (Kg/ ha)	How much of the nutrient applied did the plant take up?

applied in the soil. The nutrients absorption capacity of a genotype from the soil is known as uptake efficiency; however, it is not necessary that the genotype with high absorption ability is also shown high nutrient use efficiency. Jiang (2006) and Chatzistathis et al. (2009) have observed that the high Mn and Fe efficiency in some wheat cultivars was linked with better internal utilization, not because of higher plant Mn and Fe accumulation. Incorporation efficiency (transports to shoot and leaves) is another factor that is based on shoot parameters, and the third factor is utilization efficiency (remobilization) which relies on root and shoots parameters (the whole plant parameters).

Many researchers have recorded significant differences in nutrient utilization efficiency among different plant species along with within species genotypes (cultivars). Genetic parameters (genotypic capacity to absorb and utilize nutrients efficiently or inefficiently), the ability of the root to exudates organic compounds (that facilitates nutrient uptake and different rhizospheric properties), the mycorrhizal colonization of the root system, etc. could be the most probable reason due to which nutrient utilization efficiency varied among genotypes and species (Chatzistathis et al., 2013). Genotypes and/or species that have greater efficiency of nutrient absorption and utilization can efficiently utilize applied fertilizers, therefore, help in the reduction of input cost and nutrients loss to the ecosystem. Hence there is a need for the development of a focused breeding program for the development of high nutrient use efficient cultivars. Identification of traits that are related to nutrient absorption, transportation, utilization, and remobilization and their regulatory elements is a potential target for NUE improvement. In association with best management practices (BMPs), the development of high nutrient efficient cultivars will definitely facilitate a sustainable agricultural production system that protects environment (soil, air, and water) quality degradation.

2. Physiological processes involved in NUE

Plant NUE is a complex, polygenic trait comprised of various interrelated physiological processes, namely nutrient acquisition, storage, and remobilization or reallocation. The various physiological processes are described under the following subheadings:

2.1 Nutrient acquisition

Nutrient uptake is one of the critical components involved in nutrient acquisition efficiency (NAcE), which is one of the principal components of NUE (Reich et al., 2014). However, some nutrients can be obtained from the atmosphere, but major portion of plant nutrient requirement is based on nutrients acquired from the soil. The soil mineral nutrients can be obtained either from the parental material weathering or directly through breakdowns of organic matter via biological means. Root interception, mass flow and diffusion are three physiological mediums that facilitate the transfer of nutrients to the surface of root hairs. All three mechanisms can operate in the uptake of a single nutrient, but their significance is relay on the relative portion of nutrients in the plants and soil. After the acquisition, the absorbed nutrients either directly or indirectly participate in the biomass production as well as final yield thereafter.

2.2 Root interception

Various soil particles (sand, silt, and clay) formed soil aggregates of variable sizes and shapes in combination with organic matter. Most of the plant nutrients are present in these soil aggregates. Root interception is the process by which roots directly contact soil aggregates. However, within soil aggregates, only 1%–2% of soil surface came in contact with plant roots as roots grow around the soil aggregates, not within aggregates. The extended portion of soil aggregates remains untouched by roots which makes root interception a minor nutrient uptake mechanism.

2.3 Mass flow

Mass flow is defined as the transpiration (water loss through the leaves) driven passive transportation of nutrient from soil pores into the plant root. For bringing nutrients to the root surface, mass flow serves the major role when the solution contains comparatively larger concentration of nutrients. Alike diffusion, it not only work in the immediate volume of soil that surrounds root but can operate in wider volume of soil.

2.4 Diffusion

Diffusion is the movement of a nutrient from a high concentration area to a low concentration area. In the soil system, the clay surface in the soil aggregates is considered to be the area of high concentration, whereas, in contrast, the root surface is usually thought to be the area of low concentration. Hence, in the process of diffusion, positively charged nutrient ions may move to the plants root surface from the clay surface in the soil aggregates.

2.5 Storage

Another important physiological process is nutrient storage that can be further subdivides into three processes such accumulation of nutrients, reserve building, and nutrients recycling (Chapin et al., 1990). Those nutrients get accumulated that is not related to growth directly as excessive availability of these nutrients may increase the metabolic demand of plants for these compounds. In contrast, reserve formation is related to those nutrients that facilitate plant growth hence stored in a form that does not support growth directly, i.e., in metabolically controlled storage. Hence reserve formation in the form of these metabolically inactive compounds directly affects the growth and other plant processes that would be based on the original form of the substrate. The compounds that is originally involved in growth and physiological functions of plant but which would be lost are broken down for use in future plant activities by another physiological process, i.e., recycling. For efficient NUE, the importance of storage or remobilization of nutrients depends on the availability of nutrients.

2.6 Reallocation

Reallocation of the nutrient is another important process that involves the redistribution of nutrients between different plant organs during plant development. Under this process, either the nutrient is relocated to young leaves from old senescing leaves or to finally harvested sink organ, i.e., grain. Like in the case of nitrogen, older leaves with decreased photosynthetic efficiency will be utilized as a source of nitrogen for reallocation into young leaves for promoting plant growth. Hence, in this fashion, at the whole-plant level, the nitrogen can be used competently for photosynthesis, and consequently, NUE can be enhanced due to loss reduction (Escudero et al., 2003). Likewise, the relocation of nutrients from old senescing leaves to young leaves had also been observed for micronutrients such as Cu (Benatti et al., 2014) and Mo (Shinmachi et al., 2010). The cost of this process is generally assumed to be low; hence reallocation of the nutrient is significantly important for the improvement of NUE for plants. It has been reported that the translocation of nutrients from vegetative organs, i.e., from shoot to the harvested sink organ, i.e., grains are very important, particularly in cereals, which resulted in enhancing yield and thereby improved NUE.

3. Breeding efforts for traits related to NUE

NUE of a plant is decided by its nutrient uptake efficiency (soil to root and internal transport), utilization efficiency (take up the nutrients, storage and remobilization), which might be mediated by the coordination of gene networks. Screening and breeding for low nutrient-tolerant genotypes are the major criteria for improving the NUE of fertilizers. Breeding approaches mostly focus on the identification of cultivars accumulating high harvestable biomass per unit nutrient available as well as supplied through fertilizers (Moll et al., 1982). Higher nutrient use efficiency relies upon the exploration of available useful variations, identification of beneficial traits, and its incorporation into elite local or commercial varieties. Thus, understanding the genetic control and then breeding plants for higher N, P, K acquisition and utilization potential through modern biotechnological and bioinformatics approaches is gaining global importance as it can address the constraint of biomass loss due to poor nutrient management. Further, choosing a right plant according to the phenotypes and utilization of molecular markers for selecting the background and foreground of genetic architecture improves the accuracy and predictability of breeding for nutrient use efficiency.

In last few decades, numerous studies have been attempted in important food crops to demonstrate the available genetic variation for NUE and associated traits (Balint et al., 2009; Graham et al., 1981; Pujarula et al., 2021; Stangoulis et al., 2000). Nitrogen is one of the primary macronutrients required for crop growth and yield. Breeding for increasing N use efficiency depends on N uptake, mobilization, accumulation related traits, and microbe-rhizosphere associations (Ciampitti et al., 2013; Masclaux-Daubresse et al., 2010). Worldwide, one-third of arable land is P-deficient and affects the requirement for ideal plant growth and development. Most of the P added to the soil is fixed and becomes inaccessible for the plants. Improving P use efficiency depends on solubilizing immobilized P and then its uptake and accumulation in the plant system. Similarly, breeding for enhanced K use efficiency rely on the screening of genotypes for genetic variation in K uptake and utilization efficiency (White et al., 2017) under current and reduced input conditions. Connection among total K supplied, K content accumulated in plants, and biomass formed under varied environments has to be studied to understand the genotypic, management, and environmental potentials. Identification of linked genes/QTLs and incorporation into elite lines will enhance the nutrient use efficiency in plants under unbalanced or deficient nutrient conditions.

3.1 Molecular physiology of nutrient uptake and assimilation

Root system architecture (RSA) has been considered as an important and crucial factor in determining nutrient acquisition efficiency in plants (Den Herder et al., 2010). Plants modify their RSA by stimulating lateral rooting under low availability of N, P, and K (Drew, 1975; Giehl et al., 2014). NO_3^- and NH_4^+ are the most preferred forms of inorganic N taken up by most plants root systems (Reguera et al., 2015) and RSA plays a very crucial role in N uptake and assimilation under both normal and deficient conditions (Mandal et al., 2018). These forms are actively transported by roots, distributed though xylem, and assimilated in roots and leaves by involving several important enzymes (Feng et al., 2020). NH_4^+ transport is basically regulated by Ammonium Transporters (AMTs) and low-affinity uptake systems like Aquaporins whereas, NO_3^- transport is facilitated by Nitrate Reductase or Peptide Transporters (NRTs/NPTs), Slow Anion Associated Channel Homolog (SLC/SLAH) and Chloride Channel (CLC) family, and Aluminum-activated Malate Transporters (ALMT) (Khademi et al., 2004; Léran et al., 2014; Li et al., 2017). First, NO_3^- is reduced to NH_4^+ and is then assimilated into amino acids by involving several enzymes. Similarly, Pi transporters from PHT1 family in case of P (Nussaume et al., 2011), K^+ transporters from KUP/HAK, HKT, NHX, and CHX gene families, and channel proteins from Tandem-Pore K^+ (TPK), and K^+ Inward Rectifier families in case of K (Wang et al., 2013) are involved in the transmembrane transport, distribution and homeostasis mechanisms. Designing a robust RSA i.e., length, density, lateral distribution, thickness, volume, dry weight of roots, etc. as well as the microbial-rhizosphere associations, nutrient uptake, and assimilation through molecular tools may enhance the overall NUE in the plants. For instance, gene/protein *miR444a/ANR1* encourages lateral root development under poor soil NO^{3-} status and *EL5* enhances the functioning of root apical meristem whereas, gene *OsMADS25* is involved in accelerating primary and lateral root formation in rice (Yu et al., 2015). Likewise, transcription factors such as *NAM, ATAF* and *CUC* (*TaNAC2-5A*) (He et al., 2015) and the *TaNFYA-B1* augments quicker root development in wheat (Qu et al., 2015).

3.2 QTLs controlling NUE

Many QTLs related to N, P, K use efficiency have been reported in important agricultural and horticultural crops. In rice, four QTLs controlling grain

N content have been detected on the chromosomes 8, 9, and 10 and two other QTLs governing harvest index and N uptake have been identified on the chromosomes 5 and 7, respectively (Fiaz et al., 2021). QTLs in wheat have been detected for root dry weight, shoot dry weight, and grain yield (Zhang et al., 2019). A total of 15 QTLs controlling N use efficiency have been identified in barley (Kindu et al., 2014). In bread wheat, 11 stable loci viz., *q1A, q1D, q2B-1, q2B-2, q2D, q4D, q5B-1, q5B-2, q5B-3, q6D*, and *q7A*, four independent QTLs through Meta-QTL (MQTL) analysis, and 20 candidate genes associated with major proteins which play important roles in K stress tolerance and transport mechanisms have been identified. In addition, the importance of 4B chromosome for mineral nutrition related to P, K, Ca, and Mg was understood in wheat (Safdar et al., 2020). In maize, QTLs related to root-traits (Neelam et al., 2017; Wang et al., 2019), biomass-traits (Azevedo et al., 2015), and yield-traits (Xu et al., 2018; Zhang et al., 2013) have been detected for P use efficiency, however the locus *Pup1* for P-Starvation Tolerance 1 (*PSTOL1*), has been found most successful so far, which upon over expression increased the yield and biomass in rice (Gamuyao et al., 2012) and sorghum (Hufnagel et al., 2014) as well under P deficient conditions. Zhang et al. (2013) identified a major QTL *qKN* governing kernel numbers and Qiu et al. (2013) identified QTL *AP9*, responsible for acid phosphatase activity under poor soil P conditions in maize. Kale et al. (2021) detected seven major QTLs controlling low soil P tolerance related traits in a RIL population of rice. Recently novel and well characterized and successful QTL, *Pup1* was reported in rice 3 (Aluwihare et al., 2016). Yuan et al. (2017) found 11 high-frequency QTLs and four important QTLs for P use efficiency in wheat. Gao et al. (2021) reported 2 novel QTL clusters, and 140 candidate genes associated with P use efficiency in Barley. The chromosomal loci controlling the K uptake, utilization, shoot K accumulation, and biomass formation under K deficiency have been identified in crops like rice (Fang et al., 2015; Khan et al., 2015; Miyamoto et al., 2012; Wang et al., 2012), wheat (Gong et al., 2015; Guo et al., 2012; Kong et al., 2013; Zhao et al., 2014), barley (Nguyen, Dolstra, et al., 2013, Nguyen, Ribot, et al., 2013), maize (Zdunić et al., 2014), tomato (Asins et al., 2013), *Brassica species* (White et al., 2010), apple (Fazio et al., 2013), and cotton (Liu et al., 2015). However, many candidate genes have been reported which encode K^+ transporters viz., *AtAKT1, AtHAK5, AtKUP9, AtTPK1, AtCNGC1*, and *AtSKOR* in *Arabidopsis*, *AtKUP9, AtAKT2, AtKAT2*, and *AtTPK3* in *Brassica oleracea* (White et al., 2010), *AKT2/3* and a Na^+/H^+-antiporter in cotton (Liu et al., 2015), *OsHKT8* in rice (Ren et al., 2005), *HvHKT1;5, TmHKT1;5-A*, and *TaHKT1:5-D* in barley and wheat (Nguyen, Dolstra, et al., 2013), and *SlHKT1;1* and *SlHKT1;2* in tomato (Asins et al., 2013).

3.3 Transgenic studies related to nutrient use efficiency

A considerable success has been achieved in the field of improving NUE through molecular engineering of major traits/genes/enzymes responsible for higher acquisition, uptake, redistribution, homeostasis of N, P, K nutrients. These factors have to be necessarily considered in breeding for improved NUE and subsequent yield performances of crops. Genome-editing technologies are reliable, cost-effective, and versatile approaches that are being utilized to enhance nutrient use efficiency in different crops. For instance, in rice, the up regulation of the *GS1* enzyme significantly improve the grain yield in rice (Lea et al., 2007) and root biomass, and number of grains in wheat (Habash et al., 2001). Over expression of *PHT1* genes under deficient conditions enhanced the ability to acquire P from soil and mobilize in the plants (Raghothama et al., 2005). Increased root hair development and tiller number were reported by up regulating *OsPHT1.1* (Sun et al., 2012) whereas, primary and lateral root development and low affinity uptake of P was accelerated by *OsPHT1.2* (Ai et al., 2009) in rice. Transporters like *ARF7* and *ARF19* regulate the response against low P concentration (Huang et al., 2018), *PHR1* controls the lipid remodeling under low P (Pant et al., 2015), *ZAT6* harmonizes the P balance (Devaiah et al., 2007), *PSTOL1* enhances the root development (Gamuyao et al., 2012), *BHLH32* reduces the anthocyanin and root hair development (Chandrika et al., 2013), and the *miR399-PHO2* mechanism controls the P remobilization (Chen, Jenkins, et al., 2008, Chen, Xia, et al., 2008) in plants. On the contrary, knocking out of *gln1-3* and *gln1-4* genes in maize has decreased the kernels number and yield (Martin et al., 2006). This indicated the significant importance of GS1 enzyme in enhancing N use efficiency. Different *nif* gene clusters present in the bacterial genome may be manipulated for accumulating higher N by the cereal crops (Ryu et al., 2020). Rhizosphere-associated increase in NUE is dependent on nitrogen-fixing microbial associations in cereals (Mus et al., 2016). Accelerating the level of growth hormones like auxins (Ortíz-Castro et al., 2009), cytokinins (Moubayidin et al., 2009) or gibberellins from microorganisms can also alter the RSA. This can be thoroughly explored for further improving the NUE in various crops. Though numerous studies have attempted to enhance N, P, K use efficiency through the over expression of different genes, but only few could give inconsistent results. Further, the evaluation of transgenic plants under high and low nutrient concentrations is necessarily required.

4. Molecular basis of N, P and K use efficiency

Recently emerged molecular, biotechnological, and genomic tools have dissected the multi-genic nature of complex traits associated with NUE in different crops. Gene *OsNRT1* responsible for N uptake (Lin et al., 2000), *SP1* controls panicle size (Li, Merrick, et al., 2009, Li, Qian, et al., 2009), *OsPTR6* regulates glutamine synthase and N uptake (Fan et al., 2014), *OsNRT1.1A* and *OsNRT1.1B* modulate N utilization (Wang et al., 2018) and NO^{3-} uptake, and translocation (Hu et al., 2015) in rice. Yan et al. (2011) reported *OsNRT2.1, OsNRT2.2, OsNRT2.3a, OsNRT2.3b*, and *OsNRT2.4* genes for N uptake in rice. Similarly, *OsAMT1.1, OsAMT2.1, OsAMT3.1* loci have been reported for N use efficiency with limited success in rice (Bao et al., 2015). In maize, *RTH1* and *RTH3* candidate genes controlling accelerated root hair elongation have been identified by Hochholdinger et al. (2009). P use efficiency is mostly relying on the ability of plants to solubilize the immobilized P in its rhizosphere, its acquisition and uptake from the soils and then subsequent remobilization and accumulation in different plant parts. Various genes have been identified which are regulating mobilization of P pool in different crops viz., *ALMT1* in wheat (Delhaize et al., 2009), *MDH* in *Penicillium oxalicum* (Lu et al., 2012), *PAP3* in common bean (Liang et al., 2010), *PAP15*, in Arabidopsis (Wang et al., 2009), *phyA* in *Aspergillus niger* (Ma et al., 2013), *PHY1* and *PAP1* in *Medicago truncatula* (Ma et al., 2009), *LASAP2* in white lupin (Wasaki et al., 2009), and *ACP1* in soybean (Zhang et al., 2014). Also some regulators like *NFYA-B1* (Qu et al., 2015), *NAC2-5A* (He et al., 2015), *PHO2-1A* (Ouyang et al., 2016) in wheat, *DEP1* (Sun et al., 2014), *MADS25* (Yu et al., 2015), *NAP* (Liang et al., 2014), *PTF1* (Yi et al., 2005), *PHR2, PHR3* (Guo et al., 2015), *MYB2P-1* (Dai et al., 2012) in rice, *Dof1* (Yanagisawa et al., 2004), *PTF1* (Li et al., 2011), in maize, and *NLP7* (Yu et al., 2016) in Arabidopsis were reported for improving the traits pertaining to N, P, K use efficiency. Several phosphate transporters are found responsible for P uptake and allocation in the plants. In Arabidopsis, *MYB62*, *ARF7* and *ARF19* have been reported to increase the absorption of P by root (Huang et al., 2018). Likewise, K concentration plays a very important role in plants for ensuring tolerance against diverse environmental stresses. Sufficient K assures resistances a range of biotic and abiotic stresses. However, K use efficiency is controlled by complex traits and associated network of genes for its uptake and utilization by plants. The root phenotype, its distribution in the soil, and net K^+ influx in roots decide the K^+ uptake potential of a plant. K^+ utilization is connected with its translocation into different parts, capacity to maintain K^+ concentration within cytoplasm, etc. The importance of *OsHKT8* in rice under saline soils (Ren et al., 2005), *HvHKT1; five* for shoot Na^+ and K^+ concentrations in barley and wheat (Nguyen et al., 2013), and *SlHKT1; one* and *SlHKT1; two* in tomato (Asins et al., 2013) have been reported for enhance K use efficiency. Recently, GWAS i.e., Genome-Wide Association Study has come up as a useful tool for uncovering the genetic basis and identifying the candidate genes controlling complex traits in different crops. Luo et al. (2019) identified three potential candidate genes viz., *GRMZM2G039588*, *GRMZM5G841893*, and *GRMZM2G051806* and Wang et al. (2019) identified the candidate gene *GRMZM2G009544* through GWAS in maize populations.

5. Molecular basis of micronutrient use efficiency

For governing NUE, two processes have been recognized as important such as the efficiency of nutrient acquisition (ENA) and efficiency of nutrient utilization (ENU) The activity of the root system and ability to modulate root architecture to explore the soil directly influences ENA, whereas to achieve high nutrient utilization efficiency, plants must integrate and regulate nutrient translocation, remobilization and metabolic efficiency (Fig. 6.2). NUE is a complex process. Resolving NUE into component traits for each nutrient is the first step for resolving complexity, subsequent division of each component into many physiological traits, each of which is the consequence of complex biochemical pathways, which are encoded by several genes (Hawkesford et al., 2011). In order to uncover the alteration in gene expression, the high throughput profiling methods have been performed under nutrient-deficient conditions and again under sufficient resupply of nutrient conditions. By various high throughput studies, it has been reported that induction and suppression as well as differential root and shoot expression of responsive genes, local or systemic transcriptional regulation and root system architecture (RSA) changes mediated by a set of genes resulted in differential NUE (López-Arredondo et al., 2017).

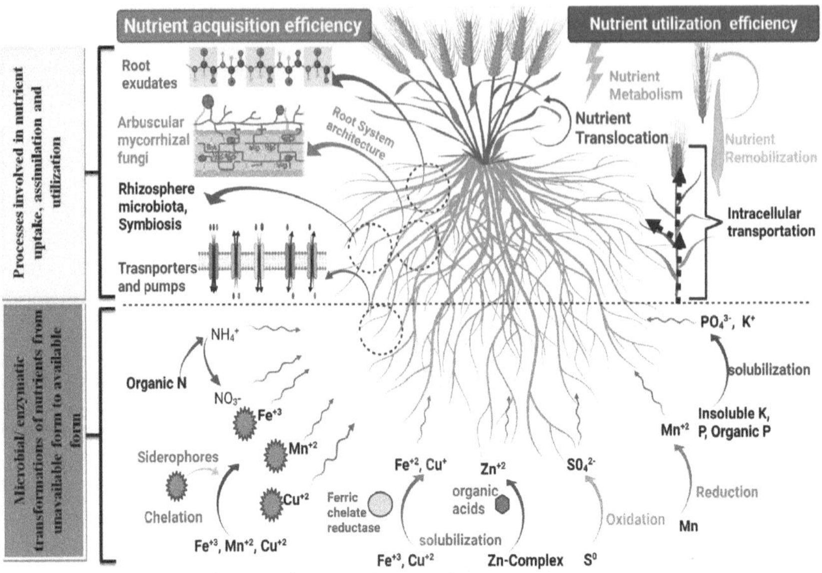

Fig. 6.2 Overview of mechanism of nutrient use efficiency.

The information regarding the molecular aspect of plant nutrition is rapidly increasing, which may facilitate the discovery and, thereby, manipulation of mineral acquisition, transportation, utilization and RSA associated key regulatory elements. Most of the earlier work is based on *Arabidopsis,* and a considerable effort has been made in major food crops, including rice, wheat, maize, barley, and soybean, etc. hence more work in this direction is needed. To date, several key regulatory elements were identified that regulates the complex process of nutrient uptake, transportation, and remobilization, and also determines root morphology. Hence by manipulation of these components, the development of plant type with enhanced ENA and ENU has become possible. Hence these regulatory elements are important targets for the improvement of the NUE of crops.

6. Molecular mechanism of nutrient uptake and transport

For metabolism and efficient growth, plants take up nutrients for which nutrients transportation across plasma membrane to the cytoplasm in the root hairs is needed first. Recent research indicates that for each nutrient, there is the active participation of a different group of genes that encodes certain specific proteins known as transporters that facilitate the entry of nutrients through the plasma membrane into the cytoplasm. For different nutrients, there is the involvement of different types of transporters (channel proteins, ATPases or ATP-powered pumps, cotransporters), and the number of genes encoding these transporters has been cataloged so far in a number of plant species.

6.1 Mechanism of iron (Fe) uptake and transport

There is limited solubility of Fe in many soils; hence before transporting it into the plant, plants first mobilize Fe in the rhizosphere. It has been studied extensively in numerous plant species, including essential food crops, namely wheat, rice, maize, and barley, that how from the soil the plant acquires Fe. To date, huge knowledge regarding the two-known uptake strategy: the "Reduction-based" strategy, and the "Chelation-based" strategy, is available (Walker & Connolly, 2008). In nongraminaceous species, the reduction-based strategy works for Fe uptake by the sequential action of proton ATPase, followed by ferric chelate reductase, and finally, by the ferrous iron transporter-mediated transportation. In this strategy, under Fe deficient condition, the activity of AHA2 H + -ATPase protons are exuded into the rhizosphere, which reduces the soil pH, results in more Fe solubilization. In reaction to Fe deficiency in the root epidermis, up-regulation of several ATPases such as *AHA*1, AHA2, and AHA7 are reported, but among which, the only primary root H + -ATPase is AHA2 (Morrissey et al., 2009). Then membrane-bound reductase, i.e., ferric chelate reductase FRO2 expressed and helps to produce the reduced form of Fe, i.e., Fe^{+2} from Fe^{+3}. In the process of Fe uptake, in *Arabidopsis* this step was considered a rate-limiting step by Connolly et al. (2003). The role of ferric chelate reductases is further confirmed by transgenic studies where overexpression of ferric chelate reductases in the roots were reported to be linked with improved tolerance to Fe deficiency in case of three crop species, i.e., rice, tobacco, and soybeans (Morrissey et al., 2009). Then Fe^{+2} becomes accessible for plant uptake and is further transported with the help of the either divalent metal transporter IRT or another group of transporters i.e., Natural Resistance Associated Macrophage Protein (NRAMP) into the root epidermis. Graminacious species follow a chelation-based strategy where under Fe deficiency, grasses modifies rhizosphere, by secreting mugineic acid family phytosiderophores (PS) (Römheld, 1991). PS can be defined as low molecular weight organic compounds; specifically, nonproteinaceous amino acids including deoxymuigenic acid, nicotianamine, avenic acid etc., that showed a high binding affinity toward respective metal ions hence facilitate their chelation and acquisition. In rice, the dramatically increased expression of transcription factor "OsIRO2" during initial 5 days of Fe starvation confirms its role in the regulation of genes that are involved in PS synthesis, thereby Fe uptake. Then the PS directly chelates Fe^{3+} form, and further the transporters that are member of Yellow Stripe-Like (YSL) family such as OsYSL15 in rice and YS1 in barley and maize facilitates PS-Fe^{3+} complex transportation into the root.

In comparison to the reduction strategy, the sensitivity of the chelation strategy toward pH is less. Existence of very strong association between resistance toward Fe deficient soil and the volume of PS released by genotypes was recorded. The combination of both strategies has been reported in the case of rice. Rice is more susceptible to calcareous soils due to reduced production of PS in comparison to other cereals such as maize and barley. Hence in response to Fe deficiency, for compensating plant Fe demand rice follow reduction strategy by expressing ferrous transporters such as OsIRT1 and OsIRT2, which facilities Fe^{+2} transportation.

6.2 Mechanism of zinc (Zn) uptake and transport

Zn followed the same strategies, i.e., Reduction-based strategy and Chelation-based strategy, as discussed in the case of Fe. In reduction-based strategy, there is the outflow of either H^{+} ions or organic acids, which encourages solubility of Zn-complexes (Zn-phosphates, Zn-hydroxides etc.) and results in free Zn^{+2} ions. These organic acids are released either directly by root epidermal cells or indirectly in root exudates/mucilage, among which malic acid, oxalic acid, zitric acid, or tartaric acid are the few major ones. In reaction to Zn deficiency in the soil, highly Zn efficient rice and barley varieties are reported to express an increased rate of citrate, malate and mugeniec acid secretion (Mitra, 2017). Then the released Zn^{+2} transported with the help of Zinc—Iron Premeases (*ZIP*) family transport proteins (*ZIP1, ZIP3,* and *ZIP4*) across the roots plasma membranes. Among plants Zn homeostasis is reported to be modulated by the number of transcription factors (TFs), like basic-region leucine zipper proteins *bZIPs19* and *bZIPs23* (Mitra, 2017). Chelation-based strategy is depends on outflow of phytosiderophores (PS) which make stable complexes with Zn, and then this Zn-PS complex is transported across the plasma membranes by YSL family transporters. However, the chelation-based strategy is restricted to cereal roots only.

6.3 Mechanism of manganese (Mn) uptake and transport

The Mn availability to the plant is affected by its oxidation state. The only available form of Mn is Mn^{+2} (reduced form); however, rest oxidized forms such as Mn^{+4} and Mn^{+3} are not bio-available. In the case of few species, certain evidence showed that plants roots take up Mn by complexing it with ethylene diamine tetra acetic acid (Mn^{+2}-EDTA) (Laurie et al., 1995). In certain species, the deficiency of both Fe and Zn is reported to associated with enhance Mn^{+2} uptakes, whereas, in contrast, Mn^{+2} accumulations are inhibited by an excess of Fe and Cd, indicating the involvement of broad-specificity transporters in Mn^{+2} accumulations (Pittman, 2005). Various well-characterized Fe and Zn transporter that belongs to ZIP, IRT, and NRAMP families have been shown to transport Mn, suggesting the coincide Mn uptake with Fe and Zn uptake process (Hawkesford et al., 2011). In *Arabidopsis* and rice there is sufficient evidence available that suggests that the Mn uptake is mainly mediated by NRAMP *transporters* (Castaings et al., 2016). Mn in-sufficient condition up-regulates the expression level of AtNRAMP1 likewise less Mn, shoots accumulation is observed in nramp1 knockout mutants which supports its role in active Mn^{+2} uptake mechanism. This transporter was reported to localize in the epidermis (plasma membrane), roots (cortex cell), and young leaves (vascular tissues). OsNRAMP5 was the first identified Mn transporter in rice which facilitates Mn uptake as well as translocation (Sasaki et al., 2012). Additional investigation revealed the role of other transporters also such as OsMTP9 (Ueno et al., 2015). A knockout mutant of either of transporters: OsNRAMP5 or OsMTP9, showed reduced uptake along with restricted Mn translocation with in plant (Alejandro et al., 2020). In the latest investigation in rice, the active involvement of OsNRAMP1 in Cd uptake was established along with Mn uptake (Chang et al., 2020). To date, several *ZIP* and *IRT*, Mn^{2+} transporters (*MtZIP4*, *MtZIP7*-*Medicago*; *AtIRT1*-Arabidopsis; *LeIRT1* and *LeIRT2* -tomato; *PsIRT1*-pea and, *HvIRT1*-barley) have been identified of which the role of only two transporters such as *AtIRT1* and *HvIRT1* have been demonstrated in plants.

6.4 Mechanism of boron (B) uptake and transport

For boron uptake and transport in plants, two mechanisms are involved. The first one is passive diffusion, which works under adequate concentrations of B. The available form of B is boric acid, which is relatively permeable across biological membranes and are small uncharged molecule. Hence under B sufficiency the plant demand gets satisfied by means of boric acid passive diffusion. Whereas under limited B concentration second mechanism, "active uptake" is operating where plants use boric acid channels and borate exporters for B uptake and transportation respectively (Takano et al., 2008). Under boron-limiting conditions, in various plant species to support optimum plant growth, the role of boric acid channels along with borate exporters has been established. In *Arabidopsis,* a MIP subfamily member, Nodulin-26-like Intrinsic Proteins (NIP5; 1), was identified as channel protein (Takano et al., 2006). Reduced root and shoot growth and low B content are associated with nonfunctional mutants of NIP5; 1(Brown et al., 2011), whereas in low boron treated Arabidopsis, significantly increased expression of AtNIP5; one was recorded (Takano et al., 2006). These results established the transporter role of NIP5; one under B deficiency condition. BOR-1 the member of SLC_4 bicarbonate family is the first identified B transporter (Frommer et al., 2002). BOR1 overexpression is reported to link with enhanced root to shoot translocation which ultimately resulted in increased seed set under B limiting conditions (Miwa et al., 2006). Under limited B concentration, the entry of boric acid into the root is facilitated by *NIP5; one* channel then B travels simplistically to the pericycle then transported by BOR1 into the xylem.

6.5 Mechanism of copper (Cu) uptake and transport

The mechanism of Cu uptake in the plant is not revealed completely. However, it is suggested that the Cu uptake is strongly overlapped by the Fe uptake mechanism (Ryan et al., 2013). As per isotope fractionation analysis of Cu uptake, it has been observed that plants prefer uptake of light Cu isotope. The Cu uptakes across the plasma membrane occur primarily in reduced form (Cu^+), suggested the possible role of reduction-based strategy in Cu uptake. In reaction to Cu deficiency, in root tips, Cu (cupric ions) is also reduced by the activity of two Ferric Reductase Oxidases (FRO4/5) (Bernal et al., 2012). Then the high-affinity Cu transporters (COPT) direct the transportation of reduced Cu^+ ions across root plasma membrane. In *Arabidopsis,* six genes are encoding COPT transporter were discovered (COPT1-COPT6) and of which only two AtCOPT/two were categorized as high-affinity Cu transporters. The reduced and increased expression of Cu transporters (COPT1 and COPT2) under Cu sufficient and Cu deficient condition respectively supported their critical role in Cu acquisition under Cu starvation conditions (Migocka et al., 2018). Copper uptake may also occur as Cu^{2+} (oxidized form) through the ZIP family transporters, but there is only a little evidence in *Arabidopsis*. In *Arabidopsis,* gene expression studies showed induction and repression of ZIP2 and ZIP4 expression under Cu deficient and excess conditions, respectively (Wintz et al., 2003). However, the findings of recent studies are contradictory where ZIP2 and ZIP4 expression was up-regulated under Cu deprivation and access, respectively (Printz et al., 2016). The ZIP4 expression is also regulated by the Zn concentration, which supports the existence of crosstalk between Cu and Zn absorption.

6.6 Mechanism of molybdenum (Mo) uptake and transport

Plants are able to acquire only oxyanion molybdate (MoO_4^{2-}), which is the most soluble form of Mo in soils. It requires a definite system for Mo uptake in the presence of competing anions. Sulfate is a potential inhibitor of Mo uptake; hence low sulfate concentration is associated with increased uptake of Mo (Shinmachi et al., 2010). Molybdate and sulfate share physical as well as chemical similarities; therefore, it is speculated that the sulfate transporters or related systems may facilitate molybdate import along with distribution (Bittner, 2014). Comparison to bacteria, in eukaryotes the mechanism of Mo transportation is less understood. In-plant species with high affinity and specific MoO_4^{2-} transporter belongs to the MOT-1, and MOT-2 families are known. In *Arabidopsis*, two members of the *MOT1* family are known, among which AtMOT1 facilitates efficient Mo uptake (Tomatsu et al., 2007). The nonfunctional mutant of *Arabidopsis* AtMOT1 display 35%–80% lower growth rate under Mo deficient environment than plants growing under Mo sufficient environment, showed possible role in scavenging of Mo (Tomatsu et al., 2007). These transporters are the members of sulfate transporters family but lack the antisigma factor antagonist (ATAS) domain which critically directs the transport activity of sulfate ions (Shibagaki et al., 2006). In plasma membrane the location of these transporters has not been detected, but there are conflicting reports of localization to the endomembrane system (Tomatsu et al., 2007) or the mitochondrial envelope (Baxter et al., 2008). Hence the precise function of MOT1 is needed to be proven in future studies. Along with MOT1 and MOT2 members deriven high-affinity system, Mo can also move nonspecifically, more particularly by sulfate transporters. In some species of plants, the activity of Caribbean stylo, high-affinity sulfate transporter SHST1 is completely suppressed by the presence of MoO_4 suggested its possible role in mediating

MoO_4 transport (Fitzpatrick et al., 2008). It has been identified that the sulfur deficiency induces sulfate transporter up-regulation and correspondingly Mo uptake in wheat and tomato (Manuel et al., 2018).

7. Molecular mechanism of nutrient assimilation and remobilization

The efficient nutrient mobilizations in the plant begin with uptake, followed by vascular system loading and final translocation to young developing plant organs (leaves, tiller, and seed), which are associated with efficient nutrient utilization. Hence, dissection of molecular mechanism and genetic key components governing these processes is critical. In this context, in plants wider range of metabolic modifications including within plant tissues nutrients mobilization, active recycling of nutrients to young growing organs/tissues from mature/senescing organs as well as efficient reutilization of nutrients from vacuoles (which perform buffering function, i.e., when nutrients are present in excess in the cytoplasm it works as a store) when plants faces nutrient deficiency.

7.1 Iron (Fe) assimilation and remobilization

Fe is almost chelated once it is transported into root epidermal cells, although the mechanism is not clear. The transporter responsible for xylem loading is not known, but once in xylem, it is bound with citrate, which is transported via FRD3. For phloem loading, YSLs transporter is involved in the case of rice here Fe bound with nicotinamide (NA). For long-distance Fe transportation to the seed, NA is a very crucial factor, but it is not clear in which form Fe is held after loading into seeds. The transporters YSL1, YSL3, and Oligo Peptide Transporter (OPT3) are involved in seed Fe loading. For storage and sequestration of mineral elements, including Fe, vacuoles are a very important place in all the plant cells, especially in seeds. Under excess concentration, transporters FPN2 and Vacuolar Iron Transporter (VIT1) facilitates buffering of Fe into vacuoles where, it is complexed with phytate and NA. During Fe deficiency, the Fe is taken out from vacuoles via NRAMP2 and NRAMP4. For minimizing oxidative stress, in mitochondria, Fe is stored as ferritin or frataxin (Morrissey et al., 2009), but the Fe transporters for mitochondrial loading have not been identified yet. For biogenesis of Fe—S complex, mitochondria are the main compartment, and in *Arabidopsis,* the transportation of Fe—S cluster from mitochondria is reported to deriven by ABC transporter of the mitochondrion

(AtATM3) (Jeong et al., 2009). In the leaf, a major portion (90%) of Fe is sequestered within the chloroplast, where the maximum portion is localized in the stroma. For entering into the chloroplast, FRO7 *reduces* Fe^{+3} to Fe^{+2} and the reduced from is taken up by inner membrane-localized Fe^{+2} transporters, i.e., Permease In Chloroplasts (PIC1). In chloroplast, Fe is sequestered in the form of ferritin (Jeong et al., 2009). The mechanism of redistribution of nutrients from mature organs to developing organs improves NUE. In *Arabidopsis,* AtOPT3 is involved in the distribution of nutrients and also involved in phloem loading, xylem to phloem Fe circulation, and shoot-to-root Fe translocation (Zhai et al., 2014).

7.2 Zinc (Zn) assimilation and remobilization

The root plasma membrane (pericycle cells) localized Heavy Metal ATPase (HMA2 and HMA4) transporters are considered the most likely candidate for xylem loading of Zn (Eren et al., 2004; Hanikenne et al., 2008). The Zn sequestrated in xylem, in both soluble forms bound to small proteins or in the form of insoluble complexes such as Zn-phytate. The easy phloem transportation of Zn indicated the importance of phloem mobility in Zn distribution to economic plant parts (fruits and seeds) (Wu et al., 2010). In phloem and leaf cells the Zn^{+2} influxes is mediated by ZIP family members, and in addition, YSL family transporters are involved in phloem loading (Sadeghzadeh et al., 2011). In the phloem tissues, Zn is transported either freely (ionic form) or by complexing with other components (Zn-NA, Zn-histidine, or Zn-malate). However, improved concentration of chelating solutes resulted in higher mobility of Zn in phloem than xylem. Once Zn is transported in phloem, further Zn distribution toward various organs and developing sinks is regulated by several short-distance or long-distance pathways. Depending upon growth state and kind of tissue, in a vacuole, Zn is sequestered by complexing with various organic acids including Zn-citrate and Zn-malate, but not in the form of free $Zn2^{+}$ ions (Broadley et al., 2007). Various transporters that belongs to Cation Diffusion Facilitator (CDF) family, including Metal Tolerance Protein (MTP1, MTP2), Zinc-Induced Facilitator 1 (ZIF1) transporter, are associated with vacuole Zn transportation via Zn^{2+}/H^{+} antiport mechanism; however, Zn mobilization from the vacuole is linked with NRAMPs transporter (White et al., 2009). The up-regulation of HMA3 transcripts expression under excess Zn^{2+} concentrations in the cytoplasm supports its possible role in vacuole Zn sequestration (Moreira et al., 2018).

7.3 Manganese (Mn) assimilation and remobilization

After Mn^{2+} absorption by root, it needs to be translocated into the xylem, phloem, and to/among the different tissues. To date, in plant cells, most of the transporters involved in root-to-shoot transportation and the molecular mechanism of xylem and phloem loading are not clearly known. The most probable reason for this is the sharing of transporters among divalent cations such as Mn with Fe. In comparison to phloem, Mn is more mobile in xylem, but there is certain evidence supporting recycling of a small amount of Mn via phloem also (Alejandro et al., 2020). The mechanism of Mn transportation is well studied in *Arabidopsis* and rice than any other crop. Depending upon the availability of Mn in rice, the OsNRAMP3 transporter behaves differently. Under high Mn availability conditions, it transfers Mn to mature tissues; however, in an opposite situation, it facilitates Mn transportation from xylem to phloem, and ultimately to young growing organs particularly leaves and panicles (Yamaji et al., 2013). Another phloem companion cell localized, rice transporter OsYSL2, involved in phloem as well as developing seeds Mn loading (Koike et al., 2004). The expression of OsYSL2 is primarily reported in above ground plant parts particularly leaves, flowers, as well as in developing seeds but not in roots. The presence of Mn was observed in various cellular compartments (mitochondria, plastids, peroxisomes, endoplasmic reticulum, and Golgi body), where it is actively involved in specific cellular functions. Majority of the transporters involved in Mn transfer in or out of these organelles remain unidentified except for *AtECA3* (endosome), *AtMTP11* (golgi apparatus), *PAM71* (thylakoids) (Feng Shao et al., 2017), and *AtMTM1* and *AtMTM2* (mitochondria) (Hu et al., 2021). To regulate cellular Mn homeostasis, vacuoles serve as an efficient nutrient reservoir. Major portion of Mn taken up by roots is allocated to the shoots, but some fraction is also sequestrated into the root cells vacuoles. In rice, OsMTP8.1 (metal tolerance proteins) is reported to involve in vacuoles sequestration of Mn (Chen et al., 2013). In *Arabidopsis* root stele cells for remobilization of Mn from vacuoles to the cytosol, the probable transporter is tonoplast localized ZIP1 (Alejandro et al., 2020). The remobililization of Mn out of senescing/mature leaves to Mn-deficient young leaves is not observed due to its less mobility in plants (Li et al., 2017).

7.4 Boron (B) assimilation and remobilization

The B deficiency in new leaf and toxicity in older leaf supported the highly immobile nature of B in most of the species. Later on, species-specific B mobility is reported, and those species transporting sugar alcohols (polyols) as the main photosynthate within the phloem showed high B phloem mobility. Later on, phloem B mobility is also reported in nonpolyol producing species, including wheat, canola, and sunflower, but the mechanism is poorly understood. The first B transporter identified in plant species is AtBOR1 which is critical for xylem loading. It is localized in the pericycle of the root stele (Hawkesford et al., 2011). Later on, it is determined that the xylem loading transporter BOR1 is also involved actively in within shoots B distribution. The prevailed distribution of B to young leaves was reported in the case of BOR-1 (wild type); however, such transport was absent in bor-1 mutant (Miwa et al., 2010). In addition to root, its expression was recorded in the shoot, but the possible mechanism of B distribution among leaves with the help of BOR1 is not known. NIP6; 1, which encodes the B channel, is another transporter involved in B distribution. It is mainly expressed in stems suggesting its possible involvement in B transport in shoots.

7.5 Copper (Cu) assimilation and remobilization

Following root uptake, the xylem loading of Cu is regulated by Heavy Metal ATPase (HMA5) transporters (Burkhead et al., 2009). After entering into the xylem, Cu makes complexes with nicotinamide (NA) or citrate, which makes it stable in mildly acidic conditions of the xylem. Few experimental evidences are available on Cu transportation in the phloem. Phloem NA abundance supported, NA mediated Cu transportation to flower and seed. The loading and unloading of phloem are thought to be achieved by YSL1 and YSL3 transporters that are localized in shoot and reproductive tissues. Reduced leaves to reproductive tissues Cu from mobilization was observed in the case of double mutants (ysl1/ysl3) (Waters et al., 2007). In rice, phloem-localized transporter OsYSL16 is reported to involve in Cu-chelate loading in the phloem. Its involvement in the reallocation of Cu from mature plants organs to the young developing plant organs including seeds has been established (Zheng et al., 2012). After phloem loading, it is transferred to various organs and metabolic pathways where it is required for further activities. Copper plays primary metabolic functions in various organs, including chloroplast, apoplast, mitochondria, and endoplasmic reticulum (ER) and exclusive transporters and metal chaperones worked in a harmonized fashion to ensure smooth delivery of Cu to these organs. In ER, the delivery of Cu to ER-localized ethylene receptor *(ETR1) is regulated* by heavy metal HMA7 transporter. In stroma-Cu/Zn

SOD (CSD2) and in thylakoid lumen-plastocyanin are two chloroplast components where the presence of Cu is critical. Transporters, HMA6 and HMA8 *ensure the Cu delivery* to the thylakoidal plastocyanin, whereas, for stomal CSD2 Cu delivery, only HMA6 is required (Brown et al., 2011). Low molecular weight metal receptors proteins (metal chaperones) are considered crucial for intracellular allocation of Cu ions and Cu positioning into the active sites of Cu-dependent enzymes (O'Halloran et al., 2000). Among metal chaperones for mitochondrial cytochrome C oxidase, Cu transportation is mediated by COX11, COX17, and COX 19 (Brown et al., 2011).

In addition, cysteine-rich proteins, i.e., Metallo Thioneins (MTs), were identified in the case of *Arabidopsis*. These MT were categorized into four groups (MT1, MT2, MT3, and, MT4) and the corresponding change in their accumulation is observed concerning changes in Cu availability. The redistribution of Cu from senescing organs to young developing organ particularly young leaves and seed is hindered in case of *Arabidopsis* MT mutant suggested active involvement of MT in remobilization of Cu (Benatti et al., 2014). *Plasma* membrane-localized protein, COPT6 has been reported to involve in Cu remobilization/redistribution to reproductive organs from vegetative plant organs (Garcia-Molina et al., 2013). During the leaf senescence stage in *A*. thaliana, accumulation of small Cu-binding proteins (CCH) in yellow-senescing leaves and detection of CCH in phloem exudates recommended its role in Cu-transportation from mature vegetative organs to young developing sink organs (Printz et al., 2016).

7.6 Molybdenum (Mo) assimilation and remobilization

Mo is enter into the plants in the form of molybdate (MoO_4^{2-}), but it remains inactive until it synthesizes pterin-based Mo cofactor (Moco). Then moco binds with various Mo-dependent enzymes including nitrate reductase (NR), sulfite oxidase (SO), xanthine oxidase/dehydrogenase (XO/XDH), and aldehyde oxidase (AO) that participates in various oxidative or reductive reactions. It has been suggested that the transporter involved in other nutrient transportation such as P-type ATPase, heavy metal transporters, and nonspecific anion transporters, including sulfate transporters, also work as cotransporter of Mo. In addition to this, some Mo specific transporters have also been uncovered in several plant species particularly *Arabidopsis, Medicago* that belongs to MOT-1 and MOT-2 family. Comparing with wild-type plants, the lack of functional mutant of

MOT-1.2 showed more elevation in leaves Mo content than seeds. Besides that, in comparison to wild-type the less activity of Mo-dependent enzymes was observed in the leaves of mutant plants. These findings suggested its involvement in intracellular transportation along with an interorgan allocation of Mo utilizing vacuolar MoO_4^{2-} (Gasber et al., 2011). MOT2 another transporter that are the member of sulfate transporter family, localized in the vacuole, was reported to involve the outflow of vacuole stored Mo to cytosol for translocation into matured leaves (Gasber et al., 2011). There are some evidences, particularly in wheat, that support the absence of remobilization of Mo from senescing/mature leaves to young reproductive organs (Shinmachi et al., 2010). Likewise, onset of Mo deficiencies at the time of flowering and seed setting in groundnut, maize, and Merlot grapevine may also reflect inadequate remobilization of Mo from senescing leaves (Brown et al., 2011).

8. Molecular mechanism of modulation of root system architecture

Root architecture is an important parameter that determines the plants' potential to access soils essential growth components such as water and nutrients. The root system is considered critical from a plant growth point of view as it carryout diverse plant functions, including the acquisition of nutrients and water, formation of symbiosis with advantageous soil microflora for obtaining improvement nutrient absorption efficiency of the plant (Hodge et al., 2009). Roots are composed of mature and elongation zone. Nutrient and water freely move through the elongation zone into the xylem then into the shoot. In contrast, nutrient entry is difficult through the mature zone due to the presence of an obstacle called a casparian strip. Hence in later growing season, the deep soil nutrient level becomes more important, particularly for deep-rooted crop species. As the plant grows, roots spread out both vertically and laterally to cover soil areas that have more water and nutrients.

In response to nutrient deprivation, plants expressed several root phenotype responses, including reduced growth of primary root, enlargement in length and number of lateral root and enhanced root hair elongation. There is a number of evidence that suggests that by modulation of RSA, the plant can cope up with severe nutrient starvation. RSA modification is a complex process, and several regulatory elements governing RSA modulation were identified, which are potential targets for the improvement of NUE. The modulation of RSA is governed by nutrient availability and symbiotic microorganism (Li et al., 2015). There are huge studies regarding RSA modulation associated with nutrient deficiency, and an arbuscular mycorrhizal fungus is conducted. Vast knowledge of molecular mechanisms which causes root system architecture alteration in response to the concentration of particular nutrients is generated. Several genes, transcription factors, microRNA which are involved in RSA modulation, facilitate the enhanced nutrient acquisition, have been discovered in many plant species, but the majority of studies are related to macronutrients. Very few evidence are available of micronutrients and discussed below; hence huge research in this direction is needed. Alteration in root architecture was also observed under low micronutrient (Fe and B) concentration. The change in root morphology associated with Fe deficiency is the same as the P deficiency response, such as significant improvement in root hair numbers as well as length (López-Bucio et al., 2003). Likewise, under B deficiency, primary root growth reduces, whereas lateral root growth strengthens by developing longer root hairs (Abreu et al., 2014). Starvation for B leads to an enhanced density of first-order lateral root without any significant effect on lateral root length. Similarly, Fe starvation causes first-order lateral root density and length reduction. Whereas under Mn deficiency, the length of lateral root inhibited along with a slight reduction in branching is also observed. However, Zn starvation stimulates lateral root branching along with lateral root growth reduction (Shahzad et al., 2017). In earlier experiments, it was confirmed that the Zn uptake in maize was positively affected by AM fungi due to improved plant growth parameters (Faber et al., 1990). In rice OsARF12, a transcription activator of the auxin response gene was detected, which facilitates Fe accumulation by elongating roots (Qi et al., 2012). The possible role of *BOR2, the* B transporter and CTR1 the negative regulator of the ethylene response pathway in B mediated root development in *Arabidopsis* were established by two different lab groups (Miwa et al., 2013; Tabata et al., 2013).

Arbuscular mycorrhizal fungi (AMF) are able to develop the profused/widespread root symbioses, associating with about 80% terrestrial plant species roots, and it supports plant growth by improving the mining and acquisition ability of roots toward essential micro as well as macronutrients (Marschner et al., 1994). In symbiotic association, by penetrating the inner cortical cell of the host root, fungal hyphae form a specialized structure called arbuscules (the organelle which facilitates nutrients exchange between the AMF and host) and vesicles (the organelle which performs nutrient storage function) which can significantly improve the water and nutrients absorption ability of the root (Parniske, 2008; Rillig et al., 2006). AMF can form a profound hyphal network extended up to 12 cm away root surface; hence plants can examine a huge volume of soil and can acquire immobile nutrients away from the rhizospheric nutrient depletion zone (Parihar et al., 2019). It is evident that the association of root with AMF facilitates the nutrient acquisition, specifically for diffusion-restricted micronutrients like Zn and Cu (Li et al., 2015). The extent of Fe and Mn availability to the plant is relay on the oxidation as well as reduction potential of soil. The concentration of various micronutrients such as Zn, Fe, and Cu was reported to be high in the shoot of plants inoculated with mycorrhiza than un-inoculated plants when sown in micronutrient deficient soils or soils with no supplementary application of micronutrient (Liu et al., 2000). On the basis of these findings, it is summarized that the effects of AMF on micronutrients uptake is significantly vary with the amount of micronutrient added to the soil. Likewise, significantly enhanced concentrations of macronutrients (N, K, P, Ca, and Mg) along with micronutrients (Mn, Cu, Fe, Zn, and Cl) were also recorded due to inoculation of AMF (Lehmann et al., 2015; Mardukhi et al., 2011).

9. Factors affecting NUE

9.1 Soil factors

Adverse physical and chemical soil constraints lead to low nutrient supply to the plant resulted in the lower production potential of many soils around the world. Among chemical constraints, nutrient deficiencies or toxicities, salinity, acidity, and low organic matter content are important. Among acidic soil, Al, Mn, Fe, and H present at a phototoxic level, whereas deficiency was recorded for macro (N, P, K, Ca, and Mg) as well as micronutrients (Mo, Zn) (Patiram, 2007); hence both the situation adversely affects NUE and ultimately plant growth. In contrast, alkaline soil contains an excess level of $NaHCO_3$ and exchangeable Na, and saline soil contains chlorides and sulfates of Na, Mg, and Ca. Therefore the predominance of salts adversely affects plant nutrient uptake by reducing the permeability of roots. Likewise, in some agriculturally cultivable soils, the level of some trace elements including Cd, Pb, Cr, Cu, Ni, As, Zn, Co, and Mn significantly increased due to the application of pesticides and soil amendments (Ha et al., 2014). Up to some level, these elements are essential, but beyond a threshold, concentration may cause phytotoxicity which eventually leads to reduced NUE (Baligar et al., 2001). The excess concentration of Co and Cd is reported induce Fe deficiency by hindering its movement from root to shoot (Morrissey et al., 2009). The degree of heavy metals availability to the plant is significantly affected by numerous soil factors, including pH, temperature, ligand-exchange mechanisms involved, redox potentials, and soil solution composition (Sarkar et al., 2017). The soil organic matter is positively correlated with NUE as it helps in amelioration of physical, chemical, and biological properties of soil, therefore enhances the soil water holding capacity along with exchangeable nutrients percent and cut downs the leaching of nutrients as well as the toxicity of Al and Mn.

High bulk density layers of soil or hardpan, poor soil texture and structure, crusting and surface sealing, water holding capacity, extreme drying, poor aeration, and water-logging are important physical constraints that adversely affect nutrient utilization efficiency. These soil physical factors can alter adsorption, precipitation, mineralization, immobilization, losses through leaching, runoff as well as gaseous losses of nutrients via denitrification or ammonification (Bhattacharyya et al., 2015). Adverse physical properties of soil affect growth, distribution, and morphology of root; hence it affects the root's ability to acquire a larger volume of soil resulted in reduced nutrient uptake and thereby low NUE. Enough soil aeration and moisture, ideal soil temperature, and absence of mechanical stress are some parameters that determine physical soil conditions and considered critical for optimum crop growth. By modifying plant and soil parameters as well as microbial activity, soil aeration, optimum temp significantly influences the efficiency of applied nutrients. Likewise, soil moisture is also an important parameter for NUE as it affects plant nutrient absorption by influencing root growth (Sarkar et al., 2017).

9.2 Fertilizer factors

Fertilizer source, rate, time, depth, and method of application are some important fertilizer factors that affect the NUE of plants. Hence there is a need for optimization of these factors according to soil type, plant species, and climatic factors for reducing nutrient loss via leaching, denitrification, runoff, and fixation. There is numerous evidence that supports that the for upgrading fertilizer use efficiency amelioration and judicious consideration of fertilizer factors is important (Baligar et al., 2001). By using the right source of nutrients, at right time with right rate, and in the right place, the NUE of any fertilizer can be upgraded. Right source means select fertilizer which matches well with the soil properties and need of the crop, whereas matching the crop need with applied fertilizer signifies the right rate. Right time signifies, fertilizer maintenance in such a way that the fertilizer remain available to crop when they needed most. Significant improvement in NUE can be reached by synchronizing the availability of nutrients with crop demand. By following split application of fertilizer, use of slow releases fertilizers, inhibitors, and stabilizers fertilizers can be better timed out for efficient plant uptake. Right place means, as much as possible try to keep the nutrients at the place where plants can use them most (Sarkar et al., 2017). In the future, site-specific technology may be helpful for designing sound fertilizer management systems thereby judicious use of fertilizer inputs, which will ultimately lead to reduced costs of cultivation and environmental degradation.

9.3 Plant factors

Genetic variability of the plant causes variation in parameters of nutrient uptake hence explains the differences in NUE. Variability in nutrients absorption and translocation, plants shoot demand, amount of dry matter produced/amount of nutrient absorbed, and interactions with environmental components are some factors which may leads to varying NUE among plants. Overall, plants' NUE is relay on the capacity of roots to take up the nutrient and their further transportation to the shoots followed by remobilization to different organs, including economic parts (grain). Various organic and inorganic-organic compounds exude from plant roots and microorganisms that may facilitate nutrient uptake either by changing soil pH directly worked as chelator or through the transformation of nutrients via reductants and oxidants. Likewise, root exudes also influence the population of the microorganisms of the rhizosphere hence improve microorganism facilitated nutrient uptake. The root is the main organ that facilitates nutrient absorption from the soil. Therefore by altering root morphology (length, thickness, surface area, volume, and root hair length), various plant species can improve NUE. By such modification, the plant can access nutrients from high-density soil layers, large coverage of soil, and also able to withstand deficiency or toxicity of nutrient and temperature or moisture extremes. Longer, thinner roots with increased root surface area are reported to be associated with increased NUE of Zn, Cu, Mn, and Fe (Sadeghzadeh, 2013). Various physiological mechanisms along with plant response toward nutrient deficiency, tolerance toward elemental toxicity, and climatic factors govern nutrient uptake and utilization efficiency. Hence by genetic improvement of plants for imparting tolerance against elemental toxicities, nutrient deficiencies, high bulk density, and climatic variables (drought, temperature extremes, and aeration) will improve the absorption and utilization efficiency of the plant.

9.4 External factors

External factors comprised of soil management practices, environmental factors (temperature, solar radiation, and rainfall), biotic factors (diseases, insects, and weeds) profoundly affect the level of available nutrients in the soil and the plant's capacity to take up the nutrient and further utilization to produce economic plant parts (grains). Soil temperature affects the nutrient release rate from the organic and inorganic reserve, followed by uptake, transport as well as utilization by plants. Solar radiation is associated with photosynthetic activity hence influences plant nutrient demands subsequently. The crop shading affects crop growth hence ion uptake is affected negatively. For crop growth, thereby NUE, the rainfall distribution pattern during crop growth season and interaction of nutrients with water at root zone is much more important than the total rainfall. The climatic variable is not under human control hence can't be managed up to a larger extent; therefore, crop management practices should be optimized for prevailing climatic conditions, and genotype selection should be performed accordingly. In breeding programs targeting improved NUE, the inclusion of physiological traits that facilitate plants tolerance ability toward climatic stresses is a vital option (Sarkar et al., 2017).

The biotic factors such as diseases, insects compete with plants for nutrients, water and weed compete for sunlight also hence reduces NUE. The allelopathic interaction between crop plants and weed exaggerates the problem further. The soil-borne pathogen creates stress which leads to alteration in morphology as well as the physiology of root and shoot, which in turn reduces the nutrient absorption capacity of plants. Leaf infecting insects and diseases negatively affect the photosynthetic capacity of plants which alters the utilization efficiency of absorbed nutrients.

10. Way forward for the improvement of crops for NUE

During the postgreen-revolution period, drastic modifications in environmental factors and the status of natural resources were recorded. The nonjudicious and uncontrolled application of chemical fertilizers caused gradual loss in soil fertility, and reduced use of organic amendments exacerbated the situation further. In addition to this, the low NUE of applied fertilizers adversely impacted agricultural productivity and sustainable production further. Development of improved crop varieties that showed efficient nutrient uptake from soil followed by their assimilation leads to reduced nutrient loss from agricultural fields hence; it will help in sustainable use of natural resources. NUE is a very complicated process comprised of signaling, acquisition, transportation, and utilization (assimilation and translocation/remobilization). Therefore deployment of information that is generated through the exploitation of existing natural genetic variation forward genetics and high-throughput sequencing is very much critical. To date, some varieties with improved ENA and ENU have been bred successfully, but all are based on traditional breeding and were mainly based on experimental knowledge. In the recent past, various sophisticated platforms were evolved that facilitates documentation of various characteristics with better precision. Gene expression data for numerous crops under a variable set of conditions are rapidly increasing. But there is a huge challenge of designing a suitable mathematical and computational approach for analyzing and integrating such huge data for effective utilization in crop improvement. By studying genetic variation, candidate genes governing ENA and ENU are identified, which confirm the existence of natural mechanisms for tolerating nutrient stresses in a variety of organisms. Genome-wide association studies are also a potential tool for the detection of QTLs governing traits associated with NUE with high statistical significance. There are only a few studies based on GWAS analysis that too for macronutrient N, P, and K exclusively (Ertiro et al., 2020; Karunarathne et al., 2020; Rakotoson et al., 2021; Reddy et al., 2021; Thudi et al., 2021) hence more such studies are needed in future for identification of genomic regions governing micronutrient use efficiency traits.

In cereals there are various open accesses genomic resources are already created; likewise, a web resource comprised of exclusive information of NUE genes or QTLs, i.e., OMIC TOOLS with unique identifier OMICS_24069 is developed by (Kumar et al., 2018). These databases work as a comprehensive repository of genes related to NUE component traits, including uptake, transportation, translocation, nitrate assimilation, signaling and regulation, reassimilation, remobilization, C/N storage, metabolism amino acid biosynthesis, DNA-binding, ion-binding, senescence, and copper homeostasis. The information regarding various QTLs related to shoot growth, relative grain and biomass yield, amino acid contents, total

N, comparative grain and biomass N, nitrate, N response, Zn and Fe homeostasis, and P starvation. The web resource also consists of information regarding various microRNAs targeting some of the NUE-related genes. The information of web resources will be critical for plant researchers, particularly for plant breeders, who may wish to improve the NUE of cereal crops. Hence more efforts are needed in the future for the development of such web resources for other crop species also. By releasing a large amount of several organic acids, plants can tolerate nutrient deficiency conditions. Hence, in the recent past, for development of transgenic plants that can acquire optimum concentration of nutrients from the soil by releasing large amounts of organic acids, work has been initiated (Chatzistathis et al., 2013). By genetic engineering of root traits that are associated with nutrients acquisition from the soil, transporters involved in uptake, mobilization, redistribution of nutrients, and enzymes linked with efficient assimilation, huge progress has been made in the direction of nutrient-efficient crop breeding (Iqrar et al., 2020; Wan et al., 2017). Recently by manipulating the genes linked with nutrients assimilation and regulation, improvements in NUE were observed. When the rice was grown under adequate nitrogen level, enhanced NUE ($\sim$40%) was recorded through the involvement of overexpression of nitrate transporter OsNRT2.3b (Fan et al., 2016). Similarly, the rice line genetically modified for barley Alanine Amino Transferase (HvAlaAT) gene showed high Nitrogen use efficiency, which was associated with overexpression of the modified gene (Tiong et al., 2021).

Some modern technologies such as is "CRISPR/Cas9" genome editing are providing great opportunities for crop improvement in various aspects, and various successful genome editing studies have been conducted for economically important crops. This technique is highly suitable for all crop species irrespective of their ploidy level or duration of life cycles. Time-consuming backcrossing and difficulties in pyramiding multiple traits in single genotypes are some disadvantages of traditional breeding, which are not associated with the CRISPR/Cas9 genome-editing tool. This tool facilitates the pyramiding of desirable alleles in an individual in a limited time period through the modification of several genes simultaneously. In addition, the replacement of genes of one species with another and more precise transgene integration in desirable locations is some complex genetic modification facilitated by this tool. Although to date, only a few efforts are made in the direction of NUE via genome editing that too in rent years. Such as, in a recent study by application of CRISPR-based cytosine base editing system (CBE) rice Nitrate Transporter gene NRT1.1B was edited by changing the amino acid positioned at 327 (Thr to Met), the resulted mutant displayed improved NUE than wild type rice (Lu et al., 2017). Recently genome editing was performed in potatoes for obtaining improved NUE genotype, which showed improved nutrient transport and use efficiency which reflected in the form of plant growth and productivity (Tiwari et al., 2020). Likewise, CRISPR-Cas9 genome editing systems could serve as an efficient tool for improving the NUE of other crop species also. It is speculated that in the coming years, we will witness intensive genome editing research in most crop species to improve ENA and ENU.

11. Conclusion

Adequate and balance application of essential nutrients is key component of improved crop productivity. Plants that are efficient nutrient absorber and utilizer enhance the efficiency of applied fertilizers hence preventing losses of nutrients resulting reduced cost of inputs. In future importance of nutrient efficiency will be increasing as under increasing fertilizer prices and the influence of a changing climate, farmers strive to maintain profitable agriculture by achieving higher productivity levels. Hence improving NUE is a key component from an agronomic, economic and environmental viewpoint. One possible direction is biotechnology through which fertilizer use and plant nutrient uptake efficiency can be increased. Despite of highly complex nature of NUE, recent improvement in biotechnological tools have facilitated the dissection of the polygenic/complex traits. The rapid development of an array of molecular and genomic techniques provides an opportunity to overcome the hurdles that have hindered progress so far. Dissection of molecular mechanism and identification of plant intrinsic factors governing nutrient absorption, transport, utilization, and mobilization in plant cultivars is cornerstone for designing molecular breeding program. By integrating SNP array with novel techniques including Next-generation sequencing (NGS) and Genotyping by sequencing (GBS) technologies, molecular breeding approaches are feasible solutions for identifying cultivars with superior NUE. The development of highly nutrient efficient cultivars, along with best management practices (BMPs) will contribute to sustainable agricultural systems. To overcome the internal and external constraints that are contributing to lower nutrient use efficiencies there is greater need of coordinated, multi-disciplinary, team effort for formulation of an effective system to make increased NUE in plants a reality.

References

Abreu, I., Poza, L., Bonilla, I., & Bolaños, L. (2014). Boron deficiency results in early repression of a cytokinin receptor gene and abnormal cell differentiation in the apical root meristem of *Arabidopsis thaliana*. *Plant Physiology and Biochemistry, 77*, 117–121.

Ai, P. H., Sun, S. B., Zhao, J. N., Fan, X. R., Xin, W. J., Guo, Q., Yu, L., Shen, Q. R., Wu, P., Miller, A. J., & Xu, G. H. (2009). Two rice phosphate transporters, *OsPht1; 2* and *OsPht1; 6*, have different functions and kinetic properties in uptake and translocation. *The Plant Journal, 57*, 798–809.

Alejandro, S., Höller, S., Meier, B., & Peiter, E. (2020). Manganese in plants: From acquisition to subcellular allocation. *Frontiers in Plant Science, 300*.

Aluwihare, C., Ishan, M., Chamikara, M. D. M., Weebadde, C. K., Sirisena, D. N., & Samarasinghe, W. L. G. (2016). Characterization and selection of phosphorus deficiency tolerant rice genotypes in Sri Lanka. Rice. *Science, 23*(4), 184–195.

Asins, M. J., Villalta, I., Aly, M. M., Olias, R., De Morales, P. A., Huertas, R., Li, J., Jaime-Perez, N., Haro, R., Raga, V., Carbonell, E. A., & Belver, A. (2013). Two closely linked tomato HKT coding genes are positional candidates for the major tomato QTL involved in Na+/K+ homeostasis. *Plant Cell Environment, 36*, 1171–1191.

Azevedo, G. C., Cheavegatti-Gianotto, A., Negri, B. F., Hufnagel, B., e Silva, L. D. C., Magalhaes, J. V., Garcia, A. A. F., Lana, U. G. P., de Sousa, S. M., & Guimaraes, C. T. (2015). Multiple interval QTL mapping and searching for *PSTOL1* homologs associated with root morphology, biomass accumulation and phosphorus content in maize seedlings under low-P. *BMC Plant Biology, 15*, 1–17.

Baligar, V. C., Fageria, N. K., & He, Z. L. (2001). Nutrient use efficiency in plants. *Communications in Soil Science and Plant Analysis, 32*(8), 921–950.

Balint, T., & Rengel, Z. (2009). Differential sulfur efficiency in canola genotypes at vegetative and grainmaturity stage. *Crop and Pasture Science, 60*, 262–270.

Bao, A., Liang, Z., Zhao, Z., & Cai, H. (2015). Overexpressing of *OsAMT1-3*, a high affinity ammonium transporter gene, modifies rice growth and carbon-nitrogen metabolic status. *International Journal of Molecular Sciences, 16*(12), 9037–9063.

Baxter, I., Muthukumar, B., Park, H. C., Buchner, P., Lahner, B., Danku, J., Zhao, K., Lee, J., Hawkesford, M. J., Guerinot, M. Lou, & Salt, D. E. (2008). Variation in molybdenum content across broadly distributed populations of *Arabidopsis thaliana* is controlled by a mitochondrial molybdenum transporter (*MOT1*). *PLOS Genetics, 4*(2), e1000004.

Benatti, M. R., Yookongkaew, N., Meetam, M., Guo, W.-J., Punyasuk, N., AbuQamar, S., & Goldsbrough, P. (2014). Metallothionein deficiency impacts copper accumulation and redistribution in leaves and seeds of *Arabidopsis*. *New Phytologist, 202*(3), 940–951.

Bernal, M., Casero, D., Singh, V., Wilson, G. T., Grande, A., Yang, H., Dodani, S. C., Pellegrini, M., Huijser, P., Connolly, E. L., Merchant, S. S., & Krämer, U. (2012). Transcriptome sequencing identifies *SPL7*-regulated copper acquisition genes *FRO4/FRO5* and the copper dependence of iron homeostasis in *Arabidopsis*. *The Plant Cell, 24*(2), 738.

Bhattacharyya, R., Ghosh, B. N., Mishra, P. K., Mandal, B., Rao, C. S., Sarkar, D., Das, K., Anil, K. S., Lalitha, M., Hati, K. M., & Franzluebbers, A. J. (2015). Soil degradation in India: Challenges and potential solutions. *Sustainability*, 7(4), 3528–3570.

Bittner, F. (2014). Molybdenum metabolism in plants and crosstalk to iron. *Frontiers in Plant Science, 28*.

Blair, G. (1993). Nutrient efficiency - what do we really mean?. In *Genetic aspects of plant mineral nutrition* (pp. 205–213). Dordrecht: Kluwer Academic.

Broadley, M. R., White, P. J., Hammond, J. P., Zelko, I., & Lux, A. (2007). Zinc in plants. *New Phytologist, 173*(4), 677–702.

Brown, P. H., & Bassil, E. (2011). Overview of the acquisition and utilization of boron, chlorine, copper, manganese, molybdenum, and nickel by plants and prospects for

improvement of micronutrient use efficiency. In *The molecular and physiological basis of nutrient use efficiency in crops* (pp. 377–428). John Wiley & Sons, Inc.
Burkhead, J. L., Reynolds, K. A. G., Abdel - Ghany, S. E., Cohu, C. M., & Pilon, M. (2009). Copper homeostasis. *The New Phytologist, 182*, 799–816.
Castaings, L., Caquot, A., Loubet, S., & Curie, C. (2016). The high-affinity metal transporters *NRAMP1* and *IRT1* team up to take up iron under sufficient metal provision. *Scientific Reports, 6*(1), 1–11.
Chandrika, N. N. P., Sundaravelpandian, K., Yu, S. M., & Schmidt, W. (2013). *ALFIN-LIKE 6* is involved in root hair elongation during phosphate deficiency in Arabidopsis. *New Phytologist, 198*, 709–720.
Chang, J.-D., Huang, S., Yamaji, N., Zhang, W., Ma, J. F., & Zhao, F. J. (2020). *OsNRAMP1* transporter contributes to cadmium and manganese uptake in rice. *Plant, Cell & Environment, 43*(10), 2476–2491.
Chapin, F. S., III, Schulze, E. D., & Mooney, H. A. (1990). The ecology and economics of storage in plants. *Annual Review of Ecology and Systematics, 21*(1), 423–447.
Chatzistathis, T., & Therios, I. (2013). *How soil nutrient availability influences plant biomass and how biomass stimulation alleviates heavy metal toxicity in soils: The cases of nutrient use efficient genotypes and phytoremediators, respectively* (pp. 427–448). Rijeka, Croatia: Biomass Now - Cultivation and Utilization Intech Open.
Chatzistathis, T., Therios, I., & Alifragis, D. (2009). Differential uptake, distribution within tissues, and use efficiency of manganese, iron, and zinc by olive cultivars kothreiki and koroneiki. *Hort Science, 44*(7), 1994–1999.
Chen, Z., Fujii, Y., Yamaji, N., Masuda, S., Takemoto, Y., Kamiya, T., Yusuyin, Y., Iwasaki, K., Kato, S., Maeshima, M., Ma, J. F., & Ueno, D. (2013). Mn tolerance in rice is mediated by *MTP8.1*, a member of the cation diffusion facilitator family. *Journal of Experimental Botany, 64*(14), 4375–4387.
Chen, Z. H., Jenkins, G. I., & Nimmo, H. G. (2008). Identification of an F-box protein that negatively regulates Pi starvation responses. *Plant & Cell Physiology, 49*, 1902–1906.
Chen, S., Xia, G., & Zhang, G. (2008). Nutrition accumulation, remobilization, and partitioning in rice on no-tillage soil. *Journal of Plant Nutrition, 31*(11), 2044–2058.
Ciampitti, I., & Vyn, T. (2013). Maize nutrient accumulation and partitioning in response to plant density and nitrogen rate: II. Calcium, magnesium, and micronutrients. *Agronomy Journal, 105*, 1645.
Connolly, E. L., Campbell, N. H., Grotz, N., Prichard, C. L., & Guerinot, M. Lou (2003). Overexpression of the *FRO2* ferric chelate reductase confers tolerance to growth on low iron and uncovers posttranscriptional control. *Plant Physiology, 133*(3), 1102.
Dai, X., Wang, Y., Yang, A., & Zhang, W. H. (2012). *OsMYB2P-1*, an R2R3 MYB transcription factor, is involved in the regulation of phosphate-starvation responses and root architecture in rice. *Plant Physiology, 159*, 169–183.
Delhaize, E., Taylor, P., Hocking, P. J., Simpson, R. J., Ryan, P. R., & Richardson, A. E. (2009). Transgenic barley (*Hordeum vulgare* L.) expressing the wheat aluminium resistance gene (*TaALMT1*) shows enhanced phosphorus nutrition and grain production when grown on an acid soil. *Plant Biotechnology Journal, 7*, 391–400.
Den, Herder, G., Van Isterdael, G., Beeckman, T., & De Smet, I. (2010). The roots of a new green revolution. *Trends in Plant Science, 15*, 600–607.
Devaiah, B. N., Nagarajan, V. K., & Raghothama, K. G. (2007). Phosphate homeostasis and root development in Arabidopsis are synchronized by the zinc finger transcription factor *ZAT6*. *Plant Physiology, 145*, 147–159.
Drew, M. C. (1975). Comparison of the effects of a localised supply of phosphate, nitrate, ammonium and potassium on the growth of the seminal root system, and the shoot, in barley. *New Phytologist, 75*, 479–490.

Eren, E., & Argüello, J. M. (2004). *Arabidopsis HMA2*, a divalent heavy metal-transporting P(IB)-type ATPase, is involved in cytoplasmic Zn^{2+} homeostasis. *Plant Physiology, 136*(3), 3712–3723.

Ertiro, B. T., Labuschagne, M., Olsen, M., Das, B., Prasanna, B. M., & Gowda, M. (2020). Genetic dissection of nitrogen use efficiency in tropical maize through genome-wide association and genomic prediction. *Frontiers in Plant Science, 474*.

Escudero, A., & Mediavilla, S. (2003). Decline in photosynthetic nitrogen use efficiency with leaf age and nitrogen resorption as determinants of leaf life span. *Journal of Ecology, 91*(5), 880–889.

Faber, B. A., Zasoski, R. J., Burau, R. G., & Uriu, K. (1990). Zinc uptake by corn as affected by vesicular-arbuscular mycorrhizae. *Plant and Soil, 129*(2), 121–130.

Fang, Y., Wu, W., Zhang, X., Jiang, H., Lu, W., Pan, J., Hu, J., Guo, L., Zeng, D., & Xue, D. (2015). Identification of quantitative trait loci associated with tolerance to low potassium and related ions concentrations at seedling stage in rice (*Oryza sativa* L.). *Plant Growth Regulation, 77*, 157–166.

Fan, X., Xie, D., Chen, J., Lu, H., Xu, Y., Ma, C., & Xu, G. (2014). Overexpression of *OsPTR6* in rice increased plant growth at different nitrogen supplies but decreased nitrogen use efficiency at high ammonium supply. *Plant Science, 227*, 1-1.

Fan, X., Tang, Z., Tan, Y., Zhang, Y., Luo, B., Yang, M., Lian, X., Shen, Q., Miller, A. J., & Xu, G. (2016). Overexpression of a pH-sensitive nitrate transporter in rice increases crop yields. *Proceedings of the National Academy of Sciences, 113*(26), 7118–7123.

Fazio, G., Kviklys, D., Grusak, M. A., & Robinson, T. (2013). Phenotypic diversity and QTL mapping of absorption and translocation of nutrients by apple rootstocks. *Aspects of Applied Biology, 119*, 37–50.

Feng Shao, J., Yamaji, N., Fang Shen, R., & Feng Ma, J. (2017). The key to Mn homeostasis in plants: Regulation of Mn transporters. *Trends in Plant Science, 22*(3), 215–224.

Feng, H., Fan, X., Miller, A. J., & Xu, G. (2020). Plant nitrogen uptake and assimilation: Regulation of cellular pH homeostasis. *Journal of Experimental Botany, 71*(15), 4380–4392.

Fiaz, S., Wang, X., Khan, S. A., Ahmar, S., Noor, M. A., Riaz, A., & Alharthi, B. (2021). *Novel plant breeding techniques to advance nitrogen use efficiency in rice: A review* (pp. 1–20). GM Crops & Food.

Fitzpatrick, K. L., Tyerman, S. D., & Kaiser, B. N. (2008). Molybdate transport through the plant sulfate transporter *SHST1*. *FEBS Letters, 582*(10), 1508–1513.

Frommer, W. B., & von Wirén, N. (2002). Ping-pong with boron. *Nature, 420*(6913), 282–283.

Gamuyao, R., Chin, J. H., Pariasca-Tanaka, J., Pesaresi, P., Catausan, S., Dalid, C., Slamet-Loedin, I., Tecson-Mendoza, E. M., Wissuwa, M., & Heuer, S. (2012). The protein kinase Pstol1 from traditional rice confers tolerance of phosphorus deficiency. *Nature, 488*, 535–539.

Gao, Y., Quan, S., Lyu, B., Tian, T., Liu, Z., Nie, Z., & Wang, Y. (2021). Barley transcription factor *HvNLP2* mediates nitrate signaling and affects nitrogen use efficiency. *Journal of Experimental Botany, 245*.

Garcia-Molina, A., Andrés-Colás, N., Perea-García, A., Neumann, U., Dodani, S. C., Huijser, P., Peñarrubia, L., & Puig, S. (2013). The *Arabidopsis COPT6* transport protein functions in copper distribution under copper-deficient conditions. *Plant and Cell Physiology, 54*(8), 1378–1390.

Gasber, A., Klaumann, S., Trentmann, O., Trampczynska, A., Clemens, S., Schneider, S., Sauer, N., Feifer, I., Bittner, F., Mendel, R. R., & Neuhaus, H. E. (2011). Identification of an *Arabidopsis* solute carrier critical for intracellular transport and inter-organ allocation of molybdate. *Plant Biology, 13*(5), 710–718.

Gerloff, G. C. (1983). Genetic basis of inorganic plant nutrition. In *Encyclopedia of plant physiology* (pp. 453–480). New York: Springer, Verlag.
Giehl, R. F., & von Wiren, N. (2014). Root nutrient foraging. *Plant Physiology, 166*, 509–517.
Gong, X. P., Liang, X., Guo, Y., Wu, C. H., Zhao, Y., Li, X. H., Li, S. S., & Kong, F. M. (2015). Quantitative trait locus mapping for potassium use efficiency traits at the seedling stage in wheat under different nitrogen and phosphorus treatments. *Crop Science, 55*, 2690–2700.
Graham, R. D. (1984). Breeding for nutritional characteristics in cereals. *Advanced Plant Nutrition, 1*, 57–102.
Graham, R. D., & Nambiar, E. K. S. (1981). Advances in research on copper deficiency in cereals. *Australian Journal of Agricultural Research, 32*, 1009–1037.
Guo, Y., Kong, F. M., Xu, Y. F., Zhao, Y., Liang, X., Wang, Y. Y., An, D. G., & Li, S. S. (2012). QTL mapping for seedling traits in wheat grown under varying concentrations of N, P and K nutrients. *Theoretical and Applied Genetics, 124*, 851–865.
Guo, M. N., Ruan, W. Y., Li, C. Y., Huang, F. L., Zeng, M., Liu, Y. Y., Yu, Y. N., Ding, X. M., Wu, Y. R., Wu, Z. C., Mao, C. Z., Yi, K. K., Wu, P., & Mo, X. R. (2015). Integrative comparison of the role of the *PHOSPHATE RESPONSE1* subfamily in phosphate signaling and homeostasis in rice. *Plant Physiology, 168*, 1762–1776.
Habash, D. Z., Massiah, A. J., Rong, H. L., Wallsgrov, R. M., & Leigh, R. A. (2001). The role of cytosolic glutamine synthetase in wheat. *Annals of Applied Biology, 138*(1), 83–89.
Hanikenne, M., Talke, I. N., Haydon, M. J., Lanz, C., Nolte, A., Motte, P., Kroymann, J., Weigel, D., & Krämer, U. (2008). Evolution of metal hyperaccumulation required cis-regulatory changes and triplication of *HMA4*. *Nature, 453*(7193), 391–395.
Ha, H., Olson, J. R., Bian, L., & Rogerson, P. A. (2014). Analysis of heavy metal sources in soil using kriging interpolation on principal components. *Environmental Science and Technology, 48*(9), 4999–5007.
Hawkesford, M. J., & Barraclough, P. (2011). *The molecular and physiological basis of nutrient use efficiency in crops*. John Wiley & Sons, Inc.
He, X., Qu, B., Li, W., Zhao, X., Teng, W., Ma, W., Ren, Y., Li, B., Li, Z., & Tong, Y. (2015). The nitrate-inducible NAC transcription factor *TaNAC2-5A* controls nitrate response and increases wheat yield. *Plant Physiology, 169*, 1991–2005.
Hochholdinger, F., & Tuberosa, R. (2009). Genetic and genomic dissection of maize root development and architecture. *Current Opinion in Plant Biology, 12*, 172–177.
Hodge, A., Berta, G., Doussan, C., Merchan, F., & Crespi, M. (2009). Plant root growth, architecture and function. *Plant and Soil, 321*(1), 153–187.
Huang, K. L., Ma, G. J., Zhang, M. L., Xiong, H., Wu, H., Zhao, C. Z., Liu, C. S., Jia, H. X., Chen, L., Kjorven, J. O., & Li, X. B. (2018). The *ARF7* and *ARF19* transcription factors positively regulate phosphate starvation response1 in Arabidopsis roots. *Plant Physiology, 178*, 413–427.
Hufnagel, B., de Sousa, S. M., Assis, L., Guimaraes, C. T., Leiser, W., Azevedo, G. C., Negri, B., Larson, B. G., Shaff, J. E., Pastina, M. M., & Barros, B. A. (2014). Duplicate and conquer: Multiple homologs of phosphorus-starvation tolerance1 enhance phosphorus acquisition and sorghum performance on low-phosphorus soils. *Plant Physiology, 166*(2), 659–677.
Hu, S. H., Lin, S. F., Huang, Y. C., Huang, C. H., Kuo, W. Y., & Jinn, T. L. (2021). Significance of *AtMTM1* and *AtMTM2* for mitochondrial MnSOD activation in *Arabidopsis*. *Frontiers in Plant Science, 1629*.
Hu, B., Wang, W., Ou, S., Tang, J., Li, H., Che, R., Zhang, Z., Chai, X., Wang, H., & Wang, Y. (2015). Variation in *NRT1.1B* contributes to nitrate-use divergence between rice subspecies. *Natural Genetics, 47*(7), 834-438.

Iqrar, S., Ashrafi, K., Kiran, U., Fatima, S., Kamaluddin, & Abdin, M. Z. (2020). Transgenic plants with improved nutrient use efficiency. *Transgenic Technology Based Value Addition in Plant Biotechnology*, 157−185.

Jeong, J., & Guerinot, M. Lou (2009). Homing in on iron homeostasis in plants. *Trends in Plant Science, 14*(5), 280−285.

Jiang, Z. (2006). Mn use efficiency in different wheat cultivars. *Environmental and Experimental Botany*, 574150.

Kale, R. R., Durga Rani, C. V., Anila, M., Mahadeva Swamy, H. K., Bhadana, V. P., Senguttuvel, P., Subrahmanyam, D., Ayyappa Dass, M., Swapnil, K., Anantha, M. S., & Punniakotti, E. (2021). Major QTLs associated with low soil phosphorus tolerance identified from the Indian rice landrace, Wazuhophek. *PLoS One, 16*(7), e0254526.

Karunarathne, S. D., Han, Y., Zhang, X.-Q., Zhou, G., Hill, C. B., Chen, K., Angessa, T., & Li, C. (2020). Genome-wide association study and identification of candidate genes for nitrogen use efficiency in barley (*Hordeum vulgare* L.). *Frontiers in Plant Science*, 1361.

Khademi, S., O'Connell, J., Remis, J., Robles-Colmenares, Y., Miericke, L. J. W., & Stroud, R. M. (2004). Mechanism of ammonia transport by amt/MEP/Rh: Structure of AmtB at 1.3.5 angstrom. *Science, 305*, 1587−1594.

Khan, M. S. K., Saeed, M., & Iqbal, J. (2015). Identification of quantitative trait loci for Na+, K+ and Ca^{++} accumulation traits in rice grown under saline conditions using F_2 mapping population. *Brazilian Journal of Botany, 38*, 555−565.

Kindu, G. A., Tang, J., Yin, X., & Struik, P. C. (2014). Quantitative trait locus analysis of nitrogen use efficiency in barley (*Hordeum vulgare* L.). *Euphytica, 199*(1−2), 207−221.

Koike, S., Inoue, H., Mizuno, D., Takahashi, M., Nakanishi, H., Mori, S., & Nishizawa, N. K. (2004). *OsYSL2* is a rice metal-nicotianamine transporter that is regulated by iron and expressed in the phloem. *The Plant Journal, 39*(3), 415−424.

Kong, F. M., Guo, Y., Liang, X., Wu, C. H., Wang, Y. Y., Zhao, Y., & Li, S. S. (2013). Potassium (K) effects and QTL mapping for K efficiency traits at seedling and adult stages in wheat. *Plant and Soil, 373*(1), 877−892.

Kumar, A., Pandeya, A., Malik, G., Sharma, M., P.,H.K., S.,A.K., Gahlaut, V., Gajula, M. N. V. P., Singh, K. P., Suravajhala, P., Balyan, H. S., & Gupta, P. K. (2018). A web resource for nutrient use efficiency-related genes, quantitative trait loci and microRNAs in important cereals and model plants. *F1000 Research, 7*, 673.

Laurie, S. H., Tancock, N. P., McGrath, S. P., & Sanders, J. R. (1995). Influence of EDTA complexation on plant uptake of manganese (II). *Plant Science, 109*(2), 231−235.

Lea, P. J., & Azevedo, R. A. (2007). Nitrogen use efficiency. 2. Amino acid metabolism. *Annals of Applied Biology, 151*(3), 269−275.

Lehmann, A., & Rillig, M. C. (2015). Arbuscular mycorrhizal contribution to copper, manganese and iron nutrient concentrations in crops - a meta-analysis. *Soil Biology and Biochemistry, 81*, 147−158.

Léran, S., Varala, K., Boyer, J. C., Chiurazzi, M., Crawford, N., Daniel-Vedele, F., David, L., Dickstein, R., Fernandez, E., Forde, B., Gassmann, W., Geiger, D., Gojon, A., Gong, J. M., Halkier, B. A., Harris, J. M., Hedrich, R., Limami, A. M., Rentsch, D., Seo, M., & Lacombe, B. (2014). A unified nomenclature of nitrate transporter 1/peptide transporter family members in plants. *Trends in Plant Science, 19*(1), 5−9.

Liang, Q. A., Cheng, X. H., Mei, M. T., Yan, X. L., & Liao, H. (2010). QTL analysis of root traits as related to phosphorus efficiency in soybean. *Annals of Botany, 106*, 223−234.

Liang, C., Wang, Y., Zhu, Y., Tang, J., Hu, B., Liu, L., Ou, S., Wu, H., Sun, X., Chu, J., & Chu, C. (2014). *OsNAP* connects abscisic acid and leaf senescence by fine-tuning abscisic acid biosynthesis and directly targeting senescence-associated genes in rice. *Proceedings of the National Academy of Sciences of the United States of America, 111*, 10013−10018.

Li, H., Huang, G., Meng, Q., Ma, L., Yuan, L., Wang, F., Zhang, W., Cui, Z., Shen, J., Chen, X., Jiang, R., & Zhang, F. (2011). Integrated soil and plant phosphorus management for crop and environment in China. A review. *Plant and Soil, 349*, 157–167.

Li, B. Z., Merrick, M., Li, S. M., Li, H. Y., Zhu, S. W., Shi, W. M., & Su, Y. H. (2009). Molecular basis and regulation of ammonium transporter in rice. *Rice Science, 16*, 314–322.

Lin, C. M., Koh, S., Stacey, G., Yu, S. M., Lin, T. Y., & Tsay, Y. F. (2000). Cloning and functional characterization of a constitutively expressed nitrate transporter gene, *OsNRT1*, from rice. *Plant Physiology, 122*(2), 379–388.

Li, S., Qian, Q., Fu, Z., Zeng, D., Meng, X., Kyozuka, J., & Wang, Y. (2009). Short panicle1 encodes a putative PTR family transporter and determines rice panicle size. *The Plant Journal, 58*(4), 592–605.

Liu, A., Hamel, C., Hamilton, R. I., Ma, B. L., & Smith, D. L. (2000). Acquisition of Cu, Zn, Mn and Fe by mycorrhizal maize (*Zea mays* L.) grown in soil at different P and micronutrient levels. *Mycorrhiza, 9*(6), 331–336.

Liu, S., Lacape, J. M., Constable, G. A., & Llewellyn, D. J. (2015). Inheritance and QTL mapping of leaf nutrient concentration in a cotton inter-specific derived RIL population. *PLoS One, 10*(5), e0128100.

Li, C., Wang, P., Menzies, N. W., Lombi, E., & Kopittke, P. M. (2017). Effects of changes in leaf properties mediated by methyl jasmonate (MeJA) on foliar absorption of Zn, Mn and Fe. *Annals of Botany, 120*(3), 405–415.

Li, X., Zeng, R., & Liao, H. (2015). Improving crop nutrient efficiency through root architecture modifications. *Journal of Integrative Plant Biology, 58*, 193–202.

López-Arredondo, D. L., Sánchez-Calderón, L., & Yong-Villalobos, L. (2017). Molecular and genetic basis of plant macronutrient use efficiency: Concepts, opportunities, and challenges. In *Plant macronutrient use efficiency*. Academic Press.

López-Bucio, J., Cruz-Ramírez, A., & Herrera-Estrella, L. (2003). The role of nutrient availability in regulating root architecture. *Current Opinion in Plant Biology, 6*(3), 280–287.

Lu, J., Gao, X., Dong, Z., Yi, J., & An, L. (2012). Improved phosphorus acquisition by tobacco through transgenic expression of mitochondrial malate dehydrogenase from. *Penicillium oxalicum. Plant Cell Reports, 31*, 49–56.

Luo, B., Ma, P., Nie, Z., Zhang, X., He, X., Ding, X., Feng, X., Lu, Q., Ren, Z., Lin, H., & Wu, Y. (2019). Metabolite profiling and genome-wide association studies reveal response mechanisms of phosphorus deficiency in maize seedling. *Plant Journal, 97*, 947–969.

Lu, Y., & Zhu, J. K. (2017). Precise editing of a target base in the rice genome using a modified CRISPR/Cas9 system. *Molecular Plant, 10*, 523–525.

Ma, Z., Liu, J., & Wang, X. (2013). Agrobacterium-mediated transformation of cotton (*Gossypium hirsutum*) shoot apex with a fungal phytase gene improves phosphorus acquisition. *Methods in Molecular Biology, 958*, 211–222.

Mandal, V. K., Sharma, N., & Raghuram, N. (2018). Molecular targets for improvement of crop nitrogen use efficiency: Current and emerging options. In *Engineering nitrogen utilization in crop plants* (pp. 77–93). Cham: Springer.

Manuel, T. J., Alejandro, C. A., Angel, L., Aurora, G., & Emilio, F. (2018). Roles of molybdenum in plants and improvement of its acquisition and use efficiency. In *Plant micronutrient use efficiency* (pp. 137–159). Academic Press.

Mardukhi, B., Rejali, F., Daei, G., Ardakani, M. R., Malakouti, M. J., & Miransari, M. (2011). Arbuscular mycorrhizas enhance nutrient uptake in different wheat genotypes at high salinity levels under field and greenhouse conditions. *Comptes Rendus Biologies, 334*(7), 564–571.

Marschner, H., & Dell, B. (1994). Nutrient uptake in mycorrhizal symbiosis. *Plant and Soil, 159*(1), 89–102.

Martin, A., Lee, J., Kichey, T., Gerentes, D., Zivy, M., Tatout, C., Dubois, F., Balliau, T., Valot, B., Davanture, M., & Tercé-Laforgue, T. (2006). Two cytosolic glutamine synthetase isoforms of maize are specifically involved in the control of grain production. *Plant Cell, 18*(11), 3252–3274.

Masclaux-Daubresse, C., Daniel-Vedele, F., Dechorgnat, J., Chardon, F., Gaufichon, L., & Suzuki, A. (2010). Nitrogen uptake, assimilation and remobilization in plants: Challenges for sustainable and productive agriculture. *Ann Bot, 105*(7), 1141–1157.

Ma, X. F., Wright, E., Ge, Y., Bell, J., Xi, Y., Bouton, J. H., & Wang, Z. Y. (2009). Improving phosphorus acquisition of white clover (*Trifolium repens* L.) by transgenic expression of plant-derived phytase and acid phosphatase genes. *Plant Science, 176*, 479–488.

Migocka, M., & Malas, K. (2018). Plant responses to copper: Molecular and regulatory mechanisms of copper uptake, distribution and accumulation in plants. In *Plant micronutrient use efficiency* (pp. 71–86). Academic Press.

Mitra, G. (2017). Essential plant nutrients and recent concepts about their uptake. In *Essential plant nutrients* (pp. 3–36). Cham: Springer.

Miwa, K., & Fujiwara, T. (2010). Boron transport in plants: Co-ordinated regulation of transporters. *Annals of Botany, 105*(7), 1103–1108.

Miwa, K., Takano, J., & Fujiwara, T. (2006). Improvement of seed yields under boron-limiting conditions through overexpression of *BOR1*, a boron transporter for xylem loading, in *Arabidopsis thaliana. The Plant Journal, 46*(6), 1084–1091.

Miwa, K., Wakuta, S., Takada, S., Ide, K., Takano, J., Naito, S., Omori, H., Matsunaga, T., & Fujiwara, T. (2013). Roles of *BOR2*, a boron exporter, in cross linking of rhamnogalacturonan II and root elongation under boron limitation in *Arabidopsis. Plant Physiology, 163*(4), 1699.

Miyamoto, T., Ochiai, K., Takeshita, S., & Matoh, T. (2012). Identification of quantitative trait loci associated with shoot sodium accumulation under low potassium conditions in rice plants. *Soil Science and Plant Nutrition, 58*, 728–736.

Moll, R. H., Kamprath, E. J., & Jackson, W. A. (1982). Analysis and interpretation of factors which contribute to efficiency of nitrogen utilization. *Agronomy Journal, 74*(3), 562–564.

Moreira, A., Moraes, L. A. C., & dos Reis, A. R. (2018). The molecular genetics of zinc uptake and utilization efficiency in crop plants. In *Plant micronutrient use efficiency* (pp. 87–108). Academic Press.

Morrissey, J., & Guerinot, M. Lou (2009). Iron uptake and transport in plants: The good, the bad, and the ionome. *Chemical Reviews, 109*(10), 4553.

Moubayidin, L., Di Mambro, R., & Sabatini, S. (2009). Cytokinin-auxin crosstalk. *Tree Plant Science, 14*, 557–562.

Mus, F., Crook, M. B., Garcia, K., Garcia Costas, A., Geddes, B. A., Kouri, E. D., Paramasivan, P., Ryu, M. H., Oldroyd, G. E., Poole, P. S., & Udvardi, M. K. (2016). Symbiotic nitrogen fixation and the challenges to its extension to non legumes. *Applied Environmental Microbiology, 82*, 3698–3710.

Neelam, K., Thakur, S., Neha, Yadav, I. S., Kumar, K., Dhaliwal, S. S., & Singh, K. (2017). Novel alleles of phosphorus-starvation tolerance 1 gene (*PSTOL1*) from *Oryza rufipogon* confers high phosphorus uptake efficiency. *Frontiers in Plant Science, 8*, 1–12.

Nguyen, V. L., Dolstra, O., Malosetti, M., Kilian, B., Graner, A., Visser, R. G. F., & van der Linden, C. G. (2013). Association mapping of salt tolerance in barley (*Hordeum vulgare* L.). *Theoretical and Applied Genetics, 126*, 2335–2351.

Nguyen, V. L., Ribot, S. A., Dolstra, O., Niks, R. E., Visser, R. G. F., & van der Linden, C. G. (2013b). Identification of quantitative trait loci for ion homeostasis and salt tolerance in barley (*Hordeum vulgare* L). *Molecular Breeding, 31*, 137–152.

Nussaume, L., Kanno, S., Javot, H., Marin, E., Pochon, N., Ayadi, A., Nakanishi, T. M., & Thibaud, M. C. (2011). Phosphate import in plants: Focus on the *PHT1* transporters. *Frontiers in Plant Science, 2*, 83.
O'Halloran, T. V., & Culotta, V. C. (2000). Metallochaperones, an intracellular shuttle service for metal ions. *Journal of Biological Chemistry, 275*(33), 25057−25060.
Ortíz-Castro, R., Contreras-Cornejo, H. A., Macías-Rodríguez, L., & LópezBucio, J. (2009). The role of microbial signals in plant growth and development. *Plant Signaling and Behavior, 4*, 701−712.
Ortiz-Monasterio, J. I., Manske, G. G. B., & van Ginkel, M. (2001). Nitrogen and phosphorus use efficiency. In *Applications of physiology in wheat breeding* (pp. 200−207). Mexico DF: CIMMYT.
Ouyang, X., Hong, X., Zhao, X., Zhang, W., He, X., Ma, W., Teng, W., & Tong, Y. (2016). Knock out of the *PHOSPHATE 2* gene *TaPHO2-A1* improves phosphorus uptake and grain yield under low phosphorus conditions in common wheat. *Scientific Reports, 6*, 29850.
Pant, B. D., Burgos, A., Pant, P., Cuadros-Inostroza, A., Willmitzer, L., & Scheible, W. R. (2015). The transcription factor *PHR1* regulates lipid remodeling and triacylglycerol accumulation in *Arabidopsis thaliana* during phosphorus starvation. *The Journal of Experimental Botany, 66*, 1907−1918.
Parihar, M., Meena, V. S., Mishra, P. K., Rakshit, A., Choudhary, M., Yadav, R. P., Rana, K., & Bisht, J. K. (2019). Arbuscular mycorrhiza: A viable strategy for soil nutrient loss reduction. *Archives of Microbiology, 201*(6), 723−735.
Parniske, M. (2008). Arbuscular mycorrhiza: The mother of plant root endosymbioses. *Nature Reviews Microbiology, 6*(10), 763−775.
Patiram. (2007). Management and future research strategies for enhancing the productivity of crops on the acid soils. *Journal of the Indian Society of Soil Science, 55*(4), 411−420.
Pittman, J. K. (2005). Managing the manganese: Molecular mechanisms of manganese transport and homeostasis. *New Phytologist, 167*(3), 733−742.
Printz, B., Lutts, S., Hausman, J. F., & Sergeant, K. (2016). Copper trafficking in plants and its implication on cell wall dynamics. *Frontiers in Plant Science, 7*, 1−16.
Pujarula, V., Pusuluri, M., Bollam, S., Das, R. R., Ratnala, R., Adapala, G., Thuraga, V., Rathore, A., Srivastava, R. K., & Gupta, R. (2021). Genetic variation for nitrogen use efficiency traits in global diversity panel and parents of mapping populations in pearl millet. *Frontiers in Plant Science, 12*, 73.
Qiu, H. B., Mei, X. P., Liu, C. X., Wang, J. G., Wang, G. Q., Wang, X., Liu, Z., & Cai, Y. L. (2013). Fine mapping of quantitative trait loci for acid phosphatase activity in maize leaf under low phosphorus stress. *Molecular Breeding, 32*, 629−639.
Qi, Y., Wang, S., Shen, C., Zhang, S., Chen, Y., Xu, Y., Liu, Y., Wu, Y., & Jiang, D. (2012). *OsARF12*, a transcription activator on auxin response gene, regulates root elongation and affects iron accumulation in rice (*Oryza sativa*). *New Phytologist, 193*(1), 109−120.
Qu, B., He, X., Wang, J., Zhao, Y., Teng, W., Shao, A., Zhao, X., Ma, W., Wang, J., Li, B., Li, Z., & Tong, Y. (2015). A wheat CCAAT box-binding transcription factor increases the grain yield of wheat with less fertilizer input. *Plant Physiology, 167*, 411−423.
Raghothama, K. G., & Karthikeyan, A. S. (2005). Phosphate acquisition. *Plant and Soil, 274*, 37−49.
Rakotoson, T., Dusserre, J., Letourmy, P., Frouin, J., Ratsimiala, I. R., Rakotoarisoa, N. V., cao, T. V., Vom Brocke, K., Ramanantsoanirina, A., Ahmadi, N., & Raboin, L. M. (2021). Genome-wide association study of nitrogen use efficiency and agronomic traits in upland rice. *Rice Science, 28*(4), 379−390.
Reddy, V. R. P., Das, S., Dikshit, H. K., Mishra, G. P., Aski, M. S., Singh, A., Tripathi, K., Pandey, R., Bansal, R., Singh, M. P., Gore, P. G., B.,M.P., Kothari, D., Rai, N., &

Nair, R. M. (2021). Genetic dissection of phosphorous uptake and utilization efficiency traits using GWAS in mungbean. *Agronomy, 11*(7), 1401.

Reguera, M., Bassil, E., Tajima, H., Wimmer, M., Chanoca, A., Otegui, M. S., Paris, N., & Blumwald, E. (2015). pH regulation by NHX-type antiporters is required for receptor-mediated protein trafficking to the vacuole in Arabidopsis. *The Plant Cell, 27*, 1200–1217.

Reich, M., Aghajanzadeh, T., & De Kok, L. J. (2014). Physiological basis of plant nutrient use efficiency-concepts, opportunities and challenges for its improvement. In *Nutrient use efficiency in plants* (pp. 1–27). Cham: Springer.

Ren, H. Z., Goa, J. P., Li, L. G., Cai, X. L., Huang, W., Chao, D. Y., Zhu, M. Z., Wang, Z. Y., Luan, S., & Lin, H. X. (2005). A rice quantitative trait locus for salt tolerance encodes a sodium transporter. *Nature Genetics, 37*, 1141–1146.

Rillig, M. C., & Mummey, D. L. (2006). Mycorrhizas and soil structure. *New Phytologist, 171*(1), 41–53.

Römheld, V. (1991). The role of phytosiderophores in acquiring iron and other micronutrients in gramineous species -an ecological approach. *Plant and Soil, 130*, 127–134.

Ryan, B. M., Kirby, J. K., Degryse, F., Harris, H., McLaughlin, M. J., & Scheiderich, K. (2013). Copper speciation and isotopic fractionation in plants: Uptake and translocation mechanisms. *New Phytologist, 199*(2), 367–378.

Ryu, M. H., Zhang, J., Toth, T., Khokhani, D., Geddes, B. A., Mus, F., Garcia-Costas, A., Peters, J. W., Poole, P. S., Ané, J. M., & Voigt, C. A. (2020). Control of nitrogen fixation in bacteria that associate with cereals. *Nature Microbiology, 5*, 314–330.

Sadeghzadeh, B. (2013). A review of zinc nutrition and plant breeding. *Journal of Soil Science and Plant Nutrition, 13*(4), 905–927.

Sadeghzadeh, B., & Rengel, Z. (2011). Zinc in soils and crop nutrition. In *The molecular and physiological basis of nutrient use efficiency in crops* (pp. 335–375). John Wiley & Sons, Inc.

Safdar, L. B., Andleeb, T., Latif, S., Umer, M. J., Tang, M., Li, X., Liu, S., & Quraishi, U. M. (2020). Genome-wide association study and QTL meta-analysis identified novel genomic loci controlling potassium use efficiency and agronomic traits in bread wheat. *Frontiers in Plant Science, 11*, 70.

Sarkar, D., & Baishya, L. K. (2017). Nutrient use efficiency. In *Essential plant nutrients* (pp. 119–146). Cham: Springer.

Sasaki, A., Yamaji, N., Yokosho, K., & Ma, J. F. (2012). *Nramp5* is a major transporter responsible for manganese and cadmium uptake in rice. *The Plant Cell, 24*(5), 2155.

Shahzad, Z., Amtmann, A., Chiou, T. J., & Fujiwara, T. (2017). Food for thought: How nutrients regulate root system architecture. *Current Opinion in Plant Biology, 39*, 80–87.

Shibagaki, N., & Grossman, A. R. (2006). The role of the *STAS* domain in the function and biogenesis of a sulfate transporter as probed by random mutagenesis. *Journal of Biological Chemistry, 281*(32), 22964–22973.

Shinmachi, F., Buchner, P., Stroud, J. L., Parmar, S., Zhao, F. J., McGrath, S. P., & Hawkesford, M. J. (2010). Influence of sulfur deficiency on the expression of specific sulfate transporters and the distribution of sulfur, selenium, and molybdenum in wheat. *Plant Physiology, 153*(1), 327.

Stangoulis, J. C. R., Grewal, H. S., Bell, R. W., & Graham, R. D. (2000). Boron efficiency in oilseed rape: I. Genotypic variation demonstrated in field and pot grown *Brassica napus* L. And *Brassica juncea* L. *Plant Soil, 225*, 243–251.

Sun, S. B., Gu, M. A., Cao, Y., Huang, X. P., Zhang, X., Ai, P. H., Zhao, J. N., Fan, X. R., & Xu, G. H. (2012). A constitutive expressed phosphate transporter, *OsPht1; 1*, modulates phosphate uptake and translocation in phosphate-replete rice. *Plant Physiology, 159*, 1571–1581.

Sun, H. Y., Qian, Q., Wu, K., Luo, J. J., Wang, S. S., Zhang, C. W., Ma, Y. F., Liu, Q., Huang, X. Z., Yuan, Q. B., Han, R. X., Zhao, M., Dong, G. J., Guo, L. B.,

Zhu, X. D., Gou, Z. H., Wang, W., Wu, Y. J., Lin, H. X., & Fu, X. D. (2014). Heterotrimeric G proteins regulate nitrogen-use efficiency in rice. *Nature Genetics, 46*, 652–656.

Tabata, R., Kamiya, T., Shigenobu, S., Yamaguchi, K., Yamada, M., Hasebe, M., Fujiwara, T., & Sawa, S. (2013). Identifying an EMS-induced causal mutation in a gene required for boron-mediated root development by low-coverage genome re-sequencing in *Arabidopsis*. *Plant Signaling & Behavior, 8*(1), 18–24.

Takano, J., Miwa, K., & Fujiwara, T. (2008). Boron transport mechanisms: Collaboration of channels and transporters. *Trends in Plant Science, 13*(8), 451–457.

Takano, J., Wada, M., Ludewig, U., Schaaf, G., von Wirén, N., & Fujiwara, T. (2006). The *Arabidopsis* major intrinsic protein *NIP5;1* is essential for efficient boron uptake and plant development under boron limitation. *The Plant Cell, 18*(6), 1498–1509.

Thudi, M., Chen, Y., Pang, J., Kalavikatte, D., Bajaj, P., Roorkiwal, M., Chitikineni, A., Ryan, M. H., Lambers, H., Siddique, K. H. M., & Varshney, R. K. (2021). Novel genes and genetic loci associated with root morphological traits, phosphorus-acquisition efficiency and phosphorus-use efficiency in chickpea. *Frontiers in Plant Science, 12*.

Tiong, J., Sharma, N., Sampath, R., MacKenzie, N., Watanabe, S., Metot, C., Lu, Z., Skinner, W., Lu, Y., Kridl, J., Baumann, U., Heuer, S., Kaiser, B., & Okamoto, M. (2021). Improving nitrogen use efficiency through overexpression of alanine aminotransferase in rice, wheat, and barley. *Frontiers in Plant Science, 29*.

Tiwari, J. K., Buckseth, T., Singh, R. K., Kumar, M., & Kant, S. (2020). Prospects of improving nitrogen use efficiency in potato: Lessons from transgenics to genome editing strategies in plants. *Frontiers in Plant Science, 11*, 597481.

Tomatsu, H., Takano, J., Takahashi, H., Watanabe-Takahashi, A., Shibagaki, N., & Fujiwara, T. (2007). An *Arabidopsis thaliana* high-affinity molybdate transporter required for efficient uptake of molybdate from the soil. *Proceedings of the National Academy of Sciences, 104*(47), 18807–18812.

Ueno, D., Sasaki, A., Yamaji, N., Miyaji, T., Fujii, Y., Takemoto, Y., Moriyama, S., Che, J., Moriyama, Y., Iwasaki, K., & Ma, J. F. (2015). A polarly localized transporter for efficient manganese uptake in rice. *Nature Plants, 1*(12), 1–8.

Walker, E. L., & Connolly, E. L. (2008). Time to pump iron: Iron-deficiency-signaling mechanisms of higher plants. *Current Opinion in Plant Biology, 11*(5), 530–535.

Wang, Z., Chen, Z., Cheng, J., Lai, Y., Wang, J., Bao, Y., Huang, J., & Zhang, H. (2012). QTL analysis of Na^+ and K^+ concentrations in roots and shoots under different levels of NaCl stress in rice (*Oryza sativa* L.). *PLoS One,* 7(12), e51202.

Wang, W., Hu, B., Yuan, D., Liu, Y., Che, R., Hu, Y., Ou, S., Liu, Y., Zhang, Z., Wang, H., & Li, H. (2018). Expression of the nitrate transporter gene *OsNRT1.1A/OsNPF6.3* confers high yield and early maturation in rice. *Plant Cell, 30*, 638–651.

Wang, X. R., Wang, Y. X., Tian, J., Lim, B. L., Yan, X. L., & Liao, H. (2009). Overexpressing *AtPAP15* enhances phosphorus efficiency in soybean. *Plant Physiology, 151*, 233–240.

Wang, Y., & Wu, W. H. (2013). Potassium transport and signaling in higher plants. *Annual Review of Plant Biology, 64*, 451–476.

Wang, Q. J., Yuan, Y., Liao, Z., Jiang, Y., Wang, Q., Zhang, L., Gao, S., Wu, F., Li, M., Xie, W., & Liu, T. (2019). Genome-wide association study of 13 traits in maize seedlings under low phosphorus stress. *Plant Genome, 12*, 1–13.

Wan, T. E. N. G., Xue, H. E., & Tong, Y. P. (2017). Transgenic approaches for improving use efficiency of nitrogen, phosphorus and potassium in crops. *Journal of Integrative Agriculture, 16*(12), 2657–2673.

Wasaki, J., Maruyama, H., Tanaka, M., Yamamura, T., Dateki, H., Shinano, T., Ito, S., & Osaki, M. (2009). Overexpression of the *LASAP2* gene for secretory acid phosphatase in white lupin improves the phosphorus uptake and growth of tobacco plants. *Soil Science and Plant Nutrition, 55*, 107–113.

Waters, B., & Grusak, M. (2007). Whole - plant mineral partitioning throughout the life cycle in *Arabidopsis thaliana* ecotypes Columbia, Landsberg erecta, Cape Verde Islands, and the mutant line *ysl1ysl3*. *The New Phytologist, 177*, 389–405.

White, P. J., & Bell, M. J. (2017). The genetics of potassium uptake and utilization in plants. In T. S. Murrell, & R. L. Mikkelsen (Eds.), *Proceedings for the frontiers of potassium science conference, 25-27 January 2017, Rome* (pp. 46–65). Peachtree Corners: International Plant Nutrition Institute.

White, P. J., & Broadley, M. R. (2009). Biofortification of crops with seven mineral elements often lacking in human diets - iron, zinc, copper, calcium, magnesium, selenium and iodine. *New Phytologist, 182*(1), 49–84.

White, P. J., Hammond, J. P., King, G. J., Bowen, H. C., Hayden, R. M., Meacham, M. C., Spracklen, W. P., & Broadley, M. R. (2010). Genetic analysis of potassium use efficiency in *Brassica oleracea*. *Annals of Botany, 105*, 1199–1210.

Wintz, H., Fox, T., Wu, Y.-Y., Feng, V., Chen, W., Chang, H.-S., Zhu, T., & Vulpe, C. (2003). Expression profiles of *Arabidopsis thaliana* in mineral deficiencies reveal novel transporters involved in metal homeostasis. *Journal of Biological Chemistry, 278*(48), 47644–47653.

Wu, C., Lu, L., Yang, X., Feng, Y., Wei, Y., Hao, H., Stoffella, P. J., & He, Z. (2010). Uptake, translocation, and remobilization of zinc absorbed at different growth stages by rice genotypes of different Zn densities. *Journal of Agricultural and Food Chemistry, 58*(11), 6767–6773.

Xu, C., Zhang, H., Sun, J., Guo, Z., Zou, C., Li, W. X., Xie, C., Huang, C., Xu, R., Liao, H., & Wang, J. (2018). Genome-wide association study dissects yield components associated with low-phosphorus stress tolerance in maize. *Theoretical and Applied Genetics, 131*, 1699–1714.

Yamaji, N., Sasaki, A., Xia, J. X., Yokosho, K., & Ma, J. F. (2013). A node-based switch for preferential distribution of manganese in rice. *Nature Communications, 4*(1), 1–11.

Yanagisawa, S., Akiyama, A., Kisaka, H., Uchimiya, H., & Miwa, T. (2004). Metabolic engineering with *Dof1* transcription factor in plants: Improved nitrogen assimilation and growth under low-nitrogen conditions. *Proceedings of the National Academy of Sciences of the United States of America, 101*, 7833–7838.

Yan, M., Fan, X., Feng, H., Miller, A. J., Shen, Q., & Xu, G. (2011). Rice *OsNAR2.1* interacts with *OsNRT2.1, OsNRT2.2* and *OsNRT2.3* nitrate transporters to provide uptake over high and low concentration ranges. *Plant Cell Environment, 34*(8), 1360–1372.

Yi, K., Wu, Z., Zhou, J., Du, L., Guo, L., Wu, Y., & Wu, P. (2005). *OsPTF1*, a novel transcription factor involved in tolerance to phosphate starvation in rice. *Plant Physiology, 138*, 2087–2096.

Yuan, Y., Gao, M., Zhang, M., Zheng, H., Zhou, X., Guo, Y., Zhao, Y., Kong, F., & Li, S. (2017). QTL mapping for phosphorus efficiency and morphological traits at seedling and maturity stages in wheat. *Frontiers in Plant Science, 8*, 614.

Yu, C. Y., Liu, Y. H., Zhang, A. D., Su, S., Yan, A., Huang, L., Ali, I., Liu, Y., Forde, B. G., & Gan, Y. B. (2015). MADS-box transcription factor *OsMADS25* regulates root development through affection of nitrate accumulation in rice. *PLOS One, 10*, e0135196.

Yu, L. H., Wu, J., Tang, H., Yuan, Y., Wang, S. M., Wang, Y. P., Zhu, Q. S., Li, S. G., & Xiang, C. B. (2016). Over expression of Arabidopsis *NLP7* improves plant growth under both nitrogen-limiting and sufficient conditions by enhancing nitrogen and carbon assimilation. *Scientific Reports, 6*, 27795.

Zdunić, Z., Grljušić, S., Ledenčan, T., Duvnjak, T., & Šimić, D. (2014). Quantitative trait loci mapping of metal concentrations in leaves of the maize IBM population. *Hereditas, 151*, 55–60.

Zhai, Z., Gayomba, S. R., Jung, H., Vimalakumari, N. K., Piñeros, M., Craft, E., Rutzke, M. A., Danku, J., Lahner, B., Punshon, T., Guerinot, M., Lou, Salt, D. E., Kochian, L. V., & Vatamaniuk, O. K. (2014). *OPT3* is a phloem-specific iron transporter that is essential for systemic iron signaling and redistribution of iron and cadmium in *Arabidopsis*. *The Plant Cell, 26*(5), 2249–2264.

Zhang, M., Gao, M., Zheng, H., Yuan, Y., Zhou, X., Guo, Y., Zhang, G., Zhao, Y., Kong, F., An, Y., & Li, S. (2019). QTL mapping for nitrogen use efficiency and agronomic traits at the seedling and maturity stages in wheat. *Molecular Breeding, 39*(5), 71.

Zhang, D., Song, H. N., Cheng, H., Hao, D. R., Wang, H., Kan, G. Z., Jin, H. X., & Yu, D. Y. (2014). The acid phosphatase-encoding gene *GmACP1* contributes to soybean tolerance to low-phosphorus stress. *PLoS Genetics, 10*, e1004061.

Zhang, G. D., Wang, X. P., Wang, B., Tian, Y. C., Li, M., Nie, Y. X., Peng, Q. C., & Wang, Z. L. (2013). Fine mapping a major QTL for kernel number per row under different phosphorus regimes in maize (*zea mays* L.). *Theoretical and Applied Genetics, 126*, 1545–1553.

Zhao, Y., Li, X. Y., Zhang, S. H., Wang, J., Yang, X. F., Tian, J. C., Hai, Y., & Yang, X. J. (2014). Mapping QTLs for potassium-deficiency tolerance at the seedling stage in wheat (*Triticum aestivum* L.). *Euphytica, 198*, 185–198.

Zheng, L., Yamaji, N., Yokosho, K., & Ma, J. F. (2012). *YSL16* is a phloem-localized transporter of the copper-nicotianamine complex that is responsible for copper distribution in rice. *The Plant Cell, 24*(9), 3767–3782.

CHAPTER SEVEN

Use of nanoparticles and fertilizers in alleviating heavy metals and improving nutrients uptake in plants

Babar Hussain[a,1], Luqman Riaz[b], Khadija Javeed[c,d], Muhammad Jawad Umer[e], Yawar Abbas[f], Shafeeq ur-Rahman[g,h], Sher Wali Khan[a], Haibat Ali[f], Qamar Abbas[a], Mohsin Zafar[i], Muhammad Nadeem Ashraf[j]

[a]Department of Plant Science, Karakoram International University, Gilgit, Pakistan
[b]College of Life Sciences, Henan Normal University, Xinxiang, China
[c]Department of Plant Pathology, Agriculture College, Guizhou University, Guiyang, China
[d]Faculty of Mountain Agriculture and Environmental Sciences, Kohsar University Murree, Pakistan
[e]State Key Laboratory of Cotton Biology/Institute of Cotton Research, Chinese Academy of Agricultural Science (ICR, CAAS), Anyang, China
[f]Department of Environmental Sciences, Karakoram International University, Gilgit, Pakistan
[g]School of Environment and Civil Engineering, Dongguan University of Technology, Dongguan, Guangdong, China
[h]MOE Laboratory for Earth Surface Processes, College of Urban and Environmental Sciences, Peking University, Beijing, China
[i]Department of Environmental Sciences, Mirpur University of Science and Technology, Mirpur-Azad Jammu and Kashmir, Pakistan
[j]Institute of Soil and Environmental Science, University of Agriculture, Faisalabad, Pakistan

1. Introduction

Nano-ferilizers and/or nano-particles (NPs) generally denotes to substances which size ranges between 1.0 and 100 nm with at least one magnitude (Abbas et al., 2020; Adeel et al., 2020; Fahad et al., 2020). Currently use of NPs have enhanced intensely in manufacturing, medication, crop cultivation and cosmetic fields (Adeel et al., 2019). NPS consist of diverse particle dimensions, geometrical shapes and roles and can be manufactured in accordance with the requirement (Adeel et al., 2020). NPs may have various significances compare to common elements, for instance they have large surface activity, greater surface reaction areas, potential catalytic efficacy, as well as distinctive visual and attractive features (Wang et al., 2019;

[1] Plan and design the entire study

Sustainable Plant Nutrition
ISBN: 978-0-443-18675-2
https://doi.org/10.1016/B978-0-443-18675-2.00008-0

Yang et al., 2017). Moreover, nano-materials are capable of adsorbing and catalysis potential widely used for environmental decontamination due to their chemical reactions, huge adsorption area, lesser temperature alteration, as well as energetic atomicity (Liang et al., 2017). Due to their small size nano-particles easily react and adsorb atoms thus reach to charge steadiness (Tosco et al., 2014). Meanwhile, greater specific surface dimension can significantly increase the adsorption capabilities of materials (Gu et al., 2018).

Nano-hydroxyapatite (NAP) is considered as a favorable P nano-fertilizer having little damage risk and higher output effectiveness than those of conventional fertilizers, because of their nano-scale dimensions, lesser leaching proportion, and slow P-discharge dynamics (Liu & Lal, 2014; Montalvo et al., 2015). Furthermore, it has been reported that rock phosphate with a minor particle magnitude (<35 μm) is more effectual in reducing the phytoavailability and geochemical firmness of soil metals as compared to bigger dimension particles (133–266 μm), this may be because of its greater specific surface sites and increased available P for the development of metal-phosphates complexes (Chen et al., 2006). Furthermore, incorporation of Fe_3O_4 nano-fertilizers on biochar enhanced its heavy metals agglomeration capability as well as refrained accumulation of iron oxide (Wang et al., 2019). Iron oxide biochar nano-fertilizer can change physico-chemical characteristics of soil and adsorb metals from soil consequently increasing plant yield and quality (Zhang et al., 2020). NPs could not only enhanced plant biomass and decreased metal content in roots and leaves, but also improve photosynthesis and diminished the oxidative stress (Ji et al., 2017). Moreover, foliar application of SiO_2-NPs enhanced the transfer of potassium (K), magnesium (Mg), and iron (Fe) from the topmost node to rachis of rice (Chen et al., 2018). Incorporation of nZVI in arsenic contaminated soil improved K translocation in lower and upper plant parts while Na in upper parts of barley (Gil-Diaz et al., 2016). Fe_2O_3 nano fertilizers enhanced P absorption in the uppermost parts of maize in arsenic exposure (Mansouri et al., 2017). Furthermore, nano-hydroxyapatite enhanced soil pH and lessen soil exchangeable acid and interchangeable aluminum (Cui et al., 2018). As a result, nanotechnology is assisting in the resolution of a wide range of environmental issues by delivering novel and practical solutions. Therefore, more research into the usage of NPs and their effects on soil and plants is required. It was necessary to investigate the role of NPs in reducing environmental toxins that pose a major hazard to human health. Thus, this study presents a summary of the uses of NPs for removing heavy metal ions from polluted soil and plant health improvement.

2. Effect of nano-fertilizers on plant growth

Nanotechnology is becoming more significant in crop production owing to its enormous potential for altering the current status of fertilizers in terms of ecological safety, sustainability, and economic stability. Nanotechnology has received special attention as a way of improving crop productivity and resource use efficiency because of large surface area, improved catalytic surface, chemical reactivity, rapid dispersibility and ability to adsorb water. These unique properties of nanomaterials enable nano-fertilizers (1–100 nm) as promising source of nutrients delivery to crop plants, which has given it priority over various types of traditional fertilizers (Naderi & Danesh-Shahraki, 2013). Nano-fertilizers are considered as tried and true method for increasing crop production and slowing nutrients release, thus reducing the frequency of fertilizer application and resulting in a substantial improvement in nitrogen use efficiency (NUE) (Kah et al., 2019), which is a useful metric in crop yield assessment. Nitrogen use efficiency concept is based on (i) plant nutrient uptake, (ii) transportation of nutrients to above ground plant parts, (iii) nutrient usage and growth of biomass, and (iv) environmental factors (Renner et al., 2020).

The occurrence of nutrients in soil as nano-fertilizer allows greater solubility, rapid absorption and uptake by the plants opposed to conventional fertilizers as demonstrated for nitrogen, phosphorous, potassium, calcium, magnesium, iron, manganese, zinc, and copper (Ditta & Arshad, 2016). Therefore, it is critical that nano-fertilizer should be applied and delivered in correct form, especially at the rate that is best suited for plants and that can avoid leaching in soil. For example size reduction may improve surface to mass ratio of nano-fertilizers, allowing considerable increase in the nutrient uptake via root. This helps to reduce application doses and cost, as well as significantly reduce nutrient loss and improve nitrogen usage efficiency by 20%–30% (Kah et al., 2018). Nano based fertilizers can be administered to plants as nano-structured formulations (nutrient elements) or by incorporating nano-structured elements into a carrier complex, which will help in nutrients delivery at right time and place within plants.

Nano-fertilizers are proven carriers of nutrients with nano-scale dimension (30–40 nm) that are capable to hold large number of nutrients and releasing them slowly, so that crop requirements can be met. The plant cell wall prevents the entry of any foreign agent that are greater than the diameter of plant cell wall (5 and 20 nm) (Fleischer et al., 1999), therefore, nano-fertilizers smaller than size of the cell wall can move in readily and penetrate plasma membrane. Plants absorb nano-fertilizers through apoplastic or symplastic pathways and transport them to various parts of plants (Rui et al., 2014). On the other hand, various types of nanoparticles were transported to plant cells via endocytosis or through various pores and channels (Gao et al., 2008). Once they are up taken and translocated to above ground plants parts, it play crucial role in various physiological (Vafa et al., 2015) and biochemical processes of plants such as chlorophyll content (Rezaei & Abbasi, 2014), auxin synthesis, enzymes production, membrane integrity, growth promoting hormones (Alloway, 2008), improved N metabolism, and protein contents (Yang et al., 2006).

Nano-fertilizers in soil first comes in contact with roots and helps to promote growth of plants by up taking nutrients that are necessary for synthesis of proteins. Additionally, nano-fertilizers play role in decomposition of carbohydrates, hormones released within plants, various types of proteins, and indole acetic acid (IAA) as reported in wheat crop (Abdel-Aziz et al., 2018), millet crop (Tarafdar et al., 2014) and cotton crop (Rezaei & Abbasi, 2014). It was reported that nano-fertilizers improves chlorophyll content and intensifies its capability to trap sunlight, increasing pigmentation, promoting Rubisco activity, enhanced CO_2 metabolism, and increasing phenomenon of photosynthesis (Gao et al., 2006; Yang et al., 2006). In addition, nano-fertilizers enhance N and P assimilation in plants that will promote the activities of photo-reduction for photosystem II, chain of electron transport, and helps significantly in detoxifying in reactive oxygen species (Janmohammadi et al., 2016; Morteza et al., 2013). Application of nano fertilizers not only improve N and P uptake but also reduce various abiotic stresses as observed in cucumber plants (Siddiqui et al., 2014) and *Phaseolus vulgaris* L (El-Saadony, Desoky, et al., 2021). Plants experience nutrients deficiency and various types of abiotic stresses at different stages of growth that can inhibit root growth, damage to membrane, and induce oxidative stress. Therefore, looking at the background information provided in this chapter and evidence from various previous studies, it is very much clear that nano-fertilizers is an eco-friendly and cost effective approach to enhance plant growth.

3. Nutrients distribution in soil after nano-fertilizers application

Growing population has resulted in significant rise in food consumption at global level. As a consequence, the agricultural production must expand by 60%–70% to fulfill food demand, which requires 67% more arable land, 65% more water, and 50% more fertilizers (N = 51%; *P* = 54%) (EC, 2019). It was expected that the use of fertilizers will rise by 137.4 million Mt for N and 52.9 million Mt for P in 2030 as compared to the current use (N = 120 million Mt and *P* = 50 million Mt) (FAO, 2017). With the current fertilization practices, the NUE for N is 30%–55% and for P is 18%–20%, indicating that only a small fraction is utilized by plants and significant amount is entering into other environmental compartments causing pollution (Fellet et al., 2021). In recent years, nanofertilizers have developed as a potential technique for crop fertilization that would reduce nutrient loss by regulating availability of nutrients according to the nutritional requirement of crops. Depending upon their use as fertilizers, they were classified as; macronutrient nanofertilizers (hydroxyapatite ($Ca_{10}(PO_4)_6(OH)_2$) as well as calcium carbonate ($CaCO_3$) along with magnesium oxide (MgO)), micronutrient nanofertilizers (copper oxide (CuO), iron oxide (Fe_3O_4), manganese oxide (MnO), zinc oxide (ZnO)) (Fellet et al., 2021), nanomaterial-enhanced fertilizers (nano chitosan and silica) (Liu & Lal, 2015), and plant growth stimulating nanomaterials (cerium oxide, titanium oxide, and carbon nanotubes) (Park et al., 2020).

Nitrogen and phosphorous are key nutrients required by plants for sustainable crop production and food security. These nutrients have high immobilization and leaching rate in soil and become unavailable for plant uptake. The efficiency of nitrogen utilization may be increased by intercalating or entrapping nitrogen in a clay matrix utilizing nanotechnology. Due to the unique qualities of zeolites, such as surface area, they may be employed to make nano-fertilizers. Similarly, nanomaterials derived from carbon based graphene oxide also serve for the slow release of potassium nitrate that helps in minimizing the leaching process (Shalaby et al., 2016). It was shown that zeolite treated with urea increased the NUE in maize by 15% when compared to urea alone (Ahmed et al., 2008). Additionally (Komarneni, 2009), also proved that modified zeolites containing NH_4NO_3 and KNO_3 released NH^{4+} more slowly than commercial fertilizers. The intercalation of NH^{4+} with nano zeolite significantly reduced nitrogen leaching and increased N content due to slow-release. This slow release of N from nano-zeolite could be attributed to: (1) prevention of urea leaching from the plant root zone; (2) slow conversion of urea by soil enzymes, that would lead to delayed formation of NH^{4+} ions; and (3) by adsorbing ammonium ions onto zeolite exchange sites, making them inaccessible to nitrifying bacteria (Subramanian & Thirunavukkarasu, 2017). It was evident from the literature N can be release for 1176 h using nano-zeolite compared to 200 h using conventional fertilizer (Subramanian & Rahale, 2012), while hydroxyapatite nanoparticles can release N for greater than 60 days (Kottegoda et al., 2011).

Soil organic matter releases P but it is immediately bound to insoluble inorganic compounds, which results in P deficiency leading to stunted plant growth. This showed that availability of phosphate in soil can be enhanced using nanoparticles having positively charged or by using modified zeolites. It was reported that synthesis of hydroxyapatite nanofertilizers can effectively play role in the controlled release of P and increase crop yield with minimum risk of P loss because it has weak interactions with soil components compared to charged ions (PO_4^{3-}, HPO_4^{2-}, $H_2PO_4^-$, or Ca^{2+}) (Liu & Lal, 2015). This result in the maximum retention of hydroxyapatite nanofertilizers in soil solution opposed to conventional P fertilizers, which will help plant roots to absorb P with maximum efficiency. On the other hand, nano-zeolite can also serve as effective sorbent for phosphate and help in slow release of P in soil. It was reported that zeolites charged with ammonium can also help to maximize solubility of minerals specifically phosphate thus increasing the availability of phosphorus in soil (Dwivedi et al., 2016). Potassium is labile in soil, and its availability is mainly determined by the soil's cation exchange capacity (CEC). This showed the capacity of soil to contain potassium and cations required by agricultural crops. The two important components of soil that promote CEC are soil organic matter and clay minerals. It was explained that greater the CEC capacity of soil higher the ability of soil to supply K to plants. Due to extensive agriculture, mostly soils are lacking clay minerals and organic matter leading to limited adsorptive sites for K. Therefore using nano-fertilizers specifically based on zeolites can increase the concentration of K by its slow release from nano-zeolites due to high capacity of ion exchange (Liu et al., 2005; Zhou & Huang, 2007).

Micronutrients also play key role in the plant development. It was noted that many developing countries are facing micronutrients deficiencies specifically Fe and Zn, which can bind in the soil. Nanotechnology contributed effectively to supply these micronutrients to plants

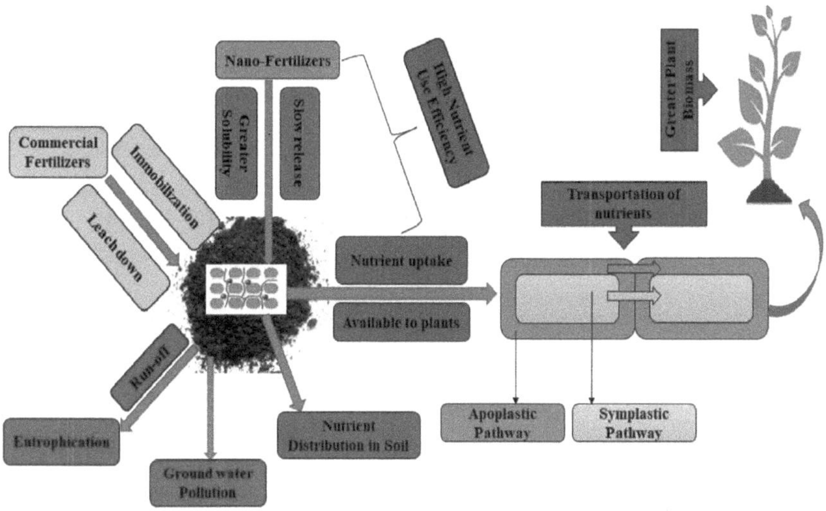

Fig. 7.1 Nutrient distribution in soil through application of nano-fertilizers and their impact on plant growth.

according to the timely requirement. This was evident from previous finding which stated that nanomaterials such as nanocalcite in combination with SiO_2, MgO, and Fe_2O_3 nanoparticles not only enhanced the absorption of Ca, Fe, and Mg but also increase P uptake with Mn and Zn (Sabir et al., 2014). This can be achieved by designing a more diffusible and soluble nano-fertilizers by adjusting size, surface area, and its reactivity (Milani et al., 2010). For example, Al and $Fe(OH)_2$ have positive charge at pH 4.5 to 7.5 and play key role in the retention of anions nutrient through adsorption and promote soil stabilization (Tarafdar & Adhikari, 2015).The food production begins with the soil and there is need to protect this resource. This can be achieved using nano-fertilizers, which will promote balanced nutrition and agricultural production and specifically encourage farmers to use them (Fig. 7.1).

4. Impact of nano-fertilizers on essential nutrients and heavy metals bioavailability

NPs improve the soil pH and enhance HM adsorption as a result HM bioavailability decreased (Zhou et al., 2021). For instance Fe_3O_4-NPs reduced the movement of Cd as well as other HM in the soil (Sebastian et al., 2019). Another study showed that Si-NPs converted Cd into more stable component in the field subsequently 3 years (Wang et al., 2020). Nano-hydroxyapatite enhanced soil pH and lessened soil interchangeable acid and interchangeable aluminum (Cui et al., 2018). In another investigation it was found that the content of Cd in brown rice were remarkably declined by 48.9%, 35.6%, and 46.5% under 0.05%, 0.2%, and 0.4% nano-Fe_3O_4-modified biochar (BC—Fe) applications respectively, meanwhile soil Cd bioavailability reduced by 6.81%—25.0% (Zhang et al., 2020). Though, Silicon-NPs enhance the nutrients phytoavailability in plants, as a result plant growth improve after NPs amendments (Liu et al., 2015; Liu & Lal, 2015). However, Peng et al. (2020) reported that application of 500 mg kg^{-1} CuO-NPs and CeO-NPs stimulated the prime agglomeration of Cu and Ce metals in belowground parts of rice with values of 235.48 mg Cu kg^{-1} and 164.84 mg Ce kg^{-1} correspondingly, whereas amount of Zn in the aboveground plant parts was found as 313.18 mg kg^{-1} treated with highest ZnO-NPs with 1.5 of translocation factor (TF). Furthermore, it was reported that foliar application of Si-NPs at the rate of 20 mg L^{-1} caused greatest reduction in metal agglomeration in brown rice (Hussain et al., 2020). In addition, grain quality and yield of brown rice was not only improve but also decreased metal accumulation. When Se and Si NPs were foliar applied at the ratio of 20:10 mg L^{-1} enhanced the Se bioavailability in grains by 86% (Hussain et al., 2020). In another study it was noted that foliage application of nano-Si markedly decreased the TF from the topmost nodes to rachises, however TFs of K, Mg and Fe from the upmost nodes to rachises enhanced remarkably. Moreover, higher level of nano-Si 25 mM was found most significant in lessening TF of Cd from belowground plant parts to the topmost nodes and from upmost nodes to rachises (Chen et al., 2018). Organic and inorganic nano-Si lessened contents and bioconcentration factor (BCF) of HMs in grain while declined the TF of HMs from belowground parts to aboveground parts and/or from shoots to grains and the most significant outcome was noted on Cd. The mean grain Cd content declined to 27.1% and 23.8% by inorganic and organic nano-Si correspondingly (Wang et al., 2016). On the other hand application of K-nano-chelate enhance the transformation of Cd roots to shoots by 76% than without fertilization treatment (control) (Zahedifar et al., 2019). Wang, Jiang, et al. (2021) reported that addition of 500 mg kg^{-1} SiO_2-NPs in alternate wetting and drying (AWD) irrigation system remarkably decreased As by 29% and Cd by 68% in rice shoots. Meanwhile, application of nano-SiO_2 at same rate under continuous flooding system lessened 70% As and 50% Cd in rice shoots. The possible mechanism for decreasing metal content in plant could be because of adsorption of metals on negatively charged areas of nano-SiO_2, which might be remained in soil by making low bioavailable masses due to homo-and/or hetero-accumulation with other colloidal elements in the soil pore water (Wang, Jiang, et al., 2021). Meanwhile, bioavailability of engineered-NPs (ENPs) such as CuO and ZnO-ENPs is an essential component for the assessment of hazard due to their accumulation in the soil system (Peng et al., 2020). Zn extraction was found to be more efficient than Cu in both soil exposed and unexposed to ENPs by using $CaCl_2$ extraction method (Jośko et al., 2021). This could be due to Zn affinity for soil interchange sites, whereas Cu is greatly held by soil organic matter and ferric oxides (Jośko et al., 2021; Sahuquillo et al., 2003). Jośko et al. (2021) observed that Zn and Cu contents in bulk soil treated with combinations of ENPs were lesser compared with sole application of ENPs. The condition of ENP dissociation was found to be primary factor of the events of the element of ENPs together with their extraction technique (Gao et al., 2017). The disintegration of Class B soft metal oxide ENPs including ZnO and CuO is frequently acid facilitated mechanism on the basis of the reaction reported by (Lowry et al., 2012; Gao et al., 2018):

$$MeO\ ENPs(s) + 2H^+(aq) \leftrightarrow Me^{2+}(aq) + H_2O$$

Inside the environment of blended pressure, ENPs might then contest for the confined amount H^+ ions (soil pH, 6.1) ensuing in lesser dissociation amount of ENPs compared with single occasion of ENPs. Secondly, much less metallic ions launched from ENPs during co-exposure can also subsequent lesser amount of $CaCl_2$-extractable Zn and Cu as compared to sole application (Jośko et al., 2021). Nano-precise fate of co-current ENPs are reported by the various effects of metal bioavailability in the situations of mixtures of metallic salts exposure. Because, soil contaminated with multi-metals exhibited competitive encounter among metallic ions for adsorption locations of soils, ensuing in a greater mobility of metals (Jośko et al., 2021; Qiu et al., 2016). Dissolution was considered a prime mechanis $MeO\ ENPs(s) + 2H^+(aq) \leftrightarrow Me^{2+}(aq) + H_2O$ m resulting the bioavailability and harmfulness of nano-CuO to plants as well as soil microbes (Xu et al., 2021). Furthermore, multiple variant stepwise retrogression examination exhibited that the dissociation of spherical nano-CuO (CuO-NPs), might be predicted soundly with pH and dissolve organic carbon (DOC) level of the soil solutions. Because the addition of four- and eight-times of DOC treatments to an acidic soil solution considerably aided the dissolving of the three sizes of CuO-NPs, tubular nano-CuO (CuO-NTs), and spherical microsized CuO (CuO-MPs) as a result Cu^{2+} enhanced from 4.55 to 11.3 and 5.67−16.2 times correspondingly. Addition of 8-flods DOC boost Cu^{2+} discharge by 2.13−16.6 times in neutral soil solution. Meanwhile, in higher pH soils encouraging result on dissolving only noted on CuOs, with Cu^{2+} raised by factor of 1.56−4.64 and 1.38−4.48 (Xu et al., 2021). Other reports showed that iron nano-particles (Fe-NPs) improved relief of Fe than that of control and $FeSO_4$ amendment. Fe-NPs increased Fe bioavailability resulting in upsurge in Fe absorption, growth, as well as improvement in photosynthetic rate in sorghum plant (Soliemanzadeh & Fekri, 2021). Moreover, the application of nano-TiO_2 agglomeration of ^{133}Cs in the shoots (731.7 $\mu g\,g^{-1}$ dw) were more as compared to roots (597.8 $\mu g\,g^{-1}$ dw). Thus it was concluded that nano-TiO_2 in Cs polluted soils can remarkably increase ^{133}Cs absorption and its agglomeration in soybean plants (Sing & Lee, 2018). Liu et al. (2021) reported that the bioavailability of Cd was markedly reduced in two different soils when treated with nano-zero-valent iron (nZVI). This was because nZVI encouraged the development of lessen accessible form of Cd such as Fe-oxide bound and organic-matter bound Cd in soils under flooding irrigation system. Also, the lessening in in Cd extractability was significantly associated with enhancement in poorly crystalline Fe(II) mineral stages among the different soils (Liu et al., 2021). Moreover, nutritional components including Fe in upper plant parts, Mg in taproots as well as Ca, Zn, and K in underground plant parts were correspondingly increased up to 66%, 64%, 41%, 143%, and 46% when nano-TiO_2 were applied (Wang et al., 2021).Nano-silica (SiO_2-SH) markedly lower Cd, Pb, and Cu absorption in pakchoi by 92.02%, 68.03%, 76.34% while in lettuce by 89.81%, 43.41%, 5.76% sequentially. Study of chemical forms of Cd, Pb, and Cu show that SiO_2-SH may convert the heavy metals in acid solvable forms into reducible and oxidizable segments, thus preventing heavy metals extractability from soil (Lian et al., 2021).

5. Improvement in antioxidant defense system in plants by nano-ferilizers

Nano-materials significantly alleviate the harmful impacts of external pressures (El-Saadony, Saad, et al., 2021). Khalid et al. (2021) noted an improvement in morphological traits of plants by 50%−92%

while chlorophyll level by 30%–80% by NPs (ZnO, FeO, and MgO-NPs). When cotton crop are stressed, foliage applications of nano-Zn at the rate of 200 ppm aid to enhance growth and productivity (Hussein & Abou-Baker, 2018). Spraying of SiO_2 while plant growth phase can improve cell wall toughness, firmness and flexibility (Yassen et al., 2017). A study showed that biological selenium nano-particles (Bio-SeNPs) repressed fungal development in the ranged of 20–40 μg mL^{-1} whereas markedly removed 88% and 92% DPPH˙ (2,2-diphenyl-1-picrylhydrazyl) and ABTS˙ (2,2-azinobis-3-ethylbenzothiazoline-6-sulfonate) radicals respectively. Under greenhouse circumstances wheat plant fed with 100 μg mL−1 of Bio-SeNPs remarkably diminished the occurrence of wheat crown and root rot disease (CRD) and dramatically improved plant growth, grain amount and value by 5%–40%. In addition photosynthetic components and gaseous exchange capability were dramatically boosted (El-Saadony, Saad, et al., 2021). Similarly, biological silicon nano-particles (Bio-SiNPs) markedly increased plant growth and productivity, chlorophyll content, carotenoids, transpiration ratio, photosynthesis level, conduction of stomata, membrane constancy indication, total soluble sugars, nutrients uptake as well as functioning of POD, CAT, SOD and APX. Meanwhile Bio-SiNPs remarkably lessened electrolyte outflow, MDA content, H_2O_2, $O_2^{\bullet -}$, and metals in upper ground parts and shells of *Phaseolus vulgaris* L than that of control (El-Saadony, Desoky, et al., 2021). Nano-Fe_3O_4 particles up to 2000 mg L^{-1} to each Pb, Zn, Cd, and Cu solution reduced growth inhibition whereas triggered defensive systems in wheat plantlets, reducing oxidative pressure caused by heavy metals. The ability of nanp-Fe_3O_4 to alleviate metal pressure might be linked to an enhancement in SOD and POD enzyme activity. Meanwhile, the reduction in MDA concentration validated their protecting function (Konate et al., 2017). Furthermore, NPs amendments mitigated the antagonistic impacts of chilling by sustaining the maximum photochemical efficacy of PSII (F_v/F_m), maximum photo-oxidizable PSI (P_m) as well as photosynthetic gas interchange. Moreover, sugarcane seedlings aided with NPs have larger levels of light absorbing components such as chlorophyll and carotenoids. The non-photochemical quenching (NPQ) of PSII was improved by maximum carotenoid level in the leaves applied by NPs (Elsheery et al., 2020). Different types and mechanisms by which various NPs affect on the absorption of heavy metals by plants are displayed in Table 7.1 .

Table 7.1 Influence of different NPs fertilizers on plants and nutrients mobilization.

Type of NPs	Size of NPs	Concentration	Plant species	Effects on plant	Effect on nutrients	References
Nano-Fe_3O_4 −modified biochar	0, 0.05%, 0.1%, 0.2%, 0.4%, 0.8%, and 1.6%; $W_{BC\text{-}Fe}/W_{soil}$	<0.149 mm	Rice	BC-Fe application promoted the formation of iron plaque and enhanced the root interception of Cd	Cd concentrations of brown rice were significantly decreased by 48.9%, 35.6%, and 46.5% by the 0.05%, 0.2%, and 0.4% BC-Fe treatments, respectively	(Zhang et al., 2020)
SiO_2	60−100 nm	0, 5, 10, 25 mM	Rice	Cd translocation to roots and other plant components is reduced	NPs reduced Cd levels in grains and rachises while increasing K, mg, and Fe levels in these parts	(Chen et al., 2018)
ZnO-NPs, CuO NPs and CeO_2 NPs	50, 100, 500 mg kg^{-1}	24.52 ± 3.97 nm (Zn-NPs), 30.73 ± 4.45 nm (Cu-NPs) and 40.28 ±	Rice	NPs significantly enhanced plant root growth. Zn-NPs can promote plants to produce more	Zn was mostly located in the root cortex while Cu was primarily accumulated in	(Peng et al., 2020)

		5.38 nm (Ce-NPs)		peroxidase, which can better eliminate excess ROS in the plant body, thereby promoting plant growth	the root exodermis and few Ce distributed in the root. Zn, Cu, Ce, K, Ca, P and S elements in the transverse sections of rice roots and leaves were highest in NPs treatments	
ZnO	20–30 nm	0, 50, 75, 100 mg L^{-1}	Rice	NPs improved plant biomass, photosynthesis	NPs increased Zn content whereas lowering Cd levels in the rice plant's root and shoot	(Ali et al., 2019)
Fe_3O_4-NPs	10 and 50 mg L^{-1}	5–10 nm	Coriander	Excellent antibacterial activity was being revealed by efficient Fe_3O_4-NPs against gram positive and	The alleviation percentage in coriander was observed to be 97.37 for Cd and 98.45 for Pb using 10 mg L^{-1} Fe_3O_4 NPs	(Fahad et al., 2020)

(Continued)

Table 7.1 Influence of different NPs fertilizers on plants and nutrients mobilization.—cont'd

Type of NPs	Size of NPs	Concentration	Plant species	Effects on plant	Effect on nutrients	References
				negative bacterium		
nZVI	~50 nm	0, 1.0, 10%	Barley	Plant height, dry weight improved	The pH of postharvest soil was raised by NPs. Concentrations of K and Na in the shoot were all enhanced by NPs, but As content in the shoots and roots was lowered	(Gil-Diaz et al., 2016)
SiO_2	5–15 nm	300 ppm	Sugarcane	By preserving optimum photochemical efficiency, the negative effects of chilling were reduced. The amount of light-harvesting pigments	–	(Elsheery et al., 2020)

				(chlorophylls and carotenoids) was enhanced; SiO_2 NPs had greater amelioration effects		
Se and Si NPs	5, 10, 20 mg L^{-1}	12.26 ± 2 nm (Se-NPs); 18.04 ± 3 nm (Si-NPs)	Rice	Rice growth and quality were increased by foliar application of both NPs	The use of Se and Si in combination was more effective in lowering Cd and Pb levels. Furthermore, Cd and Pb toxicity are reduced when Se and Si NPs are added to the foliage	(Hussain et al., 2020)
Nano-silica	3.0%	18 nm		–	The bioavailability of As, Pb, and Cd decreased by 85%, 97.1%, and 80.1% respectively	(Cao et al., 2020)

(*Continued*)

Table 7.1 Influence of different NPs fertilizers on plants and nutrients mobilization.—cont'd

Type of NPs	Size of NPs	Concentration	Plant species	Effects on plant	Effect on nutrients	References
SiO_2-NPs	150–2000 mg kg^{-1}	10 nm	Rice	Dry root and shoot biomass remarkably enhanced by SiO_2-NPs	Application of 500 mg kg^{-1} decreased the concentration of As and Cd to 70% and 50% under the water regimes in rice shoots	(Wang, Deng, et al., 2021)
ZnONPs	100 mg kg^{-1}	68.1 nm	Rice	Dimethylarsinic acid was significantly higher in ZnONPs treated soil suggests that ZnONPs stimulated the growth	ZnONPs and zinc ions (Zn^{2+}) significantly reduced the total As in rice roots (−39.5% and −83.3%) and shoots (−60.2% and −80.0%)	(Ma et al., 2020)
Nanohydroxyapatite (NAP)	25.9 tons ha^{-1} (1.16%, w/w)	<60 nm	Ryegrass	–	Increased soil pH and decreased soil exchangeable acid and	(Cui et al., 2018)

					exchangeable al. $CaCl_2$-extractable Cu and Cd in the amended soils significantly decreased compared with the control	
Se-NPs	0, 0.01, 0.1, 1, 10, 50 mg L^{-1}	39.2 ± 19.2 nm	Garlic	Plant growth was promoted to some extent under low Se-NPs exposure. Garlic growth under 10 mg L^{-1} Se-NPs exposure was not stunted obviously. Compared to the control group, the biomass was decreased by 10%, under	The total Se content was in the order of Se root ≫ Se bulb > Se leaf for the Se-NPs. Translocation factor for Se was found lowest under Se-NPs exposed garlic	(Li et al., 2020)

(*Continued*)

Table 7.1 Influence of different NPs fertilizers on plants and nutrients mobilization.—cont'd

Type of NPs	Size of NPs	Concentration	Plant species	Effects on plant	Effect on nutrients	References
				50 mg L^{-1} Se-NPs treatment		
Nano-Si fertilizers	2.5 mM	60 nm	Rice	The average spike weight of six cultivars increased to 25.3% and 24.8%, respectively, by inorganic and organic nano-Si	The average grain Cd concentration decreased to 27.1% and 23.8%, respectively, by inorganic and organic nano-Si	(Wang et al., 2016)
K-nano-chelate	0, 100, and 200 mg kg^{-1}	–	Basil	–	K-nano-chelate reduced Cd content in root and shoot. While application of 100 mg K kg^{-1} soil as K-nano-chelate increased root Zn and Cu level significantly by	(Zahedifar et al., 2019)

					about 28% and 30% respectively	
Nano-Fe_3O_4	2000 mg L^{-1}	6.8 nm	Wheat	Addition nano-Fe_3O_4 2000 mg L^{-1} in heavy metal solution (1 mM) significantly decreased the growth inhibition and activated protective mechanisms to reduce oxidative stress induced by heavy metals in the wheat seedlings	Addition of nano-Fe_3O_4 significantly reduced the absorptions of Pb, Zn, Cd and Cu by 54.04%, 23.95%, 65.51%, and 68.64% in the roots; and by 17.75%, 11.39%, 99.59% and 52.94% in the shoots, respectively	(Konate et al., 2017
NPs-Si and NPs-Se	12.5 mg L^{-1} (NPs-Si) and 6.25 mg L^{-1} (NPs-Se)		Rice	NPs-Se improved root and root volume. NPs-Se, NPs-Si and foliar application of NPs-Se at BT resulted in the highest relative	Both NPs-Si and NP-Se treatments showed positive impact on ion selectivity by reducing the Na^+ uptake and	(Badawy et al., 2021)

(Continued)

Table 7.1 Influence of different NPs fertilizers on plants and nutrients mobilization.—cont'd

Type of NPs	Size of NPs	Concentration	Plant species	Effects on plant	Effect on nutrients	References
				water content and dry matter, while foliar application of NPs-Si at booting stage gave the highest leaf area index of rice plants	raising the K^+ uptake, consequently reducing the Na^+/K^+ ratio	

6. Conclusion

Nano-fertilizers are more reactive and can reach plant cells through root hairs, according to the findings. Abiotic stress and heavy metal poisoning can be reduced by NPs in plants. As a result, NPs technology can be applied to agriculture, specifically in the transformation of nanomaterials into nano-ferilizers and/or nano-particles. This would be advantageous in terms of agricultural productivity, cost effectiveness, and carbon reduction. Heavy metal accumulation in plants is a severe food safety issue for the population consuming plant grains as a staple food such as wheat, maize, rice, soya beans etc. Nano-fertilizers can greatly reduce metal ion translocation from roots to uppermost parts and from uppermost parts to rachises. Nano-fertilizers, on the other hand, can significantly improve the translocation of critical nutrients like K, Mg, and Fe from the higher portions of the plant to the rachises. Thus, NPs have the ability to reduce heavy metal levels, MDA activity, and boost antioxidant activity and photosynthetic rate all at the same time. As a result, nanotechnology may open up new perspectives on sustainable production. Plant defense system including peroxidases, superoxide dismutase, and catalase which persistently hunt ROS could be mimicked by NPs. Thus, advancement in preparation of green nano-materials is environmental friendly, practical, and decrease the usage of toxic substances.

References

Abbas, Q., Yousaf, B., Amina Ali, M. U., Munir, M. A. M., El-Naggar, A., Rinklebe, J., Naushad, M., & Naushad, M. (2020). Transformation pathways and fate of engineered nanoparticles (ENPs) in distinct interactive environmental compartments: A review. *Environmental Interface, 138*, 105646.

Abdel-Aziz, H. M. M., Hasaneen, M. N. A. G., & Omer, A. M. (2018). Foliar application of nano chitosan NPK fertilizer improves the yield of wheat plants grown on two different soils. *Egyption Journal of Experiemntal Biology, 14*, 63–72.

Adeel, M., Ma, C., Ullah, S., Rizwan, M., Hao, Y., Chen, C., Jilani, G., Shakoor, N., Li, M., Wang, L., Tsang, D. C. W., Rinklebe, J., Rui, Y., Xing, B., et al. (2019). Exposure to nickel oxide nanoparticles insinuates physiological, ultrastructural and oxidative damage: A life cycle study on *Eisenia fetida*. *Environmental Pollution, 254*, 113032.

Adeel, M., Tingting, J., Hussain, T., He, X., Ahmad, M. M., Irshad, M. K., Shakoor, N., Zhang, P., Changjian, X., Hao, Y., Zhiyong, Z., Javed, R., Rui, Y., et al. (2020). Bioaccumulation of ytterbium oxide nanoparticles insinuate oxidative stress, inflammatory, and pathological lesions in ICR mice. *Environenmental Science and Pollution Research, 27*, 32944–32953.

Ahmed, O. H., Hussin, A., Ahmad, H. M., Rahim, A. A., & Majid, N. M. A. (2008). Enhancing the urea-N use efficiency in maize (*Zea mays*) cultivation on acid soils amended with zeolite and TSP. *The Scientific World Journal, 8*, 394–399.

Ali, S., Rizwan, M., Noureen, S., Anwar, S., Ali, B., Naveed, M., Abd_Allah, E. F., Alqarawi, A. A., & Ahmad, P. (2019). Combined use of biochar and zinc oxide nanoparticle foliar spray improved the plant growth and decreased the cadmium accumulation in rice (*Oryza sativa* L.) plant. *Environenmental Science and Pollution Research, 26*, 1128–11299.

Alloway, B. J. (2008). Zinc in soils and crop nutrition (2nd ed.). published by IZA and IFA Brussels, Belgium and Paris, France.

Badawy, S. A., Zayed, B. A., Bassiouni, S. M. A., Mahdi, A. H. A., Majrashi, A., Ali, E. F., & Seleiman, M. F. (2021). Influence of nano silicon and nano selenium on root characters, growth, ion selectivity, yield, and yield components of rice (*Oryza sativa* L.) under salinity conditions. *Plants, 10*, 1657.

Cao, P., Qiu, K., Zou, X., Lian, M., Liu, P., Niu, L., Yu, L., Li, X., & Zhang, Z. (2020). Mercapto propyltrimethoxysilane- and ferrous sulfate modified nano-silica for immobilization of lead and cadmium as well as arsenic in heavy metal-contaminated soil. *Environenmental Science, 266*, 115152.

Chen, R., Zhang, C., Zhao, Y., Huang, Y., & Liu, Z. (2018). Foliar application with nanosilicon reduced cadmium accumulation in grains by inhibiting cadmium translocation in rice plants. *Environenmental Science and Pollution Research, 25*, 2361–2368.

Chen, S. B., Zhu, Y. G., & Ma, Y. B. (2006). The effect of grain size of rock phosphate amendment on metal immobilization in contaminated soils. *Journal of Hazardous Materials, 134*, 74–79.

Cui, H., Shi, Y., Zhou, J., Chu, H., Cang, L., & Zhou, D. (2018). Effect of different grain sizes of hydroxyapatite on soil heavy metal bioavailability and microbial community composition. *Agriculture, Ecosystems and Environment, 267*, 165–173.

Ditta, A., & Arshad, M. (2016). Applications and perspectives of using nanomaterials for sustainable plant nutrition. *Nanotechnology Reviews, 5*, 209–229.

Dwivedi, S., Saquib, Q., Al-Khedhairy, A. A., & Musarrat, J. (2016). Understanding the role of nanomaterials in agriculture. In *Microbial inoculants in sustainable agricultural productivity* (pp. 271–288). Springer.

El-Saadony, M. T., Saad, A. M., Najjar, A. A., Alzahrani, S. O., Alkhatib, F. M., Shafi, M. E., Selem, E., Desoky, E. S. M., Fouda, S. E., & El-Tahan, A. M. (2021). The use of biological selenium nanoparticles to suppress *Triticum aestivum* L. crown and root rot diseases induced by *Fusarium species* and improve yield under drought and heat stress. *Saudi Journal of Biological Sciences, 28*, 4461–4471.

Elsheery, N. I., Sunoj, V. S. J., Wen, Y., Zhu, J. J., Muralidharan, G., & Cao, K. F. (2020). Foliar application of nanoparticles mitigates the chilling effect on photosynthesis and photoprotection in sugarcane. *Plant Physiology Biochemistry, 149*, 50–60.

Fahad, Balouch, A., Agheem, M. H., Memon, S. A., Baloch, A. R., Tunio, A., Abdullah, Pato, A. H., Jagirani, M. S., Panah, P., Gabole, A. A., & Qasim, S. (2020). Efficient mitigation of cadmium and lead toxicity in coriander plant utilizing magnetite (Fe_3O_4) nanofertilizer as growth regulator and antimicrobial agent. *International Journal of Environmental Analytical Chemistry*, 1–12.

FAO. (2017). *World fertilizer trends and outlook to 2020: Summary report.* Italy: Food and Agriculture Organization of the United Nations Rome.

Fellet, G., Pilotto, L., Marchiol, L., & Braidot, E. (2021). Tools for nano-enabled agriculture: Fertilizers based on calcium phosphate, silicon and chitosan nanostructures. *Agronomy, 11*, 1239.

Fleischer, A., O'Neill, M. A., & Ehwald, R. (1999). The pore size of non-graminaceous plant cell walls is rapidly decreased by borate ester cross-linking of the pectic polysaccharide rhamnogalacturonan II. *Plant Physiology, 121*, 829–838.

Gao, F., Hong, F., Liu, C., Zheng, L., Su, M., Wu, X., Yang, F., Wu, C., & Yang, P. (2006). Mechanism of nano-anatase TiO_2 on promoting photosynthetic carbon reaction of spinach. *Biological Trace Element Research, 111*, 239–253.

Gao, F., Liu, C., Qu, C., Zheng, L., Yang, F., Su, M., & Hong, F. (2008). Was improvement of spinach growth by nano-TiO_2 treatment related to the changes of Rubisco activase? *Biometals, 21*, 211–217.

Gao, X., Spielman-Sun, E., Rodrigues, S. M., Casman, E. A., & Lowry, G. V. (2017). Time and nanoparticle concentration affect the extractability of Cu from CuO NP-amended soil. *Environmental Science and Technology, 51*, 2226–2234.

Gao, X., Avellan, A., Laughton, S., Vaidya, R., Rodrigues, S. M., Casman, E. A., & Lowry, G. V. (2018). CuO nanoparticle dissolution and toxicity to wheat (Triticum aestivum) in Rhizosphere Soil. *Environmental Science & Technology, 52*, 2888–2897.

Gil-Diaz, M., Diez-Pascual, S., González, A., Alonso, J., Rodríguez-Valdés, E., Gallego, J. R., & Lobo, M. C. (2016). A nanoremediation strategy for the recovery of an As-polluted soil. *Chemosphere, 149*, 137–145.

Gu, P., Zhang, S., Li, X., Wang, X., Wen, T., Jehan, R., Alsaedi, A., Hayat, T., & Wang, X. (2018). Recent advances in layered double hydroxide-based nanomaterials for the removal of radionuclides from aqueous solution. *Environenmental Science, 240*, 493–505.

Hussain, B., Lin, Q., Hamid, Y., Sanaullah, M., Di, Liu, Hashemi, M. L. R., Khan, M. B., He, Z., & Yang, X. (2020). Foliage application of selenium and silicon nanoparticles alleviates Cd and Pb toxicity in rice (*Oryza sativa* L.). *Science of the Total Environment, 712*, 136497.

Hussein, M. M., & Abou-Baker, N. H. (2018). The contribution of nanozinc to alleviate salinity stress on cotton plants. *Revies in Society and Open Science, 5*(8), 171809.

Janmohammadi, M., Amanzadeh, T., Sabaghnia, N., & Dashti, S. (2016). Impact of foliar application of nano micronutrient fertilizers and titanium dioxide nanoparticles on the growth and yield components of barley under supplemental irrigation. *Acta Agriculturae Slovenica, 107*, 265–276.

Ji, Y., Zhou, Y., Ma, C., Feng, Y., Hao, Y., Rui, Y., Wu, W., Gui, X., Han, Y., Wang, Y., & Xing, B. (2017). Jointed toxicity of TiO_2 NPs and Cd to rice seedlings: NPs alleviated Cd toxicity and Cd promoted NPs uptake. *Plant Physiology and Biochemistry, 110*, 82–93.

Jośko, I., Kusiak, M., Xing, B., & Oleszczuk, P. (2021). Combined effect of nano-CuO and nano-ZnO in plant-related system: From bioavailability in soil to transcriptional regulation of metal homeostasis in barley. *Journal of Hazardous Materials, 416*, 126230.

Kah, M., Kookana, R. S., Gogos, A., & Bucheli, T. D. (2018). A critical evaluation of nanopesticides and nanofertilizers against their conventional analogues. *Nature Nanotechnology, 13*, 677–684.

Kah, M., Tufenkji, N., & White, J. C. (2019). Nano-enabled strategies to enhance crop nutrition and protection. *Nature Nanotechnology, 14*, 532–540.

Khalid, U., Sher, F., Noreen, S., Lima, E. C., Rasheed, T., Sehar, S., & Amami, R. (2021). Comparative effects of conventional and nano-enabled fertilizers on morphological and physiological attributes of *Caesalpinia bonducella* plants. *Journal of the Saudi Society of Agricultural Sciences, 21*, 61–72.

Komarneni, S. (2009). *Potential of nanotechnology in environmental soil science* (pp. 16–19). 한국토양비료학회 학술발표회 초록집 https://www.dbpia.co.kr/Journal/articleDetail?nodeId=NODE02059456.

Konate, A., He, X., Zhang, Z., Ma, Y., Zhang, P., Alugongo, G. M., & Rui, Y. (2017). Magnetic (Fe_3O_4) nanoparticles reduce heavy metals uptake and mitigate their toxicity in wheat seedling. *Sustainability, 9*, 790.

Kottegoda, N., Munaweera, I., Madusanka, N., & Karunaratne, V. (2011). A green slow-release fertilizer composition based on urea-modified hydroxyapatite nanoparticles encapsulated wood. *Current Science, 101*, 73–78.

Liang, S., Jin, Y., Liu, W., Li, X., Shen, S., & Ding, L. (2017). Feasibility of Pb phytoextraction using nano-materials assisted ryegrass: Results of a one-year field-scale experiment. *Journal of Environmental Management, 190*, 170–175.

Lian, M., Wang, L., Feng, Q., Niu, L., Zhao, Z., Wang, P., Song, C., Li, X., & Zhang, Z. (2021). Thiol-functionalized nano-silica for in-situ remediation of Pb, Cd, Cu contaminated soils and improving soil environment. *Environmental Pollution, 280*, 116879.

Liu, R., & Lal, R. (2014). Synthetic apatite nanoparticles as a phosphorus fertilizer for soybean (Glycine max). *Science Reports, 4*, e5686.

Liu, R., & Lal, R. (2015). Potentials of engineered nanoparticles as fertilizers for increasing agronomic productions. *Science of the Total Environment, 514*, 131–139.

Liu, M., Xu, M., Zhang, X., Zhou, J., Ma, Q., & Wu, L. (2021). Poorly crystalline Fe(II) mineral phases induced by nano zero-valent iron are responsible for Cd stabilization with different soil moisture conditions and soil types. *Ecotoxicology and Environmental Safety, 223*, 112616.

Liu, P., Yin, L., Wang, S., Zhang, M., Deng, X., & Zhang, S. (2015). Tanaka, K. Enhanced root hydraulic conductance by aquaporin regulation accounts for silicon alleviated salt-induced osmotic stress in *Sorghum nicolor* L. *Environmental Experts in Botony, 111*, 42–51.

Liu, X. M., Zhang, F. D., Zhang, S. Q., Feng, Z. B., He, X. S., Wang, Y. J., & Wang, R. F. (2005). Study on adsorption and desorption properties of nano-kaoline to nitrogen, phosphorus, potash and organic carbon. *Scientia Agricultura Sinica, 38*, 102–109.

Li, Y., Zhu, N., Liang, X., Zheng, L., Zhang, C., Li, Y. F., Zhang, Z., Gao, Y., & Zhao, J. (2020). A comparative study on the accumulation, translocation and transformation of selenite, selenate, and SeNPs in a hydroponic-plant system. *Ecotoxicology and Environmental Safety, 189*, 109955.

Lowry, G. V., Espinasse, B. P., Badireddy, A. R., Richardson, C. J., Reinsch, B. C., Bryant, L. D., Bone, A. J., Deonarine, A., Chae, S., Therezien, M., Colman, B. P., Hsu-Kim, H., Bernhardt, E. S., Matson, C. W., & Wiesner, M. R. (2012). Long-term transformation and fate of manufactured Ag nanoparticles in a simulated large scale freshwater emergent wetland. *Environmental Science & Technology, 46*, 7027–7036.

Mansouri, T., Golchin, A., & Neyestani, M. R. (2017). The effects of hematite nanoparticles on phytoavailability of arsenic and corn growth in contaminated soils. *International Journal of Environmental Science and Technology, 14*, 1525–1534.

Ma, X., Sharifan, H., Dou, F., & Sun, W. (2020). Simultaneous reduction of arsenic (As) and cadmium (Cd) accumulation in rice by zinc oxide nanoparticles. *Chemical Engineering Journal, 384*, 123802.

Milani, N., McLaughlin, M., Hettiaratchchi, G., Beak, D., Kirby, J., Stacey, S., & Gilkes, R. (2010). Fate of nanoparticulate zinc oxide fertilisers in soil. In *solubility, diffusion and solid phase speciation. World Congress of Soil Science* (pp. 172–174). Soil Solutions for a Changing World.

Montalvo, D., McLaughlin, M. J., & Degryse, F. (2015). Efficacy of hydroxyapatite nanoparticles as phosphorus fertilizer in Andisols and Oxisols. *Soil Science Society American Journal, 79*, 551–558.

Morteza, E., Moaveni, P., Farahani, H. A., & Kiyani, M. (2013). Study of photosynthetic pigments changes of maize (*Zea mays* L.) under nano TiO_2 spraying at various growth stages. *SpringerPlus, 2*, 247.

Naderi, M., & Danesh-Shahraki, A. (2013). Nanofertilizers and their roles in sustainable agriculture. *International Journal of Agriculture and Crop Sciences (IJACS), 5*, 2229–2232.

Park, S., Choi, K. S., Kim, S., Gwon, Y., & Kim, J. (2020). Graphene oxide-assisted promotion of plant growth and stability. *Nanomaterials, 10*, 758.

Peng, C., Tong, H., Shen, C., Sun, L., Yuan, P., He, M., & Shi, J. (2020). Bioavailability and translocation of metal oxide nanoparticles in the soil-rice plant system. *Science of the Total Environment, 713*, 136662.

Qiu, H., Versieren, L., Rangel, G. G., & Smolders, E. (2016). Interactions and toxicity of Cu-Zn mixtures to Hordeum vulgare in different soils can be rationalized with bioavailability-based prediction models. *Environmental Science and Technology, 50*, 1014–1022.

Renner, A., Cadillo-Benalcazar, J. J., Benini, L., & Giampietro, M. (2020). Environmental pressure of the European agricultural system: Anticipating the biophysical consequences of internalization. *Ecosystem Services, 46*, 101195.

Rezaei, M., & Abbasi, H. (2014). Foliar application of nanochelate and non-nanochelate of zinc on plant resistance physiological processes in cotton (*Gossipium hirsutum* L.). *Iranian Journal of Plant Physiology, 4*, 1137–1144.

Rui, Y., Gui, X., Li, X., Liu, S., & Han, Y. (2014). Uptake, transport, distribution and bio-effects of SiO_2 nanoparticles in Bt-transgenic cotton. *Journal of Nanobiotechnology, 12*, 1–15.

Sabir, A., Yazar, K., Sabir, F., Kara, Z., Yazici, M. A., & Goksu, N. (2014). Vine growth, yield, berry quality attributes and leaf nutrient content of grapevines as influenced by seaweed extract (*Ascophyllum nodosum*) and nanosize fertilizer pulverizations. *Scientia Horticulturae, 175*, 1–8.

Sahuquillo, A., Rigol, A., & Rauret, G. (2003). Overview of the use of leaching/extraction tests for risk assessment of trace metals in contaminated soils and sediments. *TrAC, 22*, 152–159.

Sebastian, A., Nangia, A., & Prasad, M. N. V. (2019). Cadmium and sodium adsorption properties of magnetite nanoparticles synthesized from *Hevea brasiliensis* Muell. Arg. bark: Relevance in amelioration of metal stress in rice. *Journal of Hazardous Materials, 371*, 261–272.

Shalaby, T. A., Bayoumi, Y., Abdalla, N., Taha, H., Alshaal, T., Shehata, S., Amer, M., Domokos-Szabolcsy, É., & El-Ramady, H. (2016). Nanoparticles, soils, plants and sustainable agriculture. In , *Vol. 1. Nanoscience in food and agriculture* (pp. 283–312). Springer.

Siddiqui, M. H., Al-Whaibi, M. H., Faisal, M., & Al Sahli, A. A. (2014). Nano-silicon dioxide mitigates the adverse effects of salt stress on *Cucurbita pepo* L. *Environmental Toxicology and* Chemistry, *33*, 2429–2437.

Sing, J., & Lee, B. K. (2018). Effects of Nano-TiO_2 particles on bioaccumulation of ^{133}Cs from thecontaminated soil by Soybean (*Glycine max*). *Process Safety and Environmental Protection, 116*, 301–311.

Soliemanzadeh, A., & Fekri, M. (2021). Effects of green iron nanoparticles on iron changes and phytoavailability in a calcareous soil. *Pedosphere, 31*(5), 761–770.

Subramanian, K., & Rahale, C. S. (2012). Ball milled nanosized zeolite loaded with zinc sulfate: A putative slow release Zn fertilizer. *International Journal of Innovative Horticulture, 1*, 33–40.

Subramanian, K. S., & Thirunavukkarasu, M. (2017). *Nano-fertilizers and nutrient transformations in soil* (pp. 305–319). Nanoscience and Plant–Soil Systems Springer.

Tarafdar, J., & Adhikari, T. (2015). Nanotechnology in soil science. Soil science: An introduction. In *Chapter: Nanotechnology in soil science* (pp. 775–807).

Tarafdar, J., Raliya, R., Mahawar, H., & Rathore, I. (2014). Development of zinc nanofertilizer to enhance crop production in pearl millet (*Pennisetum americanum*). *Agricultural Research, 3*, 257–262.

Tosco, T., Papini, M. P., Viggi, C. C., & Sethi, R. (2014). Nanoscale zerovalent iron particles for groundwater remediation: A review. *Journal of Cleaner Production, 77*, 10–21.

Vafa, Z. N., Sirousmehr, A. R., Ghanbari, A., Khammari, I., & Falahi, N. (2015). Effects of nano zinc and humic acid on quantitative and qualitative characteristics of savory (*Satureja hortensis* L.). *International Journal of Biosciences, 6*, 124–136.

Wang, Y., Deng, Y., Cota-Ruiz, K., Peralta-Videa, J. R., Hernandez-Viezcas, J. A., & Gardea-Torresdey, J. L. (2021). Soil-aged nano titanium dioxide effects on full-grown carrot: Dose and surface-coating dependent improvements on growth and nutrient quality. *Science of the Total Environment, 774*, 145699.

Wang, X., Jiang, J., Dou, F., Sun, W., & Ma, X. (2021). Simultaneous mitigation of arsenic and cadmium accumulation in rice (*Oryza sativa* L.) seedlings by silicon oxide nanoparticles under different water management schemes. *Paddy Water Environment, 19*, 569–584.

Wang, Y., Jiang, F., Ma, C., Rui, Y., Tsang, D. C. W., & Xing, B. (2019). Effect of metal oxide nanoparticles on amino acids in wheat grains (*Triticum aestivum*) in a life cycle study. *Journal of Environmental Management, 241*, 319–327.

Wang, Y., Liu, Y., Zhan, W., Zheng, K., Lian, M., Zhang, C., Ruan, X., & Li, T. (2020). Long-term stabilization of Cd in agricultural soil using mercapto-functionalized nano-silica (MPTS/nano-silica): A three-year field study. *Ecotoxicology Environmental Safety, 197*, 110600.

Wang, S., Wang, F., Gao, S., & Wang, X. (2016). Heavy metal accumulation in different rice cultivars as influenced by foliar application of nano-silicon. *Water Air Soil Pollution, 227*, 228.

Xu, M., Wang, Y., Mu, Z., Li, S., & Li, H. (2021). Dissolution of copper oxide nanoparticles is controlled by soil solution pH, dissolved organic matter, and particle specific surface area. *Science of the Total Environment, 772*, 145477.

Yang, J., Cao, W., & Rui, Y. (2017). Interactions between nanoparticles and plants: Phytotoxicity and defense mechanisms. *Journal of Plant Interaction, 12*, 158–169.

Yang, F., Hong, F., You, W., Liu, C., Gao, F., Wu, C., & Yang, P. (2006). Influence of nano-anatase TiO_2 on the nitrogen metabolism of growing spinach. *Biological Trace Element Research, 110*, 179–190.

Yassen, A., Abdallah, E., Gaballah, M., & Zaghloul, S. (2017). Role of silicon dioxide nano fertilizer in mitigating salt stress on growth, yield and chemical composition of cucumber (*Cucumis sativus* L.). *International Journal of Agricultural Research, 22*, 130–135.

Zahedifar, M., Moosavi, A. A., Zarei, Z., Shafigh, M., & Karimian, F. (2019). Heavy metals content and distribution in basil (*Ocimum basilicum* L.) as influenced by cadmium and different potassium sources. *International Journal of Phytoremediation, 21*, 435–447.

Zhang, J. Y., Zhou, H., Gu, J. F., Huang, F., Yang, W. J., Wang, S. L., Yuan, T. Y., & Liao, B. H. (2020). Effects of nano-Fe_3O_4-modified biochar on iron plaque formation and Cd accumulation in rice (*Oryza sativa* L.). *Environmental Pollution, 260*, 113970.

Zhou, J., & Huang, P. (2007). Kinetics of potassium release from illite as influenced by different phosphates. *Geoderma, 138*, 221–228.

EC. (2019). *EU agricultural outlook for markets and income, 2019–2030. European commission, DG agriculture and rural development brussels.*

El-Saadony, M. T., Desoky, E. S. M., Saad, A. M., Eid, R. S., Selem, E., & Elrys, A. S. (2021). Biological silicon nanoparticles improve *Phaseolus vulgaris* L. yield and minimize its contaminant contents on a heavy metals-contaminated saline soil. *Journal Environmental Science, 106*, 1–14.

Zhou, P., Adeel, M., Shakoor, N., Guo, M., Hao, Y., Azeem, I., Li, M., Liu, M., & Rui, Y. (2021). Application of nanoparticles alleviates heavy metals stress and promotes plant growth: An overview. *Nanomaterials, 11*, 26.

CHAPTER EIGHT

Role and importance of microorganisms in plant nutrition and remediation of potentially toxic elements contaminated soils

Fouzieh Beigmohammadi[a], Eisa Solgi[a], Behnam Asgari Lajayer[b], Eric D. van Hullebusch[c]

[a]Department of Environment, Faculty of Natural Resource and Environment, Malayer University, Malayer, Iran
[b]Department of Soil Science, Faculty of Agriculture, University of Tabriz, Tabriz, Iran
[c]Université Paris Cité, Institut de Physique du Globe de Paris, CNRS, Paris, France

1. Introduction

Soil serves many basic functions in the natural environment; it meets the basic needs of human society (Zhang et al., 2019), and the terrestrial ecosystem is the crossroad of ecological functions which are seen as very important, with a wide range of interactions between solid, gas and liquid phases where living organisms are developing (Giannakopoulou et al., 2012). However, soil degradation, whether through anthropogenic activities or natural processes, is a relatively slow process, but its effects are irreversible in the long run and often on a human life scale (Massas et al., 2013).

Environmental pollution is one of the most important factors affecting the balance of natural conditions of soils so that increased human activities and the resulting pollution lead to soil disturbance (Heidarpour et al., 2019). Potentially toxic elements (PTEs) are among these pollutants that not only affect soil physical and chemical properties but also reduce biological activities as well as direct access to soil nutrients (Mohammadhosseini et al., 2014).

It is also a great reservoir for contaminants because it has the ability to bind to various chemicals (Juned et al., 2018). Soil, both as a PTEs sink in the environment and as a source of PTEs, has adverse effects on human health and the environment (Acosta et al., 2015; Cai et al., 2015).

Sustainable Plant Nutrition
ISBN: 978-0-443-18675-2
https://doi.org/10.1016/B978-0-443-18675-2.00012-2

The soil contamination with PTEs is considered one of the most important challenges in the world; because these pollutants remain widespread, display bioaccumulative property and a prolonged exposure may threaten quantitative and qualitative aspects of food security and endangering the health of living things (Yousaf et al., 2017; Asgari Lajayer et al., 2017, p. 2019).

Given the increasing environmental pollution and the harmful problems they cause, new methods must be developed to protect the environment and human health, one of which is bioremediation (Raghunandan et al., 2018).

Bioremediation is a new method used to recover and reduce PTEs in contaminated soils. Microorganisms play an important role in bioremediation of PTEs (Saghafi et al., 2020). The importance of bioremediation methods is that they are more environmentally friendly than physical and chemical methods and less likely to endanger human health (Verma & Kuila, 2019). Bioremediation can be done in-situ, on-site, and or ex-situ by separating the contaminated environment and moving it to another location (Karimi et al., 2021; Raghunandan et al., 2018) and various bioremediation methods are used to remediate the transferred soil. In this chapter, while briefly referring to the methods of remediation in the soils contaminated by PTEs, we investigate the bioremediation methods and more details on the mechanisms and how it takes place.

2. Potentially toxic elements

In this chapter, we used potentially toxic elements instead of the term "heavy metals", which could better represent the properties of these metals (Duffus, 2002; Pourret & Bollinger, 2018; Pourret & Hursthouse, 2019). With increasing industrialization and urbanization, environmental pollution caused by PTEs has become a global problem (Chen et al., 2018; Liu et al., 2018).

Some PTEs play an important role in biological systems, such as zinc, copper, nickel and iron, and some other PTEs have no biological function in living systems and are toxic even at very low concentrations, including mercury, cadmium, lead and arsenic (Edelstein & Ben-Hur, 2018; Vardhan et al., 2019). The most important roles of iron, zinc and copper in biological systems are enzymatic activity, cell membrane stability and cytochrome oxidase functions, respectively (Nizamani et al., 2019). According to the classification of International Agency for Research on Cancer (IARC), arsenic and cadmium are in group one and lead is in group 2B carcinogenesis (González et al., 2019).

PTEs enter the soil in both natural and anthropogenic ways. Their natural inputs to the soil are attributed to native materials and geology, while population growth, traffic, municipal waste disposal, agricultural management and industrial activities have been identified as major human causes of environmental pollution and PTEs accumulation for various soils (Asgari Lajayer et al., 2019). Fig. 8.1 summarizes the sources of PTEs contamination in the soil. And some of the important factors and activities that lead to the entry of PTEs into the environment in each section are listed. Probably, in connection with soil pollution, the severity and extent of pollution caused by industry and agriculture is higher than other sectors. The mobility and potential risk of PTEs depends on their source (natural or anthropogenic) as well as their chemical speciation (Shaheen et al., 2017). Thus, the potential mobility of PTEs is usually higher in soils whose source of contamination is anthropogenic (Palansooriya et al., 2020).

PTEs such as Pb, Cu, Cr, Cd, Hg and As enter the soil through atmospheric precipitation and agricultural activities cause Cd, Cu, Pb, U, Se, Zn and As to enter the soil. Also, mining and smelting causes Hg, Pb, As and Cd to enter soil and industries cause entry Cd, Cr, Co, Cu, As, Ni, Zn and Hg. And Cd, Cr, Cu, Pb, Hg and Zn enter the soil through waste disposal and lead to soil contamination (Banerjee et al., 2018).

Soil cleanliness from pollutants is one of the factors affecting its health. Accordingly, different permissible and standard pollutants in the soil have been defined for different pollutants, which are presented in Table 8.1 of the United States Environmental Protection Agency (US EPA) and Food

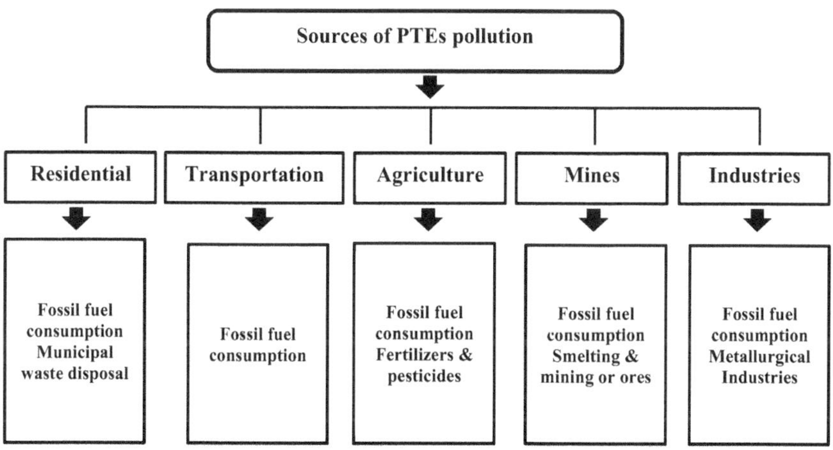

Fig. 8.1 The sources of PTEs.

Table 8.1 Range and standard of PTEs in soil.

PTEs	The range of soil (mg/kg dry wt)	Standards	
		US EPA	FAO/WHO
Cd	0.01–0.7	85	0.3
Pb	2–200	420	100
Ni	10	-	50
As	0.1–40	75	20
Cr	5–3000	3000	100
References	(Sharma et al., 2019)	(EPA, 2005)	(FAO/WHO, 2007)

and Agriculture Organization/World Health Organization (FAO/WHO) standard and range of some PTEs. In the case of PTEs with low range in the soil, indicate that these elements are less present in the earth's natural crust and enter the soil more than anthropogenic sources.

3. Methods for the remediation of PTEs contaminated soils

The remediation methods have different classifications that three important factors in these classifications include (1) the nature of the action (extraction, immobilization) performed on metals (2) the place where the correction takes place (3) the type of technology and method used (biological, physical, chemical, thermal) (Dermont et al., 2008; Xiong et al., 2019). The basis of place, remediation methods are performed in two ways: (1) In-Situ (2) Ex-Situ (Dornbusch et al., 2020).

The remediation methods for contaminated soils are summarized in Fig. 8.2. Chemical leaching involves the washing of soil contaminated with fresh water, reagents and other liquids (or gas) to remove contaminants from soils; physical cleaning mainly involves thermal desorption and soil replacement (Wang et al., 2019).

The advantages of ex-situ techniques are that they are generally: (1) fast and easy to use; (2) relatively low cost in terms of initial investment and continued operation. The disadvantages of these techniques include (1) during the soil washing process, they produce a significant volume of by-products such as contaminated solutions and solvents, which may require additional treatment and storage in special landfills (2) Permanent control is required for stored waste (3) The release of additional pollutants may occur unintentionally by altering the physiological conditions of the sites during modification (Lwin et al., 2018).

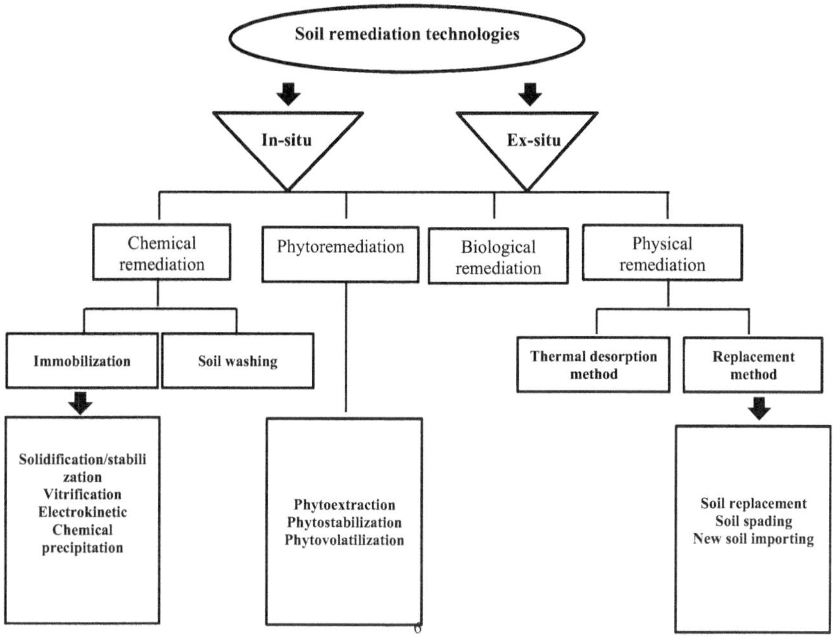

Fig. 8.2 Classification of soil remediation methods.

The most important advantages of in-situ methods: (1) site disruption (2) simplicity and speed (3) relatively low cost compared to other options (4) introduction of nutrients to the contaminated area (5) general acceptability (6) reduces the risk of contamination in a wide range of inorganic pollutants (Wuana & Okieimen, 2011). Factors affecting the applicability and selection of such technologies include (1) cost, (2) long-term effectiveness/sustainability, (3) commercial availability, (4) public acceptance, (5) target metal concentration, (6) type PTEs and materials used and (7) physicochemical properties such as toxicity and mobility (Nejad et al., 2018). In general, among the various technologies for remediation of PTEs contaminated soils, in-situ technology by adding modifiers to the soil has received much attention due to its efficient, economical and environmentally friendly benefits (Chen et al., 2020).

In the case of highly contaminated soils, most in situ remediation techniques are used, where soil is removed from its source but today, in situ techniques are preferred due to the low labor and energy requirements (Lwin et al., 2018). In fact, with the increasing global concern, in situ refining technology has emerged as a sustainable strategy (Chen et al., 2020). The method to be investigated in this chapter is bioremediation. Generally, bioremediation is defined as the technique involving the application of biological agents like the plants and microorganisms (fungi and bacteria, etc.) for the remediation of contaminated soils or water so that the environment can be reusable (Gangar et al., 2019). However, one should always bear in mind that metals cannot be destroyed by biological processes, however, microorganisms change their chemical form by adding (reducing) or removing (oxidation) atom's electrons (Tabak et al., 2005).

It is noted that some researchers classify the remediation methods dedicated to polluted soils environments, such as phytoremediation, as a subset of bioremediation, but in the present study, our main goal is to consider the bioremediation and application of microorganisms in this field. And at the end of the chapter, we briefly mention the simultaneous use of plants and microorganisms.

4. Importance of microorganisms in soil health and soil remediation

The main factors of healthy soil are enzyme activity, diversity of microorganisms and microbial biomass in soil. Each particle of healthy soil contains millions of microbes that are effective in providing nutrients and minerals and are a major component of the soil ecosystem (Das & Varma, 2010; Pankhurst et al., 1995). The most important source of enzymes in the soil is microorganisms, but plants and animals also play a minor role in the production of soil enzymes (Kumar et al., 2019). It seems, among living organisms, PTEs have the first effects on soil microorganisms after entering the soil. However, the microbial population of the soil changes due to the stress caused by the presence of PTEs in the soil (Xu et al., 2019).

Also, soil microorganisms may affect PTEs bioavailability and facilitate the uptake of various metal ions such as Cd^{2+}, Mn^{2+} and Fe^{2+} by the plant roots. Beside, in plants that grow on soils contaminated with PTEs, have grown mycorrhiza (fungal coexistence with plant roots). Fungal coexistence increases root uptake by up to 47 times and nutrients such as trace metal ions are available to plant roots (Archana & Jaitly, 2014). Therefore, the role of microorganisms in the removal of PTEs in contaminated soils can be understood, and as a result this method can be used to clean up soils contaminated with PTEs.

Generally soil microorganisms include bacteria and fungi, bacteria are the most abundant microbes in soil environment (9×10^7 in 1 gram of soil) in contrast of fungi which are reported to be two order of magnitude lower in terms of number (2×10^5 in 1 gram of soil) (Alexander, 1991, p. 467). The attendance of microorganisms in contaminated soils is very effective in reducing stress (Miransari, 2017). One way to use microorganisms for biodegradation is to cultivate strains of microorganisms resistant to PTEs contaminated soils, and after improving their survival, we use the interaction between contaminated soils and microorganisms to remediation soils contaminated with PTEs (Sun et al., 2020).

Actually, bioremediation techniques can be much more effective using growth control and improving the ability of microorganisms in a contaminated environment (Alvarez et al., 2017). By changing the form of PTEs, microorganisms affect solubility, bioavailability, and PTEs mobility (Ayangbenro & Babalola, 2017).

Metals and microbes have interactions that depend on the metal species, the soil environment, and the organism. On the other hand, the structure of compounds and metabolic activities affect speciation, and as a result, they affect the mobility, solubility, bioavailability and toxicity of metals (Gadd, 2007).

The interaction between microbes and metals depends on the soluble chemistry, the specific surface of the adsorbent, the presence or absence of organic ligands and reducing agents. Microorganisms can interact by metals through several mechanisms, some of which perhaps used as the basis for bioremediation. These mechanisms include biosorption, bioleaching, bioaccumulation, biotransformation and biomineralization (Tabak et al., 2005).

In some cases, these processes are specific biochemical pathways that protect the microbial cell from metal toxicity, and because of the specificity of this detoxification mechanism, it may be used to design biosensors to bioavailable the toxic metals concentrations. Also, microbes can produce novel mineral phases via non-specific mechanisms, trapping PTEs in the soil and making them immobile. In addition, they manufacture biogenic ligands that are complexed with metals and mobilization metals from polluted soils. These metals can be pumped from the soil and trapped in a bioreactor on the surface (Lloyd, 2002). Therefore, through these mechanisms, they prevent the absorption of PTEs into plants; limit the leaching of PTEs to groundwater, as a result, their bioavailability is reduced (Harmsen & Naidu, 2013).

Various species of microorganisms have been introduced to clean contaminated soils in various scientific sources, including Verma and Kuila (2019) stated, *Flavobacterium*, *Pseudomonas*, *Bacillus*, *Arthrobacter*, *Corynebacterium*,

Methosinus, Rhodococcus, Mycobacterium, Stereum hirsutum, Nocardia, Methanogens, Aspergilus niger, Pleurotus ostreatus, Rhizopus arrhizus, Azotobacter, Alcaligenes, Phormidium valderium, Ganoderma applantus are having an effective role in the bioremediation of PTEs. Lloyd et al. (2003) stated that various types of bacteria used to convert toxic pollutants into less toxic pollutants include *Pseudomonas* sp., *Achromobacter* sp., *Burkholderia* sp., *Rhodococcus* sp., *Ralstonia* sp., *Alcaligenes* sp., *Sphingomonas* sp., *Dehalococcoides* sp., *and Comamonas* sp. D'Annibale et al. (2007) reported that several species of fungi, including *Allescheriella* sp., *Botryosphaeria rhodina, Klebsiella oxytoca, Phlebia* sp. *Pleurotus pulmonarius,* and *Stachybotrys* sp., play important roles in metal binding processes.

Archaebacterium Sulfolobus acidocaldarius can oxidize As(III) to As(V) (Lindstrom & Sehlin, 1989). As(V) has a stronger bond with inorganic components and is less toxic than As(III), so microbial oxidation has caused the immobility and bioremediation of As(III) (Park et al., 2011). *Bacillus* sp reduces Cr(VI) to Cr(III), which has less mobility and toxicity than Cr(III). The bacterium was isolated from a chromium-contaminated landfill (Liu et al., 2006).

Therefore, the importance and role of various microorganisms (fungi, bacteria, etc.) in remediation the soil of pollutants, led to the formation of a new method in the improvement of contaminated soils and the combination of this method with other methods was used to increase efficiency.

5. Bioremediation of PTEs in contaminated soils

Bioremediation is proposed as an almost economical method that can reduce the effects of contamination and toxicity of pollutants as an efficient method in contaminated soils (Singh & Gupta, 2016). In fact, bioremediation is a method that pollutants (PTEs) through metabolic processes converts to less toxic compounds (Banerjee et al., 2018). Microorganisms utilize toxic ingredients as energy source for growth via fermentation, co-metabolism and respiration (Ayangbenro & Babalola, 2017).

The basis of place bioremediation methods are performed in two ways: (1) In-Situ Bioremediation (2) Ex-Situ Bioremediation. In-Situ Bioremediation included Bioventing, Biosparging, Anaerobic Bioremediation and Phytoremediation and Ex Situ Bioremediation included Land farming, Composting, Biopiling and Bioreactors (Gangar et al., 2019). The basic of bioremediation consist of decrease the solubility of pollutants using changing the conditions of redox and pH, and ultimately absorbing pollutants from the contaminated environment (Jain & Arnepalli, 2019).

Metals can be extracted from the contaminated environment using two mechanisms: (1) A number of heterotrophic microorganisms can increase the mobility of metals by producing organic acids, (2) Some autotrophic microorganisms, including *Thiobacillus*, can produce sulfuric acid from elmetals sulfur, which causes the leaching of metals (Lloyd, 2002). A group of bacteria are able to methylated metals, including Se, As, Pb, Sn, Te and Hg under anaerobic and anaerobic conditions. Methylated compounds are different in terms of toxicity and solubility. Methylated compounds, including $(CH_3)_2Se_2$ and $(CH_3)_2Se$, volatile from the soil (Gadd, 2007).

Table 8.2 summarizes the results of various studies on microorganisms that are effective in bioremediation PTEs. It should be noted that PTEs are more toxic (Cd, Ni, Hg, Cr, As and Pb) in this review. The mechanism used by microorganisms to remediation PTEs is also described in this table. According to the results, some microorganisms have been used to remediation contamination from more than one PTEs, which can be suitable for places that are contaminated with several PTEs.

The conditions required for the multiplication and growth of microorganisms, the ability and the method used to remediation them are among the factors that are important in the application of microorganisms. For example, the reason for the widespread use of *Pseudomonas aeruginosa* is that it uses various mechanisms to remediate polluted bodies (Agnello et al., 2016). In fact, this bacterium, in two ways, can improve the bioavailability of PTEs: (1) manufacturing metal chelating siderophores (Visca et al., 2007) (2) by producing biosurfactants (rhamnolipids), the solubility of organic compounds in water decreases and the mobility of PTEs increases (Zhang et al., 2012).

Bioremediation relative to other methods of removing PTEs contaminated soils has the main advantages of which include (1) this is done with minimal change in the environment (2) Permanent waste disposal and elimination of long-term liability (Azubuike et al., 2020). In addition to, in terms of cost, the bioremediation method is cheaper than other contaminated soil remediation methods (Gupta et al., 2020).

6. Factors influencing on the success of bioremediation of PTEs

It is important for a successful bioremediation to know the characteristics of the soil environment and to identify the contaminants as well as the microorganisms involved, these parameters are shown in Fig. 8.3. The

Table 8.2 Microorganisms used in bioremediation of PTEs.

PTEs	Microorganisms	Mechanisms	References
Lead	*S. rhizophila* *V. boronicumulans*	These microorganisms induce remediation of contaminated environment by producing metal carbonates	(Jalilvand et al., 2020)
	Alcaligenes faecalis *Brevibacterium iodonium* *Bacillus pumilus* *Pseudomonas aeruginosa*	The predominant mechanism in the remediation of lead is lead sulfide precipitation	(De et al., 2008)
	Aspergillus parasitica *Cephalosporium aphidicola*	Biosorption process by ion exchange introduced be as prevailing mechanism also complexation mechanism is effective in the remediation of Pb(II)	(Akar et al., 2007)
	Rhizopus sp. *Aspergillus* sp.	Biosorption and bioaccumulation were the predominant mechanisms in cadmium the environment from lead	(Zafar et al., 2007)
	Pseudomonas sp. *B50D*	Effective mechanisms in the remediation process include reduction, siderophores, biofilms, biosorption, and bioaccumulation	(Giovanella et al., 2017)
Mercury	*Alcaligenes faecalis* *Brevibacterium iodonium* *Bacillus pumilus* *Pseudomonas aeruginosa*	The detoxification mechanism was through volatilization	(De et al., 2008)
	Pseudomonas sp. *B50D*	The mechanism for removing mercury from the environment is volatilization	(Giovanella et al., 2017)
	Hymenoscyphus ericae *Neocosmospora vasinfecta* *Verticillium terrestre*	Mercury is converted to β-HgS and Hg (0) through the volatilization mechanism	(Kelly et al., 2006)

	Pseudomonas K-62	Using genetics, they overexpressed the gene for polyphosphates and further accumulated mercury in the microorganism.	(Kiyono & Pan-Hou, 2006)
Arsenic	*Bacillus flexus strain As-12*	The mechanism of this microorganism is arsenic oxidation of As(III) to As(V)	(Jebeli et al., 2017)
	E. coli strain	Using genetics and protein gene expression microorganisms increased As(III) bioaccumulation	(Kostal et al., 2004)
Chromium	*Pseudomonas aeruginosa*	The mechanism of this microorganism for the remdiation of chromium is to reduce Cr(VI) to Cr(III)	(Sinha et al., 2011)
Nickel	*Pseudomonas* sp. *B50D*	Effective mechanisms in the remediation process include reduction, siderophores, biofilms, biosorption, and bioaccumulation	(Giovanella et al., 2017)

PTEs	Microorganisms	Mechanisms	References
Cadmium	*Pseudomonas* sp. *B50D*	Effective mechanisms in the remediation process include reduction, siderophores, biofilms, biosorption, and bioaccumulation	(Giovanella et al., 2017)
	S. rhizophila *V. boronicumulans*	These microorganisms contribute to the remediation of contaminated environment by producing metal carbonates	(Jalilvand et al., 2020)
	Bacillus sp.	Using microbially induced carbonate precipitation caused the metals became immobile	(Zhao et al., 2016)
	A. fumigatus	The ability of the microorganism to remediation the environment, in addition to its high ability to bioaccumulate, also depends on the biological function of the microorganism	(Mohammadian Fazil et al., 2015; Priyalaxmi et al., 2014)

(Continued)

Table 8.2 Microorganisms used in bioremediation of PTEs.—cont'd

PTEs	Microorganisms	Mechanisms	References
	Klebsiella planticola	Microorganisms are able to bioremediation cadmium through the precipitation of cadmium sulfide	(Sharma et al., 2000)
	Rhizopus sp. *Aspergillus* sp.	Biosorption and bioaccumulation were the predominant mechanisms in cadmium the environment from cadmium	(Zafar et al., 2007)
	Rhizopus arrhizus	Accumulation of metal in the cell wall is cause the remediation of cadmium	(Favero et al., 1991)

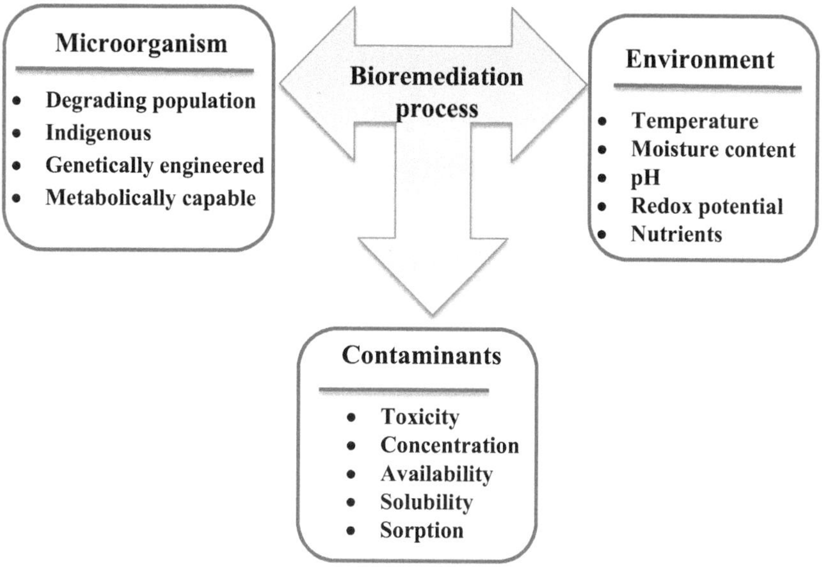

Fig. 8.3 Factors (Environmental, biological and chemical) affecting in bioremediation.

change in pH causes changes in the chemical forms of PTEs and affecting the concentration of bioavailable and mobile forms, and thus affects the adsorption of PTEs by the adsorbent cations. Temperature variations affect stability of different ions, and the equilibrium capacity, thus affecting the rate of PTEs absorption (Ayangbenro & Babalola, 2017). Also, temperature and acidity are two factors that influence the activity of microorganisms as far as some scientific sources are classified according to these two parameters. To be successful in bioremediation methods, individuals must have sufficient knowledge in the fields of physiological, genetic and ecological biochemistry of microorganisms (Gangar et al., 2019).

The use of native microorganisms provides more appropriate results for remediation of PTEs contaminated environment, however, the microorganisms that have shown their potential for soil remediation may still improve their capability but it is necessary to use molecular tools to understand the genetic processes involved (Singh et al., 2006; Valls & De Lorenzo, 2002). This increase in capability leads to widespread use of this method (Tylecote, 2019). The toxicity of PTEs in addition to the dose depends on their chemical form, which effects on the bioavailability of PTEs for organisms and their mobility in the environment (Giovanella et al., 2020). The mobility of metals associated the solubility and reactivity, which is determined by the chemical form of the metals (chemical speciation) that this word can refer to, physical state of the element, the identity of the element and its oxidation state (Ge et al., 2000; van Hullebusch et al., 2005).

On the other hand, the performance of microorganisms to clean up contaminated environments depends on the appropriate environmental characteristics for the metabolism and growth of microorganisms, such as moisture, temperature, and pH (Verma & Jaiswal, 2016). Therefore, various factors related to the environment, PTEs and microorganisms have interactions and change in one factor will change other factors.

To succeed in situ bioremediation method the four main items required are: (1) the presence of microorganisms in sufficient density and suitable for the bioremediation of pollutants (2) nutrients needed for microorganisms are available (3) the environment should be in appropriate condition (4) there is enough time to implement the bioremediation process (Ayangbenro & Babalola, 2017).

The efficiency of the bioremediation process has been proved at laboratory and field scale (Huang et al., 2015). The success rate of bioremediation and pollution elimination in this process can be monitored. Quantitatively, the rate of contaminants immobilisation/immobilization is measured using various devices such as atomic absorption (AA) and gas chromatography (GC), and in terms of quality, bioindicators are used (Azubuike et al., 2020). Some of these markers are used to monitor the elimination of pollutants such as earthworms, microorganisms, plants and Enzymes (Maila & Cloete, 2005).

7. Bioavailability of PTEs

One of the most important topics in bioremediation is the bioavailability of PTEs that its importance and relevance to the discussion of bioremediation is evident in its definition. A general definition of bioavailability is provided by the National Research Council (2003), which is: the complex mass transmission and absorption of pollutants inside soil organisms is influenced by the specific PTEs characteristics (including chemical speciation), soil properties and the biology of organisms (plants or microorganisms in the bioremediation frame) under different the conditions of environmental (van Hullebusch et al., 2005). EPA uses the concept of bioavailability to assess the risk of PTEs and risk management for their effects on living organisms and their absorption on the membrane of organisms (Park et al., 2011). Fig. 8.4 shows the reaction between PTEs with soil components and plants. These reactions eventually lead to the mobility and bioavailability of PTEs.

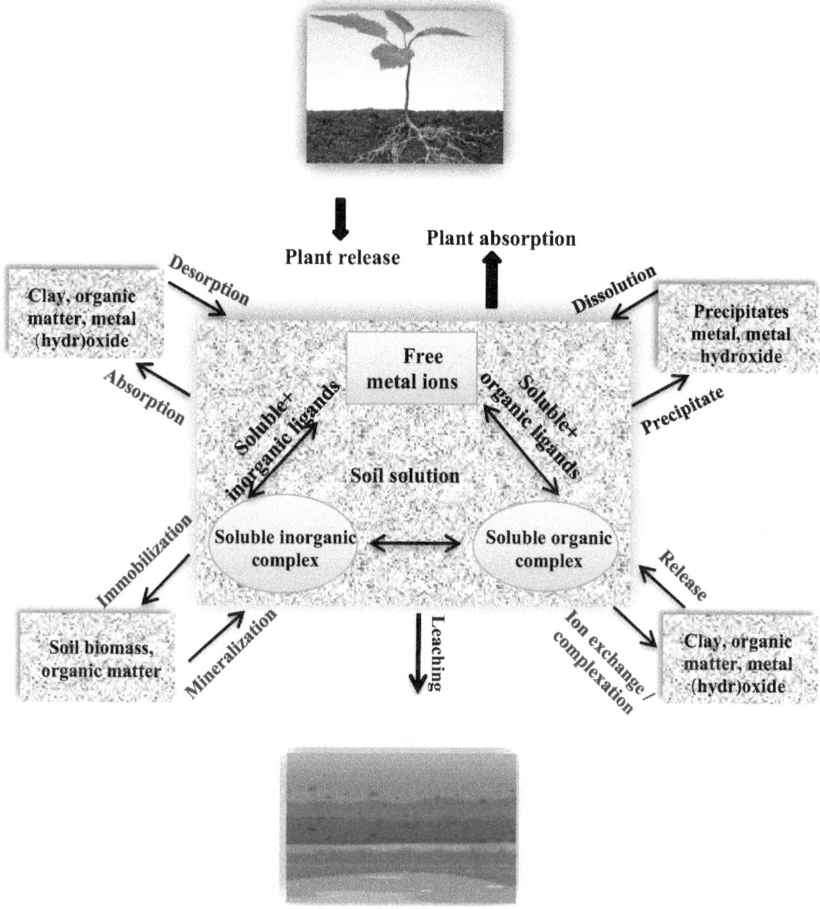

Fig. 8.4 The reaction between PETs with soil components and plants.

Monitoring frameworks around the world have moved toward risk-based approaches to assessing and correcting contaminated soils, so the bioavailability of pollutants must be studied and considered (Kim et al., 2015). Lanno et al. (2004) emphasized that bioavailability should be considered as a dynamic process involving three stages: (1) environmental availability (2) environmental bioavailability (3) toxicological bioavailability.

Environmental availability refers to the total soil content, including two parts dissolved in water in the pores of the soil (concentration of PTEs in the water of the pores of the soil) and the part that is potentially absorbed into the soil matrix (e.g., organic values and surface sedimentation values).

Environmental bioavailability refers to the metal forms that dissolve in water in soil pores and can be absorbed by plant roots or other soil organisms. And the toxicological bioavailability is defined as a physiological accumulation due to the biological effect of PTEs on plants or other organisms that depends on a number of complex processes such as transport, metabolism and detoxification (Kim et al., 2015).

Soil characteristics and the geochemical composition of each element control transferable concentrations, solutions, bioavailability and mobility (Beesley et al., 2010). Because soil chemistry and its liquid phase accommodate a large number of reactions, the effects of metals and soil are complex, so that the mobility and bioavailability of PTEs depends on soil characteristics such as pH, organic matter, cation exchange capacity, and oxidation potential (Guala et al., 2010).

PTEs do not decompose or degrade through biological processes, and can only be caused by changes in their chemical form or by their immobility in the soil (Modoi et al., 2014; Sobariu et al., 2016). This chemical deformation is closely related to the physical and chemical properties of the soil, so that the bioavailability of PTEs depends mainly on their chemical form in the soil. In fact, different chemical forms of PTEs metals in soils; makes chemical stability and different health risk a different hygiene (Tang et al., 2018).

8. Mechanism of bioremediation of PTEs from soil

Understanding the mechanisms used by microorganisms to remove PTEs from the soil is crucial to increasing the efficiency of this method as well as its extension to large-scale application. Microbes use several mechanisms to remediation of contaminated soils, including Bioleaching, Bioaccumulation, Biosorption, Biotransformation, Biomineralisation and Biosorption. Bioaccumulation involves the transfer of PTEs from the outer surfaces to the cell cytoplasm. This mechanism is similar to the way nutrients are absorbed by microorganisms. In this process, PTEs enter the cell cytoplasm after the metal ions are absorbed from the cell surface (Banerjee et al., 2018). In the process of biomineralization, a special form of the mineral is biosynthesized with microorganisms and by molecules that are useful in various functions of the biological system (Kim & Park, 2013). Biosorption is a process in which the cell structure of a microorganism can trap metal ions and then absorb from the cell wall that is autonomous of the metabolic cycle (Ayangbenro & Babalola, 2017). Economically, biosorption process is more appropriate than other methods, and the absorption capacity depends on the chemical, physical and biological properties of the biomass (Pratush et al., 2018).

Microorganisms are ubiquitous organisms and spread in places contaminated with PTEs and can convert PTEs into less toxic forms (Verma & Kuila, 2019). Fungi can grow and live in PTEs polluted environments and accumulate PTEs and nutrients. They are therefore highly absorbent and can be used on a large scale to absorb PTEs (Yin et al., 2019). There are many factors involved in the ability of fungi to absorb PTEs, the most important inclusive glucuronic acid, chitin-chitosan complex, phosphate and polysaccharides in fungi cells which operate through coordination and ion exchange (Purchase et al., 2009). Among the soil microorganisms, bacteria have the highest abundance. These microorganisms live in an extensive suffering of conditions and are used for the remediation of PTEs contaminated environments due to specific characteristics including easy cultivation, rapid growth rate and tiny size (Yin et al., 2019). In fact, PTEs are adsorbed on the polysaccharide layers of the bacteria using phosphate, amino, carboxyl and sulfate groups (Yue et al., 2015).

In the bioremediation method, microorganisms may use PTEs as a feedstock to produce energy (Raghunandan et al., 2018). Microorganisms employ various mechanisms to resist the toxicity of PTEs, including efflux, biotransformation, use of enzymes, and synthesis of metallothioneins and biosurfactants (Igiri et al., 2018). In fact microorganisms retaining the health of soil through the enzymes they produce and eliminate contaminants from contaminated environments (Kumar et al., 2019). In the way that microorganisms use enzymes to create depletion conditions for PTEs, metals acting as electron acceptors or electron donors (Banerjee et al., 2018). Fig. 8.5 illustrates the most important bioremediation mechanisms of PTEs.

Microorganisms, remediation polluted environments with the help of different methods, including: binding, immobilization, oxidation, transformation and volatizing of PTEs (Verma & Kuila, 2019). Microorganisms are complex in two ways: (1) Low molecular weight organic acids (2) High-molecular-weight ligands that occur more frequently in PTEs which is highly dependent on pH (Ayangbenro & Babalola, 2017). In relation to the type of reactions, in most cases bioremediation is a process performed by a microorganism but in some cases there are taking place in anaerobic conditions (Colberg & Young, 1995).

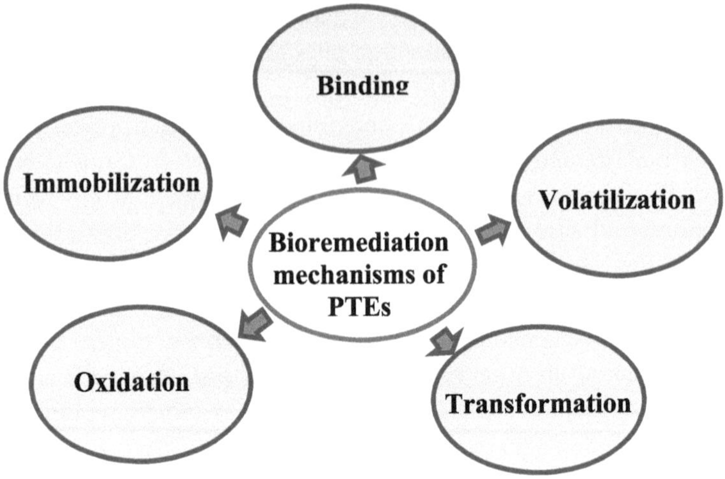

Fig. 8.5 The most important bioremediation mechanisms of PTEs.

9. Speciation and mobility of PTEs

Soil is a combination of organic and inorganic materials, including humic, clay, iron oxides, manganese and aluminum, and soil solution. Mobility and speciation of PTEs depend on the bonding to these soil components, which is related to the type and condition of compounds soil, redox conditions and soil reaction. Different forms of PTEs in connection with various soil components have surface reaction and bonding energies of different (Kushwaha et al., 2018) which are very effective on the mobility of metals. The following are some of the factors that affect on the speciation of PTEs which, are very important in the bioavailability of PTEs and mobility being possibly affected by microbial metabolic activities (Chuan et al., 1996; DeAngelis et al., 2010; Husson, 2013).

9.1 Oxidation/reductions (redox)

The redox potential (Eh) plays a key role in the solubility and bioavailability of PTEs, and these factors have a linear relationship (Seshadri et al., 2016). So that, the higher the redox potential, the higher the solubility and thus the higher the bioavailability.

Redox affects directly or indirectly the chemical properties of pollutants, including U, As, Cu, Co, Pu, Ni, Pb, Cr and Se (Alloway, 1995). According to the importance of the iron in soils and their role as sorbate of inorganic metal, the redox and microbial are important in the mobility of pollutants (van Hullebusch et al., 2005). Metals under the influence of various factors, the most important of which is redox, have different forms in the soil, and the mobility and toxicity of these forms are different. For example As(III) and As(V) are the two chemical forms of arsenic in soil that have less mobility and toxicity in As(V) redox state, and chromium is present in two chemical forms, Cr(III) and Cr(VI), which have less toxicity and mobility in Cr(III).

Arsenic, selenium, mercury, and chromium are most affected by microbial oxidation-reductions, while bimetallic metals such as cadmium, zinc, nickel, and lead are not. These reactions fall in two categories, dissimilatory and assimilatory. In assimilatory metal (loid)s as an electron receptor in and enhances bacterial growth. In dissimilatory metal (loid)s hasnot specific function in the physiology of microorganisms and occur microbial reductions of alcohols and organic acids, aromatic compounds or H_2 (Park et al., 2011).

Some microorganisms have the ability to reduce Fe(III) and in this way they produce energy for growth. In this case, the use of iron or other metals as electron acceptors is called dissimilatory (Tabak et al., 2005).

Bacteria use enzymes to convert Cr(III) to Cr(VI) and Hg(II) to Hg (0), which is less toxic and to secondary compounds such as Cr(VI) and Hg (0), and that's a kind of bioremediation strategy (Borch et al., 2010).

9.2 Soil pH

Another factor determining the chemical form PTEs in the soil is pH. Metals in different pH are of various chemical forms. Different chemical forms have various motility and toxicity. Some PTEs are more soluble in acidic pH. For example, in alkaline pH, lead is mostly present in the form of lead carbonate and lead phosphate in the environment that, these two combinations are insoluble. But in acidic pH, lead is present as free ions (Kushwaha et al., 2018). In general, in the range of $pH < 5.2 \pm 0.2$, lead solubility and bioavailability increase (Martınez & Motto, 2000).

When the pH of the soil solution is between 3 and 6, H_2AsO_4 is the most abundant chemical form of arsenic in the soil, and if the pH of the soil solution is between 7 and 8, both $H_2AsO_4^-$ and $HAsO_4^{2-}$ are present in significant concentrations (Saghafi et al., 2020; Fulladosa et al., 2004). Alkaline pH above 11 have high levels of AsO_4^{3-}, but soils with pH above 11 are very uncommon. Therefore, AsO_4^{3-} is not important in soil-soluble oxides (Koo et al., 2012). Soil pH also affect on the precipitation of cations and the bonding of cations and anions. In alkaline soils ($pH > 7$), the precipitation of PTEs with such as HPO_4^{2-}, CO_3^{2-}, OH^- and SO_4^{2-} anions is very common (Ok et al., 2010). The impact of pH perhaps results of indirect the activity of microbes that control the oxidation and reduction of iron and manganese (Olaniran et al., 2013).

9.3 Soil compounds

The effect of PTEs on soil ecology, in addition to amount in the soil, is closely related to the distribution of their various forms in the liquid and solid phase of the soil (Brümmer, 1986). Soil mineral and organic compounds are effective factors on the type of PTEs compounds and therefore in their solubility and bioavailability. Soil organic matter including of two parts, humic and non-humic (amino acids, fats, sugars etc), that 80%–90% of soil organic matter is humic (Giesking & Anderson, 1975). Humic substances are known as humic acids. Humic acids have less mobility in the soil than fulvic acids, which is due to the higher molecular weight of humic acids also; humic acids have lower carbon and oxygen content (Schulten et al., 1991). PTEs are attached to these parts. In acidic PTEs associated with organic matter have higher solubility. For example, lead complex with organic matter in pH four to six soluble and available for plants (Kushwaha et al., 2018). Soil mineral colloids that, including clay, sesquioxides, minerals hydrous oxides and oxides play an important role in the absorption of metals. Clay is classified into two groups 2:1 and 1:1. Type 2:1 have two layers, which there is a weak van der Waals bond between them. When water and chemicals, including PTEs, are placed between these two layers, they become immobile (Kushwaha et al., 2018).

On the other hand, some soil alkaline substances such as calcium carbonate and magnesium and calcium oxides are used to remediation contaminated soils and the immobilization of PTEs (Derome, 2000). Therefore, the presence of these compounds in the soil can prevent the mobility of PTEs and reduce their bioavailability through strong chemical bonds with soil components.

10. Microbe-assisted phytoremediation

Phytoremediation include the use of plants to destroy and detoxify pollutants from contaminated environments (Ahmadi et al., 2020; Bharagava et al., 2017; Kazemalilou et al., 2020). There are two strategies for phytoremediation (1) the use of hyperaccumulators plants that produce small amounts of biomass but high concentrations of PTEs accumulate (2) the use of plants that produce high biomass but accumulate PTEs at low concentrations (Tlustos et al., 2006). Simultaneous use of several methods can increase the efficiency of the bioremediation method (Cassidy et al., 2015). It is possible to combine different methods through the second strategy, which can increase the production of biomass. However, increasing biomass production can also be beneficial for the first strategy.

Phytoremediation is one of the most effective methods of remediation of contaminated soils that has been combined with various methods to increase its efficiency, including the simultaneous application of plants and microorganisms, which has increased the efficiency of phytoremediation and remediation of PTEs contaminated soils. Most plants that grow in polluted environments have little growth due to accumulation of toxins and their harmful effects (Glick, 2003; Delangiz et al., 2020). That, these harmful effects and damages can be mitigated by microorganisms (Ahemad & Khan, 2013). Microorganisms present in soils contaminated with PTEs have above resistance to high concentrations of PTEs and the positive effect of microorganisms on plant resistance to high concentrations of PTEs has been confirmed by many researchers, including Chen et al. (2014); Ma et al. (2015); Złoch et al. (2017).

In bioremediation of plant-microorganism, two basic functions are performed: immobilization of PTEs and elimination of PTEs. By increasing the resistance of plants to PTEs, microorganisms increase the absorption and transfer of PTEs. Microorganisms optimize environmental conditions through the Arbuscular mycorrhiza, symbiotic complexes among root and endophytes for plant growth and increase their resistance (Li et al., 2015).

This positive effect can improve the process of phytostabilization and phytoextraction process and generally increase plant growth under environmental conditions (Khan et al., 2009). Increased growth and biomass production is one of the most important factors in phytoremediation and can increase the efficiency of this method. Microorganisms increase phytoremediation capacity in two ways (Deb et al., 2020).

(1) Directly: Microorganisms increase the uptake and transport of PTEs (facilitation of phytoextraction) or decrease the mobility and the availability of PTEs for the rhizosphere (phytostabilization).

(2) Indirectly: Microorganisms increase plant resistance to PTEs and increase plant growth and biomass and increase removal of PTEs through accumulation in plant biomass.

To increase the capabilities and efficiency of this technique, screening should be done between microorganisms and plants to select the most suitable plants and microorganisms to design capable systems in this field.

11. Conclusions

With the increasing world industrialization, environmental pollution problems have increased, and various types of pollutants enter the food chain and endanger the human health after being exposed to the environment through various processes. PTEs enter the environment from both natural and anthropogenic sources and remain in soils for a long time. As compared to different pollutants, PTEs are widely distributed in the environment, some of which are even problematic for humans and organisms even at low concentrations. Compared to other parts of the ecosystem, the soils have high ability to spread PTEs pollution, and in fact, soil is a reservoir of various pollutants, including PTEs. Therefore, the modification of soils contaminated with PTEs has received more and more attention from researchers and they have used various methods to remediation the soil. Bioremediation is one of the efficient and useful methods in this field that, many researchers are hoping to develop so that it can replace other methods for its important benefits. And in some cases, the combination of microorganisms and plants is used to remediation contaminated soils. In these processes, microorganisms generally, use various mechanisms to convert toxic pollutants (PTEs) into less toxic compounds and, by reducing the bioavailability of PTEs, prevent their transfer to living organisms.

References

Acosta, J. A., Gabarron, M., Faz, A., Martínez-Martínez, S., Zornoza, R., & Arocena, J. M. (2015). Influence of population density on the concentration and speciation of metals in the soil and street dust from urban areas. *Chemosphere, 134*, 328–337.

Agnello, A. C., Bagard, M., van Hullebusch, E. D., Esposito, G., & Huguenot, D. (2016). Comparative bioremediation of heavy metals and petroleum hydrocarbons co-contaminated soil by natural attenuation, phytoremediation, bioaugmentation and bioaugmentation-assisted phytoremediation. *Science of the Total Environment, 563*, 693–703.

Ahemad, M., & Khan, M. S. (2013). Pesticides as antagonists of rhizobia and the legume-rhizobium symbiosis: A paradigmatic and mechanistic outlook. *Biochemistry & Molecular Biology, 1*(4), 63–75.

Ahmadi, O., Pandey, J., Khadem Moghadam, N., Asgari Lajayer, B., & Ghorbanpour, M. (2020). Phytoremediation of contaminated soils using trees. In M. Faisal, Q. Saquib, A. A. Alatar, & A. A. Al-Khedhairy (Eds.), *Cellular and molecular phytotoxicity of heavy metals. Nanotechnology in the life sciences*. Cham: Springer. https://doi.org/10.1007/978-3-030-45975-8_21

Akar, T., Tunali, S., & Cabuk, A. (2007). Study on the characterization of lead (II) biosorption by fungus *Aspergillus parasiticus*. *Applied Biochemistry and Biotechnology, 136*, 389–406.

Alexander, M. (1991). *Introduction to soil microbiology* (2nd ed., p. 467). Krieger Publishing Company Malabar. January 1, 1991.

Alloway, B. J. (1995). In B. J. Alloway (Ed.), *Heavy metals in soils*. London, UK: Blackie Academic & Professional.

Alvarez, A., Saez, J. M., Costa, J. S. D., Colin, V. L., Fuentes, M. S., Cuozzo, S. A., Benimeli, C. S., Polti, M. A., & Amoroso, M. J. (2017). Actinobacteria: Current research and perspectives for bioremediation of pesticides and heavy metals. *Chemosphere, 166*, 41–62.

Archana, H. S., & Jaitly, A. K. (2014). Bioremediation: Environmental biotechnology for heavy metal decontamination of soil and water. *Biochemical and Cellular Archives, 14*(2), 259–281.

Asgari Lajayer, B., Ghorbanpour, M., & Nikabadi, S. (2017). Heavy metals in contaminated environment: Destiny of secondary metabolite biosynthesis, oxidative status and phytoextraction in medicinal plants. *Ecotoxicology and Environmental Safety, 145*, 377–390.

Asgari Lajayer, B., Khadem Moghadam, N., Maghsoodi, M. R., Ghorbanpour, M., & Kariman, K. (2019). Phytoextraction of heavy metals from contaminated soil, water and atmosphere using ornamental plants: Mechanisms and efficiency improvement strategies. *Environmental Science and Pollution Research, 26*(9), 8468–8484.

Ayangbenro, A. S., & Babalola, O. O. (2017). A new strategy for heavy metal polluted environments: A review of microbial biosorbents. *International Journal of Environmental Research and Public Health, 14*(1). https://doi.org/10.3390/ijerph14010094

Azubuike, C. C., Chikere, C. B., & Okpokwasili, G. C. (2020). Bioremediation: An eco-friendly sustainable technology for environmental management. In G. Saxena, & R. N. Bharagava (Eds.), *Bioremediation of industrial waste for environmental safety* (pp. 19–39). Springer Singapore.

Banerjee, A., Jhariya, M. K., Yadav, D. K., & Raj, A. (2018). Micro-remediation of metals: A new frontier in bioremediation. In C. M. Hussain (Ed.), *Handbook of environmental materials management. 1-36.* Switzerland: Springer Inc.

Beesley, L., Moreno-Jiménez, E., & Gomez-Eyles, J. L. (2010). Effects of biochar and greenwaste compost amendments on mobility, bioavailability and toxicity of inorganic and organic contaminants in a multi-element polluted soil. *Environmental Pollution, 158*, 2282–2287.

Bharagava, R. N., Chowdhary, P., & Saxena, G. (2017). Bioremediation: An ecosustainable green technology: Its applications and limitations. In R. N. Bharagava (Ed.), *Environmental pollutants and their bioremediation approaches* (pp. 1–22). Taylor & Francis Inc.

Borch, T., Kretzschmar, R., Kappler, A., Cappellen PV Ginder-Vogel, M., Voegelin, A., & Campbell, K. (2010). Biogeochemical redox processes and their impact on contaminant dynamics. *Environmental Science & Technology*, 505–508.

Brümmer, G. W. (1986). Heavy metal species, mobility and availability in soils. In M. Bernhard, F. E. Brinckman, & P. J. Sadler (Eds.), *The importance of chemical "speciation" in environmental processes. 169-192.* Berlin, Heidelberg: Springer.

Cai, L., Xu, Z., Bao, P., He, M., Dou, L., Chen, L., Zhou, Y., & Zhu, Y. G. (2015). Ultivariate and geostatistical analyses of the spatial distribution and source of arsenic and heavy metals in the agricultural soils in Shunde, Southeast China. *Journal of Geochemical Exploration, 148*, 189–195.

Cassidy, D. P., Srivastava, V. J., Dombrowski, F. J., & Lingle, J. W. (2015). Combining in situ chemical oxidation, stabilization, and anaerobic bioremediation in a single application to reduce contaminant mass and leachability in soil. *Journal of Hazardous Materials, 297*, 347–355.

Chen, D., Liu, X., Bian, R., Cheng, K., Zhang, X., Zheng, J., Joseph, S., Crowley, D., Pan, G., & Li, L. (2018). Effects of biochar on availability and plant uptake of heavy metals – a meta analysis. *Journal of Environmental Management, 222*, 76–85.

Chen, L., Luo, S., Li, X., Wan, Y., Chen, J., & Liu, C. (2014). Interaction of Cd hyperaccumulator *Solanum nigrum* L. and functional endophyte *Pseudomonas* sp. Lk9 on soil heavy metals uptake. *Soil Biology and Biochemistry, 68*, 300–308.

Chen, H., Yuan, X., Xiong, T., Jiang, L., Wang, H., & Wu, Z. (2020). Biochar facilitated hydroxyapatite/calcium silicate hydrate for remediation of heavy metals contaminated soils. *Biochar Facilitated Hydroxyapatite/Calcium Silicate Hydrate for Remediation of Heavy Metals Contaminated Soils, 231*(66), 1–16. https://doi.org/10.1007/s11270-020-4425-1

Chuan, M. C., Shu, G. Y., & Liu, J. C. (1996). Solubility of heavy metals in a contaminated soil: Effects of redox potential and pH. *Water, Air, and Soil Pollution, 90*(3–4), 543–556.

Colberg, P. J. S., & Young, L. L. (1995). Anaerobic degradation of nonhalogenated homocyclic aromatic compounds coupled with nitrate, iron, or sulfate reduction. In L. L. Young, & C. E. Cerniglia (Eds.), *Microbial transformation and degradation of toxic organic chemicals* (pp. 307–330). New York: Wiley-Liss.

D'Annibale, A., Leonardi, V., Federici, E., Baldi, F., Zecchini, F., & Petruccioli, M. (2007). Leaching and microbial treatment of a soil contaminated by sulphide ore ashes and aromatic hydrocarbons. *Applied Microbiology and Biotechnology, 74*, 1135–1144.

Das, S. K., & Varma, A. (2010). Role of enzymes in maintaining soil health. In G. Shukla, & A. Varma (Eds.), *Soil enzymology* (pp. 25–42). Berlin, Heidelberg: Springer.

DeAngelis, K. M., Silver, W. L., Thompson, A. W., & Firestone, M. K. (2010). Microbial communities acclimate to recurring changes in soil redox potential status. *Environmental Microbiology, 12*(12), 3137–3149.

Deb, V. K., Rabbani, A., Upadhyay, S., Bharti, P., Sharma, H., Rawat, D. S., & Saxena, G. (2020). Microbe-Assisted phytoremediation in reinstating heavy metal contaminated sites: Concepts, mechanisms, challenges, and future perspectives. In P. K. Arora (Ed.), *Microbial technology for health and environment* (pp. 161–189). Springer Singapore.

De, J., Ramaiah, N., & Vardanyan, L. (2008). Detoxifcation of toxic heavy metals by marine bacteria highly resistant to mercury. *Marine Biotechnology, 10*(4), 471–477.

Delangiz, N., Khoshru, B., Asgari Lajayer, B., Ghorbanpour, M., & Kazemalilou, S. (2020). Molecular mechanisms of heavy metal tolerance in plants. In M. Faisal, Q. Saquib, A. A. Alatar, & A. A. Al-Khedhairy (Eds.), *Cellular and molecular phytotoxicity of heavy metals. Nanotechnology in the life sciences*. Cham: Springer.

Dermont, G., Bergeron, M., Mercier, G., & Laflèche, M. R. (2008). Metal-contaminated soils: Remediation practices and treatment technologies. *Practice Periodical of Hazardous, Toxic, and Radioactive Waste Management, 12*(3), 188–209.

Derome, J. (2000). Detoxification and amelioration of heavy-metal contaminated forest soils by means of liming and fertilisation. *Search Results Web Results Environmental Pollution, 107*, 79–88.

Dornbusch, M. J., Limb, R. F., Tomlinson, H. A. K., Daigh, A. L. M., & Sedivec, K. K. (2020). Evaluation of soil treatment techniques on remediated brine spill sites in semi-arid rangelands. *Journal of Environmental Management, 260*, 100–110.

Duffus, J. H. (2002). Heavy metals" a meaningless term? (IUPAC technical report). *Pure and Applied Chemistry, 74*(5), 793–807.

Edelstein, M., & Ben-Hur, M. (2018). Heavy metals and metalloids: Sources, risks and strategies to reduce their accumulation in horticultural crops. *Scientia Horticulturae, 234*, 431–444.

EPA. (2005). *US environmental protection agency guidance for developing ecological soil screening levels*. Washington, DC: Office of Solid Waste and Emergency Response. Available on https://www.epa.gov/chemical-research/guidance-developing-ecological-soil-screening-levels.

FAO/WHO. (2007). Joint FAO/WHO food standard programme codex alimentarius commission 13th session. In *Report of the thirty eight session of the codex committee on food hygiene*. Houston, USA, ALINORM 07/30/13.

Favero, N., Costa, P., & Massimino, M. L. (1991). In vitro uptake of cadmium by basidiomycete *Pleurotus ostreatus*. *Biotechnology Letters, 10*, 701–704.

Fulladosa, E., Murat, J. C., Martinez, M., & Villaescusa, I. (2004). Effect of pH on arsenate and arsenite toxicity to luminescent bacteria (Vibrio fischeri). *Archives of Environmental Contamination and Toxicology, 46*(2), 176–182.

Gadd, G. M. (2007). Geomycology: Biogeochemical transformations of rocks, minerals, metals and radionuclides by fungi, bioweathering and bioremediation. Search results web results. *Mycological Research, 111*, 3–49.

Gangar, T., Bhardwaj, kk, & Gupta, R. (2019). Microbes and processes in bioremediation of soil. In A. Kumar, & S. Sharma (Eds.), *Microbes and enzymes in soil health and bioremediation* (pp. 11–37). Springer Singapore.
Ge, Y., Murray, P., & Hendershot, W. H. (2000). Trace metal speciation and bioavailability in urban soils. *Environmental Pollution, 107*(1), 137–144.
Giannakopoulou, F., Gasparatos, D., Haidouti, C., & Mass, I. (2012). Sorption behavior of cesium in two Greek soil: Effects of Cs initial concentration, clay mineralogy and particle size fraction. *Soil and Sediment Contamination, 21*(8), 937–950.
Giesking, J. E., & Anderson, G. (1975). Soil components, 1. Organic components. *Soil Components and Organic Components, 1.*
Giovanella, P., Cabral, L., Costa, A. P., de Oliveira Camargo, F. A., Gianello, C., & Bento, F. M. (2017). Metal resistance mechanisms in gram-negative bacteria and their potential to remove Hg in the presence of other metals. *Ecotoxicology and Environmental Safety, 140*, 162–169.
Giovanella, P., Vieira, G. A. L., Otero, I. V. R., Pellizzer, E., Fontes, B. J., & Sette, L. D. (2020). Metal and organic pollutants bioremediation by extremophile microorganisms. *Journal of Hazardous Materials, 382.* https://doi.org/10.1016/j.jhazmat.2019.121024
Glick, B. R. (2003). Phytoremediation: Synergistic use of plants and bacteria to clean up the environment. *Biotechnology Advances, 21*(5), 383–393.
González, N., Calderón, J., Rúbies, A., Timoner, I., Castell, V., Domingo, J. L., & Nadal, M. (2019). Dietary intake of arsenic, cadmium, mercury and lead by the population of Catalonia, Spain: Analysis of the temporal trend. *Food and Chemical Toxicology, 132*, 110721.
Guala, S. D., Vega, F. A., & Covelo, E. F. (2010). Heavy metal concentrations in plants and different harvestable parts: A soil-plant equilibrium model. *Environmental Pollution, 158*, 2659–2663.
Gupta, G. S., Yadav, G., & Tiwari, S. (2020). Bioremediation of heavy metals: A new approach to sustainable agriculture. In A. Kumar, U. R. Singh, & D. P. Singh (Eds.), *Restoration of wetland ecosystem: A trajectory towards a sustainable environment* (pp. 195–226). Springer Singapore.
Harmsen, J., & Naidu, R. (2013). Bioavailability as a tool in site management. *Journal of Hazardous Materials, 261*, 840–846.
Heidarpour, A., Aliasgharzad, N., Khoshmanzar, E., khoshru, B., & Asgari Lajayer, B. (2019). Bio-removal of Zn from contaminated water by using green algae isolates. *Environmental Technology and Innovation, 16*, 100464. https://doi.org/10.1016/j.eti.2019.100464
Huang, G., Liu, F., Yang, Y., Deng, W., Li, S., Huang, Y., & Kong, X. (2015). Removal of ammonium nitrogen from groundwater using a fully passive permeable reactive barrier with oxygen releasing compound and clinoptilolite. *Journal of Environmental Management, 154*, 1–7.
Husson, O. (2013). Redox potential (eh) and pH as drivers of soil/plant/microorganism systems: A transdisciplinary overview pointing to integrative opportunities for agronomy. *Plant and Soil, 362*(1–2), 389–417.
Igiri, B. E., Okoduwa, S. I. R., Idoko, G. O., Akabuogu, E. P., Adeyi, A. O., & Ejiogu, I. K. (2018). Toxicity and bioremediation of heavy metals contaminated ecosystem from tannery wastewater: A review. *Journal of Toxicology*, 1–16, 2018.
Jain, S., & Arnepalli, D. N. (2019). Biominerlisation as a remediation technique: A critical review. In V. K. Stalin, & M. Muttharam (Eds.), *Geotechnical characterisation and geoenvironmental engineering* (pp. 155–162). Singapore: Springer.
Jalilvand, N., Akhgar, A., Alikhani, H. A., Rahmani, H. A., & Rejali, F. (2020). Removal of heavy metals zinc, lead, and cadmium by biomineralization of urease-producing bacteria isolated from Iranian mine calcareous soils. *Journal of Soil Science and Plant Nutrition, 20*, 206–219.

Jebeli, M. A., Maleki, A., Amoozegar, M. A., Kalantar, E., Izanloo, H., & Gharibi, F. (2017). *Bacillus flexus* strain As-12, a new arsenic transformer bacterium isolated from contaminated water resources. *Chemosphere, 169*, 636–641.

Juned, S., Nagargoje, B. N., Bhosle, A. B., & Yannawar, V. B. (2018). Monitoring of heavy metals and physico-chemical parameters of soil, near sudha dam bhokar, Maharashtra. *Journal of American Science, 14*(5), 35–40.

Karimi, H., Mahdavi, S., Asgari Lajayer, B., Moghiseh, E., Rajput, V. D., Minkina, T., & Astatkie, T. (2021). Insights on the bioremediation technologies for pesticide contaminated soils. *Environmental Geochemistry and Health*. https://doi.org/10.1007/s10653-021-01081-z

Kazemalilou, S., Delangiz, N., Asgari Lajayer, B., & Ghorbanpour, M. (2020). Insight into plant-bacteria-fungi interactions to improve plantperformance via remediation of heavy metals: An overview. In V. Sharma, R. Salwan, & L. K. T. Al-Ani (Eds.), *Molecular aspects of plant beneficial microbes in agriculture*. Elsevier Inc. https://doi.org/10.1016/B978-0-12-818469-1.00010-9

Kelly, D. J. A., Budd, K., & Lefebvre, D. D. (2006). The biotransformation of mercury in pH-stat cultures of microfungi. *Canadian Journal of Botany, 84*(2), 254–260.

Khan, M. S., Zaidi, A., Wani, P. A., & Oves, M. (2009). Role of plant growth promoting rhizobacteria in the remediation of metal contaminated soils. *Environmental Chemistry Letters, 7*, 1–19.

Kim, S., & Park, C. B. (2013). Bio-inspired synthesis of minerals for energy, environment and medicinal applications. *Advanced Functional Materials, 23*(1), 10–25.

Kim, R. Y., Yoon, J. K., Kim, T. S., Yang, J. E., Owens, G., & Kim, K. R. (2015). Bioavailability of heavy metals in soils: Definitions and practical implementation—a critical review. *Environmental Geochemistry and Health, 37*(6), 1041–1061.

Kiyono, M., & Pan-Hou, H. (2006). Genetic engineering of bacteria for environmental remediation of mercury. *Journal of Health Science, 52*(3), 199–204.

Koo, N., Lee, S. H., & Kim, J. G. (2012). Arsenic mobility in the amended mine tailings and its impact on soil enzyme activity. *Environmental Geochemistry and Health, 34*, 337–348.

Kostal, J., Yang, R., Wu, C. H., Mulchandani, A., & Chen, W. (2004). Enhanced arsenic accumulation in engineered bacterial cells expressing ArsR. *Applied and Environmental Microbiology, 70*(8), 4582–4587.

Kumar, A., Sharma, T., Mulla, S. I., Kamyab, H., Pant, D., & Sharma, S. (2019). Let's protect our earth: Environmental challenges and implications. In A. Kumar, & S. Sharma (Eds.), *Microbes and enzymes in soil health and bioremediation* (pp. 11–37). Singapore: Springer.

Kushwaha, A., Hans, N., Kumar, S., & Rani, R. (2018). A critical review on speciation, mobilization and toxicity of lead in soil-microbe-plant system and bioremediation strategies. *Ecotoxicology and Environmental Safety, 147*, 1035–1045.

Lanno, R., Wells, J., Conder, J., Bradham, K., & Basta, N. (2004). The bioavailability of chemicals in soil for earthworms. *Ecotoxicology and Environmental Safety, 57*, 39–47.

Li, Y. S., Feng, C. L., Wu, X. F., & Shi, R. (2015). A review on the functions of microorganisms in the phytoremediation of heavy metal-contaminated soils. *Acta Ecologica Sinica, 35*(20), 6881–6890.

Lindstrom, E. B., & Sehlin, H. M. (1989). High efficiency of plating of the thermophilic sulfur-dependent archaebacterium *Sulfolobus acidocaldarius*. *Applied and Environmental Microbiology, 55*, 3020–3021, 1989.

Liu, Y. G., Xu, W. H., Zeng, G. M., Li, X., & Gao, H. (2006). Cr (VI) reduction by *Bacillus* sp. isolated from chromium landfill. *Process Biochemistry, 41*, 1981–1986.

Liu, J., Zhanga, Y., Wang, H., & Du, Y. (2018). Study on the prediction of soil heavy metal elements content based on visible near-infrared spectroscopy. *Spectrochimica Acta A: Molecular and Biomolecular Spectroscopy, 199*, 43–49.

Lloyd, J. R. (2002). Bioremediation of metals; the application of micro-organisms that make and break minerals. *Interactions, 2*, 1–3.
Lloyd, J. R., Lovley, D. R., & Macaskie, L. E. (2003). Biotechnological application of metal-reducing microorganisms. *Advances in Applied Microbiology, 53*, 85–128.
Lwin, C. S., Seo, B. H., Kim, H. U., Owens, G., & Kim, K. R. (2018). Application of soil amendments to contaminated soils for heavy metal immobilization and improved soil quality—a critical review. *Soil Science and Plant Nutrition, 64*(2), 156–167.
Maila, M. P., & Cloete, T. E. (2005). The use of biological activities to monitor the removal of fuel contaminants – perspective for monitoring hydrocarbon contamination: A review. *International Biodeterioration & Biodegradation, 55*(1), 1–8.
Ma, Y., Oliviera, R. S., Nai, F., Rajkumar, M., Luo, Y., Rocha, I., & Freitas, H. (2015). The hyperaccumulator *Sedum plumbizincicola* harbors metal-resistant endophytic bacteria that improve its phytoextraction capacity in multi-metal contaminated soil. *Journal of Environmental Management, 156*, 62–69.
Martınez, C. E., & Motto, H. L. (2000). Solubility of lead, zinc and copper added to mineral soils. *Environmental Pollution, 107*, 153–158.
Massas, I., Kalivas, D., Ehaliotis, C., & Gasparator, D. (2013). Total and available heavy metal concentration in soils of Thriassio plain (Greece) and assessment of soil pollution indexes. *Environmental Monitoring Assessment, 185*, 6751-6466.
Miransari, M. (2017). The interactions of soil microbes affecting stress alleviation in agroecosystems. In V. Kumar, M. Kumar, S. Sharma, & R. Prasad (Eds.), *Probiotics in agroecosystem* (pp. 31–50). Springer Singapore.
Modoi, O. C., Roba, C., Török, Z., & Ozunu, A. (2014). Environmental risks due to heavy metal pollution of water resulted from mining wastes in NW Romania. *Environmental Engineering and Management Journal, 13*, 325–2336.
Mohammadhosseini, M., Bahmanpour, H., & Lotfi, S. (2014). Investigation and identification of types and amounts of heavy metals in soil of an industrial area. *Journal of Chemical Health Risk, 4*(4), 33–44.
Mohammadian Fazli, M., Soleimani, N., Mehrasbi, M., Darabian, S., Mohammadi, J., Ramazani, A., et al. (2015). Highly cadmium tolerant fungi: Their tolerance and removal potential. *Journal of Environmental Health Science and Engineering, 13*, 1–9.
Nejad, Z. D., Jung, M. C., & Kim, K. H. (2018). Remediation of soils contaminated with heavy metals with an emphasis on immobilization technology. *Environmental Geochemistry and Health, 40*, 927–953.
Nizamani, P., Afridi, H. I., Kazi, T. G., Talpur, F. N., & Baig, J. A. (2019). Essential trace elemental levels (zinc, iron and copper) in the biological samples of smoker referent and pulmonary tuberculosis patients. *Toxicology Reports, 6*, 1230–1239.
Ok, Y. S., Oh, S. E., Ahmad, M., Hyun, S., Kim, K. R., Moon, D. H., Lee, S. S., Lim, K. J., Jeon, W. T., & Yang, J. E. (2010). Effects of natural and calcined oyster shells on Cd and Pb immobilization in contaminated soils. *Environmental Earth Sciences, 61*, 1301–1308.
Olaniran, A. O., Balgobind, A., & Pillay, B. (2013). Bioavailability of heavy metals in soil: Impact on microbial biodegradation of organic compounds and possible improvement strategies. *International Journal of Molecular Sciences, 14*, 10197–10228.
Palansooriya, K. N., Shaheen, S. M., Chen, S. S., Tsang, D. C., Hashimoto, Y., Hou, D., Bolan, N. S., Rinklebeb, J., & Ok, Y. S. (2020). Soil amendments for immobilization of potentially toxic elements in contaminated soils: A critical review. *Environment International, 134*. https://doi.org/10.1016/j.envint.2019.105046
Pankhurst, C. E., Hawke, B. G., McDonald, H. J., Kirkby, C. A., Buckerfield, J. C., Michelsen, P., Brien, K. A. O., Gupta, V. V. S. R., & Doube, B. M. (1995). Evaluation of soil biological properties as potential bioindicators of soil health. *Australian Journal of Experimental Agriculture, 35*(7), 1015–1028.

Park, J. H., Lamb, D., Paneerselvam, P., Choppala, G., Bolan, N., & Chung, J. W. (2011). Role of organic amendments on enhanced bioremediation of heavy metal (loid) contaminated soils. *Journal of Hazardous Materials, 185*(2–3), 549–574.

Pourret, O., & Bollinger, J. C. (2018). Heavy metal"-What to do now: To use or not to use? *The Science of the Total Environment, 610*, 419.

Pourret, O., & Hursthouse, A. (2019). It's time to replace the term "heavy metals" with "potentially toxic elements" when reporting environmental research. *International Journal of Environmental Research and Public Health, 16*(22), 1–6.

Pratush, A., Kumar, A., & Hu, Z. (2018). Adverse effect of heavy metals (as, Pb, Hg, and Cr) on health and their bioremediation strategies: A review. *International Microbiology, 21*, 97–106.

Priyalaxmi, R., Murugan, A., Raja, P., & Raj, K. D. (2014). Bioremediation of cadmium by *Bacillus safensis* (JX126862), a marine bacterium isolated from mangrove sediments. *International Journal of Current Microbiology and Applied Sciences, 3*, 326–335.

Purchase, D., Scholes, L. N., Revitt, D. M., & Shutes, R. B. E. (2009). Effects of temperature on metal tolerance and the accumulation of Zn and Pb by metal-tolerant fungi isolated from urban runoff treatment wetlands. *Journal of Applied Microbiology, 106*, 1163–1174.

Raghunandan, K., Kumar, A., Kumar, S., Permaul, K., & Singh, S. (2018). Production of gellan gum, an exopolysaccharide, from biodiesel-derived waste glycerol by *Sphingomonas* spp. *3Biotech., 8*(71), 1–13.

Saghafi, D., Bagherifam, S., Hatami, M., & Asgari Lajayer, B. (2020). The role of microbes in detoxification and availability of metalloids. In R. Deshmukh, D. K. Tripathi, & G. Guerriero (Eds.), *Metalloids in plants: Advances and future prospects*. John Wiley & Sons. https://doi.org/10.1002/9781119487210.ch5

Schulten, H. R., Plage, B., & Schnitzer, M. (1991). A chemical structure for humic substances. Search Results Web results. *The Science of Nature, 78*, 311–312.

Seshadri, B., Bolan, N. S., Wijesekara, H., Kunhikrishnan, A., Thangarajan, R., Qi, F., Matheyarasu, R., Rocco, C., Mbene, K., & Naidu, R. (2016). Phosphorus–cadmium interactions in paddy soils. *Geoderma, 270*, 43–59.

Shaheen, S. M., Balbaa, A. A., Khatab, A. M., & Rinklebe, J. (2017). Compost and sulfur affect the mobilization and phyto-availability of Cd and Ni to sorghum and barnyard grass in a spiked fluvial soil. *Environmental Geochemistry and Health, 39*(6), 1305–1324.

Sharma, P. K., Balkwill, D. L., Frenkel, A., & Vairavamurthy, M. A. (2000). New Klebsiella planticola strain (Cd-1) grows anaerobically at high cadmium concentrations and precipitates cadmium sulfide. *Applied and Environmental Microbiology, 66*, 3083–3087.

Sharma, B., Singh, P., Chauhan, P. S., Singh, S., & Singh, R. P. (2019). Microbes-Assisted remediation of metal polluted soils. In V. K. Gupta (Ed.), *New and future developments in microbial biotechnology and bioengineering:* Aspergillus *system properties and applications* (pp. 223–232). Elsevier Inc.

Singh, S., & Gupta, V. K. (2016). Biodegradation and bioremediation of pollutants: Perspectives strategies and applications. *International Journal of Pharmacology and Biological Sciences, 10*(1), 53–65.

Singh, R., Paul, D., & Jain, R. K. (2006). Biofilms: Implications in bioremediation. *Trends in Microbiology, 14*(9), 389–397.

Sinha, S. N., Biswas, M., Paul, D., & Rahaman, S. (2011). Biodegradation potential of bacterial isolates from tannery effluent with special reference to hexavalent chromium. *Biotechnology Bioinformatics and Bioengineering, 1*(3), 381–386.

Sobariu, D. L., Fertu, D. I. T., Diaconu, M., Pavel, L. V., Hlihor, R. M., Drăgoi, E. N., Curteanu, S., Lenz, M., Corvini, P. F., & Gavrilescu, M. (2016). Rhizobacteria and plant symbiosis in heavy metal uptake and its implications for soil bioremediation. *New Biotechnology, 25*(39), 125–134.

Sun, X., Meng, J., Huo, S., Zhu, J., & Zheng, S. (2020). Remediation of heavy metal pollution in soil by microbial immobilization with carbon microspheres. *International Journal of Environmental Science and Development, 11*(1), 43–47.

Tabak, H. H., Lens, P., van Hullebusch, E. D., & Dejonghe, W. (2005). Developments in bioremediation of soils and sediments polluted with metals and radionuclides–1. Microbial processes and mechanisms affecting bioremediation of metal contamination and influencing metal toxicity and transport. *Reviews in Environmental Science and Bio/Technology, 4*(3), 115–156.

Tang, W., Xia, Q., Shan, B., & Ng, J. C. (2018). Relationship of bioaccessibility and fractionation of cadmium in long-term spiked soils for health risk assessment based on four in vitro gastrointestinal simulation models. *Science of the Total Environment, 631*, 1582–1589.

Tlustos, P., Szakova, J., Hruby, J., Hartman, I., Najmanova, J., Nedelnik, J., Pavlikova, D., & Batysta, M. (2006). Removal of As, Cd, Pb and Zn from contaminated soil by high biomass producing plants. *Plant Soil Environment, 52*, 413–423.

Tylecote, A. (2019). Biotechnology as a new techno-economic paradigm that will help drive the world economy and mitigate climate change. *Research Policy, 48*(4), 858–868.

Valls, M., & De Lorenzo, V. (2002). Exploiting the genetic and biochemical capacities of bacteria for the remediation of heavy metal pollution. *FEMS Microbiology Reviews, 26*(4), 327–338.

van Hullebusch, E. D., Lens, P. N., & Tabak, H. H. (2005). Developments in bioremediation of soils and sediments polluted with metals and radionuclides. 3. Influence of chemical speciation and bioavailability on contaminants immobilization/mobilization bioprocesses. *Reviews in Environmental Science and Bio/Technology, 4*(3), 185–212.

Vardhan, K. H., Kumar, P. S., & Panda, R. C. (2019). A review on heavy metal pollution, toxicity and remedial measures: Current trends and future perspectives. *Journal of Molecular Liquids, 290*(15), 1–22.

Verma, J. P., & Jaiswal, D. K. (2016). Book review: Advances in biodegradation and bioremediation of industrial waste. *Frontiers in Microbiology*. https://doi.org/10.3389/fmicb.2015.01555

Verma, S., & Kuila, A. (2019). Bioremediation of heavy metals by microbial process. *Environmental Technology & Innovation, 14*, 1–11.

Visca, P., Imperi, F., & Lamont, I. L. (2007). Pyoverdine siderophores: From biogenesis to biosignificance. *Trends in Microbiology, 15*, 22–30.

Wang, Y., Wang, H. S., Tang, C. S., Gu, K., & Shi, B. (2019). Remediation of heavy-metal-contaminated soils by biochar: A review. *Environmental Geotechnics, 1–14*. https://doi.org/10.1680/jenge.18.00091

Wuana, R. A., & Okieimen, F. E. (2011). Heavy metals in contaminated soils: A review of sources, chemistry, risks and best available strategies for remediation. *International Scholarly Research Notices*. https://doi.org/10.5402/2011/402647

Xiong, T., Yuan, X., Wang, H., Wu, Z., Jiang, L., Leng, L., Xi, K., Cao, X., & Zeng, G. (2019). Highly efficient removal of diclofenac sodium from medical wastewater by Mg/Al layered double hydroxidepoly(m-phenylenediamine) composite. *Chemical Engineering Journal, 366*, 83–91.

Xu, Y., Seshadri, B., Bolan, N., Sarkar, B., Ok, Y. S., Zhang, W., & Dong, Z. (2019). Microbial functional diversity and carbon use feedback in soils as affected by heavy metals. *Environment International, 125*, 478–488.

Yin, K., Wang, Q., Lva, M., & Chen, L. (2019). Microorganism remediation strategies towards heavy metals. *Chemical Engineering Journal, 360*, 1553–1563.

Yousaf, B., Liu, G., Abbas, Q., Ullah, H., Wang, R., Zia-ur-Rehman, M., & Niu, Z. (2017). Addition of biochar nanosheets to soil alleviate health risks of potentially toxic elements via consumption of wheat grown in an industrially contaminated soil. *Chemosphere, 1*, 161–170.

Yue, Z. B., Li, Q., Li, C. C., Chen, T. H., & Wang, J. (2015). Component analysis and heavy metal adsorption ability of extracellular polymeric substances (EPS) from sulfate reducing bacteria. *Bioresource Technology, 194*, 399–402.

Zafar, S., Aqil, F., & Ahmad, I. (2007). Metal tolerance and biosorption potential of filamentous fungi isolated from metal contaminated agricultural soil. *Bioresource Technology, 98*(13), 2557–2561.

Zhang, X., Sun, W., Cena, Y., Zhang, L., & Wang, N. (2019). Predicting cadmium concentration in soils using laboratory and field reflectance spectroscopy. *Science of the Total Environment, 650*, 321–334.

Zhang, X., Xu, D., Zhu, C., Lundaa, T., & Scherr, K. E. (2012). Isolation and identification of biosurfactant producing and crude oil degrading *Pseudomonas aeruginosa* strains. *Chemical Engineering Journal, 209*, 138–146.

Zhao, Y., Yao, J., Yuan, Z., Wang, T., Zhang, Y., & Wang, F. (2016). Bioremediation of Cd by strain GZ-22 isolated from mine soil based on biosorption and microbially induced carbonate precipitation. *Environmental Science and Pollution Research, 24*(1), 372–380.

Złoch, M., Kowalkowski, T., Tyburski, J., & Hrynkiewicz, K. (2017). Modeling of phytoextraction efficiency of microbially stimulated *Salix dasyclados* L. *International Journal of Phytoremediation, 19*(12), 1150–1164.

CHAPTER NINE

Plant nutrition, transport, mechanism and sensing in plants

Farheen Naz
Department of Biosciences, Jamia Millia Islamia, New Delhi, India

1. Introduction

Nutrient is an essential for every living creature to stay alive. Same as with plants require nutrients for growth and development to complete their full life span. Living beings can move from one place to another for the source of food. But it's clear that plants can't move from one place to another. So, it's necessary for them to adapt according to the availability of nutrient in the soil and environmental conditions. To maintain nutrient homeostasis, they uptake nutrients and also respond to change in soil and also modification take place within the plants. For example; primary root elongation inhibits and in place of that lateral root start elongating.

All mineral nutrients are needed in great abundance by plants. Plants, however, must compete for nutrients in the soil with biotic and abiotic processes such as leaching, erosion, and microbial consumption. Soil minerals are also lost when crops are harvested and crop products are removed from the soil. To be competitive, higher plants evolves many mechanisms to acquire nutrients at low levels and to use a various form of nutrients. Plants can absorb inorganic forms, such as nitrate (NO_3^-), ammonium (NH_4^+), and organic forms, such as urea (Owen, 2001). Some plants, such as legumes, can fix dinitrogen gas by nitrogen fixation process in association with symbiotic bacteria (Mylona & Pawlowski, 1995) (see Mylona et al., 1995, this issue). This chapter focuses on the nutrient sensing of plants and its mechanism.

2. Plants essential nutrients

Soil is the major source of nutrients for plants. For better plant growth, development and reproduction various nutrients are essential which is called as plant essential nutrients. Researchers reported a list of macro- and micronutrients that are essential for plants (Arnon, 1939).

Sustainable Plant Nutrition
ISBN: 978-0-443-18675-2
https://doi.org/10.1016/B978-0-443-18675-2.00002-X

There are three types of nutrients: Macronutrients, Secondary nutrients and Micronutrients.

3. Macronutrients

Soil contains large number of major elements. Major elements present in soil are nitrogen (N), potassium(K) and phosphorous (P) and. Together which is denoted as NPK trio. And other carbon, hydrogen and oxygen. In Table 9.1 describes macronutrient (Nitrogen, potassium and phosphorous) form, mode of uptake and major functions of the plant essential nutrients (Miller, 1943).

4. Nitrogen

Nitrogen is the most abundant element found on the earth. It is eventually used by plants from soil either through nitrogen containing minerals released from decomposition of minerals or from atmosphere. Nitrogen in atmosphere exists in the form of inert gas N_2 Approximately 20 lb nitrogen/acre per year exists from this source. It is the major source of element present in the soil for plant growth, development and reproduction (Miller, 1943). Maximum amount is found in an organic form of nitrogen (Hans, 1958). It is not the direct source of plants. It goes in the form of nitrate (NO_3), ammonium (NH_4^+) as a food source. Nitrate mainly is used for the growth of roots such as lateral roots. For example; Legume nodule formation explains in Fig. 9.1 that the fixation of nitrogen for plants. This

Table 9.1 Macronutrient (Nitrogen, potassium and phosphorous) form, mode of uptake and major functions of the plant essential nutrients.

Nutrient	Taken up by plants (form)	Mode of uptake	Functions
Nitrogen	Nitrate (NO_3^+) and ammonium (NH_4^+)	Mass flow	Formation of protein/ nucleic acids, cell formation
Potassium	Potassium (K^+)	Mass flow	Regulation of water, increases enzyme activity
Phosphorous	Dihydrogen phosphate ion (H_2PO_4 −, HPO_4^{2-}), phosphate (PO_4^{2-})	Root interception	Formation of protein/ nucleic acids, cell formation, fat and carbohydrates metabolism

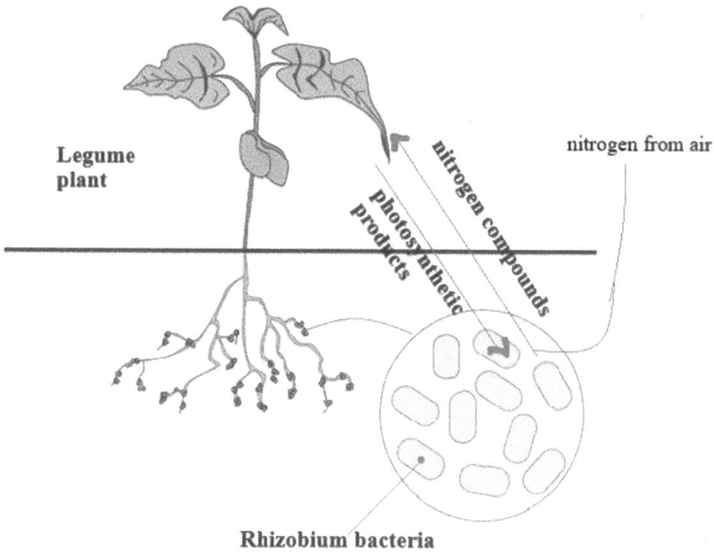

Fig. 9.1 Mechanism of nitrogen fixation.

fixation process demonstrate an orchestrated interaction between the host plant and bacteria (Napoli, 1975). The rhizobium bacteria colonize in the legume plant roots which is called as host. These colonies cause the plant roots to form nodules which becomes the house of bacteria. Due to which bacteria fixes the nitrogen that is required by the host plant. This allows the plant to produce leaves, increases photosynthetic capacity, which in turn produces nitrogen – rich seeds. If the legume plant is growing in nitrogen deficiency soil, this fixation compensates the plant. Most importantly, nitrogen act as a fertilizer a plant nutrient for some field crops for optimum development and growth in especially desert environments (Fuller, 1975; Hendricks, 1985).

4.1 Transport and uptake of nitrogen

Transport and uptake of nitrogen as a nutrient is an important task for the development and growth plants and crop. This activity is influenced and controlled by many factors which interplay among each other. For the same process three key classes are involved which are named as following:

I. Transporters
II. Receptors
III. Transceptors

Transporters are also named as selective carrier proteins which carry the nutrient from one part of the plant to other part (Forde, 2002). Transceptors is said to those proteins which control both functions transport as well as nutrient sensing function process (Alain, 2011). Lastly it has been noted that the final concentration of nitrogen in the plant always depend upon the availability of nitrogen salts in the soil. And also depends on the absorbance by roots as well as maintaining the balance of net uptake (influx and efflux phenomenon), vacuolar accumulation, root metabolization and xylem/phloem (un)loading (Hawkesford et al., 2012).

Nitrogen is transported or uptake in different forms which was described earlier also. Detailed explanation was below:

a. Nitrate Uptake
b. Ammonium Uptake
c. Amino acids
d. Urea uptake

5. Nitrate uptake

In higher plants, NO_3^- uptake is influenced and controlled by transporters which belong to NPF family and of the NRT2 family, which in *Arabidopsis* plant consist of 53 and seven members, respectively (Léran et al., 2014; Little et al., 2005). In roots, different function of NRT2 was reported. However, there are no significance sequence homologies between the both two families. But these transporters share the same topology having 12 transmembrane domains and consisting both the C- and N- and terminal lying on the cytosolic side of the membrane (Brain, 2000). The control of these transporter proteins is strictly mediated by NO_3^- availability in plant and plant nitrogen nutritional status. In deficiency of N which are called as nitrogen starved plants, reduced availability of NO_3^- triggers the typical NO_3^- primary response (NPR), which facilitates the rapid induction of NO_3^- transporters and nitrogen assimilating enzymes. An essential role for the transcriptional control of NO_3^- transporters was reported in *Arabidopsis* and crops for the better growth and development.

For example, in Arabidopsis, three members of the NPF family are situated in the plasma membrane of root cells and participate in NO_3^- uptake.

6. Ammonium uptake

Ammonium is also an important nitrogen nutrient which is rapidly absorbed and assimilated by higher plants, as well as a nutrient signaling molecule influencing plant development and growth and root system architecture. Abundance of NH_4^+ concentration absorbed by plants also causes serious problem and leads to toxicity. It results to the reduced growth, depletion of root growth and leaf chlorosis. It has been examined that there should be balance between the availability and transport of NO_3^- and NH_4^+ nutrients for better crop yield and plant growth. Indeed, for ideal growth many crops and higher plants require the contemporaneous availability of NO_3^- and NH_4^+, even if the best ratio depends on developmental phase, plant species, and environmental factors (Miller & Cramer, 2005).

There are three sub family for transporters which are AMT1; 1, AMT1; 3 and AMT1; 5 present in the epidermis and root hairs, which contribute to the uptake of NH_4^+ uptake through symplastic route system (Loqué et al., 2006).

7. Uptake of amino acids

For plant growth the content of total amino acids in plant tissues usually ranges within 1 and 10 mM, but this value varies and can be increase, especially under high NH_4^+ nutrition (Miller & Cramer, 2005). Similarly, the transportation of total amino acids in plant cell organelles changes from one part to another, with the highest level typically found in cytosol and plastids (Noctor & Novitskaya, 2002).

8. Urea uptake

Now the utilization of urea fertilizers is more than 50% of the world nitrogen fertilizer applications. As it has high nitrogen concentration and have low costs. Urea is rapidly hydrolyzed to release of NH(Arnon, 1939) by ubiquitous microbial urease and converted into NO(Arnon, 1939) by nitrifying bacteria (Glibert et al., 2006). In addition, urea is an important molecule in nitrogen recycling and translocation in plants. In higher plants first urea transporter discovered was DUR3 in *Arabidopsis* (Liu et al., 2003).

9. Factors determining the quantity of nitrogen contributing by the soil

1. Contribution of nitrogen from organic matter in the soil
2. Contribution of nitrogen decomposes of previous crop residues

3. Contribution of nitrogen by previous organic waste
4. Contribution of nitrogen by previous fertilizers

Nitrogen is very essential for maximum plant development and growth. So, supply should be adequate. Deficiency as well excess of nitrogen for a plant is the case of serious matter. In both condition it can affect the development and growth of plants. Excess of nitrogen available concentration in the soil can leads to the growth of foliage, leaves and stems and it promotes the vegetative growth at the level of fruit, flower and root growth in most plants. An unusual growth of the leaves which causes increase in leaf area is due to excess of nitrogen supply to the plants roots from the soil. It is one of the most critical issue or drawbacks in excess of nitrogen supply.

No doubt nitrogen is one of the most essential factors in the development and growth rate of the leaves. But excess causes the difficulty in proper growth. If the size of some plants

is thus greater a measure of the rate of nitrogen metabolism. With an increase of nitrogen, the water holding of the tissues of those plants also increased.

10. Deficiency of nitrogen in plants

Deficiency of nitrogen salt in the soil effects the growth of plant and yield of crops (Fuller, 1975; National Plant Food Institute, 1962; Hans, 1958). If there is deficiency in soil plants are unable to uptake from root system. It limits the quantity of stem, leaves, stem, seed and flowers. Most plants face deficit in nitrogen which exhibit the following distinct and well-characterized symptoms:

1. Leaves become yellow or light green
2. Growth of plants stunted
3. Limitation in annual branching
4. Lower leaves from the stem becomes dry and drop in early stage than usual time
5. Reduction of amount of protoplasm
6. Reduction in quantity of leaves and stems
7. Yield of crop reduced rapidly

11. Abundance of nitrogen in plants

Abundance of nitrogen soil also a serious matter for the development and growth of plant as well as crop. It leads seriously affect the growth of leaves, stem, seed, flowers and stem. And sometimes it leads to toxicity and finally plant dies.

1. Abundance of nitrogen gas may even kill the plants
2. Plants wither become dry and turn brown
3. Toxicity due to excess of salts
4. Increases lodging process in grain crops, which instantly reduces the yield of crop
5. Limits the growth and development of root
6. Delayed in maturation of plants
7. Pre maturation of ripening process
8. Seeds of plants become shriveled and become light in weight
9. Increase in leaf area

12. Potassium

Potassium is vital for confirming the optimal growth of plant. It is crucial for cell growth by which development of plant takes place (Hepler et al., 2001). It is generally reported that K controls and stimulates ATPase in the plasma membrane to develop acid stimulation, which further triggers hydrolase activation and cell wall loosening which helps in promoting cell growth (Oosterhuis & Loka, 2014). Cell elongation, ability of stomatal opening and closing, maintaining ions balance in the cytoplasm, regulating osmotic pressure and other various physiological processes (Hu et al., 2016). It is an activator of several important enzymes, like photosynthesis, sugar transport, protein synthesis, N and C metabolism. It is reported that K affects notably the utilization and absorption of other nutrients by plants, and the appropriate K amount differs in various crops.

Some ways potassium aids in plant development and health:

1. Produces sturdy, robust stalk
2. Reduces lodging
3. Imparts enlarge vigor
4. Increases disease resistance in plants
5. Facilitates the plumpness of the seeds and grains
6. Promotes the formation and transfer of sugars, starches and oils
7. Increases winter hardiness in some plants such as legume plants and various crops.

13. Deficiency of potassium in plants

Most plants face deficit in potassium which exhibit the following distinct and well-characterized symptoms:

1. Growth of plants becomes very slow like dwarf plants
2. Older leaves margins start becoming brown from green color
3. The fruits and seeds become small and shrink
4. Weakening of the plant stalks
5. The alfalfa and clovers have developed white spots on their leaves margin

All plants have different requirements of potassium concentration overall and have different ability to absorb it from the soil. In one stage of physiological life span of plants requirement of potassium may be crucial than the other stage of growth. For examples, in early stage of potato plants (*Solanum tuberosum* L.) there is no deficiency of potassium but in later stage requirement of it in crucial stage and needs more concentration of it to maintain growth and development. This critical stage wise situation was only seen in the case of potatoes plant not in case of other plants or crops.

Many fertilizers furnish K in the form of potash (K20). Potassium chloride (KCl), frequently called muriate of potash (60%–62% K20) is the most popular potassium fertilizer grade and it accounts for about 95% of the potassium fertilizers used in commercial agriculture (Fuller, 1975).

14. Phosphorous

Phosphorous is essential to plant growth, reproduction and is found in every plant cell. It is involved in various key plant functions, including photosynthesis, energy transfer, transformation of starches and sugars, nutrient movement in the plant system and transfer of genetic material from one generation to the other. The best source of phosphorous in soil occurs from natural sources such as bone meal, feces, hair, shrimp waste, mushroom compost and burn cucumber skin. Nutritional value can vary in many cases. Phosphorous has less content than N and K but apart from that it is an important content. Deficiency of phosphorous leads to unsatisfactory growth and development for the plant. Some deficiencies are such as stunted roots which decreases their absorbing and feeding zone, depletion of ripening system etc. In sufficient concentration it improves the quality of whole plant, seeds formation started in early period, vigorous growth of flowers, leaves that makes the plant resistant to diseases.

15. Transport and uptake of phosphorus

Phosphorus directly enters the plant via root tip, lateral roots, outer layer of root cells, root hairs. Uptake is also improved by mycorrhizal fungi that grow in association with the roots of various crops. It is taken up as the primary orthophosphate ion (H_2PO_4-), but some of it is also absorbed as secondary orthophosphate (HPO_4-), this later form of P increasing as the soil pH increases. It is stored in the plant root system for later activities and used up by upper part of the plants later when required. By various chemical reactions, it is moved into organic compounds, such as nucleic acids (DNA and RNA), phospholipids, energy-rich phosphate compounds phosphoproteins, phosphates, sugar, enzymes, and for example, ATP. In these organic forms as well as the inorganic phosphate ion P moved throughout the plant system, wherever it is required for further physiological functions.

Some ways phosphorus aids in plant development and health:

1. Promotes photosynthesis
2. Builds proteins, nucleic acids and enzyme
3. Increases root system growth
4. Improves the quantity of flower and seed production
5. Improves plant quality
6. Strengthens stalks and stems
7. Promotes growth of the crop
8. Contributes to earlier period maturity
9. Improves disease resistance
10. Stimulate overall plant quality
11. Increases nitrogen fixation process of legumes

16. Deficiency of phosphorous in plants

Most plants face deficit in phosphorus which exhibit the following distinct and well-characterized symptoms:

1. Reduction of shoot and root growth
2. Shoot growth becomes spindle and upright
3. Limitations in lateral shoots and lateral buds or they may go in dormant stage or may die.
4. Older leaves start showing premature defoliation
5. Depletion of blossoming
6. Foliage start becoming purple in color

7. Leaf margins start becoming brown

Various plant species have dissimilar requirements of phosphorous from fertiliser or soil content. For example, in case of beans *(Phaseolus vulgaris* L.) it has been seen that often these plants need more phosphorus whereas in case of cotton *(Gossypium hirsutum* L.) it rarely counter to phosphorus in low P yield, during development and growth. Usually, vegetable crops respond to P but ornamental shrubs have no such respond. Young citrus trees often respond to phosphorus; however, mature trees show seldom response. However, in the early stages of plant need for P from soil is relatively high for the plant development and growth. Primary as well as lateral roots are small in early stage of plant life and absorbing area is also limited and a greater level of P is needed as per root volume. And it is sufficient for plant development and growth. Early-stage plants also demand high phosphorus as compared to mature plants. Many higher plants have obtained 50% of their P needs by the time they have attain 20% of their entire life growth. During summer in warmer period or high temperature fast-growing plants respond more P applications at the time of maturation. And in the period of winters or low temperature they grow slowly at the same level of P for maturation (Fuller, 1975).

17. Carbon, hydrogen and oxygen

Carbon, hydrogen and oxygen are other vital macronutrients for the plant development, growth and reproduction. Carbon is taken as carbon dioxide (CO_2) and bicarbonate ($HCO^-{}_3$) as a source of plant. These both are taken up by the process of photosynthesis. There main function is to maintain the structure of plants.

Hydrogen is taken up in the form of water H_2O. The main function is water retention, formation of carbohydrates and pH regulation.

Oxygen is also taken up in the form of H_2O. Main functions are maintaining the plant structure, respiration and energy production.

Table 9.2 describes macronutrient (Carbon, hydrogen and oxygen) form, mode of uptake and major functions of the plant essential nutrients.

18. Secondary nutrients

Secondary nutrients are usually needed in moderate amount as compared to macronutrients. These are magnesium, calcium and sulfur.

Calcium is taken up in the form of Ca^{2+} as mass flow. Its main function is to activate enzyme activity and maintaining root permeability.

Table 9.2 Macronutrient (Carbon, hydrogen and oxygen) form, mode of uptake and major functions of the plant essential nutrients.

Nutrient	Taken up by plants (form)	Mode of uptake	Functions
Carbon	Carbon dioxide (CO_2)	Opening stomatal pores	Maintain structure of plant
Hydrogen	Water (H_2O)	Mass flow	Water retention process, synthesis of carbohydrates, pH regulation
Oxygen	Water (H_2O)	Mass flow	Energy production, maintain plant structure, respiration process

Magnesium is taken up in the form of Mg^{2+} as mass flow. The functions include the process of metabolism, formation of fat and chlorophyll.

Sulfur is taken up in the form of SO_4^{2-} as mass flow. The functions are the formation of proteins, nucleic acid, amino acids, vitamin and oil.

19. Micronutrients/trace nutrients

Micronutrients/trace nutrients are needed in minute amounts as compared to macro or secondary nutrients. These are boron, copper, chlorine, iron, molybdenum, manganese and zinc (Janick & Schery, 1969). Table 9.3 will describe micronutrients, form, mode of uptake and major functions of the plant essential nutrients.

Table 9.3 Micronutrient, form, mode of uptake and major functions of the plant essential nutrients.

Nutrient	Taken up by plants (form)	Mode of uptake	Functions
Boron	Boric acid (H_3BO_3), borate (BO_3^{3-}), tetraborate (B_4O_7)	Root interception	Increase enzyme activity
Copper	Copper ion (Cu^{2+})	Mass flow	Increase enzyme activity
Chlorine	Chloride (Cl^-)	Root interception	Increase enzyme activity, chlorophyll formation and cellular development
Iron	Iron ion (Fe^{2+}, Fe^{3+})	Root interception	Increase enzyme activity and development
Molybdenum		Mass flow	

(Continued)

Table 9.3 Micronutrient, form, mode of uptake and major functions of the plant essential nutrients.—cont'd

Nutrient	Taken up by plants (form)	Mode of uptake	Functions
	Molybdenum ions ($HMoO^{4-}$, MoO_4^{2-})		Increase enzyme activity and nitrogen fixation process of legumes
Manganese	Manganese (Mn^{2+})	Root interception	Increase enzyme activity and pigmentation
Zinc	Zinc ion (Zn^{2+})	Root interception	Increase enzyme activity

20. Nutrients of plants essential for growth, development and reproduction

In this modern agriculture management is required for the control of plant nutrition. Plant growth, development and reproduction depends on the sufficient availability of nutrient in the soil. Nutrition supplied to plants or crops depends on the physical and chemical properties of soil. Properties such as texture of soil, type of soil, composition of soil, ability of soil to hold water, presence of minerals. Fertility of soil depends on the composition of organic and inorganic substances present in the soil. These factors are the basic maintenance of nutritional value. Many factors involve in the availability of nutrients which includes soil pH, ion-exchange, soil texture, capacity of soil, and the amount of organic and inorganic matter present in the soil.

21. Nutrients uptake by plants

The nutrients uptake system in plants must be vigorous and versatile because higher plants have to transport sufficient level of nutrients to satisfy the total demand for all types of nutrients required for the growth and development of plants. The external nutrients concentrations can face limitations to plant requirement and that can vary by five orders of magnitude. To function physiological and morphological activities efficiently in the transit of such environmental conditions or variations, higher plants have provided a transport system that is regulated, active, and multiphasic. The internal energy that drives nutrients uptake comes from the proton H+ gradient maintained across the plasma membrane by the H+ ATPase. The uptake system of individual components for nutrition are just now being reported, and they are explaining interesting parallels with uptake systems from fungi, bacteria algae, and mammals. The initial uptake of nutrients occurs across the plasma membrane of epidermal and cortical cells of the root (Larsson, 1989). Subsequently, transport can occur across the tonoplast membrane of the plant cell (Granstedt, 1982) and the plasma membrane of cells in the vascular system and leaf. This nutrients uptake discussion is restricted to the initial uptake process, that is, transport of nutrients across the plasma membrane of root cells.

22. Mechanism of nitrate uptake by plants

Nitrogen is taken up by plants in two forms in which one is nitrate. Energy which drives nitrate uptake by plants comes from the proton H^+ gradient maintained across the plasma membrane by the H^+-ATPase. Initially nitrate uptake occurs across the plasma membrane of cortical cells and epidermal of the root (Larsson, 1989). Subsequent transport can occur across the plasma membrane of the plant cells in the vascular system and leaf and by tonoplast membrane (Granstedt, 1982). The transport of anionic nitrate by plants into a root cell must occur an electrical gradient with potentials that are normally between −100 and −250 mV (negatively charge inside the cell). Given an external nitrate level of concentration of 2 mM and an electrical potential of −110 mV is seen. It would be expected that the concentration of NO_3^- in cytoplasm would be 0.028 mM, based on the Nernst equation. However, in barley root cells, cytoplasmic NO_3^- concentrations have been recorded between 1 and 5 mM[27]. This proves that energy is required to uptake the transport of NO_3^- into the plant cell. All these current evidences strongly proves that NO_3^- uptake in plants cells is driven by electrogenic H^+ co-transport. Similarly the depolarization and proton co-transport phenomenon have been observed for chloride and phosphate uptake (Ullrich, 1981). Higher plants have distinct NO_3^- transport systems with varying different concentration levels of NO_3^-. Fig. 9.2 explains the nitrogen sensing and uptake of nitrogen in stress conditions.

The cloning of an immense affinity NO_3^- transporter gene has not been discovered for higher plants, but three such genes have been isolated from algae and fungi. First identified gene was the *Aspergillus* ***CRNA*** in a search for chlorate-resistant fungi (Cove, 1979). ***CRNA*** is part of a gene cluster that incorporate ***NIIA and NIAD*** and, which encode NiR and NR respectively (Tomsett, 1979). In addition with NO_3^-, light is also an important regulatory signal for NiR and NR expression ***CRNA*** was cloned and found to encode a membrane protein with 10−12 putative membrane-spanning segments, as would be expected for a cotransporter (Unkles & Hawker, 1991).

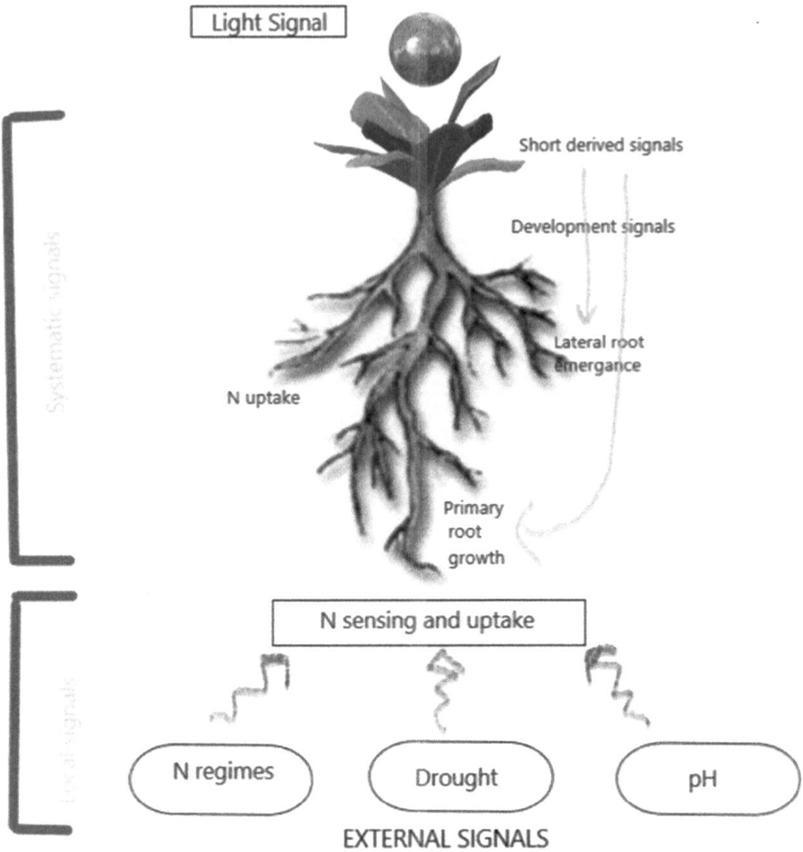

Fig. 9.2 Nitrogen sensing and uptake process.

It was examines that ***CRNA*** encodes a higher affinity NO_3^- transporter (Unkles & Hawker, 1991). No research data on the enzymatic activity or localization of the ***CRNA*** gene product are available yet now to prove this conclusion. Hypothetically, there are more *Aspergillus* proteins that take up NO_3^-, but none has been explained.

For example, if tomato plants are depleted from nitrogen source, the level of NR protein falls to 20% of the original concentration level within 24 h, thus, NR mRNA concentration remain up for 12 days (Galangau & Daniel-Vedele, 1988). Similarly, if the NR coding region is constitutively expressed with a CaMV 35s promoter in plant of tobacco, NR protein levels decreases four-to fivefold after 72 h of dark treatment, even though NR mRNA levels remain to increase.

Two other related transporter genes have been identified in *Chlamydomonas*. This alga also has a cluster of genes that includes the ***NIT-7,*** NR gene (Quesada et al., 1993). If three of the algal genes from five in the cluster ***(NAR-2, NAR-3,*** and ***NAR-4)*** are removed, NO_3^- uptake at **50** pM will be blocked and cells can no longer grow on NO_3^- (Quesada et al., 1994). Accompanying, of the triple mutant needed two of the three algal genes: ***NAR-2*** and ***NAR-3*** or ***NAR-2*** and ***NAR-4.*** Sequence analysis described that the NAR-3 and NAR-4 proteins are nearly **80%** identical to each other where the region noted (the C-terminal 250 amino acids) and are 27%–31% nearly identical to the CRNA protein of *Aspergillus* over same region. Therefore, *Chlamydomonas* appears to be having two high affinity NO_3^- transporter genes that are greatly related, are located within the same algal gene cluster, and shares similarity with fungal NO_3^- transporters. No heterologous expression or localization research data for the ***NAR-3*** and ***NAR-4*** genes are yet available. The function of ***NAR-2*** is unknown, but it may play a vital role.

Thus, the sequences of these three nutrients uptake genes are conserved between algae and fungi explains that higher plants may also have related uptake genes.

23. Mechanism for nutrient sensing in plants

Till now, sensing mechanism for the nutrients of plants is unknown. Nutrition uptake of plants from the soil relies on some of the sensing system. It depends on the uptake of ions from the soil. And availability of these inorganic ions pivot on the space and time. Nutrient – limiting factors are also faced by the plants for general development. To control this constraint, plants adapt responses which are triggered by some sensing systems that maintain nutrient availability to plants (Gojon & Nacry, 2009; Microbiology, 2000; Schachtman & Shin, 2007). Nitrate is taken up for nitrogen source in plants as well as a signal molecule for regulating plant growth. For example, in *Arabidopsis thaliana* plant, NRT1.1 nitrate transporter is vital for nitrate signaling for root growth and development and same has been derived to play a function as a nitrate sensor. Generally, this mineral nutrient sensing systems are uncharacterized in higher plants. But recently it has been reported that in case of *Arabidopsis thaliana*, the plasma membrane nitrate (NO_3^-) transporter NRT1.1 (CHL1) has been discovered. CHL1 is involved in the primary nitrate response. Ions act as an essential nutrient in plants but can also play an important role as signaling molecules. How higher plants sense modification in soil nutrient concentrations is described here.

Main function is to uptake NO_3^- by the root system from soil solution as an influx carrier (Ho et al., 2009). Nitrate plays an important role not only as a nitrogen source but also act as a major signaling molecule promoting plant growth and metabolism (cell, 1995). Particularly signaling system of NO_3^- is strong related to the growth of lateral roots of plants. This clarifies the emergence of branching of roots post embryonically in the root system. NRT1.1 is an important for NO_3^- regulation in root system shown in Fig. 9.3. The sensor stimulates a specific NO_3^- signaling that triggers lateral root growth in supply of NO_3^-. Apart from this function NRT1.1 plays a crucial role in the adaptive response of the higher plants to nitrogen limitation as it directs preferential growth of lateral roots in NO_3^- rich patches of the external medium. It has been reported that NRT1.1 not only act as a transports nitrate but also increases uptake of the phytohormone auxin (Krouk et al., 2010). Herein, nitrate inhibits NRT1.1-dependent auxin uptake, revealing that transduction of nitrate signal via NRT1.1 is associated with a moderation of auxin transport. Along with other effects, auxin promotes lateral root growth and development. Mutation of NRT1.1 increases both auxin concentration in lateral roots and growth of these roots at low, but not high, nitrate concentration. Thus, we can say that the NRT1.1 represses lateral root growth at low nitrate level availability by promoting basipetal auxin transport out of these roots. This defines a mechanism connecting nutrient and hormone signaling during organ development (Krouk et al., 2010).

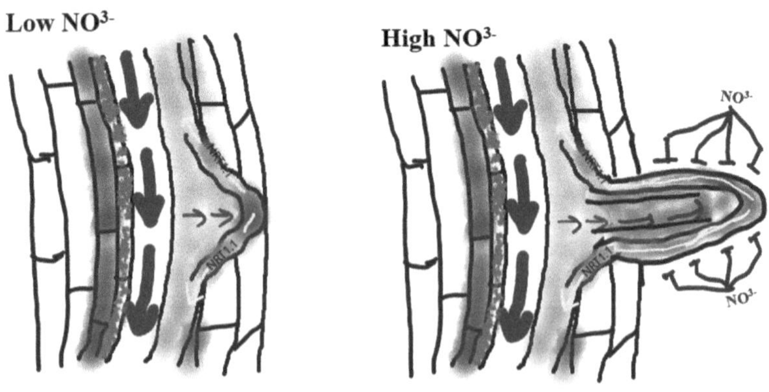

Fig. 9.3 NRT1.1-dependent mechanism of lateral root outgrowth. *Adapted (Beeckman, Tom, and Jiří Friml. "Nitrate contra auxin: nutrient sensing by roots."* Developmental cell *18.6 (2010): 877–878).*

24. Nitrate as a signal for transporting in plant cells

Really one of the interesting features of NO_3^- assimilation is its regulation in plants. NO_3^- serves as the prime signal for regulating NO_3^- assimilation, but cytokinin, light, CO_2 levels, nitrogen metabolites, circadian rhythms, carbon, and such as glutamine and sucrose all play immense roles as well. A complex regulatory network of transport involves both transcriptional and post-transcriptional phenomenon serves to integrate NO_3^- assimilation, that consumes carbon skeletons, energy and, with carbon metabolism and photosynthesis. The explanation of the plant regulatory machinery has been guided by the high work with funga1 systems, such as *Aspergillus and Neurospora.*

25. Phosphate as a signal for transporting in plant cells

Phosphorus (P) is one of the crucial structural elements for the production of organic molecules such as ATP, nucleic acids, and phospholipids. Though P is highly available in the soil but plants can absorb only its inorganic phosphate (Pi), which has low mobility. In case of *Arabidopsis*, Pi deficiency is noticed by the root tip which has the most sensing power. Higher plants have Pi transporters (PHT) including Pi/H+ Symporters to absorb Pi.

26. Potassium as a signal for transporting in plant cells

Potassium (K^+) is an essential element, the most abundant cation presents in plants, whereas

its concentration in the soil is limited. Plants absorb K^+ through root tips and lateral roots from soils through both K^+ channels and carriers situated at the plasma membrane of root cells, especially in the epidermal and root hair cells. At the time of K^+ deficiency, transport of K^+ is possible by transporters that are available in plant cells that are proposed to be K+ sensors (Wang & plant, 2010). In *Arabidopsis*, a large collection of K+ transporters is encoded by the genome. Under K^+ deficiency the AKT1 (*Arabidopsis K+ t*ransporter 1), an inward K+ channel, and other transporters are AtCHX13, HAK5, and AtCHX17, play a key role in K+ acquisition and homeostasis (Wang & plant, 2010).

27. Conclusion

At the condition of stress, it seems difficult to transport nutrients from the soil to plants. It needs sufficient energy for the uptake of essential elements. To maintain homeostasis plants, establish a special process that is known as signaling mechanism. Transporters, regulators, carrier protein or channel work under the same process. Several functions had been identified and several are left to identify the mechanism of signaling other nutrients.

Sufficient amount of nutrient supply to plants from soil is necessary for better growth. There should not be the deficiency of any nutrient for the optimal growth. All physiological and morphological conditions depend on the sufficient available concentration of nutrients. New findings related to understand the mechanism of nutrient sensing are the future prospective.

References

Arnon, D. (1939). The essentiality of certain elements in minute quantity for plants with special reference to copper. *Plant Physiology* Accessed November 9, 2021 https://www.ncbi.nlm.nih.gov/pmc/articles/PMC437744/.

Brain, G. (2000). Nitrate transporters in plants: Structure, function and regulation. *Biomembranes*. Elsevier. Accessed October 29, 2021 https://www.sciencedirect.com/science/article/pii/S0005273600001401.

Cove, D. J. (1979). Genetic studies of nitrate assimilation in Aspergillus nidulans. *Biological Reviews of the Cambridge Philosophical Society, 54*(3), 291–327. https://doi.org/10.1111/J.1469-185X.1979.TB01014.X

Forde, B. G. (2002). Local and long-range signaling pathways regulating plant responses to nitrate. *Annual Review of Plant Biology, 53*, 203–224. https://doi.org/10.1146/ANNUREV.ARPLANT.53.100301.135256

Fuller, W. (1975). *Management of southwestern desert soils* Accessed October 28, 2021 https://agris.fao.org/agris-search/search.do?recordID=US201300524157.

Galangau, F., & Daniel-Vedele, F. (1988). *Expression of leaf nitrate reductase genes from tomato and tobacco in relation to light-dark regimes and nitrate supply* (Vol. 88, pp. 383–0388) Accessed October 28, 2021 https://academic.oup.com/plphys/article-abstract/88/2/383/6083279.

Glibert, P. M., Harrison, J., Heil, C., & Seitzinger, S. (2006). Escalating worldwide use of urea - a global change contributing to coastal eutrophication. *Biogeochemistry, 77*(3), 441–463. https://doi.org/10.1007/S10533-005-3070-5

Gojon, A., & Nacry, P. (2009). *BiologyJD-C opinion in plant, 2009 undefined. Root uptake regulation: a central process for NPS homeostasis in plants*. Elsevier Accessed October 27, 2021 https://www.sciencedirect.com/science/article/pii/S1369526609000375.

Granstedt, R. (1982). *Identification of the leaf vacuole as a major nitrate storage pool* (Vol. 70, pp. 410–413). Accessed October 27, 2021 https://academic.oup.com/plphys/article-abstract/70/2/410/6077523.

Hawkesford, M., Horst, W., Kichey, T., & plants, H. L. (2012). *Functions of macronutrients*. Elsevier Accessed October 28, 2021 https://www.sciencedirect.com/science/article/pii/B9780123849052000066.

Hendricks, D. M. (1985). *Arizona soils*. Tucson: University of Arizona Accessed November 8, 2021 https://scholar.google.com/scholar?hl=en&as_sdt=0%2C5&q=Hendricks+DM+

%281985%29+Arizona+soils.+University+of+Arizona%2C+College+of+Agriculture %2C+Tucson%2C+244+pp&btnG=.
Hepler, P. K., Vidali, L., & Cheung, A. Y. (2001). Polarized cell growth in higher plants. *Annual Review of Cell and Developmental Biology, 17*, 159–187. https://doi.org/10.1146/ANNUREV.CELLBIO.17.1.159
Ho, C., Lin, S., Hu, H., & Cell, Y. T.- (2009). *CHL1 functions as a nitrate sensor in plants.* Elsevier Accessed October 27, 2021 https://www.sciencedirect.com/science/article/pii/S0092867409008423.
Hu, W., Jiang, N., Yang, J., et al. (2016). Potassium (K) supply affects K accumulation and photosynthetic physiology in two cotton (Gossypium hirsutum L.) cultivars with different K sensitivities. *Elsevier* Accessed October 27, 2021 https://www.sciencedirect.com/science/article/pii/S0378429016301824.
Janick, R. W., & Schery, F. W. W. (1969). *WHF& C. Plant science: An introduction to world crops. Plant science* (p. 740). https://ideas.repec.org/a/eee/agisys/v9y1982i4p325-328.html Accessed November 8, 2021.
Microbiology JH-C opinion. (2000). *Two-component and phosphorelay signal transduction.* Elsevier Accessed October 27, 2021 https://www.sciencedirect.com/science/article/pii/S1369527400000709.
Krouk, G., Lacombe, B., & Bielach, A. (2010). *Nitrate-regulated auxin transport by NRT1. 1 defines a mechanism for nutrient sensing in plants.* Elsevier Accessed October 27, 2021 https://www.sciencedirect.com/science/article/pii/S1534580710002169.
Larsson, C., & BI-M and genetic aspects. (1989). *Molecular aspects of nitrate uptake in higher plants.* Accessed October 27, 2021 https://www.cabdirect.org/cabdirect/abstract/19901949253.
Léran, S., Varala, K., & Boyer, J. (2014). *A unified nomenclature of nitrate transporter 1/peptide transporter family members in plants.* Elsevier Accessed October 29, 2021 https://www.sciencedirect.com/science/article/pii/S1360138513001702.
Little, D. Y., Rao, H., Oliva, S., Daniel-Vedele, F., Krapp, A., & Malamy, J. E. (2005). *The putative high affinity nitrate transporter NRT2. 1 represses lateral root initiation in response to nutritional cues.* National Acad Sciences Accessed November 9, 2021 https://www.pnas.org/content/102/38/13693/.
Liu, L., Ludewig, U., & Frommer, W. (2003). *AtDUR3 encodes a new type of high-affinity urea/H+ symporter in Arabidopsis* Accessed October 29, 2021 https://academic.oup.com/plcell/article-abstract/15/3/790/6009927.
Loqué, D., Yuan, L., Kojima, S., & Gojon, A. (2006). Additive contribution of AMT1; 1 and AMT1; 3 to high-affinity ammonium uptake across the plasma membrane of nitrogen-deficient Arabidopsis roots. *Wiley Online Library, 48*(4), 522–534. https://doi.org/10.1111/j.1365-313X.2006.02887.x
Miller, C. E. T. L. (1943). *Fundamentals of soil science : Soil science* Accessed November 8, 2021 https://journals.lww.com/soilsci/Citation/1943/07000/Fundamentals_of_Soil_Science.9.aspx.
Miller, A. J., & Cramer, M. D. (2005). Root nitrogen acquisition and assimilation. *Plant and Soil, 274*(1), 1–36. https://doi.org/10.1007/S11104-004-0965-1, 2005 274:1.
Mylona, P., & Pawlowski, K. (1995). Symbiotic nitrogen fixation. *Cell,* 7, 869–885 Accessed October 27, 2021 https://www.ncbi.nlm.nih.gov/pmc/articles/PMC160880/.
Napoli, C. (1975). Ultrastructure of Rhizobium-induced infection threads in clover root hairs. *American Society of Microbiology.*, 1003–1009 Accessed October 26, 2021 https://journals.asm.org/doi/abs/10.1128/am.30.6.1003-1009.1975.
National Plant Food Institute. (1962). *Our land and its care.* Google book Accessed November 8, 2021 https://books.google.co.in/books/about/Our_Land_and_Its_Care.html?id=86bxV_qfJmQC&redir_esc=y.

Crawford, N. M. (1995). Nitrate: Nutrient and signal for plant growth. *Cell, 7*, 859–868 Accessed October 27, 2021 https://www.ncbi.nlm.nih.gov/pmc/articles/PMC160877/.

Alain, G. (2011). *Nitrate transceptor (s) in plants* Accessed October 28, 2021 https://academic.oup.com/jxb/article-abstract/62/7/2299/581270.

Noctor, G., & Novitskaya, L. (2002). *Co-ordination of leaf minor amino acid contents in crop species: Significance and interpretation* Accessed October 29, 2021 https://academic.oup.com/jxb/article-abstract/53/370/939/537246.

Oosterhuis, D., & Loka, D. (2014). *The physiology of potassium in crop production*. Elsevier Accessed October 27, 2021 https://www.sciencedirect.com/science/article/pii/B9780128001325000031.

Owen, A., & biochemistry DJ-S biology. (2001). *Competition for amino acids between wheat roots and rhizosphere microorganisms and the role of amino acids in plant N acquisition*. Elsevier Accessed November 9, 2021 https://www.sciencedirect.com/science/article/pii/S0038071700002091.

Quesada, A., Galvan, A., & Fernandez, E. (1994). Identification of nitrate transporter genes in Chlamydomonas reinhardtii. *The Plant Journal, 5*(3), 407–419. https://doi.org/10.1111/J.1365-313X.1994.00407.X

Quesada, A., Galván, A., Schnell, R. A., Lefebvre, P. A., & Fernández, E. (1993). Five nitrate assimilation-related loci are clustered in Chlamydomonas reinhardtii. *MGG Molecular & General Genetics, 240*(3), 387–394. https://doi.org/10.1007/BF00280390

Schachtman, D. P., & Shin, R. (2007). Nutrient sensing and signaling: NPKS. *Annual Review of Plant Biology, 58*, 47–69. https://doi.org/10.1146/ANNUREV.ARPLANT.58.032806.103750

Hans, P. (1958). Soil-plant relationships. *Journal of Science* Accessed November 8, 2021 https://journals.lww.com/soilsci/Citation/1958/03000/Soil_Plant_Relationships.23.aspx.

Tomsett, A. (1979). Deletion mapping of the niiA niaD gene region of Aspergillus nidulans. *Research, 34*, 19–32. https://doi.org/10.1017/S001667230001925X

Ullrich, W., & Letters AN-PS. (1981). *Nitrate-dependent membrane potential changes and their induction in Lemna gibba G 1*. Elsevier Accessed October 28, 2021 https://www.sciencedirect.com/science/article/pii/0304421181902339.

Unkles, S., & Hawker, K. (1991). crnA encodes a nitrate transporter in Aspergillus nidulans. *National Acad Sciences, 88*, 204–208 Accessed October 28, 2021 https://www.pnas.org/content/88/1/204.short.

Wang, Y., & Plant, W. W. (2010). *Plant sensing and signaling in response to K+-deficiency*. Elsevier Accessed October 29, 2021 https://www.sciencedirect.com/science/article/pii/S1674205214604135.

CHAPTER TEN

Role of inorganic bio stimulant elements in plant growth

Muhammad Ashar Ayub[a,b,c], Misbah Abbas[a], Muhammad Zia ur Rehman[a]

[a]Institute of Soil and Environmental Sciences, University of Agriculture Faisalabad, Faisalabad, Punjab, Pakistan
[b]Horticultural Sciences Department, University of Florida, Institute of Food and Agricultural Sciences, Indian River Research and Education Centrer, Fort Pierce, FL, United States
[c]Institute of Agro-Industry and Environment, The Islamia University of Bahawalpur, Bahawalpur, Punjab, Pakistan

1. Introduction

Plant nutrition is an important subject of discussion in the modern era where agriculture sustainability is at stake (Barker and Pilbeam 2005). There are 17 essential plant nutrients reported to have primary roles in plant metabolism and without a sufficient supply of each, the plant's sustainability is at stake (Taiz et al., 2015; Riaz et al., 2020). Besides these essential nutrients, there is a list of inorganic biostimulants and beneficial nutrients which take part in nutritional and metabolism. The inorganic biostimulants include nanoparticles, polymers, and elements like Silicon, Sodium, Selenium, and many lanthanides (Ahmad et al., 2018; du Jardin, 2015; EL Boukhari et al., 2020; Ma et al., 2010). The term biostimulant was devised by horticulturalists using plant growth-promoting substances excluding nutrients, soil conditioners, and pesticides and used in literature for elements (other than nutrients) having beneficial effects for plants (Kauffman et al., 2007). During current changing climatic conditions, plants are being widely exposed to various kinds of biotic and abiotic stresses for which mare nutritional modulation is not enough. For the same, plant nutrition only sometimes is not enough, and some biostimulants are to be applied in an adequate amount as they have an ameliorative effect on the plant under stress (Borhannuddin Bhuyan et al., 2020; Bulgari et al., 2019) as they can enhance N assimilation and better nutrition acquisition (Kunicki et al., 2010; du Jardin, 2015).

Sustainable Plant Nutrition
ISBN: 978-0-443-18675-2
https://doi.org/10.1016/B978-0-443-18675-2.00014-6

1.1 Biostimulants categories and role in plants

In the modern era, Agriculture is under immense pressure owing to the huge influx of abiotic and biotic stresses compromising the world's food security, and emerging biostimulants can be a very helpful asset to tackle these issues (Rouphael & Colla, 2020). The exogenous introduction of biostimulants in stressed agriculture systems can be a very helpful boost for crop productivity which not only can inverse negative effects of environmental stresses but also can increase crop productivity (Szparaga et al., 2019). The biostimulants though have a significant effect on plants but cannot replace nutrients in any condition (Bulgari et al., 2019; Van Oosten et al., 2017). In changing era of global climate change, plants are more prone to various natural as well as anthropogenic driven stresses including biotic as well as abiotic factors. Among prominent classes of inorganic biostimulants, major classes are Beneficial elements Al, Se, Si, Na Co (Pilon-Smits et al., 2009), nanoparticles (Jayarambabu et al., 2015; Nair, 2016; Landa, 2021; Sohail et al., 2021), silicon-based acids (Ma & Takahashi, 1989; Neeru et al., 2016) and inorganic chelating agents (Flora & Pachauri, 2010). The bioavailability of each of these stimulants primarily depends upon its chemical nature, fate in soil, soil-plant interaction, transport mechanism, and assimilation mechanisms in plants.

The beneficial elements are specific in their effects on plants and primarily depend upon dose, chemical speciation, the necessity for plants, and bioavailable concentration for heavy metals like Aluminum (Bojórquez-Quintal et al., 2017) and Cobalt (Pilon-Smits et al., 2009; Vatansever et al., 2016). Silicon is overall beneficial for plants (Richmond & Sussman, 2003) and sodium is only beneficial for beats (Kaur et al., 2016; Takahashi & Maejima, 1998). The role of inorganic acids and nanoparticles as biostimulants is also well reported and studied in the literature (Jayarambabu et al., 2015; Landa, 2021; Ma & Takahashi, 1989; Nair, 2016; Neeru et al., 2016; Sohail et al., 2021). The beneficial nutrient category is assigned for elements that do not meet the essentiality criteria for mineral nutrients, but their significant amount is helpful in yield enhancement but has no significant requirement for plant survival (Kirkby, 2012). Among beneficial nutrients, Na, Si, Al, and Se are suggested only for hyperaccumulator plants while Co is reported to be significant for plants having symbiotic association with the microbial community (Pilon-Smits et al., 2009). The mechanism of action of these elements is not much studied but in general, they are involved in osmoregulation, transpiration modification, enzyme activation, pathogen suppression, defense system activation, stress mitigation, help in nutritional acquisition, yield improvement, and post-harvest quality assurance (Trejo-Téllez & GómezMerino 2012; Anitha et al., 2020). The silicic acids (Exley et al., 2019), metallic and nutrient element nanoparticles (Jampílek & Králová, 2019; Pacheco & Buzea, 2018) are also involved in similar kind of effects in plants facing environmental stresses.

Rare earth metals are another category of elements being significantly studied for their benefits as well as a detrimental role on agriculture. The rare earth metal's specific category is of lanthanides including lanthanum (La), cerium (Ce), praseodymium (Pr), neodymium (Nd), promethium (Pm), samarium (Sm), europium (Eu), gadolinium (Gd), terbium (Tb), dysprosium (Dy), holmium (Ho), erbium (Er), thulium (Tm), ytterbium (Yb), and lutetium (Lu) (Hu et al., 2006). Among these, La, Ce, and Nd are reported to have enormous tested upon uses for agriculture application (Kotelnikova et al., 2020). These elements are reported to have effects on water absorption, cellular permeability, seed germination, enzyme activation, hormones production, metal solubilization, and enhanced N assimilation (Buckingham et al., 1999; Hu et al., 2004; Kovaříková et al., 2019; Liang et al., 2008; Tyler, 2004).

2. Inorganic plant biostimulants: fate in soil and plants

The mode of action of biostimulants in a plant varies as per the type of biostimulants. Silicon, an important beneficial element is well reported to have significant positive effects on plants facing abiotic and biotic stresses. Among major inorganic biostimulants, silicon (Si), selenium (Se), cobalt (Co), aluminum (Al), and a few lanthanides are important.

The major role of Si is reported in tackling salinity (Coskun et al., 2016; Rios et al., 2017), drought (Coskun et al., 2016; Malik et al., 2021), heavy metals (Adrees et al., 2015; Bhat et al., 2019; Kim et al., 2014; Rehman et al., 2019). The Na is beneficial for plants in certain conditions as reported for plants (Subbarao et al., 2003) but its higher concentration is highly toxic for plants as a major constraint in salt-affected soils (Kronzucker et al., 2013) and silicon can be very beneficial in countering Na stress (Flam-Shepherd et al., 2018). The plant takes up Si from the soil as orthosilicic acid and upon its translocation to roots, then into shoos due to transpiration pull as colloidal silica gel (SiO_2. nH20) as well as via forming other biomolecules (Harrison 1996; Yoshida et al., 1962). The uptake of silicon in plants is primarily controlled by the genetics of plants as specie specific accumulators/ excluders are present in plants for Si transport and due to the same effect, Si translocation is different in different. The apoplastic and symplastic pathways have been found involved in the translocation of Si into plant tissue along with the involvement of transporter proteins The accumulation and translocation of Si in plants primarily depend upon plant species, root architecture, and soil chemistry The ways of translocation of Si from soil to plants can be active (transported based) as well as passive depending upon the need of plant (Ma et al., 2001; Ma & Takahashi, 1989; Ma & Yamaji, 2006). After absorption from the soil, Si is translocated to xylem parenchyma cells in the stem by LSi6, which is a homolog of LSi1 transporter (Yamaji et al., 2008). Other transporters involved in cellular and subcellular translocation of Si are LSi1-3 and LSi6having a coordinated network facilitating Si uptake, translocation, and accumulation in plants (Feng et al., 2011), and the majority of Si is deposited in epidermal cells of the leaf, inflorescence brackets, root endodermis as silica and once deposited, its usually non translocated (Currie & Perry, 2007; Mabagala et al., 2020). The Si uptake, absorption, and assimilation potential of plants is an important factor in the stress mitigating potential of plants (Deshmukh et al., 2020; Hodson et al., 2005; Sonah et al., 2017).

The role of Si in stress mitigation of plants is well-reported and in addition to that, it helps in plant growth, regulation of enzymes, gene expression regulation, improvement in plant's physiological responses as well as in antioxidant defense system regulation under stress conditions. It is reported to provide protection against oxidative stress under drought in *Solanum lycopersicum* (Cao et al., 2015), *Sorghum bicolor* (Galvez & Clark, 1991), *Spinacia oleracea* (Gunes et al., 2007), *Triticum aestivum* (Alzahrani et al., 2018; Kim et al., 2014, 2016; Tayyab et al., 2018), Mangifera indica (Helaly et al., 2017). The role of Si in salinity tolerance is also well reported as given by Liang et al. (2007), Zhu et al. (2004, 2019, 2020), Tuna et al. (2008) in *Hordeum vulgare, Cucumis sativus,* and *Triticum aestivum* respectively. Well studied roles of Si for heavy metal stress mitigation (Kollárová et al., 2019; Singh et al., 2019), temperature stress tolerance (Nwugo & Huerta, 2010; Rehman et al., 2019), and biotic stress management (Deshmukh et al., 2020; Prabhu et al., 2001, 2015) are also well reported. A detailed review of all is given in Table 10.1.

The physical and physiological effects of silicon in plants under biotic and abiotic stress are more of molecular level (Coskun et al., 2018; Deshmukh et al., 2020; Ma & Yamaji, 2015) as these stresses initiate oxidative stress

Table 10.1 Detailed review on the role of inorganic biostimulants on plant growth under various environmental conditions.

Stimulants	Sources, dose, and method of application	Stress	Crop	Effects on crop	References
Silicon (Si)	W-14 (drought susceptible) and w-20 (drought resistant) rice varieties were exogenously applied with Si as K_2SiO_3	Drought stress	Rice (*Oryza sativa* L.)	Application of Si resulted in higher water utilization efficiency, improved chlorophyll contents, and nutritional acquisition of both plants	(Chen et al., 2011)
	Silicon applied from sodium silicate (2 mM sodium silicate per kg of soil)	Drought stress	Wheat (*Triticum aestivum* L.)	Si fertilizer application helped wheat enhance antioxidants (SOD, CAT, APX, and POD) activity helping it escape drought stress to some extent	(Tale Ahmad & Haddad, 2011)
	(0, 200, 400, 800 mg L^{-1}) of Si as sodium silicate ($Na_2SiO_3.9H_2O$) was applied on a 1-year-old crop plant under greenhouse	Drought stress	Kentucky bluegrass *(Poa pratensis)*	Si application resulted in more activity of Rubisco enzyme, stable proteins and resulted in higher carbon assimilation.	(Saud et al., 2016)
	1 mM of silicon dioxide (SiO_2)	Drought stress (10% and 20% PEG)	Rapeseed (*Brassica napus* L.)	Si helped plants upregulate various enzymatic and nonenzymatic antioxidants and assisted plants in higher chlorophyll contents	(Hasanuzzaman et al., 2018)

(Continued)

Table 10.1 Detailed review on the role of inorganic biostimulants on plant growth under various environmental conditions.—cont'd

Stimulants	Sources, dose, and method of application	Stress	Crop	Effects on crop	References
				production helping plants withstand drought stress	
	Si priming treatments (0, 4 mM, 6 mM)	Drought stress (plant kept at 60% field capacity for 3 weeks)	Maize (*Zea mays* subsp. *Mays*)	Si priming helped plants alter several antioxidants as well as manage oxidative stress in a better way compared to control	(Parveen et al., 2019)
	0, 100, 200 and 400 kg ha $^{-1}$ as H_4SiO_4, (20% Si content)	Water stress (100%, 75% and 50% field capacity)	Cantaloupe (*Cucumis melo* var. cantalupensis)	H_4SiO_4 application rates of 200 and 400 kg ha^{-1} have shown an incremental behavior on crop growth	Ali et al., 2021)
	1.5 mM Si in form of Na_2SiO_3	Salinity stress (50 mM NaCl in hydroponic condition)	Cucumber *Cucumis sativus*	Si supplementation improved plants' photosynthetic rate, fluorescence parameters, and chlorophyll contents.	(Harizanova & Koleva-Valkova, 2019)
	0.3 mM Si from $Na_2SiO_3 \cdot 9H_2O$	Salinity/Na stress was applied @ 50 mM NaCl	Cucumber *Cucumis sativus*		(Zhu et al., 2020)
	SiO_2-NPs at (0, 200, 400 and 600 mg/L).	Salinity and drought stress (PEG-8000)	Banana *(Musa acuminata 'Grand Nain')*	Helped plant tissue K: Na ratio rise, helped plant with higher chlorophyll contents under in vitro study	(Mahmoud et al., 2020)

Si (2 mM)	Salinity stress (100 mM NaCl)	Soybean (*Glycine max*)	Si application helped plant better manage reactive oxygen and reactive nitrogen species	(Chung et al., 2020)
Foliar and basal application of 100 ppm Si	6000 mg Na per kg dry soil.	Sweet basil (*Ocimum basilicum* L)	Si application helped plants mitigate salinity stress via helping in water balance, managing ion homeostasis, increasing oil yield and plant growth	(Farouk et al., 2020)
Si (Na_2SiO_3; 1 mM)	Salinity stress (80 and 160 mM)	Maize (*Zea mays* L.) P1574 and Hycorn 11 cultivars	The effect of Si salinity mitigation was more prominent on P1574 cultivar. Overall, Si helped in water balance, antioxidants metabolism and nutrients acquisitions by rice plants	(Ali et al., 2021)
2.5 mM Si dose application via organic and inorganic sources	Cd, Pb, Cu and Zn	rice (*Oryza sativa* L.)	Both sources of Si helped plant lowering bioaccumulation of all metals as well as improve plant growth and production.	(Wang et al., 2016)
Si @ 42 mg kg^{-1} using $K_2SiO_3 \cdot nH_2O$	Cd contaminated soils	rice (*Oryza sativa* L.)	Si application helped plant upregulate antioxidant enzymes as well as helped it lowed Cd uptake and	(Huang et al., 2018)

(Continued)

Table 10.1 Detailed review on the role of inorganic biostimulants on plant growth under various environmental conditions.—cont'd

Stimulants	Sources, dose, and method of application	Stress	Crop	Effects on crop	References
				accumulation which helped in overall healthy and safe crop	
	Si was applied in the split as K_2SiO_3	Cd stress	rice (*Oryza sativa* L.)	Split application of Si at various growth stages has proven to be beneficial in decreasing soil bioavailable Cd as well as its translocation into plant tissues.	(Rehman et al., 2019)
	Silicon and nitric oxide application	Cd stress	Maize (*Zea mays* subsp. *mays*)	The application of Si increased plant growth, lower Cd uptake, translocation, and bioaccumulation	(Liu et al., 2020)
	Si nanoparticles application	Pb stress	Coriander (*Coriandrum sativum* L.)	The application of Si has helped plants increase vitamin C and flavonoid contents as well as helped increase antioxidants activity in plant leaves also helped lower Pb uptake and translocation.	(Fatemi et al., 2020)
Selenium (Se)	Se	Drought stress	Rice (Oryza sativa L.)	Application of Se overall improved grain yield	(Emam et al., 2014)

		cv. Giza 177 and cv. IET 1444	quality and yield and helped mitigate severe effects of drought to some extent.	
0.03 mM Se	Drought stress	Wheat (*Triticum aestivum* L.) Seedlings	The Se application enhanced the expression of genes like TFs, DREB2A, NAC5, OsRDCP1, OsCMO, and OsRAB16b helpful in the provision of drought resistance to plants.	(Khattab et al., 2014)
Foliar applied Se @ 40 mg L using Na_2SeO_4	Drought stress	Wheat (*Triticum aestivum* L.)	Improved turgor pressure, water use efficiency, sugar, and free amino acid activity as well as of antioxidant system helping overall reversal of drought stress.	(Nawaz et al., 2015)
Foliar application of Se in field conditions	Drought stress	Wheat (*Triticum aestivum* L.)	The Se application increased grain yield, helped in overall nutritional homeostasis and antioxidants metabolism resulting in the overall escape of wheat crop from	(Rady et al., 2020)

(*Continued*)

Table 10.1 Detailed review on the role of inorganic biostimulants on plant growth under various environmental conditions.—cont'd

Stimulants	Sources, dose, and method of application	Stress	Crop	Effects on crop	References
				adverse effects of drought	
	Se (2 μM Na_2SeO_4)	Cd stress 600 μM $CdCl_2$	*Rape (Brassica napus oleifera);* wheat (*Triticum aestivum* L.)	Se application has shown a protective role in both crops against Cd stress via modulating macro and micronutrients uptake and translocation in plants.	(Zembala et al., 2010)
	Se applied @ 0, 5, or 10 μM	Salinity stress NaCl (0, 25, or 50 mM)	Tomato (*Solanum lycopersicum*)	The Se application have shown a net oxygen scavenging capability helping in overall physiological and biochemical uplift of plant helping it out of salinity stress.	(Mozafariyan et al., 2016)
	Se (0, 1, 5 and 25 μM Na_2SeO_3)	Salinity stress (Na^+ ionic toxicity)	Maize (*Zea mays* L.)	Se application enhanced photosynthetic activity, Na + accumulation and translocation, and reduced oxidative stress via modulating	(Jiang et al., 2017)

				antioxidants enzymes production.	
	Se application (20 and 40 mg/L)	Salinity stress	Maize (*Zea mays* L.)	Helped improve plant growth, enhanced oxidative defense and ionic homeostasis in tested plants.	
	Se solution application @ 0, 5, and 20 μM	Cd stress under the hydroponic condition	Bok choy (*Brassica chinensis*)	Se application helped plant enhance antioxidants thus helping in overall reactive oxygen species quenching and reversal of Cd mediated stress.	(Yu et al., 2018)
	selenium (0, 4, 6, 8 mgL −1)	Cd (0, 20, 25 μM) and Pd (0, 60, 100 μM) stress	Cucumber (*Cucumis sativus* L.)	Se application has shown a protective role on flowering indices and total yield via controlling of metal translocation in plants.	(Shekari et al., 2019)
	5 μmol L^{-1} Se application	5 μmol L^{-1} Cd	rice (*Oryza sativa* L.)	Se application helped seedlings withstand oxidative stress via the production of antioxidants as well as helped lower Cd uptake in plant tissues.	(Wan et al., 2019)
Cerium	0.75 μM $Ce(NO_3)_3$	80 mM NaCl	Maize (*Zea mays* subsp. *mays*)	Ce (III) partially provided salinity mitigation to	(Qu et al., 2013)

(Continued)

Table 10.1 Detailed review on the role of inorganic biostimulants on plant growth under various environmental conditions.—cont'd

Stimulants	Sources, dose, and method of application	Stress	Crop	Effects on crop	References
				crop plants but cannot substitute the role of K.	
	Cerium oxide nanoparticles	100 mM NaCl stress	Rapeseed (*Brassica napus* L.)	Cerium oxide nanoparticles changed growth and physiology to some extent but didn't not countersalinity stress entirely.	(Rossi et al., 2016)
	CeO_2NPs (0, 500 mg kg $^{-1}$ dry sand)	50 mM NaCl stress	Rapeseed (*Brassica napus* L.)	Modified root barriers to enhance Na translocation toward upper plant tissues.	(Rossi et al., 2017)
	Ce from cerium nitrate	Salinity	Maize (*Zea mays* subsp. *mays*)	Ce helped plant in mitigation of salt stress induced oxidative stress via modulating ascorbate-glutathione cycle in maize roots.	(Hu & Shan, 2018)
	CeO_2 NPs (25, 50 and 100 mg L^{-1})	25 and 75 mM NaCl	Grapevine (*Vitis vinifera* L.) *cv.*	The NPs helped plant escape salinity stress via enhancing antioxidant activity and altering elemental uptake in plants.	(Gohari et al., 2021)

	Ce (+3)	Drought stress (PEG6000)	Common bean (*Phaseolus vulgaris*)	Ce application alleviated drought stress to some extent.	(Salgado et al., 2020)
Praseodymium (Pr)	20–80 mg L^{-1} Pr	200 μmol L^{-1} Cd	Wheat (*Triticum aestivum* L.)	40 ppm Pr solution application increase wheat seedling growth and potentially mitigated Cd toxicity.	(He et al., 2010)
	100 mg/L Pr	Cd stress	rice (*Oryza sativa* L.)	The Pr application helped the plant overcome Cd stress via mitigation of oxidative stress.	(Ren et al., 2010)

via the production of (ROS) in the cell causing oxidative damage to cellular proteins, lipids, DNA and other macromolecules (Naeem et al., 2018; Verma et al., 2020). The main effects to detoxify/neutralize the oxidative stress, enzymatic and non-enzymatic antioxidants production as of superoxide dismutase (SOD), catalase (CAT), ascorbate peroxidase (APX), carotenoids, tocopherols, ascorbate, and glutathione etc via upregulation of gene expression involved in activation of aquaporins (AQP) and antioxidant production as well as pathogenesis-related proteins (PR1) (Aghabary et al., 2004; Kim et al., 2017; Zhang et al., 2017). The ROS produced in biotic and abiotic stress tend to disrupt AQP production and signal transduction involved in it opening and closing and can act as stress signal but in higher concentration is highly disruptive for plant growth (Ranjan et al., 2021).

Selenium is found in trace amounts in the environment and in special circumstances can act as an essential nutrient for human and animal growth or as a pollutant (depending upon source and concentration) making its chemistry in the environment special for researchers (Combs & Combs, 1986; Fan et al., 2002; Hasanuzzaman et al., 2020; Salhani et al., 2003). The role of Se in the human body is well reported as an important antioxidant and the same role is expected to be present in plants as well as its reported to delay leaf senescence and help promote seedling growth (Ekelund & Danilov, 2001; Yang et al., 2022). In well-reported works, the exogenous application of Se has shown promising effects on the growth of various plants under stressed environmental conditions (Das et al., 2021; Hossain et al., 2021). Though the beneficial effects of Se are superficial, its toxicity is also a concern for the plants (Brown & Shrift, 1982; Hasanuzzaman et al., 2021). Detailed effects of Se application in plants are reported in Table 10.1. The applied Se activates an antioxidants mechanism in plants (both enzymatic and non-enzymatic) helping plants withstand stresses (Banerjee & Roychoudhury, 2019).

Cobalt is a heavy metal element with proven beneficial effects in plants in lower concentrations while being toxic in higher amounts (Hanson et al., 2001; Pilon-Smits et al., 2009; Rauser & Samarakoon, 1980) but its very critical for plant growth owing to its multidisciplinary applications (Banerjee & Bhattacharya, 2021; Bhattacharya et al., 2016). The most-reported beneficial effects of cobalt in plants are for leguminous crops (Roychoudhury & Chakraborty, 2022; Banerjee and Bhattacharya et al., 2021). The Co translocation in the plant from soil to upper tissue is via various physiological pathways and transporters like iron-regulated transporter proteins (IRT1) along with zinc-iron permease (ZIP) family (Lange, Ent, et al., 2016, Lange,

Pourret et al., 2016; Vatansever et al. 2016; Yamaguchi et al., 2017). Aluminum is another heavy metal element having proven beneficial and toxic effects depending upon its concentration and bioavailability to plants. The Al hyperaccumulators can use it as a weapon against pests and assistance in disease control and also help plants in nutritional uptake but can be toxic in higher concentrations as well (Bojórquez-Quintal et al., 2017). The other elements which are sometimes categorized as beneficial belong to the heavy metal group of lanthanides commonly know as rare earth elements or REEs having diverse effects on plants and animals and are widely being used in China as fertilizer for ages (Hu et al., 2004; Tommasi et al., 2020; Wang et al., 2003; Zhang & Shan, 2001). Despite their name being rare, their concentration in soil is well over 0.015% and their sufficiency and toxicity also depend upon the type of element, bioavailable concentrations, and organism type (Balaram, 2019; Goecke et al., 2015; Malhotra et al., 2020).

3. Inorganic plant biostimulants: molecular aspects of the role in plant's metabolism

3.1 Silicon

The plant needs a total of 17 essential nutrients (macro-C, H, O, N, P, K, Ca, Mg, S and Micro-Fe, Mn, B, Zn, Ni, Cu, Cl) which are crucial for plant growth and development to ensure agriculture sustainability (Riaz et al., 2020; Taiz et al., 2015; Umar et al., 2020; Naveed et al., 2020). Under stressed conditions, it becomes very difficult for plants to ensure the acquisition of balanced nutrition and in these circumstances, biostimulant elements can play their role. The role of Si in enhanced plant nutrition is well reported in sorghum and barley (Hurtado et al., 2019), sugar beet (Artyszak et al., 2019), melons (do Nascimento et al., 2020), gerbera (Kamenidou et al., 2010) and tomato (Zhang et al., 2019). The main role of Si in plant nutrition acquisition is due to its role in upregulating gene expression of macro and micronutrient transporters in roots involved in enhanced uptake of nutrients from root to shoot helping in enhanced metabolism. While under HM nutrient toxicity, Si supplementation helps in the net chelation of HMNs in root cells restricting its translocation toward shoot and helping plants escape toxicity (Ali et al., 2020). It is well reported that Si supplementation controls the genes and transporters like N (OsAMT1; 1OsGS1, 1OsNRT1:1, OsGS2, OsFd-GOGAT, OsNADH-GOGAT2, and OsGDH2) in rice plants (Wu et al., 2017). The P transporters TaPHT1.1 and TaPHT1.2 and P transporter gene OsPT6 are also found to be in direct control of Si supplementation under P deficiency conditions (Hu et al., 2018; Kostic et al., 2017). The three important macronutrients to be taken from soil is potassium, and it's a key component of a plant's solute and food transport system, and Si—K cross talk in plants is well reported (Miao et al., 2010). In a hydroponic investigation, Chen et al. (2016) reported that Si supplementation in water-stressed sorghum helped alter gene expression of SKOR channels mediating K secretion from root cortex to xylem as well as down-regulated HAK5 and AKT1 transporters resulting in higher K retention in xylem, higher aquaporins activity and higher hydraulic conductance in vascular bundles under K deficiency. Similarly, Si cross-talk with other macronutrients is also reported under various conditions like for Mg and S (Hosseini et al., 2019; Maillard et al., 2018). The main effect of Si—S interaction is via hormonal homeostasis and ABA is important in this regard and Si supplementation plays a vital role in S metabolism in plants (Réthoré et al., 2020). The Si exogenous application helps in decreasing the concentration of stress hormones in the plant while helping the plant have sustainable and higher production under S-deprived conditions (Hosseini et al., 2017). Similarly, Si—Mg crosstalk in Mg deficiency condition is also interesting as it influences the contents of organic acid, amino acids like (GABA, Ser, and Gly), and contributes toward phytohormones metabolism like cytokinin, jasmonic acid as well as their derivatives like JA-I1e to help plant overcome Mg deficiency (Maillard et al., 2018; Hosseini et al., 2017, 2019). The Si application is also reported to enhance uptake of Mg as well as help forage plants increase their production owing to Si—S—Mg cross talk (Buchelt et al., 2020).

The role of Si in micronutrient-related nutrition management in plants is also well studied on a molecular level. Pavlovic and coworkers provided evidence that Si application can help ameliorate the iron (Fe) deficiency in cucumber via assisting Fe translocation into younger leaves via production of secretions (citrate and malate) helpful in Fe solubilization and translocation (Pavlovic et al., 2013, 2016). The role of Si in the upregulation of Fe-related gene expression is also well reported and NAS1 and YSL1 are important Fe genes in this regard. The application of Si has been shown to enhance Fe contents in young leaves of cucumber and barley under Fe deficient conditions thus can actively help young plants escape Fe deficiency-related stress (Bityutskii et al., 2018; Nikolic et al., 2019) which sometimes can be pH-dependent (Gonzalo et al., 2013). Other metallic micronutrients like Mn and Zn are also essential for plant growth and their deficiency (and toxicity) can disrupt normal physiology and biochemical balance in plants. In plants,

Zn translocation can be changed by Si application as reported in various studies and the main reason for that was the co-existence of both elements in the same plant tissues suggesting their synergistic transport (Bityutskii et al., 2014, Gu et al., 2012). While in Zn toxicity conditions, Si application has shown changes in photosynthesis-related gene expression (Ali et al., 2020; Gu et al., 2012; Song et al., 2011). The work conducted by Anwaar et al. (2015) has shown that Si application has shown a decrease in Zn mediated toxicity in cotton by reducing Zn translocation as well as changing cellular oxidation stress of cells via modulating antioxidants enzymes production. At the molecular level, Si application helped mediate Zn toxicity via altering organelle structure and improving root endodermis and pericycle structure via Si deposition (Patrícia Vieira da Cunha et al., 2010). Under Mn deficiency conditions, the plant's photosynthetic activity is disrupted due to the deactivation of various enzymes (Tripathi et al., 2015). Exogenous application of sodium and potassium silicates in sorghum has shown an improved molecular activity in sorghum helping attain higher dry matter production, improved photosynthetic activity, better antioxidants production, and improved root development (de Oliveira et al., 2019). While under Mn toxicity to various plants, the role of Si is also well studied and has proven to have ameliorative effects (Iwasaki et al., 2001, 2002). The effect of Si on plants under Mn deficiency and toxicity is also related to crop type as variable effects have been reported for variable Mn contents in different crops. In rice, the Si—Mn complex in root cells has proven to contribute toward lower translocation of Mn which is further assisted by reduced expression of OsNramp5 (Mn transporter) upon long-term Si application (Doncheva et al., 2009; Li et al., 2012). In cucumber, Mn toxicity has resulted in oxidative stress and Si application helps in reducing oxidative stress via enhancing antioxidants enzyme activity and reducing reactive oxygen species (ROS) as well as Mn contents in the apoplast (Dragišić Maksimović et al., 2007, 2012). For boron, the deficiency-toxicity fringe is very minute and a very little over the presence of B can lead to toxicity (Camacho-Cristóbal et al., 2008). The well-documented role of Si under B deficiency and excess conditions in cereals are present in literature (Gunes et al., 2007; Inal et al., 2009; Liu et al., 2017; Farooq et al., 2019; Pereira de Souza Junior et al., 2019). In recent work, different modes of Si application in cotton crops have helped plants withstand B deficiency with the foliar application being most beneficial via improving the plant's physiology and reducing B uptake (Pereira de Souza Junior et al., 2019). Under B toxicity conditions, exogenous application of B has proven to help quench ROS, via upregulating antioxidants production and reducing leaf B translocation (Farooq et al., 2019; Celikkol Akcay & Erkan, 2015).

Besides the well-versed beneficial effects of Si in plant nutrition, the role of Si as a beneficiary plant nutrient under biotic and abiotic stress conditions is also well reported. The role of Si in cell wall strengthening, mechanical support, suberization/lignification enhancement and silicification is an interesting phenomenon that helps a lot in plants' defense against drought conditions (He et al., 2013; Guntzer et al., 2012; He et al., 2015). The exogenous application of Si has also been shown to have root barriers (with thickening Casparian bands) in root endo and exodermis. In rice, Si has shown excessive suberization and lignification of sclerenchyma in endodermis and exodermis and triggered genes related to lignin and suberin synthesis (Fleck et al., 2011, 2015). These barriers help the plant to make apoplastic barriers that check Na^+ transport in roots providing salt tolerance in rice (Gong et al., 2006; Krishnamurthy et al., 2011). Similarly, the role of Si in assisting plants to tolerate pollutants is also well reported for mitigation of heavy metal stress which includes antioxidants production, cellular protection, lowering HMs uptake, and enhancing nutritional homeostasis (WU et al., 2013; Kim et al., 2014; Adrees et al., 2015; Emamverdian et al., 2018; Bhat et al., 2019; Rehman et al., 2019).

3.2 Selenium

The selenium (Se) present in soil commonly has four oxidation states: Se (+6)-selenate, Se (+4), Se (0)-elemental, and Se (−2)-selenide. The role of Se in plant nutrition and growth under abiotic and biotic stresses is a widely explored phenomenon (Kleiber et al., 2020; Sieprawska et al., 2015). The beneficial role of Se under drought stress is well reported due to its potential in the enhancement of both enzymatic as well as non-enzymatic antioxidants as well as help in water balance (Ahmad et al., 2021; Emam et al., 2014; Habibi, 2013; Ibrahim, 2014). The beneficial effects of a low concentration of Selenium have been reported for plants under stresses like drought (Ahmad et al., 2016; Adnan, 2020), salinity (Kamran et al., 2020), and heavy metals (Babula et al., 2009; Feng et al., 2021; Shekari et al., 2019) via managing oxidative stress and mediating transporters for facilitated.

The main molecular mechanism of Se mediated drought tolerance in plants is associated with better energy conversion in PS(II) as expressed by higher effective quantum energy in potatoes (Germ et al., 2007). In wheat, the application of Se has shown potential to assist the plant in the maintenance of turgor pressure and improved gas exchange (Nawaz et al., 2015). In rapeseed, Se pretreatment has proven to be beneficial for the plant as was found to be managing the metabolism of GSH, glutathione, APX and monodehydroascorbate reductase (MDHAR) activities resulting in a significant decrease in drought stress (Hasanuzzaman & Fujita, 2011). The application of Se in salinity stressed garlic has shown to have a supplementation effect and enhanced Chl a, b, total Chl, and increased antioxidants activity (Astaneh et al., 2019). In salinity stressed Canola crop, exogenous application of Se has shown an incremental effect on soluble sugar, potassium, enhanced photosynthesis, chlorophyll contents, antioxidants activity, and improved plant growth (Bybordi, 2016). In salinity stressed grapes Vitis vinifera L. cv. Sultana, Se application resulted in improved nutrition and enzymatic activity in grapevine leaves (Karimi et al., 2020). Under heavy metal stress, Se application has also been shown to improve plant physiology, biochemical attributes, and molecular parameters (Choppala et al., 2014; Feng et al., 2021; Riaz et al., 2021; Shekari et al., 2019; Zwolak, 2020).

3.3 Cobalt and aluminum

The Co in the soil system is highly labile and bioavailable but its transition from residual to a labile fraction is dependent on the soil types as well as soil conditions. The bioavailable fraction of Co in the soil is of main concern regarding its uptake into the plant as well as its translocation into different tissues (Jalali et al., 2019; Liu et al., 2019; Perez-Espinosa et al., 2005). The translocation of Co into the plant is via iron-regulated transporters and ferroproteins (IRT1 and FPN1) along with organic chelating agents (Lange et al., 2017; Yamaguchi et al., 2017). Besides this, there is not much information about exact Co translocation mechanisms in plants (Lange, Ent, et al., 2016). The use of Co on plants for potential mitigation of stresses is well researched and reported. The role of Co application on yellow passion fruits under NO_3^- starvation/NH_4^+ deprivation stress has shown an incremental effect on photosynthetic pigment and plant growth (Nuncio et al., 2021). The Co role in the reversal of reactive oxygen species (ROS) and nutrition provision enhancement under salinity stress conditions resulting in redox homeostasis is also reported in cucumber (*Cucumis sativus* L.) by Brengi et al. (2021). Aluminum being another metallic can also act as a biostimulant in a small concentration as it can help plant nutrition, help promote growth, quench oxidative stress, and help plants in sustainable metabolism (Hodson & Evans, 2020). The detailed and specific effects of these bio-stimulants are presented in Table 10.1.

4. Future prospects and summary

Inorganic bio stimulating elements can help plants withstand abiotic stresses and achieve better growth, yield, and production via enhancing nutritional acquisition and suppressing stress. A lot of publications are available regarding the beneficial role of Si and Se but of Co and Al not much research is available. The role of lanthanides in plant growth promotion is also well reported but comprehensive research is needed to better understand the behaviour of these elements in plan's metabolism. Authors encourage readers to explore these biostimulants for effective insurance of sustainable agriculture.

References

Adnan, M. (2020). Application of selenium A useful way to mitigate drought stress: A review. *Open Access Journal of Biogeneric Science and Research, 3*(1). https://doi.org/10.46718/jbgsr.2020.03.000064

Adrees, M., Ali, S., Rizwan, M., Zia-ur-Rehman, M., Ibrahim, M., Abbas, F., Farid, M., Qayyum, M. F., & Irshad, M. K. (2015). Mechanisms of silicon-mediated alleviation of heavy metal toxicity in plants: A review. *Ecotoxicology and Environmental Safety, 119*, 186–197. https://doi.org/10.1016/j.ecoenv.2015.05.011

Aghabary, A. L., Zhu, K., & Shi, Q. H. (2004). Influence of silicon supply on chlorophyll content, chlorophyll fluorescence, and antioxidative enzyme activities in tomato under salt stress. *J. Plant Nutr, 27*, 2101–2115.

Ahmad, Z., Anjum, S., Skalicky, M., Waraich, E. A., Muhammad Sabir Tariq, R., Ayub, M. A., Hossain, A., Hassan, M. M., Brestic, M., Sohidul Islam, M., Habib-Ur-Rahman, M., Wasaya, A., Aamir Iqbal, M., & El Sabagh, A. (2021). Selenium alleviates the adverse effect of drought in oilseed crops camelina (camelina sativa L.) and canola (Brassica napus L.). *Molecules, 26*(6), 1699. https://doi.org/10.3390/molecules26061699

Ahmad, H. R., Zia-ur-Rehman, M., Sohail, M. I., Anwar ul Haq, M., Khalid, H., Ayub, M. A., & Ishaq, G. (2018). Effects of rare earth oxide nanoparticles on plants. In *Nanomaterials in plants, algae, and microorganisms* (pp. 239–275). Elsevier. https://doi.org/10.1016/b978-0-12-811487-2.00011-6

Ali, M., Afzal, S., Parveen, A., Kamran, M., Javed, M. R., Abbasi, G. H., Malik, Z., Riaz, M., Ahmad, S., Chattha, M. S., Ali, M., Ali, Q., Uddin, M. Z., Rizwan, M., & Ali, S. (2021). Silicon mediated improvement in the growth and ion homeostasis by decreasing Na+ uptake in maize (Zea mays L.) cultivars exposed to salinity stress. *Plant Physiology and Biochemistry, 158*, 208–218. https://doi.org/10.1016/j.plaphy.2020.10.040

Ali, N., Réthoré, E., Yvin, J.-C., & Hosseini, S. A. (2020). The regulatory role of silicon in mitigating plant nutritional stresses. *Plants (Basel, Switzerland), 9*(12), 1779. https://doi.org/10.3390/plants9121779

Alzahrani, Y., Kuşvuran, A., Alharby, H. F., Kuşvuran, S., & Rady, M. M. (2018). The defensive role of silicon in wheat against stress conditions induced by drought, salinity or cadmium. *Ecotoxicology and Environmental Safety, 154*, 187–196. https://doi.org/10.1016/j.ecoenv.2018.02.057

Anitha, K. V., Chikkaramappa, T., Basavaraja, P. K., Prakash, N. B., Murali, K., & Boraiah, B. (2020). Effect of different levels of fertilizers on productivity and nutrient dynamics of browntop millet. *International Research Journal of Pure and Applied Chemistry*, 308–320. https://doi.org/10.9734/irjpac/2020/v21i2430368

Anwaar, S. A., Ali, S., Ali, S., Ishaque, W., Farid, M., Farooq, M. A., Najeeb, U., Abbas, F., & Sharif, M. (2015). Silicon (Si) alleviates cotton (Gossypium hirsutum L.) from zinc (Zn) toxicity stress by limiting Zn uptake and oxidative damage. *Environmental Science and Pollution Research, 22*(5), 3441–3450. https://doi.org/10.1007/s11356-014-3938-9

Artyszak, A., Gozdowski, D., & Kucińska, K. (2019). Impact of foliar fertilization on the content of silicon and macronutrients in sugar beet. *Plants (Basel, Switzerland), 8*(5), 136. https://doi.org/10.3390/plants8050136

Astaneh, R. K., Bolandnazar, S., Nahandi, F. Z., & Oustan, S. (2019). Effects of selenium on enzymatic changes and productivity of garlic under salinity stress. *South African Journal of Botany, 121*, 447–455. https://doi.org/10.1016/j.sajb.2018.10.037

Babula, P., Adam, V., Opatrilova, R., Zehnalek, J., Havel, L., & Kizek, R. (2009). Uncommon heavy metals, metalloids and their plant toxicity: A review. In *Sustainable agriculture reviews* (pp. 275–317). Springer Netherlands. https://doi.org/10.1007/978-1-4020-9654-9_14

Balaram, V. (2019). Rare earth elements: A review of applications, occurrence, exploration, analysis, recycling, and environmental impact. *Geoscience Frontiers, 10*(4), 1285–1303. https://doi.org/10.1016/j.gsf.2018.12.005

Banerjee, P., & Bhattacharya, P. (2021). Investigating cobalt in soil-plant-animal-human system: Dynamics, impact and management. *Journal of Soil Science and Plant Nutrition, 21*(3), 2339–2354. https://doi.org/10.1007/s42729-021-00525-w

Banerjee, A., & Roychoudhury, A. (2019). Role of selenium in plants against abiotic stresses: Phenological and molecular aspects. In *Molecular plant abiotic stress* (pp. 123–133). Wiley. https://doi.org/10.1002/9781119463665.ch7

Barker, A. V., & Pilbeam, D. J. (2015). *Handbook of plant nutrition*. CRC Press. https://doi.org/10.1201/b18458

Bhat, J. A., Shivaraj, S. M., Singh, P., Navadagi, D. B., Tripathi, D. K., Dash, P. K., Solanke, A. U., Sonah, H., & Deshmukh, R. (2019). Role of silicon in mitigation of heavy metal stresses in crop plants. *Plants (Basel, Switzerland), 8*(3), 71. https://doi.org/10.3390/plants8030071

Bhattacharya, P. T., Misra, S. R., & Hussain, M. (2016). Nutritional aspects of essential trace elements in oral health and disease: An extensive review. *Scientifica*, 5464373. https://doi.org/10.1155/2016/5464373

Bityutskii, N., Pavlovic, J., Yakkonen, K., Maksimović, V., & Nikolic, M. (2014). Contrasting effect of silicon on iron, zinc and manganese status and accumulation of metal-mobilizing compounds in micronutrient-deficient cucumber. *Plant Physiology and Biochemistry, 74*, 205–211. https://doi.org/10.1016/j.plaphy.2013.11.015

Bityutskii, N. P., Yakkonen, K. L., Petrova, A. I., Lukina, K. A., & Shavarda, A. L. (2018). Silicon ameliorates iron deficiency of cucumber in a pH-dependent manner. *Journal of Plant Physiology, 231*, 364–373. https://doi.org/10.1016/j.jplph.2018.10.017

Bojórquez-Quintal, E., Escalante-Magaña, C., Echevarría-Machado, I., & Martínez-Estévez, M. (2017). Aluminum, a friend or foe of higher plants in acid soils. *Frontiers in Plant Science, 8*, 1767. https://doi.org/10.3389/fpls.2017.01767

Borhannuddin Bhuyan, M. H. M., Mohsin, S. M., Mahmud, J. Al, & Hasanuzzaman, M. (2020). Use of biostimulants for improving abiotic stress tolerance in brassicaceae plants. In *The plant family brassicaceae* (pp. 497–531). Singapore: Springer. https://doi.org/10.1007/978-981-15-6345-4_19

Brengi, S. H., Khedr, A. A. E. M., & Abouelsaad, I. A. (2021). Effect of melatonin or cobalt on growth, yield and physiological responses of cucumber (Cucumis sativus L.) plants under salt stress. *Journal of the Saudi Society of Agricultural Sciences*. https://doi.org/10.1016/j.jssas.2021.06.012

Brown, T. A., & Shrift, A. (1982). Selenium: Toxicity and tolerance in higher plants. *Biological Reviews, 57*(1), 59–84. https://doi.org/10.1111/j.1469-185x.1982.tb00364.x

Buchelt, A. C., Teixeira, G. C. M., Oliveira, K. S., Rocha, A. M. S., de Mello Prado, R., & Caione, G. (2020). Silicon contribution via nutrient solution in forage plants to mitigate nitrogen, potassium, calcium, magnesium, and sulfur deficiency. *Journal of Soil Science and Plant Nutrition, 20*(3), 1532–1548. https://doi.org/10.1007/s42729-020-00245-7

Buckingham, S., Maheswaran, J., Meehan, B., & Peverill, K. (1999). The role of applications of rare earth elements in enhancement of crop and pasture production. *Materials Science Forum, 315–317*, 339–347. https://doi.org/10.4028/www.scientific.net/msf.315-317.339

Bulgari, R., Franzoni, G., & Ferrante, A. (2019). Biostimulants application in horticultural crops under abiotic stress conditions. *Agronomy, 9*(6), 306. https://doi.org/10.3390/agronomy9060306

Bybordi, A. (2016). Influence of zeolite, selenium and silicon upon some agronomic and physiologic characteristics of canola grown under salinity. *Communications in Soil Science and Plant Analysis, 47*(7), 832–850. https://doi.org/10.1080/00103624.2016.1146898

Calero Hurtado, A., Aparecida Chiconato, D., de Mello Prado, R., da Silveira Sousa Junior, G., & Felisberto, G. (2019). Silicon attenuates sodium toxicity by improving nutritional efficiency in sorghum and sunflower plants. *Plant Physiology and Biochemistry, 142*, 224–233. https://doi.org/10.1016/j.plaphy.2019.07.010

Camacho-Cristóbal, J. J., Rexach, J., & González-Fontes, A. (2008). Boron in plants: Deficiency and toxicity. *Journal of Integrative Plant Biology, 50*(10), 1247–1255. https://doi.org/10.1111/j.1744-7909.2008.00742.x

Cao, B., Ma, Q., Zhao, Q., Wang, L., & Xu, K. (2015). Effects of silicon on absorbed light allocation, antioxidant enzymes and ultrastructure of chloroplasts in tomato leaves under simulated drought stress. *Scientia Horticulturae, 194*, 53–62. https://doi.org/10.1016/j.scienta.2015.07.037

Celikkol Akcay, U., & Erkan, I. E. (2015). Silicon induced antioxidative responses and expression of BOR2 and two PIP family aquaporin genes in barley grown under boron toxicity. *Plant Molecular Biology Reporter, 34*(1), 318–326. https://doi.org/10.1007/s11105-015-0923-5

Chen, D., Cao, B., Wang, S., Liu, P., Deng, X., Yin, L., & Zhang, S. (2016). Silicon moderated the K deficiency by improving the plant-water status in sorghum. *Scientific Reports, 6*, 22882. https://doi.org/10.1038/srep22882

Choppala, G., Saifullah, Bolan, N., Bibi, S., Iqbal, M., Rengel, Z., Kunhikrishnan, A., Ashwath, N., & Ok, Y. S. (2014). Cellular mechanisms in higher plants governing tolerance to cadmium toxicity. *Critical Reviews in Plant Sciences, 33*(5), 374–391. https://doi.org/10.1080/07352689.2014.903747

Chung, Y. S., Kim, K.-S., Hamayun, M., & Kim, Y. (2020). Silicon confers soybean resistance to salinity stress through regulation of reactive oxygen and reactive nitrogen species. *Frontiers in Plant Science, 10*, 1725. https://doi.org/10.3389/fpls.2019.01725

Combs, G. F., & Combs, S. B. (1986). Selenium and cancer. In *The role of selenium in nutrition* (pp. 413–461). Elsevier. https://doi.org/10.1016/b978-0-12-183495-1.50015-4

Coskun, D., Britto, D. T., Huynh, W. Q., & Kronzucker, H. J. (2016). The role of silicon in higher plants under salinity and drought stress. *Frontiers in Plant Science, 7*. https://doi.org/10.3389/fpls.2016.01072

Coskun, D., Deshmukh, R., Sonah, H., Menzies, J. G., Reynolds, O., Ma, J. F., Kronzucker, H. J., & Bélanger, R. R. (2018). The controversies of silicon's role in plant biology. *New Phytologist, 221*(1), 67–85. https://doi.org/10.1111/nph.15343

Currie, H. A., & Perry, C. C. (2007). Silica in plants: Biological, biochemical and chemical studies. *Annals of Botany, 100*(7), 1383–1389. https://doi.org/10.1093/aob/mcm247

Das, S., Majumder, B., & Biswas, A. K. (2021). Selenium alleviates arsenic induced stress by modulating growth, oxidative stress, antioxidant defense and thiol metabolism in rice seedlings. *International Journal of Phytoremediation*, 1–15. https://doi.org/10.1080/15226514.2021.1975639

Deshmukh, R., Sonah, H., & Belanger, R. R. (2020). New evidence defining the evolutionary path of aquaporins regulating silicon uptake in land plants. *Journal of Experimental Botany, 71*(21), 6775–6788. https://doi.org/10.1093/jxb/eraa342

Doncheva, S., Poschenrieder, C., Stoyanova, Z., Georgieva, K., Velichkova, M., & Barceló, J. (2009). Silicon amelioration of manganese toxicity in Mn-sensitive and Mn-tolerant maize varieties. *Environmental and Experimental Botany, 65*(2–3), 189–197. https://doi.org/10.1016/j.envexpbot.2008.11.006

Dragišić Maksimović, J., Bogdanović, J., Maksimović, V., & Nikolic, M. (2007). Silicon modulates the metabolism and utilization of phenolic compounds in cucumber (Cucumis sativusL.) grown at excess manganese. *Journal of Plant Nutrition and Soil Science, 170*(6), 739–744. https://doi.org/10.1002/jpln.200700101

Dragišić Maksimović, J., Mojović, M., Maksimović, V., Römheld, V., & Nikolic, M. (2012). Silicon ameliorates manganese toxicity in cucumber by decreasing hydroxyl radical accumulation in the leaf apoplast. *Journal of Experimental Botany, 63*(7), 2411–2420. https://doi.org/10.1093/jxb/err359

Ekelund, N. G., & Danilov, R. A. (2001). The influence of selenium on photosynthesis and" light-enhanced dark respiration"(LEDR) in the flagellate Euglena gracilis after exposure to ultraviolet radiation. *Aquatic Sciences, 63*(4), 457–465.

El Boukhari, M. E. M., Barakate, M., Bouhia, Y., & Lyamlouli, K. (2020). Trends in seaweed extract based biostimulants: Manufacturing process and beneficial effect on soil-plant systems. *Plants (Basel, Switzerland), 9*(3), 359. https://doi.org/10.3390/plants9030359

Emam, M. M., Khattab, H. E., Helal, N. M., & Deraz, A. E. (2014). Effect of selenium and silicon on yield quality of rice plant grown under drought stress. *Australian Journal of Crop Science, 8*(4), 596–605.

Emamverdian, A., Ding, Y., Xie, Y., & Sangari, S. (2018). Silicon mechanisms to ameliorate heavy metal stress in plants. *BioMed Research International*. https://doi.org/10.1155/2018/8492898

Exley, C., Guerriero, G., & Lopez, X. (2019). Silicic acid: The omniscient molecule. *Science of The Total Environment, 665*, 432–437. https://doi.org/10.1016/j.scitotenv.2019.02.197

Fan, T. W.-M., Teh, S. J., Hinton, D. E., & Higashi, R. M. (2002). Selenium biotransformations into proteinaceous forms by foodweb organisms of selenium-laden drainage waters in California. *Aquatic Toxicology, 57*(1–2), 65–84. https://doi.org/10.1016/s0166-445x(01)00261-2

Farooq, M. A., Saqib, Z. A., Akhtar, J., Bakhat, H. F., Pasala, R.-K., & Dietz, K.-J. (2019). Protective role of silicon (Si) against combined stress of salinity and boron (B) toxicity by improving antioxidant enzymes activity in rice. *Silicon, 11*(4), 2193–2197. https://doi.org/10.1007/s12633-015-9346-z

Farouk, S., Elhindi, K. M., & Alotaibi, M. A. (2020). Silicon supplementation mitigates salinity stress on Ocimum basilicum L. via improving water balance, ion homeostasis, and antioxidant defense system. *Ecotoxicology and Environmental Safety, 206*, 111396. https://doi.org/10.1016/j.ecoenv.2020.111396

Feng, R., Wang, L., Yang, J., Zhao, P., Zhu, Y., Li, Y., Yu, Y., Liu, H., Rensing, C., Wu, Z., Ni, R., & Zheng, S. (2021). Underlying mechanisms responsible for restriction of uptake and translocation of heavy metals (metalloids) by selenium via root application in plants. *Journal of Hazardous Materials, 402*, 123570. https://doi.org/10.1016/j.jhazmat.2020.123570

Feng, J., Yamaji, N., & Mitani-Ueno, N. (2011). Transport of silicon from roots to panicles in plants. *Proceedings of the Japan Academy, Series B, 87*(7), 377–385.

Flam-Shepherd, R., Huynh, W. Q., Coskun, D., Hamam, A. M., Britto, D. T., & Kronzucker, H. J. (2018). Membrane fluxes, bypass flows, and sodium stress in rice:

The influence of silicon. *Journal of Experimental Botany, 69*(7), 1679–1692. https://doi.org/10.1093/jxb/erx460

Fleck, A. T., Nye, T., Repenning, C., Stahl, F., Zahn, M., & Schenk, M. K. (2011). Silicon enhances suberization and lignification in roots of rice (Oryza sativa). *Journal of Experimental Botany, 62*(6), 2001–2011. https://doi.org/10.1093/jxb/erq392

Fleck, A. T., Schulze, S., Hinrichs, M., Specht, A., Waßmann, F., Schreiber, L., & Schenk, M. K. (2015). Silicon promotes exodermal casparian band formation in Si-accumulating and Si-excluding species by forming phenol complexes. *PloS One, 10*(9). https://doi.org/10.1371/journal.pone.0138555. e0138555–e0138555.

Flora, S. J. S., & Pachauri, V. (2010). Chelation in metal intoxication. *International Journal of Environmental Research and Public Health, 7*(7), 2745–2788. https://doi.org/10.3390/ijerph7072745

Galvez, L., & Clark, R. B. (1991). Effects of silicon on growth and mineral composition of sorghum (Sorghum Bicolor) grown with toxic levels of aluminium. In *Plant-soil interactions at low pH* (pp. 815–823). Netherlands: Springer. https://doi.org/10.1007/978-94-011-3438-5_92

Germ, M., Kreft, I., Stibilj, V., & Urbanc-Berčič, O. (2007). Combined effects of selenium and drought on photosynthesis and mitochondrial respiration in potato. *Plant Physiology and Biochemistry, 45*(2), 162–167. https://doi.org/10.1016/j.plaphy.2007.01.009

Goecke, F., Jerez, C. G., Zachleder, V., Figueroa, F. L., Bišová, K., Řezanka, T., & Vítová, M. (2015). Use of lanthanides to alleviate the effects of metal ion-deficiency in Desmodesmus quadricauda (Sphaeropleales, Chlorophyta). *Frontiers in Microbiology, 6*, 2. https://doi.org/10.3389/fmicb.2015.00002

Gohari, G., Zareei, E., Rostami, H., Panahirad, S., Kulak, M., Farhadi, H., Amini, M., Martinez-Ballesta, M., del, C., & Fotopoulos, V. (2021). Protective effects of cerium oxide nanoparticles in grapevine (Vitis vinifera L.) cv. Flame Seedless under salt stress conditions. *Ecotoxicology and Environmental Safety, 220*, Article 112402. https://doi.org/10.1016/j.ecoenv.2021.112402

Gong, H. J., Randall, D. P., & Flowers, T. J. (2006). Silicon deposition in the root reduces sodium uptake in rice (Oryza sativa L.) seedlings by reducing bypass flow. *Plant, Cell and Environment, 29*(10), 1970–1979. https://doi.org/10.1111/j.1365-3040.2006.01572.x

Gonzalo, M. J., Lucena, J. J., & Hernández-Apaolaza, L. (2013). Effect of silicon addition on soybean (Glycine max) and cucumber (Cucumis sativus) plants grown under iron deficiency. *Plant Physiology and Biochemistry, 70*, 455–461. https://doi.org/10.1016/j.plaphy.2013.06.007

Gu, H.-H., Zhan, S.-S., Wang, S.-Z., Tang, Y.-T., Chaney, R. L., Fang, X.-H., Cai, X.-D., & Qiu, R.-L. (2012). Silicon-mediated amelioration of zinc toxicity in rice (Oryza sativa L.) seedlings. *Plant and Soil, 350*(1–2), 193–204. https://doi.org/10.1007/s11104-011-0894-8

Gunes, A., Inal, A., Bagci, E. G., Coban, S., & Sahin, O. (2007). Silicon increases boron tolerance and reduces oxidative damage of wheat grown in soil with excess boron. *Biologia Plantarum, 51*(3), 571–574. https://doi.org/10.1007/s10535-007-0125-6

Guntzer, F., Keller, C., & Meunier, J. D. (2012). Benefits of plant silicon for crops: a review. *Agronomy for Sustainable Development, 32*(1), 201–213.

Habibi, G. (2013). Effect of drought stress and selenium spraying on photosynthesis and antioxidant activity of spring barley/Učinek sušnega stresa in škropljenja s selenom na fotosintezo in antioksidativno aktivnost jarega ječmena. *Acta Agriculturae Slovenica, 101*(1). https://doi.org/10.2478/acas-2013-0004

Hansen, H., Larssen, T., Seip, H. M., & Vogt, R. D. (2001). Trace metals in forest soils at four sites in southern China. In *Acid rain 2000* (pp. 1721–1726). Netherlands: Springer. https://doi.org/10.1007/978-94-007-0810-5_134

Harizanova, A., & Koleva-Valkova, L. (2019). Effect of Silicon on photosynthetic rate and the chlorophyll fluorescence parameters at hydroponically grown cucumber plants under salinity stress. *Journal of Central European Agriculture, 20*(3), 953–960. https://doi.org/10.5513/jcea01/20.3.2312

Harrison (formerly Perry), C. C. (1996). Evidence for intramineral macromolecules containing protein from plant silicas. *Phytochemistry, 41*(1), 37–42. https://doi.org/10.1016/0031-9422(95)00576-5

Hasanuzzaman, M., Bhuyan, M. H. M. B., Raza, A., Hawrylak-Nowak, B., Matraszek-Gawron, R., Mahmud, J. Al, Nahar, K., & Fujita, M. (2020). Selenium in plants: Boon or bane? *Environmental and Experimental Botany, 178*, 104170. https://doi.org/10.1016/j.envexpbot.2020.104170

Hasanuzzaman, M., Bhuyan, M. H. M. B., Raza, A., Hawrylak-Nowak, B., Matraszek-Gawron, R., Nahar, K., & Fujita, M. (2020). Selenium toxicity in plants and environment: Biogeochemistry and remediation possibilities. *Plants (Basel, Switzerland), 9*(12), 1711. https://doi.org/10.3390/plants9121711

Hasanuzzaman, M., & Fujita, M. (2011). Selenium pretreatment upregulates the antioxidant defense and methylglyoxal detoxification system and confers enhanced tolerance to drought stress in rapeseed seedlings. *Biological Trace Element Research, 143*(3), 1758–1776. https://doi.org/10.1007/s12011-011-8998-9

Hasanuzzaman, M., Nahar, K., Anee, T. I., Khan, M. I. R., & Fujita, M. (2018). Silicon-mediated regulation of antioxidant defense and glyoxalase systems confers drought stress tolerance in Brassica napus L. *South African Journal of Botany, 115*, 50–57. https://doi.org/10.1016/j.sajb.2017.12.006

Helaly, M. N., El-Hoseiny, H., El-Sheery, N. I., Rastogi, A., & Kalaji, H. M. (2017). Regulation and physiological role of silicon in alleviating drought stress of mango. *Plant Physiology and Biochemistry, 118*, 31–44. https://doi.org/10.1016/j.plaphy.2017.05.021

He, C., Ma, J., & Wang, L. (2015). A hemicellulose-bound form of silicon with potential to improve the mechanical properties and regeneration of the cell wall of rice. *New Phytologist, 206*(3), 1051–1062. https://doi.org/10.1111/nph.13282

He, C., Wang, L., Liu, J., Liu, X., Li, X., Ma, J., Lin, Y., & Xu, F. (2013). Evidence for 'silicon' within the cell walls of suspension-cultured rice cells. *New Phytologist, 200*(3), 700–709. https://doi.org/10.1111/nph.12401

Hodson, M. J., & Evans, D. E. (2020). Aluminium-silicon interactions in higher plants: An update. *Journal of Experimental Botany, 71*(21), 6719–6729. https://doi.org/10.1093/jxb/eraa024

Hodson, M. J., White, P. J., Mead, A., & Broadley, M. R. (2005). Phylogenetic variation in the silicon composition of plants. *Annals of Botany, 96*(6), 1027–1046. https://doi.org/10.1093/aob/mci255

Hossain, A., Skalicky, M., Brestic, M., Maitra, S., Sarkar, S., Ahmad, Z., Vemuri, H., Garai, S., Mondal, M., Bhatt, R., Kumar, P., Banerjee, P., Saha, S., Islam, T., & Laing, A. M. (2021). Selenium biofortification: Roles, mechanisms, responses and prospects. *Molecules (Basel, Switzerland), 26*(4), 881. https://doi.org/10.3390/molecules26040881

Hosseini, S. A., Maillard, A., Hajirezaei, M. R., Ali, N., Schwarzenberg, A., Jamois, F., & Yvin, J.-C. (2017). Induction of barley silicon transporter HvLsi1 and HvLsi2, increased silicon concentration in the shoot and regulated starch and ABA homeostasis under osmotic stress and concomitant potassium deficiency. *Frontiers in Plant Science, 8*, 1359. https://doi.org/10.3389/fpls.2017.01359

Hosseini, S. A., Naseri Rad, S., Ali, N., & Yvin, J.-C. (2019). The ameliorative effect of silicon on maize plants grown in Mg-deficient conditions. *International Journal of Molecular Sciences, 20*(4), 969. https://doi.org/10.3390/ijms20040969

Huang, F., Wen, X. H., Cai, Y. X., & Cai, K. Z. (2018). Silicon-mediated enhancement of heavy metal tolerance in rice at different growth stages. *International Journal of Environmental Research and Public Health, 15*(10). https://doi.org/10.3390/ijerph15102193

Hu, Z., Haneklaus, S., Sparovek, G., & Schnug, E. (2006). Rare earth elements in soils. *Communications in Soil Science and Plant Analysis, 37*(9–10), 1381–1420. https://doi.org/10.1080/00103620600628680

Hu, Z., Richter, H., Sparovek, G., & Schnug, E. (2004). Physiological and biochemical effects of rare earth elements on plants and their agricultural significance: A review. *Journal of Plant Nutrition, 27*(1), 183–220. https://doi.org/10.1081/pln-120027555

Hu, H., & Shan, C. (2018). Effect of cerium (Ce) on the redox states of ascorbate and glutathione through ascorbate-glutathione cycle in the roots of maize seedlings under salt stress. *Cereal Research Communications, 46*(1), 31–40. https://doi.org/10.1556/0806.45.2017.057

Ibrahim, H. M. (2014). Selenium pretreatment regulates the antioxidant defense system and reduces oxidative stress on drought-stressed wheat (*Triticum aestivum* L.) plants. *Asian Journal of Plant Sciences, 13*(3), 120–128. https://doi.org/10.3923/ajps.2014.120.128

Inal, A., Pilbeam, D. J., & Gunes, A. (2009). Silicon increases tolerance to boron toxicity and reduces oxidative damage in barley. *Journal of Plant Nutrition, 32*(1), 112–128. https://doi.org/10.1080/01904160802533767

Iwasaki, K., Fecht, M., Maier, P., & Horst, W. J. (2001). Can leaf apoplastic manganese and silicon concentrations explain Si-enhanced manganese tolerance of Vigna unguiculata (L.) Walp. In *Plant nutrition* (pp. 246–247). Netherlands: Springer. https://doi.org/10.1007/0-306-47624-x_118

Iwasaki, K., Maier, P., Fecht, M., & Horst, W. J. (2002). Leaf apoplastic silicon enhances manganese tolerance of cowpea (Vigna unguiculata). *Journal of Plant Physiology, 159*(2), 167–173. https://doi.org/10.1078/0176-1617-00691

Jalali, M., Majeri, M., & Najafi, S. (2019). Kinetic release and fractionation of cobalt in some calcareous soils. *Journal of Geochemical Exploration, 204*, 131–141. https://doi.org/10.1016/j.gexplo.2019.02.006

Jampílek, J., & Králová, K. (2019). Beneficial effects of metal- and metalloid-based nanoparticles on crop production. In *Nanotechnology for agriculture* (pp. 161–219). Singapore: Springer. https://doi.org/10.1007/978-981-32-9370-0_11

Jardin, du P. (2015). Plant biostimulants: Definition, concept, main categories and regulation. *Scientia Horticulturae, 196*, 3–14. https://doi.org/10.1016/j.scienta.2015.09.021

Jiang, C., Zu, C., Lu, D., Zheng, Q., Shen, J., Wang, H., & Li, D. (2017). Effect of exogenous selenium supply on photosynthesis, Na(+) accumulation and antioxidative capacity of maize (Zea mays L.) under salinity stress. *Scientific Reports*, 7, 42039. https://doi.org/10.1038/srep42039

Kamenidou, S., Cavins, T. J., & Marek, S. (2010). Silicon supplements affect floricultural quality traits and elemental nutrient concentrations of greenhouse produced gerbera. *Scientia Horticulturae, 123*(3), 390–394. https://doi.org/10.1016/j.scienta.2009.09.008

Karimi, R., Ghabooli, M., Rahimi, J., & Amerian, M. (2020). Effects of foliar selenium application on some physiological and phytochemical parameters ofVitis viniferaL. cv. Sultana under salt stress. *Journal of Plant Nutrition, 43*(14), 2226–2242. https://doi.org/10.1080/01904167.2020.1766072

Kauffman, G. L., Kneivel, D. P., & Watschke, T. L. (2007). Effects of a biostimulant on the heat tolerance associated with photosynthetic capacity, membrane thermostability, and polyphenol production of perennial ryegrass. *Crop Science, 47*(1), 261–267. https://doi.org/10.2135/cropsci2006.03.0171

Kaur, S., Kaur, N., Siddique, K. H., & Nayyar, H. (2016). Beneficial elements for agricultural crops and their functional relevance in defence against stresses. *Archives of Agronomy and Soil Science, 62*(7), 905–920.

Khattab, H. I., Emam, M. A., Emam, M. M., Helal, N. M., & Mohamed, M. R. (2014). Effect of selenium and silicon on transcription factors NAC5 and DREB2A involved in drought-responsive gene expression in rice. *Biologia Plantarum, 58*(2), 265–273. https://doi.org/10.1007/s10535-014-0391-z

Kim, Y.-H., Khan, A. L., Kim, D.-H., Lee, S.-Y., Kim, K.-M., Waqas, M., Jung, H.-Y., Shin, J.-H., Kim, J.-G., & Lee, I.-J. (2014). Silicon mitigates heavy metal stress by regulating P-type heavy metal ATPases, Oryza sativa low silicon genes, and endogenous phytohormones. *BMC Plant Biology, 14*, 13. https://doi.org/10.1186/1471-2229-14-13

Kim, Y. H., Khan, A. L., Waqas, M., & Lee, I. J. (2017). Silicon regulates antioxidant activities of crop plants under abiotic-induced oxidative stress: a review. *Frontiers in Plant Science, 8*, 510.

Kim, Y. H., Khan, A. L., Waqas, M., Shahzad, R., & Lee, I. J. (2016). Silicon-mediated mitigation of wounding stress acts by up-regulating the rice antioxidant system. *Cereal Research Communications, 44*(1), 111–121.

Kirkby, E. (2012). Introduction, definition and classification of nutrients. In *Marschner's mineral nutrition of higher plants* (pp. 3–5). Elsevier. https://doi.org/10.1016/b978-0-12-384905-2.00001-7

Kleiber, T., Borowiak, K., Kosiada, T., Breś, W., & Ławniczak, B. (2020). Application of selenium and silicon to alleviate short-term drought stress in French marigold (Tagetes patula L.) as a model plant species. *Open Chemistry, 18*(1), 1468–1480. https://doi.org/10.1515/chem-2020-0183

Kollárová, K., Kusá, Z., Vatehová-Vivodová, Z., & Lišková, D. (2019). The response of maize protoplasts to cadmium stress mitigated by silicon. *Ecotoxicology and Environmental Safety, 170*, 488–494. https://doi.org/10.1016/j.ecoenv.2018.12.016

Kostic, L., Nikolic, N., Bosnic, D., Samardzic, J., & Nikolic, M. (2017). Silicon increases phosphorus (P) uptake by wheat under low P acid soil conditions. *Plant and Soil, 419*(1–2), 447–455. https://doi.org/10.1007/s11104-017-3364-0

Kotelnikova, A., Fastovets, I., Rogova, O., & Volkov, D. S. (2020). La, Ce and Nd in the soil-plant system in a vegetation experiment with barley (Hordeum vulgare L.). *Ecotoxicology and Environmental Safety, 206*, 111193. https://doi.org/10.1016/j.ecoenv.2020.111193

Kovarikova, M., Tomaskova, I., & Soudek, P. (2019). Rare earth elements in plants. *Biologia Plantarum, 63*(1), 20–32. https://doi.org/10.32615/bp.2019.003

Krishnamurthy, P., Ranathunge, K., Nayak, S., Schreiber, L., & Mathew, M. K. (2011). Root apoplastic barriers block Na+ transport to shoots in rice (Oryza sativa L.). *Journal of Experimental Botany, 62*(12), 4215–4228. https://doi.org/10.1093/jxb/err135

Kronzucker, H. J., Coskun, D., Schulze, L. M., Wong, J. R., & Britto, D. T. (2013). Sodium as nutrient and toxicant. *Plant and Soil, 369*(1–2), 1–23. https://doi.org/10.1007/s11104-013-1801-2

Kunicki, E., Grabowska, A., Sękara, A., & Wojciechowska, R. (2010). The effect of cultivar type, time of cultivation, and biostimulant treatment on the yield of spinach (Spinacia oleracea L.). *Folia Horticulturae, 22*(2), 9–13. https://doi.org/10.2478/fhort-2013-0153

Landa, P. (2021). Positive effects of metallic nanoparticles on plants: Overview of involved mechanisms. *Plant Physiology and Biochemistry, 161*, 12–24. https://doi.org/10.1016/j.plaphy.2021.01.039

Lange, B., Ent, A., Baker, A. J. M., Echevarria, G., Mahy, G., Malaisse, F., Meerts, P., Pourret, O., Verbruggen, N., & Faucon, M. (2017). Copper and cobalt accumulation in plants: A critical assessment of the current state of knowledge. *New Phytologist, 213*(2), 537–551. https://doi.org/10.1111/nph.14175

Lange, B., Pourret, O., Meerts, P., Jitaru, P., Cancès, B., Grison, C., & Faucon, M.-P. (2016). Copper and cobalt mobility in soil and accumulation in a metallophyte as influenced by experimental manipulation of soil chemical factors. *Chemosphere, 146*, 75–84. https://doi.org/10.1016/j.chemosphere.2015.11.105

Li, P., Song, A., Li, Z., Fan, F., & Liang, Y. (2012). Silicon ameliorates manganese toxicity by regulating manganese transport and antioxidant reactions in rice (Oryza sativa L.). *Plant and Soil, 354*(1–2), 407–419. https://doi.org/10.1007/s11104-011-1076-4

Liang, T., Ding, S., Song, W., Chong, Z., Zhang, C., & Li, H. (2008). A review of fractionations of rare earth elements in plants. *Journal of Rare Earths, 26*(1), 7–15. https://doi.org/10.1016/s1002-0721(08)60027-7

Liang, Y., Sun, W., Zhu, Y., & Christie, P. (2007). *Mechanisms of Silicon-Mediated Alleviation of Abiotic Stresses in Higher Plants : A Review, 147*, 422–428. https://doi.org/10.1016/j.envpol.2006.06.008

Liu, B., Huang, Q., Su, Y., Xue, Q., & Sun, L. (2019). Cobalt speciation and phytoavailability in fluvo-aquic soil under treatments of spent mushroom substrate from Pleurotus ostreatus. *Environmental Science and Pollution Research, 26*(8), 7486–7496. https://doi.org/10.1007/s11356-018-04080-3

Liu, C., Lu, W., Ma, Q., & Ma, C. (2017). Effect of silicon on the alleviation of boron toxicity in wheat growth, boron accumulation, photosynthesis activities, and oxidative responses. *Journal of Plant Nutrition, 40*(17), 2458–2467. https://doi.org/10.1080/01904167.2017.1380817

Liu, X., Yin, L., Deng, X., Gong, D., Du, S., Wang, S., & Zhang, Z. (2020). Combined application of silicon and nitric oxide jointly alleviated cadmium accumulation and toxicity in maize. *Journal of Hazardous Materials, 395*, 122679. https://doi.org/10.1016/j.jhazmat.2020.122679

Mabagala, F. S., Geng, Y., Cao, G., Wang, L., Wang, M., & Zhang, M. (2020). Silicon accumulation, partitioning and remobilization in spring maize (Zea mays L.) under silicon supply with straw return in Northeast China. *Journal of Plant Nutrition, 44*(10), 1498–1514. https://doi.org/10.1080/01904167.2020.1867576

Mahmoud, L. M., Dutt, M., Shalan, A. M., El-Kady, M. E., El-Boray, M. S., Shabana, Y. M., & Grosser, J. W. (2020). Silicon nanoparticles mitigate oxidative stress of in vitro-derived banana (Musa acuminata 'Grand Nain') under simulated water deficit or salinity stress. *South African Journal of Botany, 132*, 155–163. https://doi.org/10.1016/j.sajb.2020.04.027

Maillard, A., Ali, N., Schwarzenberg, A., Jamois, F., Yvin, J.-C., & Hosseini, S. A. (2018). Silicon transcriptionally regulates sulfur and ABA metabolism and delays leaf senescence in barley under combined sulfur deficiency and osmotic stress. *Environmental and Experimental Botany, 155*, 394–410. https://doi.org/10.1016/j.envexpbot.2018.07.026

Ma, Y., Kuang, L., He, X., Bai, W., Ding, Y., Zhang, Z., Zhao, Y., & Chai, Z. (2010). Effects of rare earth oxide nanoparticles on root elongation of plants. *Chemosphere*. https://doi.org/10.1016/j.chemosphere.2009.10.050

Malhotra, N., Hsu, H.-S., Liang, S.-T., Roldan, M. J. M., Lee, J.-S., Ger, T.-R., & Hsiao, C.-D. (2020). An updated review of toxicity effect of the rare earth elements (REEs) on aquatic organisms. *Animals : An Open Access Journal from MDPI, 10*(9), 1663. https://doi.org/10.3390/ani10091663

Malik, M. A., Wani, A. H., Mir, S. H., Rehman, I. U., Tahir, I., Ahmad, P., & Rashid, I. (2021). Elucidating the role of silicon in drought stress tolerance in plants. *Plant Physiology and Biochemistry, 165*, 187–195. https://doi.org/10.1016/j.plaphy.2021.04.021

Ma, J. F., Miyake, Y., & Takahashi, E. (2001). Chapter 2 Silicon as a beneficial element for crop plants. *Studies in Plant Science, 8*(C), 17–39. https://doi.org/10.1016/S0928-3420(01)80006-9

Ma, J., & Takahashi, E. (1989). Effect of silicic acid on phosphorus uptake by rice plant. *Soil Science and Plant Nutrition, 35*(2), 227–234. https://doi.org/10.1080/00380768.1989.10434755

Ma, J. F., & Yamaji, N. (2006). Silicon uptake and accumulation in higher plants. *Trends in Plant Science, 11*(8), 392–397. https://doi.org/10.1016/j.tplants.2006.06.007

Ma, J. F., & Yamaji, N. (2015). A cooperative system of silicon transport in plants. *Trends in Plant Science, 20*(7), 435–442. https://doi.org/10.1016/j.tplants.2015.04.007

Miao, B.-H., Han, X.-G., & Zhang, W.-H. (2010). The ameliorative effect of silicon on soybean seedlings grown in potassium-deficient medium. *Annals of Botany, 105*(6), 967–973. https://doi.org/10.1093/aob/mcq063

Mozafariyan, M., Kamelmanesh, M. M., & Hawrylak-Nowak, B. (2016). Ameliorative effect of selenium on tomato plants grown under salinity stress. *Archives of Agronomy and Soil Science, 62*(10), 1368–1380. https://doi.org/10.1080/03650340.2016.1149816

Naeem, A., Saifullah, Zia-ur-Rehman, M., Akhtar, T., Zia, M. H., & Aslam, M. (2018). Silicon nutrition lowers cadmium content of wheat cultivars by regulating transpiration rate and activity of antioxidant enzymes. *Environmental Pollution, 242*, 126–135. https://doi.org/10.1016/j.envpol.2018.06.069

Nair, R. (2016). Effects of nanoparticles on plant growth and development. In *Plant nanotechnology* (pp. 95–118). Springer International Publishing. https://doi.org/10.1007/978-3-319-42154-4_5

Naveed, M., Khalid, H., Ayub, M. A., Rehman, M. Z. ur, Rizwan, M., Rasul, A., & Haq, M. A. ul (2020). Biofortification of cereals with zinc and iron: Recent advances and future perspectives. In *Resources use efficiency in agriculture* (pp. 615–646). Springer Singapore. https://doi.org/10.1007/978-981-15-6953-1_17

Nawaz, F., Ahmad, R., Ashraf, M. Y., Waraich, E. A., & Khan, S. Z. (2015). Effect of selenium foliar spray on physiological and biochemical processes and chemical constituents of wheat under drought stress. *Ecotoxicology and Environmental Safety, 113*, 191–200. https://doi.org/10.1016/j.ecoenv.2014.12.003

Nikolic, D. B., Nesic, S., Bosnic, D., Kostic, L., Nikolic, M., & Samardzic, J. T. (2019). Silicon alleviates iron deficiency in barley by enhancing expression of strategy II genes and metal redistribution. *Frontiers in Plant Science, 10*, 416. https://doi.org/10.3389/fpls.2019.00416

Nuncio Junior, R. A., Baron, D., & Ferreira, I. E. de P. (2021). Does molybdenum and cobalt foliar spray mitigate nitrate starvation and ammonium deprivation stress? *Revista Brasileira de Fruticultura, 43*(4). https://doi.org/10.1590/0100-29452021163

Nwugo, C. C., & Huerta, A. J. (2010). The effect of silicon on the leaf proteome of rice (oryza sativa L.) plants under cadmium-stress. *Journal of Proteome Research, 10*(2), 518–528. https://doi.org/10.1021/pr100716h

Oliveira, de R. L. L., de Mello Prado, R., Felisberto, G., Checchio, M. V., & Gratão, P. L. (2019). Silicon mitigates manganese deficiency stress by regulating the physiology and activity of antioxidant enzymes in sorghum plants. *Journal of Soil Science and Plant Nutrition, 19*(3), 524–534. https://doi.org/10.1007/s42729-019-00051-w

Pacheco, I., & Buzea, C. (2018). Nanoparticle uptake by plants: Beneficial or detrimental?. In *Phytotoxicity of nanoparticles* (pp. 1–61). Springer International Publishing. https://doi.org/10.1007/978-3-319-76708-6_1

Parveen, A., Liu, W., Hussain, S., Asghar, J., Perveen, S., & Xiong, Y. (2019). Silicon priming regulates morpho-physiological growth and oxidative metabolism in maize under drought stress. *Plants (Basel, Switzerland), 8*(10), 431. https://doi.org/10.3390/plants8100431

Pavlovic, J., Samardzic, J., Kostic, L., Laursen, K. H., Natic, M., Timotijevic, G., Schjoerring, J. K., & Nikolic, M. (2016). Silicon enhances leaf remobilization of iron in cucumber under limited iron conditions. *Annals of Botany, 118*(2), 271–280. https://doi.org/10.1093/aob/mcw105

Pavlovic, J., Samardzic, J., Maksimović, V., Timotijevic, G., Stevic, N., Laursen, K. H., Hansen, T. H., Husted, S., Schjoerring, J. K., Liang, Y., & Nikolic, M. (2013). Silicon

alleviates iron deficiency in cucumber by promoting mobilization of iron in the root apoplast. *New Phytologist, 198*(4), 1096–1107. https://doi.org/10.1111/nph.12213
Pereira de Souza Junior, J., de Mello Prado, R., Machado dos Santos Sarah, M., & Felisberto, G. (2019). Silicon mitigates boron deficiency and toxicity in cotton cultivated in nutrient solution. *Journal of Plant Nutrition and Soil Science, 182*(5), 805–814. https://doi.org/10.1002/jpln.201800398
Perezespinosa, A. (2005). Co phytoavailability for tomato in amended calcareous soils. *Bioresource Technology, 96*(6), 649–655. https://doi.org/10.1016/j.biortech.2004.07.002
Pilon-Smits, E. A. H., Quinn, C. F., Tapken, W., Malagoli, M., & Schiavon, M. (2009). Physiological functions of beneficial elements. *Current Opinion in Plant Biology, 12*(3), 267–274. https://doi.org/10.1016/j.pbi.2009.04.009
Qu, C., Liu, C., Guo, F., Hu, C., Ze, Y., Li, C., Zhou, Q., & Hong, F. (2013). Improvement of cerium on photosynthesis of maize seedlings under a combination of potassium deficiency and salt stress. *Biological Trace Element Research, 155*(1), 104–113. https://doi.org/10.1007/s12011-013-9767-8
Rady, M. O., Semida, W. M., Abd El-Mageed, T. A., Howladar, S. M., & Shaaban, A. (2020). Foliage applied selenium improves photosynthetic efficiency, antioxidant potential and wheat productivity under drought stress. *International Journal of Agriculture and Biology, 24*, 1293–1300.
Ranjan, A., Sinha, R., Bala, M., Pareek, A., Singla-Pareek, S. L., & Singh, A. K. (2021). Silicon-mediated abiotic and biotic stress mitigation in plants: Underlying mechanisms and potential for stress resilient agriculture. *Plant Physiology and Biochemistry, 163*, 15–25. https://doi.org/10.1016/j.plaphy.2021.03.044
Rauser, W. E., & Samarakoon, A. B. (1980). Vein loading in seedlings of Phaseolus vulgaris exposed to excess cobalt, nickel, and zinc. *Plant Physiology, 65*(4), 578–583. https://doi.org/10.1104/pp.65.4.578
Rehman, M. Z. U., Rizwan, M., Rauf, A., Ayub, M. A., Ali, S., Qayyum, M. F., Waris, A. A., Naeem, A., & Sanaullah, M. (2019). Split application of silicon in cadmium (Cd) spiked alkaline soil plays a vital role in decreasing Cd accumulation in rice (Oryza sativa L.) grains. *Chemosphere*. https://doi.org/10.1016/j.chemosphere.2019.03.182
Ren, Y., He, J., Zhou, G., & Chang, H. (2010). Alleviated effects of cadmium stress on rice roots and physiological characteristics with rare earth praseodymium treatment. *Journal of Hunan Agricultural University, 36*(5), 519–523. https://doi.org/10.3724/sp.j.1238.2010.00519
Réthoré, E., Ali, N., Yvin, J.-C., & Hosseini, S. A. (2020). Silicon regulates source to sink metabolic homeostasis and promotes growth of rice plants under sulfur deficiency. *International Journal of Molecular Sciences, 21*(10), 3677. https://doi.org/10.3390/ijms21103677
Riaz, M. U., Ayub, M. A., Khalid, H., ul Haq, M. A., Rasul, A., ur Rehman, M. Z., & Ali, S. (2020). Fate of micronutrients in alkaline soils. In *Resources use efficiency in agriculture*. https://doi.org/10.1007/978-981-15-6953-1_16
Riaz, M., Kamran, M., Rizwan, M., Ali, S., Parveen, A., Malik, Z., & Wang, X. (2021). Cadmium uptake and translocation: Selenium and silicon roles in Cd detoxification for the production of low Cd crops: A critical review. *Chemosphere, 273*, 129690. https://doi.org/10.1016/j.chemosphere.2021.129690
Richmond, K. E., & Sussman, M. (2003). Got silicon? The non-essential beneficial plant nutrient. *Current Opinion in Plant Biology, 6*(3), 268–272. https://doi.org/10.1016/s1369-5266(03)00041-4
Rios, J. J., Martínez-Ballesta, M. C., Ruiz, J. M., Blasco, B., & Carvajal, M. (2017). Silicon-mediated improvement in plant salinity tolerance: The role of aquaporins. *Frontiers in Plant Science, 8*, 948. https://doi.org/10.3389/fpls.2017.00948

Rossi, L., Zhang, W., Lombardini, L., & Ma, X. (2016). The impact of cerium oxide nanoparticles on the salt stress responses of Brassica napus L. *Environmental Pollution, 219*, 28–36. https://doi.org/10.1016/j.envpol.2016.09.060

Rossi, L., Zhang, W., & Ma, X. (2017). Cerium oxide nanoparticles alter the salt stress tolerance of Brassica napus L. by modifying the formation of root apoplastic barriers. *Environmental Pollution, 229*, 132–138. https://doi.org/10.1016/j.envpol.2017.05.083

Rouphael, Y., & Colla, G. (2020). Editorial: Biostimulants in agriculture. *Frontiers in Plant Science, 11*, 40. https://doi.org/10.3389/fpls.2020.00040

Roychoudhury, A., & Chakraborty, S. (2022). Cobalt and molybdenum: Deficiency, toxicity, and nutritional role in plant growth and development. In *Plant nutrition and food security in the era of climate change* (pp. 255–270). Elsevier. https://doi.org/10.1016/b978-0-12-822916-3.00021-4

Salgado, O. G. G., Teodoro, J. C., Alvarenga, J. P., de Oliveira, C., de Carvalho, T. S., Domiciano, D., Marchiori, P. E. R., & Guilherme, L. R. G. (2020). Cerium alleviates drought-induced stress in Phaseolus vulgaris. *Journal of Rare Earths, 38*(3), 324–331. https://doi.org/10.1016/j.jre.2019.07.014

Salhani, N., Boulyga, S. F., & Stengel, E. (2003). Phytoremediation of selenium by two helophyte species in subsurface flow constructed wetland. *Chemosphere, 50*(8), 967–973.

Saud, S., Yajun, C., Fahad, S., Hussain, S., Na, L., Xin, L., & Alhussien, S. A. A. F. E. (2016). Silicate application increases the photosynthesis and its associated metabolic activities in Kentucky bluegrass under drought stress and post-drought recovery. *Environmental Science and Pollution Research, 23*(17), 17647–17655. https://doi.org/10.1007/s11356-016-6957-x

Shekari, L., Aroiee, H., Mirshekari, A., & Nemati, H. (2019). Protective role of selenium on cucumber (Cucumis sativusL.) exposed to cadmium and lead stress during reproductive stage role of selenium on heavy metals stress. *Journal of Plant Nutrition, 42*(5), 529–542. https://doi.org/10.1080/01904167.2018.1554075

Sieprawska, A., Kornaś, A., & Filek, M. (2015). Involvement of selenium in protective mechanisms of plants under environmental stress conditions – review. *Acta Biologica Cracoviensia S. Botanica, 57*(1), 9–20. https://doi.org/10.1515/abcsb-2015-0014

Singh, S., Singh, V. P., Prasad, S. M., Sharma, S., Ramawat, N., Dubey, N. K., Tripathi, D. K., & Chauhan, D. K. (2019). Interactive effect of silicon (Si) and salicylic acid (SA) in maize seedlings and their mechanisms of cadmium (Cd) toxicity alleviation. *Journal of Plant Growth Regulation, 38*(4), 1587–1597. https://doi.org/10.1007/s00344-019-09958-1

Sohail, M. I., Ayub, M. A., Zia ur Rehman, M., Azhar, M., Farooqi, Z. U. R., Siddiqui, A., Umar, W., Iftikhar, I., Nadeem, M., & Fatima, H. (2021). Sufficiency and toxicity limits of metallic oxide nanoparticles in the biosphere. In *Nanomaterials: Synthesis, characterization, hazards and safety* (pp. 145–221). Elsevier. https://doi.org/10.1016/b978-0-12-823823-3.00002-1

Sonah, H., Deshmukh, R. K., Labbé, C., & Bélanger, R. R. (2017). Analysis of aquaporins in Brassicaceae species reveals high-level of conservation and dynamic role against biotic and abiotic stress in canola. *Scientific Reports, 7*(1), 2771. https://doi.org/10.1038/s41598-017-02877-9

Song, A., Li, P., Li, Z., Fan, F., Nikolic, M., & Liang, Y. (2011). The alleviation of zinc toxicity by silicon is related to zinc transport and antioxidative reactions in rice. *Plant and Soil, 344*(1–2), 319–333. https://doi.org/10.1007/s11104-011-0749-3

Subbarao, G. V., Ito, O., Berry, W. L., & Wheeler, R. M. (2003). Sodium—a functional plant nutrient. *Critical Reviews in Plant Sciences, 22*(5), 391–416. https://doi.org/10.1080/07352680390243495

Szparaga, Kuboń, Kocira, Czerwińska, Pawłowska, Hara, Kobus, & Kwaśniewski. (2019). Towards sustainable agriculture—agronomic and economic effects of biostimulant use in common bean cultivation. *Sustainability, 11*(17), 4575. https://doi.org/10.3390/su11174575

Taiz, L., Zeiger, E., Møller, I. M., & Murphy, A. (2015). Plant physiology and development. In . *No. Ed.* *6*Sinauer Associates Incorporated.

Takahashi, E., & Maejima, K. (1998). *Comparative research on sodium as a beneficial element for crop plants.* Memoirs of the Faculty of Agriculture of Kinki University (Japan).

Tale Ahmad, S., & Haddad, R. (2011). Study of silicon effects on antioxidant enzyme activities and osmotic adjustment of wheat under drought stress. *Czech Journal of Genetics and Plant Breeding, 47*(No. 1), 17–27. https://doi.org/10.17221/92/2010-cjgpb

Tayyab, M., Islam, W., & Zhang, H. (2018). Promising role of silicon to enhance drought resistance in wheat. *Communications in Soil Science and Plant Analysis, 49*(22), 2932–2941. https://doi.org/10.1080/00103624.2018.1547394

Tommasi, F., Thomas, P. J., Pagano, G., Perono, G. A., Oral, R., Lyons, D. M., Toscanesi, M., & Trifuoggi, M. (2020). Review of rare earth elements as fertilizers and feed additives: A knowledge gap analysis. *Archives of Environmental Contamination and Toxicology, 81*(4), 531–540. https://doi.org/10.1007/s00244-020-00773-4

Tripathi, D. K., Singh, S., Singh, S., Mishra, S., Chauhan, D. K., & Dubey, N. K. (2015). Micronutrients and their diverse role in agricultural crops: Advances and future prospective. *Acta Physiologiae Plantarum, 37*(7). https://doi.org/10.1007/s11738-015-1870-3

Tuna, A. L., Kaya, C., Higgs, D., Murillo-Amador, B., Aydemir, S., & Girgin, A. R. (2008). Silicon improves salinity tolerance in wheat plants. *Environmental and Experimental Botany, 62*(1), 10–16. https://doi.org/10.1016/j.envexpbot.2007.06.006

Tyler, G. (2004). Rare earth elements in soil and plant systems - a review. *Plant and Soil, 267*(1–2), 191–206. https://doi.org/10.1007/s11104-005-4888-2

Umar, W., Ayub, M. A., Rehman, M. Z. ur, Ahmad, H. R., Farooqi, Z. U. R., Shahzad, A., Rehman, U., Mustafa, A., & Nadeem, M. (2020). Nitrogen and phosphorus use efficiency in agroecosystems. In *Resources use efficiency in agriculture* (pp. 213–257). Springer Singapore. https://doi.org/10.1007/978-981-15-6953-1_7

Van Oosten, M. J., Pepe, O., De Pascale, S., Silletti, S., & Maggio, A. (2017). The role of biostimulants and bioeffectors as alleviators of abiotic stress in crop plants. *Chemical and Biological Technologies in Agriculture, 4*(1). https://doi.org/10.1186/s40538-017-0089-5

Verma, K. K., Anas, M., Chen, Z., Rajput, V. D., Malviya, M. K., Verma, C. L., Singh, R. K., Singh, P., Song, X.-P., & Li, Y.-R. (2020). Silicon supply improves leaf gas exchange, antioxidant defense system and growth in saccharum officinarum responsive to water limitation. *Plants (Basel, Switzerland), 9*(8), 1032. https://doi.org/10.3390/plants9081032

Wang, X. P., Shan, X. Q., Zhang, S. Z., & Wen, B. (2003). Distribution of rare earth elements among chloroplast components of hyperaccumulator Dicranopteris dichotoma. *Analytical and Bioanalytical Chemistry, 376*(6), 913–917.

Wang, S., Wang, F., Gao, S., & Wang, X. (2016). Heavy metal accumulation in different rice cultivars as influenced by foliar application of nano-silicon. *Water, Air, & Soil Pollution, 227*(7). https://doi.org/10.1007/s11270-016-2928-6

Wan, Y., Wang, K., Liu, Z., Yu, Y., Wang, Q., & Li, H. (2019). Effect of selenium on the subcellular distribution of cadmium and oxidative stress induced by cadmium in rice (Oryza sativa L.). *Environmental Science and Pollution Research, 26*(16), 16220–16228. https://doi.org/10.1007/s11356-019-04975-9

WU, J.-W., SHI, Y., ZHU, Y.-X., WANG, Y.-C., & GONG, H.-J. (2013). Mechanisms of enhanced heavy metal tolerance in plants by silicon: A review. *Pedosphere, 23*(6), 815–825. https://doi.org/10.1016/s1002-0160(13)60073-9

Wu, X., Yu, Y., Baerson, S. R., Song, Y., Liang, G., Ding, C., Niu, J., Pan, Z., & Zeng, R. (2017). Interactions between nitrogen and silicon in rice and their effects on resistance toward the Brown planthopper nilaparvata lugens. *Frontiers in Plant Science, 8*, 28. https://doi.org/10.3389/fpls.2017.00028

Yamaguchi, T., Tomioka, R., & Takenaka, C. (2017). Accumulation of cobalt and nickel in tissues of Clethra barbinervis in a metal dosing trial. *Plant and Soil, 421*(1–2), 273–283. https://doi.org/10.1007/s11104-017-3455-y

Yamaji, N., Mitatni, N., & Ma, J. F. (2008). A transporter regulating silicon distribution in rice shoots. *The Plant Cell, 20*(5), 1381–1389. https://doi.org/10.1105/tpc.108.059311

Yang, H., Yang, X., Ning, Z., Kwon, S. Y., Li, M.-L., Tack, F. M. G., Kwon, E. E., Rinklebe, J., & Yin, R. (2022). The beneficial and hazardous effects of selenium on the health of the soil-plant-human system: An overview. *Journal of Hazardous Materials, 422*, 126876. https://doi.org/10.1016/j.jhazmat.2021.126876

Yoshida, S., Ohnishi, Y., & Kitagishi, K. (1962). Chemical forms, mobility and deposition of silicon in rice plant. *Soil Science and Plant Nutrition, 8*(3), 15–21. https://doi.org/10.1080/00380768.1962.10430992

Yu, Y., Yuan, S., Zhuang, J., Wan, Y., Wang, Q., Zhang, J., & Li, H. (2018). Effect of selenium on the uptake kinetics and accumulation of and oxidative stress induced by cadmium in Brassica chinensis. *Ecotoxicology and Environmental Safety, 162*, 571–580. https://doi.org/10.1016/j.ecoenv.2018.07.041

Zembala, M., Filek, M., Walas, S., Mrowiec, H., Kornaś, A., Miszalski, Z., & Hartikainen, H. (2010). Effect of selenium on macro- and microelement distribution and physiological parameters of rape and wheat seedlings exposed to cadmium stress. *Plant and Soil, 329*(1–2), 457–468. https://doi.org/10.1007/s11104-009-0171-2

Zhang, Y., Liang, Y., Zhao, X., Jin, X., Hou, L., Shi, Y., & Ahammed, G. (2019). Silicon compensates phosphorus deficit-induced growth inhibition by improving photosynthetic capacity, antioxidant potential, and nutrient homeostasis in tomato. *Agronomy, 9*(11), 733. https://doi.org/10.3390/agronomy9110733

Zhang, S., & Shan, X. (2001). Speciation of rare earth elements in soil and accumulation by wheat with rare earth fertilizer application. *Environmental Pollution, 112*(3), 395–405. https://doi.org/10.1016/s0269-7491(00)00143-3

Zhang, W., Xie, Z., Wang, L., Li, M., Lang, D., & Zhang, X. (2017). Silicon alleviates salt and drought stress of Glycyrrhiza uralensis seedling by altering antioxidant metabolism and osmotic adjustment. *Journal of Plant Research, 130*(3), 611–624. https://doi.org/10.1007/s10265-017-0927-3

Zhu, Y., Jiang, X., Zhang, J., He, Y., Zhu, X., Zhou, X., Gong, H., Yin, J., & Liu, Y. (2020). Silicon confers cucumber resistance to salinity stress through regulation of proline and cytokinins. *Plant Physiology and Biochemistry, 156*, 209–220. https://doi.org/10.1016/j.plaphy.2020.09.014

Zhu, Z., Wei, G., Li, J., Qian, Q., & Yu, J. (2004). Silicon alleviates salt stress and increases antioxidant enzymes activity in leaves of salt-stressed cucumber (Cucumis sativus L.). *Plant Science, 167*(3), 527–533. https://doi.org/10.1016/j.plantsci.2004.04.020

Zhu, Y., Yin, J., Liang, Y., Liu, J., Jia, J., Huo, H., Wu, Z., Yang, R., & Gong, H. (2019). Transcriptomic dynamics provide an insight into the mechanism for silicon-mediated alleviation of salt stress in cucumber plants. *Ecotoxicology and Environmental Safety, 174*, 245–254. https://doi.org/10.1016/j.ecoenv.2019.02.075

Zwolak, I. (2020). The role of selenium in arsenic and cadmium toxicity: An updated review of scientific literature. *Biological Trace Element Research, 193*(1), 44–63. https://doi.org/10.1007/s12011-019-01691-w

CHAPTER ELEVEN

Role of microorganisms in plant nutrition and soil health

Masarat Nabi
Department of Environmental Science, University of Kashmir, Srinagar, Jammu and Kashmir, India

1. Introduction

Soil serves as a medium for root development, and plants rely on it for nutrients and water (Bhatia, 2008). Previously, only the physiochemical aspects of soil were considered significant. But, the importance of soil biological diversity in sustaining fertility, as well as the interaction of soil biological activity with physiochemical properties, is now well acknowledged (Biswal, 2022). Plant growth and development are highly influenced by the combination and proportion of nutrients in the soil (Morgan and Connolly, 2013). Plant development, fertility, and yield are affected by insufficient supply, constraints, imbalance, and excessive nutrients in the soil system (Osvalde, 2011). There are two types of essential nutrients for plants, depending on their requirements (1) macronutrients and (2) micronutrients (Grusak, 2001) (Fig. 11.1). Macronutrients are required in large quantities and act as building blocks for important biological components such as proteins and nucleic acids (Chen et al., 2018). Some of the essential mineral and non-mineral macronutrients include nitrogen, phosphorus, magnesium, potassium, calcium, sulfur, carbon, hydrogen, and oxygen (Kathpalia & Bhatla, 2018). Pertaining to the role, carbon, hydrogen, and oxygen are non-mineral macronutrients that are considered structural elements since they are needed to build the larger cellular organic molecules, whereas nitrogen, phosphorus, and potassium are primary nutrients, and calcium, magnesium, and sulfur are secondary nutrients (Barker & Pilbeam, 2015). Micronutrients such as iron, zinc, manganese, boron, molybdenum, cobalt, and copper are required in minute quantities (Siddiqui et al., 2014) and are required as co-factors for enzymatic activity (Grusak, 2001). The bioavailability and absorption of nutrients from the soil are driven by a multitude of characteristics, such as physiochemical properties of soil, environmental factors, and the microbial population (Rajwar et al., 2021). Plants absorb soluble ions from the

Sustainable Plant Nutrition
ISBN: 978-0-443-18675-2
https://doi.org/10.1016/B978-0-443-18675-2.00016-X

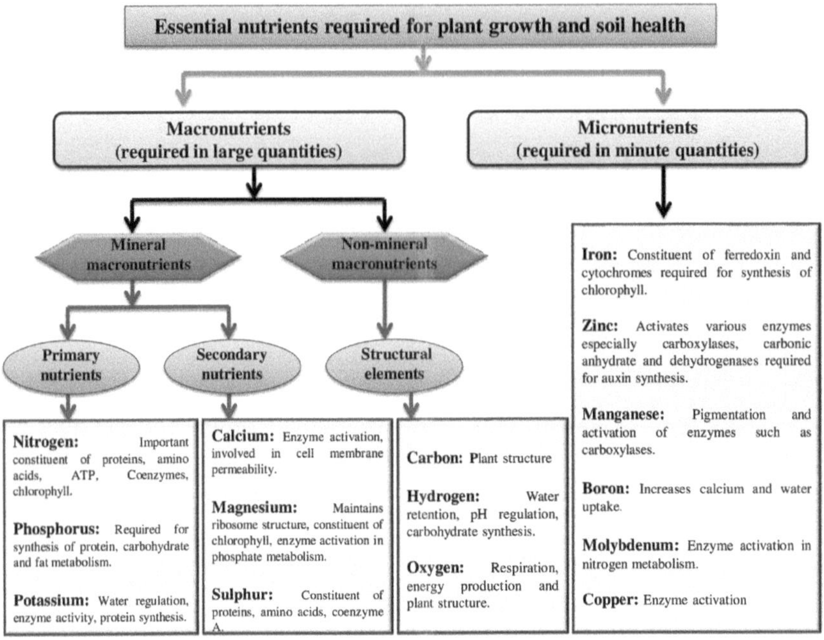

Fig. 11.1 Role of essential nutrients in improving plant nutrition and soil health.

soil *via* a variety of pathways, such as direct contact between the plant root and the nutrient, mass flow of dissolved nutrient toward the plant root, and diffusion of nutrient from a zone of higher concentration to lower concentration (Cramer & Hawkins, 2009). There has been a lot of interest in identifying the beneficial ecological interactions that stimulate plant growth, for instance, mycorrhizal fungi and bacteria found in nodulated legumes are both regarded as root symbionts (Jacoby et al., 2017). Several bacterial strains, primarily *Pseudomonas*, *Azospirillum*, and others, have been reported to promote plant growth (Dobbelaere, 2003). To promote growth and productivity, agricultural seeds have been coated with bacterial strains including *Azotobacter chroococcum* (Yousefi et al., 2017) or *Bacillus megaterium* (Turan et al., 2014). The emphasis of studies has now turned away from the particular microbial strain and toward using genome sequencing to explore the quantity and root microbiome diversity (Jacoby et al., 2017). As per the findings of such a sequencing study, the rhizospheric niche is a hotspot of biological diversity, with plant roots housing a diverse array of microbial species (Singh, 2021). This chapter aims to provide a basic knowledge of how soil microorganisms improve soil health and plant growth.

2. Microbial strains and plant nutrient bioavailability

Different microbial species (e.g., bacterial and fungal species) perform key functions in maintaining soil health. These microorganisms increase the fertility of soil *via* enhancing nutrient bioavailability, such as nitrogen, phosphorus, potassium, iron, etc (Rashid et al., 2016). There are three basic pathways involved in microbial activity-assisted plant growth viz. modulating plant hormone signaling, resisting or outpacing pathogenic microbes, and enhancing the bioavailability of soil-borne nutrients, *via* which soil microorganisms breakdown resistant types of soil-borne nutrients to release them for plant nutrition (Jacoby et al., 2017). The majority of nutrients in biological habitats, like nitrogen, phosphorus, and sulfur, are bound in organic compounds and therefore have limited bioavailability to plants. Plants rely on the proliferation of soil microorganisms including bacteria and fungi to depolymerize and mineralize organic forms of nitrogen, phosphorus, and sulfur (Mahala et al., 2020). The constituents of these microbial cells are then discharged, either *via* cycling and cell lysis or by protozoic predation. This releases inorganic nitrogen, phosphorus, and sulfur molecules into the soil, particularly ionic species like ammonium nitrate, phosphate, and sulfate, which are the desirable nutrition sources for plants (Kathpalia & Bhatla, 2018; Sharma et al., 2013).

3. Role of microbes in maintaining soil health

Soil microorganisms are active soil architects (Kaviya et al., 2019) preparing it for plant growth *via* increasing bioavailability and producing required growth factors (Bhattacharyya et al., 2012). Natural microbiomes perform various functions in adhering and desorbing inorganic nutrients to physical surfaces, as well as decomposing organic wastes making them soil-friendly (Priya et al., 2021). Plants and microorganisms play a combined function in the agricultural and farming efficiency of soil (Duchene et al., 2017). Even little anthropogenic impacts, like the incorporation of sewage sludge, increased the soil resident microbial community of Proteobacteria and Bacteroidetes in bauxite overburdened dumping regions, aiding the soil development process (Rebello et al., 2021). Microorganisms in the various biogeochemical cycles direct the process of nutrient cycling, which is an important aspect of maintaining soil fertility (Andrade, 2008). To achieve long-term plant growth, it is recommended that rhizosphere bacteria be used to boost soil fertility rather than synthetic pesticides (Gupta et al.,

2015). Plant efficiency enhancement is now well understood to be a complicated process involving interactions with particular microorganisms or consortiums (Artursson et al., 2006). Engineering symbiotic interactions to enable non-leguminous and other staple crops to fix nitrogen and transforming them into soil fertility-contributing plants is one of the new techniques (Pankievicz et al., 2019). This will substantially increase the worldwide food supply while also assisting in the achievement of environmental goals (Soumare et al., 2020). Another factor that contributes to soil health is the potential of soil microbes to detoxify and neutralize harmful pollutants that enter the soil *via* diverse pathways (Abdullah et al., 2020). Pesticides such as endosulfan (Jha et al., 2015), λ-cypermethrin (Bhatt et al., 2019), deltamethrin (Naeem et al., 2020), and pyrethroid (Wołejko et al., 2020) are broken down by microbes including *Aspergillus* sp. and *Raoultella ornithinolytica*. Plastics, hydrocarbons, surfactants, polychlorinated biphenyls, radioactive waste, heavy metals, and other xenobiotics may all be efficaciously eliminated through certain microbes (Mishra et al., 2020). However, several microorganisms known as 'Superbugs' exist that can degrade a broad spectrum of xenobiotics (Rangasamy et al., 2018). Many studies involving six isolates for tetrachlorvinphos degradation and an actinobacteria consortium consisting of strains of *Pseudomonas*, *Enterobacter*, *Aspergillus*, and *Rhodotorula* that efficiently eliminated seven pesticides have shown that microbial consortiums are more efficient in xenobiotic deterioration than individual isolates (Rebello et al., 2021). The native microbes provide ecological benefits to plants, particularly in the rhizosphere, and they can also assist plants to cope with stress (Barea et al., 2005). These rhizosphere-dwelling microorganisms are great at deteriorating xenobiotics and improving soil health (Tripathi & Gaur, 2021). Plant growth-promoting microorganisms have been used to disintegrate second-generation pesticides, particularly organophosphate insecticides (Sundari et al., 2019). Pesticides can be degraded by rhizosphere microorganisms, resulting in better plant growth (Khatoon et al., 2020). The existence and longevity of the microorganisms in the soil are determined by interactions with other biotic factors in the ecosystem (Moënne-Loccoz et al., 2015), and plant interactions with microbial consortium are frequently more successful than individual organisms (Sarma et al., 2015). Therefore, soil fertility is certainly dependent on microbial diversity and growth-promoting abilities. The role of various microbial strains (bacterial and fungal) involved in the improvement of soil health is presented in Tables 11.1 and 11.2.

Table 11.1 Various bacterial strains involved in the improvement of soil health.

Nutrient parameter	Bacterial strain(s)	References
Nitrogen fixers	*Azotobacter* sp., *Azospirillum* sp., *Cyanobacteria, Rhizobium, Bradyrhizobium, Azorhizobium, Mesorhizobium*, and *Sinorhizobium*	(Mitter et al., 2021)
Phosphorous solubilizers	*Pseudomonas aeruginosa, Azotobacter chroococcum, Azospirillum brasilense, Bacillus megateriu, Bacillus mucilaginosus, Rhizobium leguminosarum*	(Seenivasagan & Babalola, 2021)
Potassium solubilizers	*Bacillus, Rhizobium, Acidithiobacillus, Paenibacillus, Pseudomonas*, and *Burkholderia*	(Kour et al., 2020)
Organic matter	*Bradyrhizobium diazoefficiens* USDA110 and *Bradyrhizobium japonicum* THA6, *Stylosanthes guianensis*, and *Paspalun natatu*, Rhizobacteria	(Cui et al., 2015, Prakamhang et al., 2015)
Iron	*Pseudomonas fluorescens*	(Abbaszadeh-Dahaji et al., 2020)
Sulfur oxidation	*Xanthobacter, Alcaligenes, Bacillus, Pseudomonas, Streptomyces*, and *Thiobacillus*	(Sahu et al., 2018)

Table 11.2 Various fungal strains involved in the improvement of soil health.

Nutrient parameter	Fungal strain(s)	References
Nitrogen fixers	*Claroideoglomus etunicatum*, *Colletotrichum tropicale*	(Cely et al., 2016 Christian et al., 2019)
Phosphorous solubilizers	*Glomus mosseae*, *Claroideoglomus etunicatum*, *Claroideoglomus claroideum*	(Xu et al., 2008, Cely et al., 2016, Majewska et al., 2017)
Potassium solubilizers	*Aspergillus*, *Cladosporium*, *Macrophomina*, *Sclerotinia*, *Trichoderma*, *Glomus*, and *Penicillium*	(Kour et al., 2020)
Organic matter	*Glomus mosseae*, *Paraglomus occultum*, *Rhizophagus irregularis*	(Rashid et al., 2016)
Iron	*Aspergillus phytase*	(Drakakaki et al., 2005)
Sulfur oxidation	*Fusarium*, *Aspergillus*, and *Penicillium*	(Macik et al., 2020)
Manganese	*Epichloe* sp.	(Waghunde et al., 2017)
Aggregate formation/ Glomalin protein	*Rhizophagus irregularis*, *Glomus mosseae*, *Agaricus lilaceps*. *Laccaria bicolor*, *Laccaria laccata*, *Lactarius theiogalus*, *Paxillus involutus*, *Suillus bovinus*	(Rashid et al., 2016)

4. Role of microorganisms in plant nutrition

Plants have the unique ability to produce their food *via* the process of photosynthesis (Demmig-Adams et al., 2022), but they often need the assistance of soil microbes to obtain the nutrients they require to function effectively (Elnahal et al., 2022). Bacterial and fungal species have developed symbiotic interactions with plants, allowing them to survive in different environments (Sandrini et al., 2022). Plant nutrition and productivity are affected by several factors. Apart from plant genotype, conditions including availability and uptake of nutrients and water in the soil play a significant role in plant growth and development (Fageria & Moreira, 2011). The role of microbes in plant nutrient mobility and uptake is presented in Fig. 11.2.

4.1 Role of arbuscular mycorrhizal fungi in nutrient mobility and uptake

Arbuscular mycorrhizal fungi (AMF) is the most prevalent plant-microbe symbiotic relationship (Begum et al., 2019). AMF are found in almost all

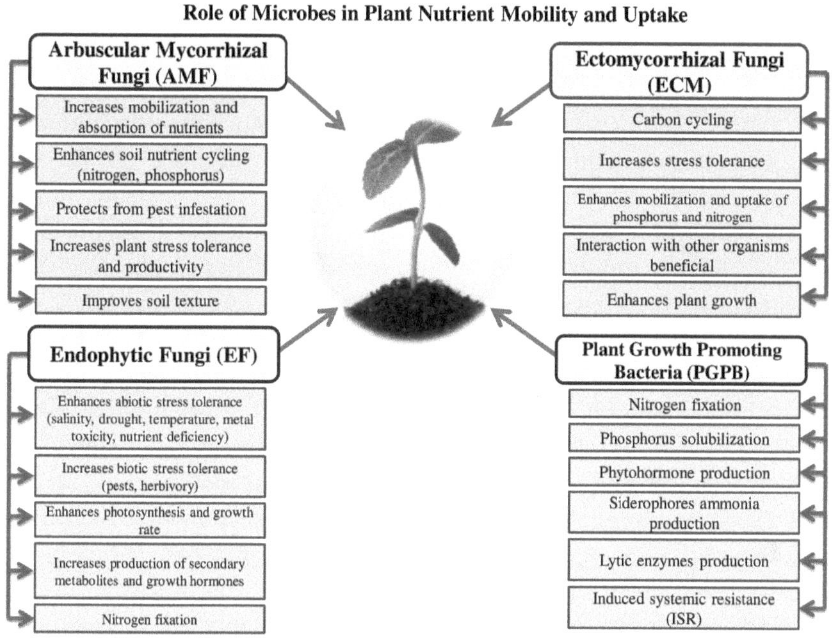

Fig. 11.2 The role of microbes in plant nutrient mobility and uptake.

natural environments and perform several key ecological functions, including increased plant nutrition, stress tolerance, and soil health (Chen et al., 2018). Since AMF associates with the majority of agricultural plants, they are gaining much attention for their potential application in sustainable agriculture (Devi et al., 2021). They have the potential to absorb considerable amounts of nitrogen from dead and decayed matter, which improves their ability to grow and survive. They play a crucial role in the nitrogen cycle as they are capable of transferring roughly 20%–75% nitrogen uptake of arbuscular mycorrhiza (AM) plants to their hosts (Begum et al., 2019). Furthermore, AMF's massive biomass and growing nitrogen demand make them the dominant participant in the global nitrogen pool, on a scale equivalent to fine roots. AMF assists host plants develop more strongly under stress *via* facilitating a sequence of the intricate network of interactions between the plant-fungus, resulting in improved photosynthetic activity, water uptake, and other gas exchange-associated attributes (Balestrini et al., 2020). AMF provides nutrition to plants even when the environment within the root cells is unfavourable (Bahadur et al., 2019). AMF forms symbiotic relationships with the roots of the host plant to acquire nutrients in exchange for mineral nutrients such as nitrogen, phosphorus, potassium, calcium, zinc, and sulfur (Bhantana et al., 2021). After developing a symbiotic association, AMF produces large underground extra-radical mycelium (ERM) extending from the roots to the surrounding rhizosphere (de Novais et al., 2020), assisting in enhancing nutrient absorption, particularly nitrogen (a major growth-limiting factor), and hence improving plant growth and development (Garcia et al., 2016). Furthermore, AMF develops fungal structures known as arbuscules, which facilitate the interchange of inorganic minerals and macronutrients like carbon and phosphorus, allowing host plants to grow more vigorously (Bargaz et al., 2018). As a result, they can considerably increase the level of phosphorus in both root and shoot systems (Chen et al., 2007). In phosphorus-limited conditions, mycorrhizal association enhances phosphorus availability to damaged roots of host plants (Chahal et al., 2021). Phosphate ion absorption, for example, has dramatically enhanced in AMF-colonized maize plants (Garcés-Ruiz et al., 2017). AMF has been shown to promote phosphorus and nitrogen absorption, resulting in better plant development at low and high phosphorus concentrations under different irrigation schemes. For instance, in the absence of water, mycorrhizal symbiosis increased the nitrogen, phosphorus, and iron levels of *Pelargonium graveolens* L (Amiri et al., 2017). The interplay of salt stress and AMF has a profound impact on the levels of phosphorus (P) and nitrogen (N), as well as the N:P ratio in plant shoots (Kazemi et al., 2019). Many studies suggest that inoculating plants with AMF positively affects nutrient levels under stress conditions. For example, AMF-inoculated *Euonymus japonica* improved levels of phosphorus, calcium, and potassium under salt stress due to quick fungal attachment (Chahal et al., 2021), while *Pistachio* spp. displayed elevated levels of phosphorus, potassium, zinc, and manganese under drought conditions (Fattahi et al., 2021), improved phosphorus and nitrogen rates in *Chrysanthemum morifolium* plant tissues (Begum et al., 2019), and increased seedling weight *via* improving water content and intercellular carbon dioxide (CO_2), phosphorus, and nitrogen levels in *Leymus chinensis* (Liu et al., 2020) have also been reported.

4.2 Role of endophytic fungi in soil nutrient mobility and uptake

Fungal endophytes are a diverse group of organisms with varying morphology, physiology as well as symbiotic interactions with their hosts (Sonowal et al., 2022). The endophytic fungus lives asymptomatically within the tissue voids of plants (Van Overbeek & Saikkonen, 2016).

They complete their life cycle by growing intercellularly or intracellularly in a specific host plant (Sarkar et al., 2021). Fungal endophytes can impact the overall plant growth *via* modulating photosynthesis, nitrogen uptake, plant hormone production, and other biophysiological processes (Qiang et al., 2019). Furthermore, they have proven to be extremely beneficial to their host plants in adversities such as drought, salt, heat, heavy metal toxicity, and so on (Chouhan et al., 2022), as well as enhancing their tolerance to biotic stress, such as herbivory and pest invasion (Grabka et al., 2022). Various root endophytic fungi, such as Dark Septate Endophytic (DSE) fungi, *Serendipita indica, Trichoderma* sp., and others, have been reported to play a key role in nutrient uptake and distribution to different parts of the host plant (García-Latorre et al., 2021). DSE, for example, has gained a lot of interest attributed to their growth-promoting benefits on various plant species including coniferous sp., (Kageyama et al., 2008), barley, cabbages (Card et al., 2016), etc. DSE can directly stimulate plant growth by increasing nutrient absorption or indirectly *via* protecting against abiotic stressors by synthesizing plant hormones and similar compounds (Hidayat, 2019). They have been shown to assist plant growth *via* improving both nutrient recovery and uptake (Vergara et al., 2017). DSE can increase the absorption rate of two essential plant macronutrients viz, nitrogen and phosphorus in several plant species (Eid et al., 2019). Besides nitrogen and phosphorus, another essential macronutrient such as potassium has also been recovered in tomato plants with significant efficiency by DSE *via* the translocation of potassium deprivation zones and potassium capture in sites inaccessible by the root surface (Vergara et al., 2017). In addition, DSE assisted in the absorption of micronutrients such as iron, manganese, and zinc by their host tomato plants. Therefore, it may have a broader influence on the photosynthetic activity of host plants, promoting plant growth (Rashid et al., 2021). Furthermore, a range of different root endophytes has proved their potential to absorb nutrients for the host plant. *Serendipita indica* is another prominent root endophytic fungus involved in nitrogen absorption (Jangir et al., 2021). It stimulated the growth of tobacco and *Arabidopsis thaliana* seedlings by enabling massive nitrogen translocation from agar plates into the aerial regions of the seedlings (Sarkar et al., 2021). Also, it can increase growth in the same way as mycorrhizal symbiotic associations do by increasing phosphorus uptake, for instance, *S. indica* has shown a significant rise in corn plant biomass sulfur (Narayan et al., 2020), and phosphorus levels (Mosaddeghi et al., 2021). *Trichoderma* strains are other beneficial fungal endophytes and are typical soil and plant root dwellers (Haris et al., 2021). They play a vital function in allowing *Trichoderma* cells to colonize roots and in regulating defense responses (Nogueira-Lopez et al., 2018). Furthermore, certain *Trichoderma* species have other advantages for plants, including enhancing plant growth and nutrient uptake, fertilizer efficiency, and abiotic stress tolerance. *Trichoderma asperellum* can help white lupin with iron deficit *via* creating siderophores that bind iron molecules (Fe^{3+}) and act as Fe donors to the plant (De Santiago et al., 2009). Some *Trichoderma* strains can solubilize large quantities of phosphate as well as various complex nutritional forms such as manganese dioxide (MnO_2), zinc, and rock phosphate (Altomare et al., 1999).

4.3 Role of ectomycorrhizal fungi in nutrient mobility and uptake

Ectomycorrhizal fungi (ECM) are facultative saprotrophs with a dual lifestyle (Martin & Nehls, 2009). Ectomycorrhizae are symbiotic interactions developed by the woody plant roots and soil-dwelling fungus that can assist the host plant in nutrient intake (Taylor & Peterson, 2005). Ectomycorrhizal interactions are developed by plant families such as the pinaceae, fabaceae, betulaceae, and fagaceae (Frąc et al., 2018). They are known to provide several advantages to their hosts, such as drought and salt tolerance (Wen et al., 2022), and are usually classified as advantageous for host nutrient absorption (Netherway et al., 2021). They compete for nutrients in the soil, secreting various hydrolytic enzymes that allow them to breakdown litter polymers and consume organic nutrition sources (Stuart & Plett, 2020). Simultaneously, they reside as symbionts within plant roots, which necessitate a series of adaptive processes to avoid plant parasitism. For instance, ECM have lost their ability to breakdown plant cell wall polysaccharides such as cellulose, pectins, pectates, and so on, restricting their access into the root to intercellular spaces (Bücking et al., 2012). Enzymes including chitinases, phosphatases, and proteases are excreted in substantial quantities by the fungal symbionts of ectomycorrhizal plants, enabling extraction of nutrients such as nitrogen and phosphorus from organic waste (Tibbett & Sanders, 2002). These fungi colonize the lateral roots of host trees, forming interlacing mycelial structures that penetrate between and surround root epidermal cells (Stuart & Plett, 2020). The Hartig net is a distinctive morphological arrangement that acts as a nutrition exchange point and enables a wide contact area between the two symbiotic partners (Balestrini & Bonfante, 2014). The fungus receives carbon sources from the host in return for restricted nutrients, which it may obtain either outside the nutrient depletion area around the host's root system or from immobilized sources that are generally inaccessible to the plant (Vaario & Matsushita, 2021). As a result, ECM play an important role in the health of forest ecosystems (Looney et al., 2018). These fungi assist forest soil processes such as organic matter decomposition, nitrogen cycling, and carbon sequestration (Han et al., 2020), by developing a favourable symbiotic interaction with the roots of 80%–90% of all temperate and boreal forest trees (Van Der Heijden et al., 2008). ECM fungi absorb and release a variety of macronutrients (such as phosphorus, potassium, calcium, magnesium, and sulfur) as well as micronutrients (including iron, zinc, copper, and manganese) to their host plants (Mahala et al., 2020). Furthermore, they specialize in providing nitrogen, the most critical growth-limiting component in most forest ecosystems (Guo et al., 2021). Most of the nitrogen in forest soils is bound up in soil organic matter or leaf litter that accumulates on the forest surface (Prescott & Vesterdal, 2021). Although the mineralization processes of these organic substances are slow to provide enough nitrogen, ECM fungi can degrade organic compounds, making them significant participants in the soil nitrogen cycle (Kluber et al., 2010).

4.4 Role of plant growth promoting bacteria (PGPBs) in soil nutrient mobility and uptake

PGPBs are beneficial microbial communities prevalent in plants that thrive in a wide range of environments. Organic compounds from the rhizosphere, including glutamine, betaine, and trehalose, are used by these microbes to help them grow. PGPBs are classified as intracellular or extracellular. Extracellular PGPBs are bacteria that increase growth factors and colonize the root surface area or the cortex's intercellular space. In contrast, intracellular PGPBs are bacteria that promote plant growth but dwell inside the root surface/rhizoplant root cell nodular structure. *Bacillus, Serratia, Azotobacter, Azospirillum, Micrococcus, Arthrobacter, Erwinia,* and *Pseudomonas* are among the extracellular PGPBs, whereas other intracellular PGPB belong to *Rhizobia*, including *Allorhizobium, Mesorhizobium, Bradyrhizobium, Frankia.* etc. PGPBs are important for improving the physiochemical and biological properties of soil by enabling nutrient absorption and flow of water, which are key constraints to crop growth and productivity, particularly under abiotic stress regimes. Through their inhibitory action against plant pathogens, PGPBs can function as biological control agents and indirectly help in the growth and development of plants (Skinder et al., 2022). By increasing available mineral nutrients or regulating phytohormone rates, PGPBs, on the other hand, may directly boost the growth and development of plants (Gamalero & Glick, 2011). They can produce plant hormones, including auxins, cytokinins, and gibberellins (Cohen et al., 2008) which can help plants, developing at different levels produce siderophores, which are iron chelators that improve plant Fe nutrition or suppress the activity of phytopathogens, synthesize 1-aminocyclopropane-1-carboxylate (ACC) deaminase, which can lower ethylene concentration in plants (Glick, 2005), and dissolve minerals such as phosphorous, making them easily available for plant development (Olanrewaju et al., 2017). Either some or all of these mechanisms could be employed by PGPBs to aid the growth and development of plants, and a single PGPB could use various mechanisms depending on the environment. When plants grow in inadequate or environmental stresses, such as salinity and heavy metals, the positive results of PGPBs on the growth of plants are usually more prominent (Shilev, 2020). Therefore, PGPBs as biofertilizers provide a low, ecologically sound strategy to enhance plant growth and development under drought circumstances, making them a significant and necessary technique for facilitating self-sustaining agriculture.

5. Conclusions

Fertilizers and other synthetic chemicals used extensively to increase plant growth, soil fertility, and crop productivity have a detrimental effect on soil health, aquatic ecosystems, soil beneficial organisms, human health, and the environment. Microbes, on the other hand, such as mycorrhizal fungi (MF), PGPBs, etc. which include various fungal and bacterial species can be effectively used to enhance plant growth and productivity in a diverse array of environments *via* suppressing the negative effects of abiotic factors such as heavy metals, salinity, and drought, which ultimately result in the reactive oxygen species (ROS) formation resulting in oxidative stress. Various microbial species have been shown to benefit plants through a range of physiological activities, including phosphate solubilization, nitrogen fixation, nutrient absorption, plant hormone, and growth regulator synthesis. Therefore, these beneficial microbes promote plant growth and crop productivity in response to various stressors and the growing demands of an increasing population. As a result, they may be used as a replacement for fertilizers and other chemical substances and may serve an important function in enhancing soil properties, crop productivity, and sustainable agriculture under stress. However, more research is needed on the application of microorganisms as biofertilizers, plant growth regulators under stress conditions, determining the most efficient fungal and bacterial strains, and identifying beneficial treatment approaches to promote soil health and crop productivity.

References

Abbaszadeh-Dahaji, P., Masalehi, F., & Akhgar, A. (2020). Improved growth and nutrition of sorghum (*Sorghum bicolor*) plants in a low-fertility calcareous soil treated with plant growth–promoting rhizobacteria and Fe-EDTA. *Journal of Soil Science and Plant Nutrition, 20*, 31–42. https://doi.org/10.1007/s42729-019-00098-9

Abdullah, S. R. S., Al-Baldawi, I. A., Almansoory, A. F., Purwanti, I. F., Al-Sbani, N. H., & Sharuddin, S. S. N. (2020). Plant-assisted remediation of hydrocarbons in water and soil: Application, mechanisms, challenges and opportunities. *Chemosphere, 247*, 125932.

Altomare, C., Norvell, W. A., Bjorkman, T. H. O., & Harman, G. (1999). Solubilization of phosphates and micronutrients by the plant-growth-promoting and biocontrol fungus Trichoderma harzianum Rifai 1295-22. *Applied and Environmental Microbiology, 65*(7), 2926–2933.

Amiri, R., Nikbakht, A., Etemadi, N., & Sabzalian, M. R. (2017). Nutritional status, essential oil changes and water-use efficiency of rose geranium in response to arbuscular mycorrhizal fungi and water deficiency stress. *Symbiosis, 73*(1), 15–25.

Andrade, G. (2008). Role of functional groups of microorganisms on the rhizosphere microcosm dynamics. In *Plant surface microbiology* (pp. 51–69). Berlin, Heidelberg: Springer.

Artursson, V., Finlay, R. D., & Jansson, J. K. (2006). Interactions between arbuscular mycorrhizal fungi and bacteria and their potential for stimulating plant growth. *Environmental Microbiology, 8*(1), 1–10.

Bahadur, A., Batool, A., Nasir, F., Jiang, S., Mingsen, Q., Zhang, Q., Pan, J., Liu, Y., & Feng, H. (2019). Mechanistic insights into arbuscular mycorrhizal fungi-mediated drought stress tolerance in plants. *International Journal of Molecular Sciences, 20*(17), 4199. https://doi.org/10.3390/ijms20174199

Balestrini, R., & Bonfante, P. (2014). Cell wall remodeling in mycorrhizal symbiosis: A way towards biotrophism. *Frontiers in Plant Science, 5*, 237.

Balestrini, R., Brunetti, C., Chitarra, W., & Nerva, L. (2020). Photosynthetic traits and nitrogen uptake in crops: Which is the role of arbuscular mycorrhizal fungi? *Plants, 9*(9), 1105.

Barea, J. M., Pozo, M. J., Azcon, R., & Azcon-Aguilar, C. (2005). Microbial co-operation in the rhizosphere. *Journal of Experimental Botany, 56*(417), 1761–1778.

Bargaz, A., Lyamlouli, K., Chtouki, M., Zeroual, Y., & Dhiba, D. (2018). Soil microbial resources for improving fertilizers efficiency in an integrated plant nutrient management system. *Frontiers in Microbiology, 9*, 1606.

Barker, A. V., & Pilbeam, D. J. (Eds.). (2015). *Handbook of plant nutrition*. CRC press.

Begum, N., Qin, C., Ahanger, M. A., Raza, S., Khan, M. I., Ashraf, M., Ahmed, N., & Zhang, L. (2019). Role of arbuscular mycorrhizal fungi in plant growth regulation: implications in abiotic stress tolerance. *Frontiers in Plant Science, 10*, 1068. https://doi.org/10.3389/fpls.2019.01068

Bhantana, P., Rana, M. S., Sun, X. C., Moussa, M. G., Saleem, M. H., Syaifudin, M., Shah, A., Poudel, A., Pun, A. B., Bhat, M. A., Mandal, D. L., Shah, S., Zhihao, D.,

Tan, Q., & Hu, C. X. (2021). Arbuscular mycorrhizal fungi and its major role in plant growth, zinc nutrition, phosphorous regulation and phytoremediation. *Symbiosis, 84*(1), 19–37. https://doi.org/10.1007/s13199-021-00756-6
Bhatia, C. R. (2008). Role of microbial diversity for soil, health and plant nutrition. In *Molecular mechanisms of plant and microbe coexistence* (pp. 53–74). Berlin, Heidelberg: Springer.
Bhattacharyya, P. N., & Jha, D. K. (2012). Plant growth-promoting rhizobacteria (PGPR): Emergence in agriculture. *World Journal of Microbiology and Biotechnology, 28*(4), 1327–1350.
Bhatt, P., Huang, Y., Zhan, H., & Chen, S. (2019). Insight into microbial applications for the biodegradation of pyrethroid insecticides. *Frontiers in Microbiology*, 1778.
Biswal, D. (2022). Nematodes as ghosts of land use past: Elucidating the roles of soil nematode community studies as indicators of soil health and land management practices. *Applied Biochemistry and Biotechnology*, 1–61.
Bücking, H., Liepold, E., & Ambilwade, P. (2012). The role of the mycorrhizal symbiosis in nutrient uptake of plants and the regulatory mechanisms underlying these transport processes. *Plant Science, 4*, 108–132.
Card, S., Johnson, L., Teasdale, S., & Caradus, J. (2016). Deciphering endophyte behaviour: The link between endophyte biology and efficacious biological control agents. *FEMS Microbiology Ecology, 92*(8).
Cely, M. V., Siviero, M. A., Emiliano, J., Spago, F. R., Freitas, V. F., Barazetti, A. R., Erika, T., Goya, E. T., Lamberti, G. D., dos Santos, I. M. O., De Oliveira, A. G., & Andrade, G. (2016). Inoculation of Schizolobium parahyba with mycorrhizal fungi and plant growth-promoting rhizobacteria increases wood yield under field conditions. *Frontiers in Plant Science,* 7, 1708. https://doi.org/10.3389/fpls.2016.01708
Chahal, K., Gupta, V., Verma, N. K., Chaurasia, A., & Rana, B. (2021). Arbuscular mycorrhizal (AM) fungi as a tool for sustainable agricultural system. *Mycorrhizal Fungi: Utilization in Agriculture and Forestry, 19.*
Chen, M., Arato, M., Borghi, L., Nouri, E., & Reinhardt, D. (2018). Beneficial services of arbuscular mycorrhizal fungi—from ecology to application. *Frontiers in Plant Science*, 1270.
Chen, B., Xiao, X., Zhu, Y. G., Smith, F. A., Xie, Z. M., & Smith, S. E. (2007). The arbuscular mycorrhizal fungus Glomus mosseae gives contradictory effects on phosphorus and arsenic acquisition by Medicago sativa Linn. *Science of the Total Environment, 379*(2–3), 226–234.
Chouhan, S., Agrawal, L., & Prakash, A. (2022). Amelioration in traditional farming system by exploring the different plant growth-promoting attributes of endophytes for sustainable agriculture. *Archives of Microbiology, 204*(2), 1–21.
Christian, N., Herre, E. A., & Clay, K. (2019). Foliar endophytic fungi alter patterns of nitrogen uptake and distribution in Theobroma cacao. *New Phytologist, 222*(3), 1573–1583.
Cohen, A. C., Bottini, R., & Piccoli, P. N. (2008). Azospirillum brasilense Sp 245 produces ABA in chemically-defined culture medium and increases ABA content in arabidopsis plants. *Plant Growth Regulation, 54*(2), 97–103.
Cramer, M. D., & Hawkins, H. J. (2009). A physiological mechanism for the formation of root casts. *Palaeogeography, Palaeoclimatology, Palaeoecology, 274*(3–4), 125–133.
Cui, H., Zhou, Y., Gu, Z., Zhu, H., Fu, S., & Yao, Q. (2015). The combined effects of cover crops and symbiotic microbes on phosphatase gene and organic phosphorus hydrolysis in subtropical orchard soils. *Soil Biology and Biochemistry, 82*, 119–126. https://doi.org/10.1016/j.soilbio.2015.01.003
de Novais, C. B., Sbrana, C., da Conceição Jesus, E., Rouws, L. F. M., Giovannetti, M., Avio, L., Siqueira, J. O., Júnior, O. J. S., da Silva, E. M. R., & de Faria, S. M. (2020).

Mycorrhizal networks facilitate the colonization of legume roots by a symbiotic nitrogen-fixing bacterium. *Mycorrhiza, 30*(2), 389–396. https://doi.org/10.1007/s00572-020-00948-w

De Santiago, A., Quintero, J. M., Avilés, M., & Delgado, A. (2009). Effect of Trichoderma asperellum strain T34 on iron nutrition in white lupin. *Soil Biology and Biochemistry, 41*(12), 2453–2459.

Demmig-Adams, B., Polutchko, S. K., Zenir, M. C., Fourounjian, P., Stewart, J. J., López-Pozo, M., & Adams, W. W., III (2022). Intersections: Photosynthesis, abiotic stress, and the plant microbiome. *Photosynthetica, 60*(SI).

Devi, S. H., Bhupenchandra, I., Sinyorita, S., Chongtham, S. K., & Devi, E. L. (2021). Mycorrhizal fungi and sustainable agriculture. In *Nitrogen in agriculture-physiological, agricultural and ecological aspects*. IntechOpen.

Dobbelaere, S., Vanderleyden, J., & Okon, Y. (2003). Plant growth-promoting effects of diazotrophs in the rhizosphere. *Critical Reviews in Plant Sciences, 22*(2), 107–149.

Drakakaki, G., Marcel, S., Glahn, R. P., Lund, E. K., Pariagh, S., Fischer, R., … Stoger, E. (2005). Endosperm-specific co-expression of recombinant soybean ferritin and Aspergillus phytase in maize results in significant increases in the levels of bioavailable iron. *Plant Molecular Biology, 59*(6), 869–880.

Duchene, O., Vian, J. F., & Celette, F. (2017). Intercropping with legume for agroecological cropping systems: Complementarity and facilitation processes and the importance of soil microorganisms. A review. *Agriculture, Ecosystems & Environment, 240*, 148–161.

Eid, A. M., Salim, S. S., Hassan, S. E. D., Ismail, M. A., & Fouda, A. (2019). Role of endophytes in plant health and abiotic stress management. In *Microbiome in plant health and disease* (pp. 119–144). Singapore: Springer.

Elnahal, A. S., El-Saadony, M. T., Saad, A. M., Desoky, E. S. M., El-Tahan, A. M., Rady, M. M., … El-Tarabily, K. A. (2022). The use of microbial inoculants for biological control, plant growth promotion, and sustainable agriculture: A review. *European Journal of Plant Pathology*, 1–34.

Fageria, N. K., & Moreira, A. (2011). The role of mineral nutrition on root growth of crop plants. *Advances in Agronomy, 110*, 251–331. Academic Press.

Fattahi, M., Mohammadkhani, A., Shiran, B., Baninasab, B., Ravash, R., & Gogorcena, Y. (2021). Beneficial effect of mycorrhiza on nutritional uptake and oxidative balance in pistachio (Pistacia spp.) rootstocks submitted to drought and salinity stress. *Scientia Horticulturae, 281*, 109937.

Frąc, M., Hannula, S. E., Bełka, M., & Jędryczka, M. (2018). Fungal biodiversity and their role in soil health. *Frontiers in Microbiology, 9*, 707.

Gamalero, E., & Glick, B. R. (2011). Mechanisms used by plant growth-promoting bacteria. In *Bacteria in agrobiology: Plant nutrient management* (pp. 17–46). Berlin, Heidelberg: Springer.

Garcés-Ruiz, M., Calonne-Salmon, M., Plouznikoff, K., Misson, C., Navarrete-Mier, M., Cranenbrouck, S., & Declerck, S. (2017). Dynamics of short-term phosphorus uptake by intact mycorrhizal and non-mycorrhizal maize plants grown in a circulatory semi-hydroponic cultivation system. *Frontiers in Plant Science, 8*, 1471.

García-Latorre, C., Rodrigo, S., & Santamaría, O. (2021). Endophytes as plant nutrient uptake-promoter in plants. *Endophytes: Mineral Nutrient Management, 3*, 247–265. Springer, Cham.

Garcia, K., Doidy, J., Zimmermann, S. D., Wipf, D., & Courty, P. E. (2016). Take a trip through the plant and fungal transportome of mycorrhiza. *Trends in Plant Science, 21*(11), 937–950.

Glick, B. R. (2005). Modulation of plant ethylene levels by the bacterial enzyme ACC deaminase. *FEMS Microbiology Letters, 251*(1), 1–7.

Grabka, R., d'Entremont, T. W., Adams, S. J., Walker, A. K., Tanney, J. B., Abbasi, P. A., & Ali, S. (2022). Fungal endophytes and their role in agricultural plant protection against pests and pathogens. *Plants, 11*(3), 384.

Grusak, M. A. (2001). *Plant macro-and micronutrient minerals.* e LS.

Guo, W., Ding, J., Wang, Q., Yin, M., Zhu, X., Liu, Q., … Yin, H. (2021). Soil fertility controls ectomycorrhizal mycelial traits in alpine forests receiving nitrogen deposition. *Soil Biology and Biochemistry, 161*, 108386.

Gupta, G., Parihar, S. S., Ahirwar, N. K., Snehi, S. K., & Singh, V. (2015). Plant growth promoting rhizobacteria (PGPR): Current and future prospects for development of sustainable agriculture. *Journal of Microbiology and Biochemical Technology, 7*(2), 096–102.

Han, M., Sun, L., Gan, D., Fu, L., & Zhu, B. (2020). Root functional traits are key determinants of the rhizosphere effect on soil organic matter decomposition across 14 temperate hardwood species. *Soil Biology and Biochemistry, 151*, 108019.

Haris, M., Shakeel, A., Ansari, M. S., Hussain, T., Khan, A. A., & Dhankar, R. (2021). Sustainable crop production and improvement through bio-prospecting of fungi. In *Fungi bio-prospects in sustainable agriculture, environment and nano-technology* (pp. 407–428). Academic Press.

Hidayat, I. (2019). Dark septate endophytes and their role in enhancing plant resistance to abiotic and biotic stresses. In *Plant growth promoting Rhizobacteria for sustainable stress management* (pp. 35–63). Singapore: Springer.

Jacoby, R., Peukert, M., Succurro, A., Koprivova, A., & Kopriva, S. (2017). The role of soil microorganisms in plant mineral nutrition—current knowledge and future directions. *Frontiers in Plant Science, 8*, 1617.

Jangir, P., Shekhawat, P. K., Bishnoi, A., Ram, H., & Soni, P. (2021). Role of Serendipita indica in enhancing drought tolerance in crops. *Physiological and Molecular Plant Pathology, 116*, 101691.

Jha, S. K., Jain, P., & Sharma, H. P. (2015). Xenobiotic degradation by bacterial enzymes. *International Journal of Current Microbiology and Applied Sciences, 4*(6), 48–62.

Kageyama, S. A., Mandyam, K. G., & Jumpponen, A. (2008). Diversity, function and potential applications of the root-associated endophytes. In *Mycorrhiza* (pp. 29–57). Berlin, Heidelberg: Springer.

Kathpalia, R., & Bhatla, S. C. (2018). Plant mineral nutrition. In *Plant physiology, development and metabolism* (pp. 37–81). Singapore: Springer.

Kaviya, N., Upadhayay, V. K., Singh, J., Khan, A., Panwar, M., & Singh, A. V. (2019). Role of microorganisms in soil genesis and functions. In *Mycorrhizosphere and pedogenesis* (pp. 25–52). Singapore: Springer.

Kazemi, R., Ronaghi, A., Yasrebi, J., Ghasemi-Fasaei, R., & Zarei, M. (2019). Effect of shrimp waste–derived biochar and arbuscular mycorrhizal fungus on yield, antioxidant enzymes, and chemical composition of corn under salinity stress. *Journal of Soil Science and Plant Nutrition, 19*(4), 758–770.

Khatoon, Z., Huang, S., Rafique, M., Fakhar, A., Kamran, M. A., & Santoyo, G. (2020). Unlocking the potential of plant growth-promoting rhizobacteria on soil health and the sustainability of agricultural systems. *Journal of Environmental Management, 273*, 111118.

Kluber, L. A., Tinnesand, K. M., Caldwell, B. A., Dunham, S. M., Yarwood, R. R., Bottomley, P. J., & Myrold, D. D. (2010). Ectomycorrhizal mats alter forest soil biogeochemistry. *Soil Biology and Biochemistry, 42*(9), 1607–1613.

Kour, D., Rana, K. L., Kaur, T., Yadav, N., Halder, S. K., Yadav, A. N., … Saxena, A. K. (2020). Potassium solubilizing and mobilizing microbes: Biodiversity, mechanisms of solubilization, and biotechnological implication for alleviations of abiotic stress. In *New and future developments in microbial biotechnology and bioengineering* (pp. 177–202). Elsevier. https://doi.org/10.1016/B978-0-12-820526-6.00012-9

Liu, L., Zhou, C., Pei, X., Guo, L., Li, J., Wu, R., & Huang, D. (2020). Nitrogen deposition regulates the clonal growth of Leymus chinensis, a typical clonal plant in arid and semi-arid regions. *International Journal of Climate Change Strategies and Management.*

Looney, B. P., Meidl, P., Piatek, M. J., Miettinen, O., Martin, F. M., Matheny, P. B., & Labbé, J. L. (2018). Russulaceae: A new genomic dataset to study ecosystem function and evolutionary diversification of ectomycorrhizal fungi with their tree associates. *New Phytologist, 218*(1), 54–65.

Macik, M., Gryta, A., & Frac, M. (2020). Biofertilizers in agriculture: An overview on concepts, strategies and effects on soil microorganisms. *Advanced Agron, 162*, 31–87. https://doi.org/10.1016/bs.agron.2020.02.001

Mahala, D. M., Maheshwari, H. S., Yadav, R. K., Prabina, B. J., Bharti, A., Reddy, K. K., … Ramesh, A. (2020). Microbial transformation of nutrients in soil: An overview. *Rhizosphere Microbes*, 175–211.

Majewska, M. L., Rola, K., & Zubek, S. (2017). The growth and phosphorus acquisition of invasive plants Rudbeckia laciniata and Solidago gigantea are enhanced by arbuscular mycorrhizal fungi. *Mycorrhiza, 27*(2), 83–94.

Martin, F., & Nehls, U. (2009). Harnessing ectomycorrhizal genomics for ecological insights. *Current Opinion in Plant Biology, 12*(4), 508–515.

Mishra, B., Varjani, S., Agrawal, D. C., Mandal, S. K., Ngo, H. H., Taherzadeh, M. J., … Guo, W. (2020). Engineering biocatalytic material for the remediation of pollutants: A comprehensive review. *Environmental Technology & Innovation, 20*, 101063.

Mitter, E. K., Tosi, M., Obregón, D., Dunfield, K. E., & Germida, J. J. (2021). Rethinking crop nutrition in times of modern microbiology: Innovative biofertilizer technologies. *Frontiers in Sustainable Food Systems, 5*, 29. https://doi.org/10.3389/fsufs.2021.606815

Moënne-Loccoz, Y., Mavingui, P., Combes, C., Normand, P., & Steinberg, C. (2015). Microorganisms and biotic interactions. In *Environmental microbiology: Fundamentals and applications* (pp. 395–444). Dordrecht: Springer.

Morgan, J.Á., & Connolly, E.Á. (2013). Plant-soil interactions: nutrient uptake. *Nature Education Knowledge, 4*(8), 2.

Mosaddeghi, M. R., Hosseini, F., Hajabbasi, M. A., Sabzalian, M. R., & Sepehri, M. (2021). Epichloë spp. and Serendipita indica endophytic fungi: Functions in plant-soil relations. *Advances in Agronomy, 165*, 59–113.

Naeem, H., Ahmad, K. S., & Jaffri, S. B. (2020). Biotechnological tools based lithospheric management of toxic pyrethroid pesticides: A critical evaluation. *International Journal of Environmental Analytical Chemistry*, 1–24.

Narayan, O. P., Verma, N., Jogawat, A., Dua, M., & Johri, A. K. (2020). *Role of sulphate transporter (PiSulT) of endophytic fungus Serendipita indica in plant growth and development.* BioRxiv.

Netherway, T., Bengtsson, J., Krab, E. J., & Bahram, M. (2021). Biotic interactions with mycorrhizal systems as extended nutrient acquisition strategies shaping forest soil communities and functions. *Basic and Applied Ecology, 50*, 25–42.

Nogueira-Lopez, G., Greenwood, D. R., Middleditch, M., Winefield, C., Eaton, C., Steyaert, J. M., & Mendoza-Mendoza, A. (2018). The apoplastic secretome of Trichoderma virens during interaction with maize roots shows an inhibition of plant defence and scavenging oxidative stress secreted proteins. *Frontiers in Plant Science, 9*, 409.

Olanrewaju, O. S., Glick, B. R., & Babalola, O. O. (2017). Mechanisms of action of plant growth promoting bacteria. *World Journal of Microbiology & Biotechnology, 33*(11), 197. https://doi.org/10.1007/s11274-017-2364-9

Osvalde, A. (2011). Optimization of plant mineral nutrition revisited: The roles of plant requirements, nutrient interactions, and soil properties in fertilization management. *Environmental and Experimental Biology, 9*(3), 1–8.

Pankievicz, V., Irving, T. B., Maia, L. G., & Ané, J. M. (2019). Are we there yet? The long walk towards the development of efficient symbiotic associations between nitrogen-fixing bacteria and non-leguminous crops. *BMC Biology, 17*(1), 1–17.

Prakamhang, J., Tittabutr, P., Boonkerd, N., Teamtisong, K., Uchiumi, T., Abe, M., & Teaumroong, N. (2015). Proposed some interactions at molecular level of PGPR coinoculated with Bradyrhizobium diazoefficiens USDA110 and B. japonicum THA6 on soybean symbiosis and its potential of field application. *Applied Soil Ecology, 85*, 38–49. https://doi.org/10.1016/j.apsoil.2014.08.009

Prescott, C. E., & Vesterdal, L. (2021). Decomposition and transformations along the continuum from litter to soil organic matter in forest soils. *Forest Ecology and Management, 498*, 119522.

Priya, P., Aneesh, B., & Harikrishnan, K. (2021). Genomics as a potential tool to unravel the rhizosphere microbiome interactions on plant health. *Journal of Microbiological Methods, 185*, 106215.

Qiang, X., Ding, J., Lin, W., Li, Q., Xu, C., Zheng, Q., & Li, Y. (2019). Alleviation of the detrimental effect of water deficit on wheat (*Triticum aestivum* L.) growth by an indole acetic acid-producing endophytic fungus. *Plant and Soil, 439*(1), 373–391.

Rajwar, J., Joshi, D., Suyal, D. C., & Soni, R. (2021). Factors affecting soil ecosystem and productivity. In *Microbiological activity for soil and plant health management* (pp. 437–457). Singapore: Springer.

Rangasamy, K., Athiappan, M., Devarajan, N., Samykannu, G., Parray, J. A., Aruljothi, K. N., … Abd_Allah, E. F. (2018). Pesticide degrading natural multidrug resistance bacterial flora. *Microbial Pathogenesis, 114*, 304–310.

Rashid, M. I., Mujawar, L. H., Shahzad, T., Almeelbi, T., Ismail, I. M., & Oves, M. (2016). Bacteria and fungi can contribute to nutrients bioavailability and aggregate formation in degraded soils. *Microbiological Research, 183*, 26–41.

Rashid, T. S., Qadir, S. A., & Awla, H. K. (2021). Induction of defence related enzymes and biocontrol efficacy of Trichoderma harzianum in tomato plants infected with Fusarium oxysporum and Fusarium solani. *Acta Agriculturae Slovenica, 117*(1), 1–6.

Rebello, S., Nathan, V. K., Sindhu, R., Binod, P., Awasthi, M. K., & Pandey, A. (2021). Bioengineered microbes for soil health restoration: Present status and future. *Bioengineered, 12*(2), 12839–12853. https://doi.org/10.1080/21655979.2021.2004645

Sahu, A., Bhattacharjya, S., Mandal, A., Thakur, J. K., Atoliya, N., Sahu, N., … Patra, A. K. (2018). Microbes: A sustainable approach for enhancing nutrient availability in agricultural soils. In *Role of Rhizospheric microbes in soil* (pp. 47–75). Singapore: Springer.

Sandrini, M., Nerva, L., Sillo, F., Balestrini, R., Chitarra, W., & Zampieri, E. (2022). Abiotic stress and belowground microbiome: The potential of omics approaches. *International Journal of Molecular Sciences, 23*(3), 1091.

Sarkar, S., Dey, A., Kumar, V., Batiha, G. E. S., El-Esawi, M. A., Tomczyk, M., & Ray, P. (2021). Fungal endophyte: An interactive endosymbiont with the capability of modulating host physiology in myriad ways. *Frontiers in Plant Science, 12*.

Sarma, B. K., Yadav, S. K., Singh, S., & Singh, H. B. (2015). Microbial consortium-mediated plant defense against phytopathogens: Readdressing for enhancing efficacy. *Soil Biology and Biochemistry, 87*, 25–33.

Seenivasagan, R., & Babalola, O. O. (2021). Utilization of microbial consortia as biofertilizers and biopesticides for the production of feasible agricultural product. *Biology, 10*(11), 1111.

Sharma, S. B., Sayyed, R. Z., Trivedi, M. H., & Gobi, T. A. (2013). Phosphate solubilizing microbes: Sustainable approach for managing phosphorus deficiency in agricultural soils. *SpringerPlus, 2*(1), 1–14.

Shilev, S. (2020). Plant-growth-promoting bacteria mitigating soil salinity stress in plants. *Applied Sciences, 10*(20), 7326.

Siddiqui, K., Bawazeer, N., & Scaria Joy, S. (2014). Variation in macro and trace elements in progression of type 2 diabetes. *The Scientific World Journal*, 2014.
Singh, D. (2021). The role of soil microorganisms in plant mineral Nutrition.international. *Journal of Innovative Research in Science and Engineering*, 7(12), 114–124.
Skinder, M. B., Nabi, M., Gojree, B. A. S., Dar, G. H., & Ganai, B. A. (2022). *Endophytic microbes:Bioremediation of soil contaminants. Microbial consortium and biotransformation for pollution decontamination* (pp. 243–256). Elsevier. https://doi.org/10.1016/B978-0-323-91893-0.00002-X.
Sonowal, S., Ahmed, R., Chikkaputtaiah, C., Basar, E., & Velmurugan, N. (2022). A comprehensive characterization of culturable endophytic bacteria of Paris polyphylla and their potential role in microalgal growth in co-culture. *Applied Soil Ecology, 174*, 104410.
Soumare, A., Diedhiou, A. G., Thuita, M., Hafidi, M., Ouhdouch, Y., Gopalakrishnan, S., & Kouisni, L. (2020). Exploiting biological nitrogen fixation: A route towards a sustainable agriculture. *Plants, 9*(8), 1011.
Stuart, E. K., & Plett, K. L. (2020). Digging deeper: In search of the mechanisms of carbon and nitrogen exchange in ectomycorrhizal symbioses. *Frontiers in Plant Science, 1658*.
Sundari, S. K., Prakash, A., Yadav, P., & Kumari, A. (2019). Plant growth-promoting microbes as front-runners for on-site remediation of organophosphate pesticide residues in agriculture soils. In *Phyto and rhizo remediation* (pp. 249–285). Singapore: Springer.
Taylor, J. H., & Peterson, C. A. (2005). Ectomycorrhizal impacts on nutrient uptake pathways in woody roots. *New Forests, 30*(2), 203–214. https://doi.org/10.1007/s11056-005-0489-4
Tibbett, M., & Sanders, F. (2002). Ectomycorrhizal symbiosis can enhance plant nutrition through improved access to discrete organic nutrient patches of high resource quality. *Annals of Botany, 89*(6), 783–789.
Tripathi, M., & Gaur, R. (2021). Bioactivity of soil microorganisms for agriculture development. In *Microbes in land use change management* (pp. 197–220). Elsevier.
Turan, M., Ekinci, M., Yildirim, E., Güneş, A., Karagöz, K., Kotan, R., & Dursun, A. (2014). Plant growth-promoting rhizobacteria improved growth, nutrient, and hormone content of cabbage (*Brassica oleracea*) seedlings. *Turkish Journal of Agriculture and Forestry, 38*(3), 327–333.
Vaario, L. M., & Matsushita, N. (2021). Conservation of edible ectomycorrhizal mushrooms: Understanding of the ECM fungi mediated carbon and nitrogen movement within forest ecosystems. *Nitrogen in Agriculture-Physiological, Agricultural and Ecological Aspects*.
Van Der Heijden, M. G., Bardgett, R. D., & Van Straalen, N. M. (2008). The unseen majority: Soil microbes as drivers of plant diversity and productivity in terrestrial ecosystems. *Ecology Letters, 11*(3), 296–310.
Van Overbeek, L. S., & Saikkonen, K. (2016). Impact of bacterial–fungal interactions on the colonization of the endosphere. *Trends in Plant Science, 21*(3), 230–242.
Vergara, C., Araujo, K. E., Urquiaga, S., Schultz, N., Balieiro, F. D. C., Medeiros, P. S., … Zilli, J. E. (2017). Dark septate endophytic fungi help tomato to acquire nutrients from ground plant material. *Frontiers in Microbiology, 8*, 2437.
Waghunde, R. R., Shelake, R. M., Shinde, M. S., & Hayashi, H. (2017). Endophyte microbes: A weapon for plant health management. In *Microorganisms for green revolution* (pp. 303–325). Singapore: Springer.
Wen, Z., Xing, J., Liu, C., Zhu, X., Zhao, B., Dong, J., … Hong, L. (2022). The effects of ectomycorrhizal inoculation on survival and growth of Pinus thunbergii seedlings planted in saline soil. *Symbiosis*, 1–10.

Wołejko, E., Jabłońska-Trypuć, A., Wydro, U., Butarewicz, A., & Łozowicka, B. (2020). Soil biological activity as an indicator of soil pollution with pesticides—a review. *Applied Soil Ecology, 147*, 103356.

Xu, P., Christie, P., Liu, Y., Zhang, J., & Li, X. (2008). The arbuscular mycorrhizal fungus Glomus mosseae can enhance arsenic tolerance in Medicago truncatula by increasing plant phosphorus status and restricting arsenate uptake. *Environmental Pollution, 156*(1), 215—220. https://doi.org/10.1016/j.envpol.2008.01.003

Yousefi, S., Kartoolinejad, D., Bahmani, M., & Naghdi, R. (2017). Effect of Azospirillum lipoferum and Azotobacter chroococcum on germination and early growth of hopbush shrub (Dodonaea viscosa L.) under salinity stress. *Journal of Sustainable Forestry, 36*(2), 107—120.

CHAPTER TWELVE

Integrated nutrient management strategies for improving crop yield

Najeebul Tarfeen[a], Khair-Ul-Nisa[b], Shahnawaz Hassan[b], Auqib Manzoor[c], Humaira[a], Zahara Sultan[d]
[a]Centre of Research for Development, University of Kashmir, Srinagar, India
[b]Department of Environmental Sciences, University of Kashmir, Srinagar, India
[c]Department of Biochemistry, University of Kashmir, Srinagar, India
[d]Department of Botany, University of Kashmir, Srinagar, India

1. Introduction

With the unprecedented growth in the population and reduction in the accessible land and other yielding units, are putting tremendous strain on the current natural and agricultural resources to meet the increasing demand of food. In the developing world, attaining food security poses challenge and is highly crucial for reducing poverty. To overcome this challenge, producers of food crops tended to exploit chemical pesticides and fertilizers which in turn have started worsening the environment. To meet the sustainability needs and future security of food crops, production in the food must enhance substantially, while environmental effect of agriculture must reduce drastically at the same time (Foley et al., 2011). By 2050, food production must grow by 70% at global level to keep pace with rising demands (Bruinsma, 2009). To meet this target, an annual rise in average production of cereals of 43MMT/yr is required. It has been analyzed that there is an increase in the yield of cereals per unit area instead from crop harvested area.

Further enhancement in the crop yield poses challenge because of alterations in the global environmental conditions and related uncertainties (Wheeler & Von Braun, 2013). Besides food supply, there is a growing demand for energy, land area, water and wood products for infrastructure, urbanization, disposal of industrial and urban wastes (Lal, 2008). Agricultural sector being the main contributor to the emissions of CO_2-GHGs as well as non CO_2-GHGs. The increase in global emissions of these gases are

Sustainable Plant Nutrition
ISBN: 978-0-443-18675-2
https://doi.org/10.1016/B978-0-443-18675-2.00003-1

mainly caused by synthetic pesticide and fertilizer use in recent past and are swiftly rising in recent years. This excessive use of nitrogen fertilizer, is a main cause of environmental pollution such as GHG emissions and eutrophication (Davidson et al., 2014).

Agricultural techniques that enhance agricultural sustainability and quality of soil have received attention by farmers and researchers from both developing and developed nations. In plant nutrition, the role of organic manures is focusing the attention of soil scientists and agriculturists throughout the world. Despite the fact, that the fertilizers have increased the productivity of crops, but has also lead to the soil deterioration. Organic manures not only provide nutrients to the soil but also replenish organic matter content of soils. This would stress the need to employ organic manures only or in combination with chemical fertilizers so to retain fertility of soil for viable crop production. Therefore judicious application of fertilizers and organic manures are important to enhance the productivity of soil, replenish the soil health and input use efficiency. Combination of varied sources of nutrients has a significant impact on soil health as well as on sustained productivity. There is a need to implement the sustainable agricultural development in which agro-ecological methods, which have a part in mitigating the environmental impact as well as climate change role in conservation of resources are aimed at enhancing the production.

Integrated nutrient management (INM) has proved to improve the production of rice by managing the supply of nutrients and minimizing the nutrient loss to the surroundings thereby leading to improved resistance to environmental stresses, and high resource-use efficiency (Zhang et al., 2012) (Parkinson, 2013). INM can be regarded as an effective agricultural practice to enhance the quality of environment as well as to ensure food security worldwide, especially in developing countries.

2. Concept of INM

Integrated nutrient management (INM) is a method of amalgamation of traditional and modern practices of nutrient management into eco-friendly and cost effective farming system that utilizes the interests from all achievable sources of inorganic, organic and biological elements into wise, productive and integrated manner (Janssen, 1993). It optimizes all facets of nutrient cycling consisting of N, P and K and other micro and macronutrient inputs and outputs with the goal of concurring the nutrient requirement by the crop and discharge into surrounding environment.

Under INM techniques, the losses via emissions, runoff, leaching, volatilization and immobilization are curtailed, While enhanced nutrient use efficacy is attained (Zhang et al., 2012). Additionally, it tends to improve the soil conditions by optimizing its hydrological, biological, physical and chemical attributes to amplify farm output and lesser land degradation (Esilaba et al., 2005). Besides enhancing crop productivity, INM also conserves soil resources. It employs application of farm wastes, farmyard manures, crop residues, green manures, chemical and natural fertilizers, fallows, irrigation and crop rotations (Janssen, 1993). This practice also includes many techniques, like employment of inhibitors, deep installment of fertilizers that have been evolved to lessen the loss of nutrients and enhanced intake by plants (Zhang et al., 2012; Parkinson, 2013). Such strategies persuade farmers to focus on persistent program and have pronounced attention for ecological impacts, in lieu only focusing on yield-scaled financial gain.

A set of INM practices consists of various steps

(i) Determine availability of soil nutrients and insufficiency of nutrients in crop plants. There are two ways to determine the deficiency of nutrients, (a) Visual clues through the analysis of plant symptoms (b). Analysis of soil samples and post harvest tissue and its comparison with reference sample (healthy plant).

(ii) Assessment of limitations and opportunities in the present management practices related to soil fertility, and their relationship with diagnosis of nutrients like excessive or insufficient use of N fertilizers.

(iii) Determine the technologies and practices of farming that balance the nutrients which are important under varied soil types and climates.

(iv) Estimate the sustainability and productivity of INM methods, INM practices need farmers participation and appropriate technologies in the testing and analysis.

3. Main principles of INM

3.1 Utilizing all achievable sources to enhance the input

The aim of INM is to enhance the use of nutrients in the soil to improve the productivity of crops. In China, excessive use of Nitrogen enriched fertilizer associated with increased pollution results in the significant crop dependency on N derived from soil (Peng et al., 2002). Experimentations suggested that in intensive wheat-maize systems, the N supply provides approx. 274 kg N/ha/yr, which accounts for crop nitrogen absorption of 75% (Cui et al., 2010). Thus enhanced crop yields can be attained even without the employment of increased nitrogen rates (Peng et al., 2010). Elevated increase in the nitrate-N buildup in the soil profile seems to be a "N resource" that could burst off, conversely, it will be lost finally to the environment via leaching or denitrification (Ju et al., 2009). For crop growth, deposition of atmospheric nitrogen as well as water irrigation also provides a significant quantity of nitrogen. In china, the inputs of nitrogen from the deposition of atmospheric nitrogen and irrigated water were respectively up to 33 and 12 kg N/ha/yr; the quantity of both is likely to elevate continuously, making them paramount contributors to the environmental nitrogen (Ju et al., 2009). Hence, the nitrogen resources from water irrigation and deposition of atmospheric nitrogen can be believed as sources for important nutrient inputs for the INM approach.

3.2 Harmonizing the nutrient supply of soil with the demand of crop temporally and spatially

The amount of nutrient application and the timing to be in consistent with the requirements of crop nutrients which is necessary to attain maximum yield and enhance the efficiency of nutrient use (Cassman et al., 2002). Nitrogen fertilizers utilized during the time of crop demand at lesser quantities and with periodic application can decrease the nitrogen losses while enhancing the quantity and the yield of crop (Witt & Dobermann, 2004). Farmers employ large quantities of nitrogen fertilizer prior to planting. Mostly, 80% of total nitrogen fertilizer is utilized as a basal dressing, and the rest nitrogen is employed within 10 days after transferring the plant for rice cultivation or within the starting 30 days following the seedling date for winter wheat (Chen et al., 2011). This application pattern of nitrogen has lead to the poor harmony between the supply of soil nitrogen and the demand of crop, leading to the large quantity of inorganic nitrogen being useable in the soil prior to be needed for rapid growth of crop (Chen et al., 2006).

3.3 Diminishing the loss of nitrogen while enhancing the crop yield

Excessive use of nitrogen fertilizer leads to accelerated nitrate leaching into groundwater and enhanced emission loses to the surrounding environment i.e., atmosphere. The INM regulates the loss of nitrogen and its harmful impact on environment while attaining more crop productivity (Gruhn et al., 2000). The effect of nitrogen in a field is a combined outcome of nitrogen uptake, its immobilization and their residues in the soil and finally nitrogen emission to the surroundings, like run off, nitrogen leaching, ammonia volatilization, denitrification and NO_X release (Jambert et al., 1997). The fate of nitrogen is determined by various factors which includes soil characteristics, properties of crop, management methods and climatic conditions. Thus, INM promotes deep positioning of fertilizers such as ammonium carbonate or urea, which can notably enhance the nitrogen use efficiency with poor volatilization of ammonia and diminished leaching of nitrates (Jambert et al., 1997). The use of inhibitors of nitrification can also decrease the emissions of N_2O (Ma et al., 2010). Combination of both organic manure as well as other management strategies, like advancement of conservation tillage and establishment of crop residues, enhance the soil quality and diminish GHG emissions accompanying increased yield of crops (Huang & Sun, 2006) (Table12.1).

4. Components of INM

Main components of INM are (i) Use of crops that restore soil fertility like legumes in combination with green manures; (ii) Use of manures like slurry, biogas, vermicompost, biological composts, FYM compost and phosphate-compost. (iii) Use of biological agent. (iv) residual crop recycling (v) efficient genotypes (vi) optimal use of fertilizer as per need and target yields.

4.1 Green manures and legumes

Green manures and legumes can fix atmospheric nitrogen prior to the beginning of flowering (40–60 days growth) exception being soya-bean. After the harvest of crops, the remaining residual part of legumes contain 25–100 kg of nitrogen per hectare which is discharged at a steady rate. Additionally nitrogen, the grain mobilizes potassium and phosphorus that are less available in soil and can be recovered into system. GM pile up 100–200 kg of nitrogen per hectare almost in 50 days, among them 60%–80% is contributed by atmospheric fixation (Rao, 2014). The deep-rooted legumes have the capability to recycle the nutrients of soil to benefit the cereal crops (Maize, wheat, rice) in the cropping system, GM crops can meet nitrogen demand entirely under many situations more effectively than urea (Rao, 2014). Many studies demonstrated that in many rice–rice and rice-wheat systems, 50% nitrogen need of rice is supplied by GM crops, with significant effect on NPK and SOC status of soils in addition to the improvement in physical conditions of soil.

Table 12.1 Impact of integrated nutrient management in crop improvement.

Cotton	INM enhanced the uptake of NPK and sustain fertility status of soil.	(Marimuthu et al., 2014)
Soybean	INM reduced the chemical fertilizer rate by upto 50%, while enhancing Zn and S uptake for grain yield	(Chander et al., 2013)
Wheat	Fertilizer/manure application enhanced the grain yield.	(Nawab et al., 2011)
Rice	INM with green manure combination (100 kg N ha^{-1}) results to increased rice yields (3.9 t ha^{-1}).	(Balasubramaniyan, 2004)
Sorghum/Wheat	INM enhanced fodder and grain yields of sorghum by 9.4% and 18.5%. Nutrient uptake was also increased in sorghum. It also enhanced the yield of straw and grain in wheat.	(Patidar & Mali, 2001)
Cereal crops	INM enhanced N accumulation, cereal yield and dry matter	(Poutala et al., 1994)
Pearl millet	INM increased yield characters, millet yield and net returns	(Verma, 1996)
Maize/groundnut	INM enhanced pod yield of groundnut and yield of maize by 9.2% and 8.8% respectively.	(Palled et al., 2000)
Sunflower	INM enhanced the yield of seed by 22.1%–26.4% when 0.2% boron and azophos were applied.	(Paulpandi et al., 2008)

Table 12.1 Impact of integrated nutrient management in crop improvement.—cont'd

Hybrid maize	INM increased stover and grain yield.	(Shobana & Imyavaramban, 2008)
Potato/sunflower	Use of inorganic manures enhanced C, P and K content in soil and increased seed yield, phenology and growth of sunflower.	(Thind et al., 2007)
Fenugreek	INM enhanced yield components, physiological characters, straw and seed yield increased 25% and 71% over the control.	(Dubey et al., 2011)
Boro rice	INM increased the yield of boro rice and application of *Azotobacter* and PSB enhanced available N and P in soil.	(Garai et al., 2014)
Chickpea	INM enhanced the number of seeds per pod (1.2), pods per plant (66.4), grain yield (2,4t ha^{-1}).	(Patil et al., 2013)

4.2 Organic source of nutrients

Organic nutrient sources in cropping system comprises of growing of legumes, crop residues, biofertilizers, green manures and organic manures (compost, bio-compost, vermi-compost, phosphor-compost, oil cakes, press mud and FYM etc). Combined use of fertilizers and organic manures supports high yield of crops over long periods as compared to use of only fertilizers (Prasad, 2008). These results revealed that in intensive cropping, more than 25% of RDF is replaced by organic sources. Legumes and green manures when in combination in cropping system has an ability to meet nitrogen demand of rice crops more than 50%. Besides, introduction of organic manures as a supplementary, improves both plant health as well as soil fertility.

Biofertilizers (cellulolytic microbes, mineral solubilization and nitrogen-fixing) assists economizing use of fertilizer nutrients through the use of BNF systems, mobilizing less mobile nutrients from immovable elements and nutrient recycling from crop residues. Blending of these systems leads to sustainable and more stable production system (Meena, Mayura, Verma, 2014, Meena, Mayura, Bahadur, 2014).

4.3 Use of fertilizers

India follows China and USA and maintains third position in the world as producer and consumer of chemical fertilizers. On one side intensive agriculture has increased the production of food crops, while on the other side has created nutrient imbalance. Problems include, depletion of nutrients in soil approx 10 MT every year; lowering of water table and its water quality; rising deficiencies of micronutrients; diminishing organic content and finally affecting soil condition (Figs. 12.1 and 12.2).

5. Quality of soil and agricultural productivity

The efficiency of soil for crop production is dependent on soil environment that provides surroundings for root growth and root requires, water, air, nutrients and space to flourish. Characteristics of soil, like texture, depth, soil reaction, density and the potential to store water shows how roots will develop well. Alterations in the characteristics of soil affect the plant health. For instance, bulk density of soil affects the productivity of crops. When bulk density is increased above its critical level, roots will have difficulty to penetrate the soil, and thus restricting root growth. Erosion and loss

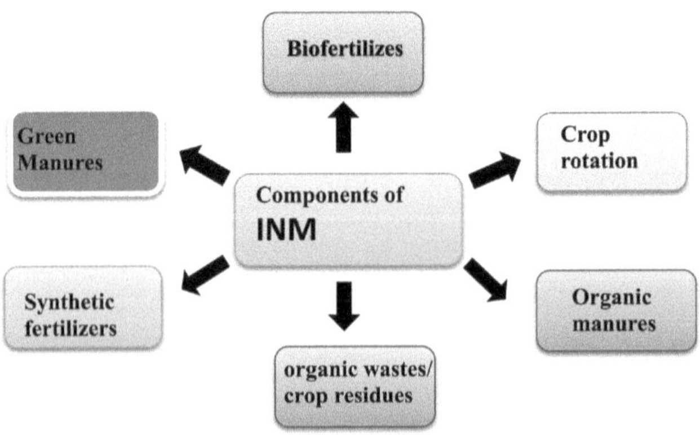

Fig. 12.1 The components of integrated nutrient management.

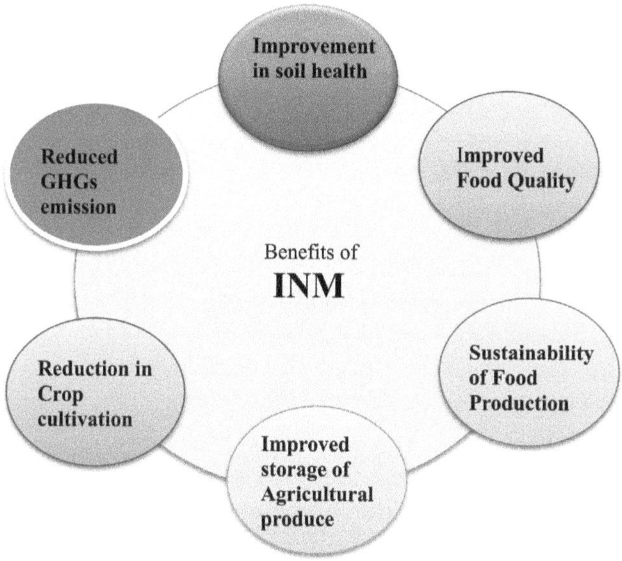

Fig. 12.2 Benefits of integrated nutrient management.

of organic matter from soil may result in enhancement in bulk density. These alterations in the quality of soil may affect the productivity as well as health of plant and can lead to the low crop yield or higher production costs.

6. Impact of INM on physical properties of soil

Physical characteristics of soil are related with organic matter and soil organic carbon, thus, any management strategies of soil increases organic matter of soil has direct bearing on microbial biomass as well as physical characteristics, for this collective use of inorganic and organic nutrient sources might be the right proposition for this soils, mainly for the enhancement of physical health of soil. Notable improvement in the physical characteristics of soil was found by many researchers under integrated use of inorganic fertilizers and organic manures. The combination of NPK fertilizers and biofertilizers, lime and manure enhanced WSA, SOC, rate of soil infiltration and moisture-retention capacity. Integration of organic either in the form of organic manure, crop residue has a significant impact of BD of agricultural soils, infiltration rate, soil aggregation and soil structure. The WSA, BD, SOC, WHC and fertility condition of soil bettered by combination of inorganic and organics (Hasanain et al.), microbial biomass carbon and organic carbon enhanced in the treatment by the usage of green manures, organic manures and biofertilizers in combination with inorganic fertilizers (Singh, 2010; Bhattacharyya et al., 2007).

7. Impact of INM on fertility of soil

Day by day, agricultural soil fertility is declining because of the disproportionate use of fertilizers, a notable enhancement in the condition of soil fertility was observed by many researchers (Bhattacharyya et al., 2007; Jat et al., 2013; Mahanta et al., 2013). By the use of organic manure with RDF, SOC of the soil gets enhanced, this combination affected the development, growth and crop productivity (Jat et al., 2013; Meena, Mayura, Bohra, et al., 2013, Meena, Mayura, Verma, et al., 2013).

Various studies revealed that the INM increases the crop yield above achievable levels by the application of fertilizers (Liang et al., 2014; Kannan et al., 2013; Mahanta et al., 2013; Hemmat et al., 2010) and results in better concurrence of requirement of crop nitrogen because of (1) decreased losses of nitrogen via nitrate leaching and denitrification (2) Gradual mineralization of organics (3) Enhancement in health conditions and productivity of soil(4) Improvement in the efficiency of nutrient use and restoration by crops and hence could support increased crop yield in many cropping systems assuring persistent supportability of the system (Aulakh, 2010; Yaduvanshi et al., 2013).

Judicious use of organic manure as well as mineral fertilizers in combination with micronutrients and biofertilizers supply with highest nitrogen, potassium and phosphate in the soil in comparison with other treatments (Jat et al., 2013; Meena, Mayura, Bohra, et al., 2013, Meena, Mayura, Verma, et al., 2013). Studies by Kusro et al. (2014) suggested that the NH_4^+ $-N$, mineralisable nitrogen and organic carbon showed significant enhancement over control. Introduction of BGA, GM and FYM via an inorganic source in the treatment enhanced the mineralisable NH_4^+ $-N$, nitrogen, organic carbon and lowered the bulk density (Kumar et al., 2012; Rahman et al., 2013).

8. Impact of biofertilizer on the productivity of crops

Application of biofertilizer to soil, plant surfaces, soil, colonizes the interior or rhizosphere of plant, facilitates growth by raising the supply or accessibility of macronutrients to the host plants such as in apple (increase in soil health and crop yield) (Mosa et al., 2015), soyabean (Nutrient balance, soil fertility, productivity) (Akbari et al., 2011), Mung bean (soil fertility and crop productivity) (Rana et al., 2011), and Lentil (soil health) (Bhavana et al., 2014). Application of biofertilizers is among the components of INM, as they are renewable and cost effective to enhance the mineral fertilizers for sustainable agriculture. Various microbes in combination with crops are being used in the biofertilizers. Various microorganisms used as biofertilizers are; *Azospirillum, Azolla, Rhizobium, Azotobacter*, Cyanobacteria, Potassium and phosphate solubilizing microbes (Meena, Mayura, Bohra, et al., 2013; Meena, Mayura, Verma, et al., 2013; Meena, Mayura, Verma, 2014; Meena, Mayura, Bahadur, 2014; Verma et al., 2014). Rhizobacteria for promoting growth of plants, bacteria for solubilizing bacteria and these are also employable as liquid biofertilizers (Verma et al., 2014; Zhang & Kong, 2014; Parewa et al., 2014).

9. Methods for the further advancement of INM

There are various beneficial effects of INM practices, that are advantageous for both farmers as well as for environment. While observing various research reports, here we have presented some ways and recent opportunities that can be analyzed and further improved by adjustment and alteration in the adoption of INM methods that are site -specific. For future advancement of INM various points are as (i) Amalgamation of both plant and soil analysis (ii) Adaptation to the local environmental circumstances (iii) Rainwater harvesting technologies and conservation tillage (iv) Organic nutrient recycling (v) New innovative technologies.

10. Conclusion

Integrated nutrient management are using many nutrient sources to enhance the nutrient input, temporal as well as spatial matching of nutrient supply of soil with crop demand and minimizing the loss of nitrogen while improving the yield of crop. Various experimentations have revealed that INM results in the considerable increase in the yield of crop while decreasing the loss of nitrogen and emission of GHGs. The overuse of fertilizers (chemical) in combination with low resource-use efficiency causes environmental pollution; the nutrient inputs in the form of organic fertilizers will be essential for the growth and development of the plant and poses less threat to environment which should be the prime concern for INM practices.

INM provides an opportunity to enhance the production of crops as well as simultaneously decrease the environmental impact. The INM methodology should be site specific and also attuned to local surroundings.

References

Akbari, P., Ghalavand, A., Sanavy, A. M., & Alikhani, M. A. (2011). The effect of biofertilizers, nitrogen fertilizer and farmyard manure on grain yield and seed quality of sunflower (Helianthus annus L.). *Journal of Agricultural Technology, 7*(1), 173–184.

Aulakh, M. (2010). *Integrated nutrient management for sustainable crop production, improving crop quality and soil health, and minimizing environmental pollution.* Paper presented at the 19th world congress of soil science, soil solutions for a changing world.

Balasubramaniyan, P. (2004). Yield potential of fine-grain rice (Oryza sativa) under integrated nutrient management. *Indian Journal of Agronomy, 49*(3), 157–159.

Bhattacharyya, R., Chandra, S., Singh, R., Kundu, S., Srivastva, A., & Gupta, H. (2007). Long-term farmyard manure application effects on properties of a silty clay loam soil under irrigated wheat–soybean rotation. *Soil and Tillage Research, 94*(2), 386–396.

Bhavana, G., Gabhane, V., Nagdeve, M., Sonune, B., & Ganvir, M. (2014). Effect of integrated nutrient management on soil fertility, nutrient balance, productivity and economics of soybean in an Inceptisol of semi arid region of Maharashtra. *Karnataka Journal of Agricultural Sciences, 27*(3), 303–307.

Bruinsma, J. (2009). *The resource outlook to 2050: By how much do land, water and crop yields need to increase by 2050.* Paper presented at the Expert meeting on how to feed the world in.

Cassman, K. G., Dobermann, A., & Walters, D. T. (2002). Agroecosystems, nitrogen-use efficiency, and nitrogen management. *AMBIO: A Journal of the Human Environment, 31*(2), 132–140.

Chander, G., Wani, S. P., Sahrawat, K. L., Pal, C., & Mathur, T. (2013). Integrated plant genetic and balanced nutrient management enhances crop and water productivity of rainfed production systems in Rajasthan, India. *Communications in Soil Science and Plant Analysis, 44*(22), 3456–3464.

Chen, X.-P., Cui, Z.-L., Vitousek, P. M., Cassman, K. G., Matson, P. A., Bai, J.-S., ... Römheld, V. (2011). Integrated soil–crop system management for food security. *Proceedings of the National Academy of Sciences, 108*(16), 6399–6404.

Chen, X., Zhang, F., Römheld, V., Horlacher, D., Schulz, R., Böning-Zilkens, M., ... Claupein, W. (2006). Synchronizing N supply from soil and fertilizer and N demand of winter wheat by an improved N min method. *Nutrient Cycling in Agroecosystems, 74*(2), 91–98.

Cui, Z., Chen, X., & Zhang, F. (2010). Current nitrogen management status and measures to improve the intensive wheat–maize system in China. *Ambio, 39*(5), 376–384.

Davidson, E., Galloway, J., Millar, N., & Leach, A. (2014). N-related greenhouse gases in North America: Innovations for a sustainable future. *Current Opinion in Environmental Sustainability, 9*, 1–8.

Dubey, P. K., Pandey, C., & Bhardwaj, S. (2011). Effect of integrated nutrient management on growth, yield attributes and yield of fenugreek (trigonella foenum-graecum L.). *International Journal of Agricultural Statistics and Science, 7*(1), 327–339.

Esilaba, A. O., Byalebeka, J., Delve, R. J., Okalebo, J., Ssenyange, D., Mbalule, M., & Ssali, H. (2005). On farm testing of integrated nutrient management strategies in eastern Uganda. *Agricultural Systems, 86*(2), 144–165.

Foley, J. A., Ramankutty, N., Brauman, K. A., Cassidy, E. S., Gerber, J. S., Johnston, M., ... West, P. C. (2011). Solutions for a cultivated planet. *Nature, 478*(7369), 337–342.

Garai, T. K., Datta, J. K., & Mondal, N. K. (2014). Evaluation of integrated nutrient management on boro rice in alluvial soil and its impacts upon growth, yield attributes, yield and soil nutrient status. *Archives of Agronomy and Soil Science, 60*(1), 1–14.

Gruhn, P., Goletti, F., & Yudelman, M. (2000). *Integrated nutrient management, soil fertility, and sustainable agriculture: Current issues and future challenges.* Intl Food Policy Res Inst.

Hasanain, M., Fatima, A., & Yadav, R. Integrated nutrient management for improve soil health.

Hemmat, A., Aghilinategh, N., Rezainejad, Y., & Sadeghi, M. (2010). Long-term impacts of municipal solid waste compost, sewage sludge and farmyard manure application on organic carbon, bulk density and consistency limits of a calcareous soil in central Iran. *Soil and Tillage Research, 108*(1–2), 43–50.

Huang, Y., & Sun, W. (2006). Changes in topsoil organic carbon of croplands in mainland China over the last two decades. *Chinese Science Bulletin, 51*(15), 1785–1803.

Jambert, C., Serça, D., & Delmas, R. (1997). Quantification of N-losses as NH_3, NO, and N_2O and N_2 from fertilized maize fields in southwestern France. *Nutrient Cycling in Agroecosystems, 48*(1), 91–104.

Janssen, B. (1993). Integrated nutrient management: The use of organic and mineral fertilizers. In *The role of plant nutrients for sustainable food crop production in sub-Saharan Africa* (pp. 89–105). Ver. Kunstmest Producenten.

Jat, L. K., Singh, S., Latare, A., Singh, R., & Patel, C. (2013). Effect of dates of sowing and fertilizer on growth and yield of wheat (*Triticum aestivum*) in an Inceptisol of Varanasi. *Indian Journal of Agronomy, 58*(4), 611–614.

Ju, X.-T., Xing, G.-X., Chen, X.-P., Zhang, S.-L., Zhang, L.-J., Liu, X.-J., … Zhu, Z.-L. (2009). Reducing environmental risk by improving N management in intensive Chinese agricultural systems. *Proceedings of the National Academy of Sciences, 106*(9), 3041–3046.

Kannan, R. L., Dhivya, M., Abinaya, D., Krishna, R. L., & Krishnakumar, S. (2013). Effect of integrated nutrient management on soil fertility and productivity in maize. *Bulletin of Environment, Pharmacology and Life Sciences, 2*(8), 61–67.

Kumar, M., Baishaya, L., Ghosh, D., Gupta, V., Dubey, S., Das, A., & Patel, D. (2012). Productivity and soil health of potato (Solanum tuberosum L.) field as influenced by organic manures, inorganic fertilizers and biofertilizers under high altitudes of eastern Himalayas. *Journal of Agricultural Science, 4*(5), 223.

Kusro, P., Singh, D., Paikra, M., Kumar, D., & Viashwavidyalaya, I. G. K. (2014). Effect of organic and inorganic additions on physico-chemical properties in Vertisol. *American International Journal of Research Formal, Applied & Natural Science, 5*(1), 51–53.

Lal, R. (2008). Soils and sustainable agriculture. A review. *Agronomy for Sustainable Development, 28*(1), 57–64.

Liang, Q., Chen, H., Gong, Y., Yang, H., Fan, M., & Kuzyakov, Y. (2014). Effects of 15 years of manure and mineral fertilizers on enzyme activities in particle-size fractions in a North China Plain soil. *European Journal of Soil Biology, 60*, 112–119.

Mahanta, D., Bhattacharyya, R., Gopinath, K., Tuti, M., Jeevanandan, K., Chandrashekara, C., … Mishra, P. (2013). Influence of farmyard manure application and mineral fertilization on yield sustainability, carbon sequestration potential and soil property of gardenpea–French bean cropping system in the Indian Himalayas. *Scientia Horticulturae, 164*, 414–427.

Ma, L., Ma, W., Velthof, G., Wang, F., Qin, W., Zhang, F., & Oenema, O. (2010). Modeling nutrient flows in the food chain of China. *Journal of Environmental Quality, 39*(4), 1279–1289.

Marimuthu, S., Surendran, U., & Subbian, P. (2014). Productivity, nutrient uptake and post-harvest soil fertility as influenced by cotton-based cropping system with integrated nutrient management practices in semi-arid tropics. *Archives of Agronomy and Soil Science, 60*(1), 87–101.

Meena, V., Maurya, B., & Bahadur, I. (2014). Potassium solubilization by bacterial strain in waste mica. *Bangladesh Journal of Botany, 43*(2), 235–237.

Meena, V., Maurya, B., Bohra, J., Verma, R., & Meena, M. (2013). Effect of concentrate manure and nutrient levels on enzymatic activities and microbial population under submerged rice in alluvium soil of Varanasi. *Crop Research, 45*(1to3), 6–12.

Meena, V. S., Maurya, B., & Verma, J. P. (2014). Does a rhizospheric microorganism enhance K+ availability in agricultural soils? *Microbiological Research, 169*(5–6), 337–347.

Meena, V., Maurya, B., Verma, R., Meena, R., Jatav, G., Meena, S. K., ... Meena, S. (2013). Soil microbial population and selected enzyme activities as influenced by concentrate manure and inorganic fertilizer in alluvium soil of Varanasi. *Bioscan, 8*(3), 931–935.

Mosa, W. F. A. E.-G., Paszt, L. S., Frąc, M., & Trzciński, P. (2015). The role of biofertilization in improving apple productivity–a review. *Advances in Microbiology, 5*(01), 21.

Nawab, K., Amanullah, S. P., Rab, A., Arif, M., Khan, M. A., Mateen, A., & Munsif, F. (2011). Impact of integrated nutrient management on growth and grain yield of wheat under irrigated cropping system. *Pakitan Journal of Botany, 4*, 1943–1947.

Palled, Y., Desai, B., & Prabhakar, A. (2000). Integrated nutrient management in alley cropped maize (Zea mays)-groundnut (*Arachis hypogaea*) system with subabul (Leucaena leucocephala). *Indian Journal of Agronomy, 45*(3), 520–525.

Parewa, H. P., Yadav, J., Rakshit, A., Meena, V., & Karthikeyan, N. (2014). Plant growth promoting rhizobacteria enhance growth and nutrient uptake of crops. *Agricultural Sustainable Development, 2*(2), 101–116.

Parkinson, R. (2013). In B. Gangwar, & V. K. Singh (Eds.), *System based integrated nutrient management* (p. 371). New Delhi, India: Publishing Agency, Wiley Online Library. US $90; INR2250. Hardback (ISBN 978-93-81450-05-5) www.nipabooks.com

Patidar, M., & Mali, A. (2001). Integrated nutrient management in sorghum (Sorghum bicolor) and its residual effect on wheat (*Triticum aestivum*). *The Indian Journal of Agricultural Sciences, 71*(9).

Patil, S., Halikatti, S., Gurumurthy, S., & Lokesh, M. (2013). Impact of weather on performance of chickpea (Cicer arietinum L.) with integrated organic nutrient management practices grown in vertisol of northern dry zone of Karnataka. *Research on Crops, 14*(3), 777–785.

Paulpandi, V., Ganesaraja, V., & Kavitha, M. (2008). Studies on the effect of insitu moisture conservation and integrated nutrient management on growth and yield of sunflower (Helianthus annus L.) in rainfed vertisol. *Legume Research-An International Journal, 31*(3), 174–178.

Peng, S., Buresh, R. J., Huang, J., Zhong, X., Zou, Y., Yang, J., ... Tang, Q. (2010). Improving nitrogen fertilization in rice by sitespecific N management. A review. *Agronomy for Sustainable Development, 30*(3), 649–656.

Peng, S.-B., Huang, J.-L., Zhong, X.-H., Yang, J.-C., Wang, G.-H., Zou, Y.-B., ... Witt, C. (2002). Challenge and opportunity in improving fertilizer-nitrogen use efficiency of irrigated rice in China. *Agricultural Sciences in China, 1*(7), 776–785.

Poutala, R., Kuoppamäki, O., Korva, J., & Varis, E. (1994). The performance of ecological, integrated and conventional nutrient management systems in cereal cropping in Finland. *Field Crops Research, 37*(1), 3–10.

Prasad, R. (2008). Integrated plant nutrient supply system (IPNS) for sustainable agriculture. *Indian Journal of Fertilizer, 4*(12), 71–90.

Rahman, M., Islam, M., Jahiruddin, M., Rafii, M., Hanafi, M., & Malek, M. (2013). Integrated nutrient management in maize-legume-rice cropping pattern and its impact on soil fertility. *Journal of Food, Agriculture & Environment, 11*(1), 648–652.

Rana, M., Chowdhury, A., & Bhuiya, M. (2011). Effects of plant population and biofertilizer on the growth parameters of three summer mungbean (Vigna radiata l.) cultivars. *Bangladesh Journal of Agricultural Research, 36*(3), 537–542.

Rao, D. (2014). *Recent advances in biological nitrogen fixation in agricultural systems*. Paper presented at the Proc Indian Natl Sci Acad.

Shobana, R., & Imyavaramban, V. (2008). Integrated nutrient management in hybrid maize (Zea mays L.) cv. Pionner 30V 92. *Plant Archives, 8*(1), 433–434.

Singh, A. K. (2010). *Soil quality parameters as influenced by management practices in rice-wheat and maize-wheat cropping systems*. Brisbane, Australia: Paper presented at the Proceedings of the 19th World Congress of Soil Science: Soil solutions for a changing world.

Thind, S., Sidhu, A., Sekhon, N., & Hira, G. (2007). Integrated nutrient management for sustainable crop production in potato-sunflower sequence. *Journal of Sustainable Agriculture, 29*(4), 173–188.

Verma, O. (1996). Integrated nutrient management in pearlmillet (Pennisetum glaucum) under rainfed conditions. *Indian Journal of Agronomy, 41*(1), 58–60.

Verma, J. P., Yadav, J., Tiwari, K. N., & Jaiswal, D. K. (2014). Evaluation of plant growth promoting activities of microbial strains and their effect on growth and yield of chickpea (Cicer arietinum L.) in India. *Soil Biology and Biochemistry, 70*, 33–37.

Wheeler, T., & Von Braun, J. (2013). Climate change impacts on global food security. *Science, 341*(6145), 508–513.

Witt, C., & Dobermann, A. (2004). 17 toward a decision support system for site-specific nutrient management. *Increasing Productivity of Intensive Rice Systems Through Site Rice Systems Through Site-specific Nutrient Management, 359*.

Yaduvanshi, N., Sharma, D., & Swarup, A. (2013). Impact of integrated nutrient management on soil properties and yield of rice and wheat in a long-term experiment on a reclaimed sodic soil. *Journal of the Indian Society of Soil Science, 61*(3), 188–194.

Zhang, F., Cui, Z., Chen, X., Ju, X., Shen, J., Chen, Q., … Fan, M. (2012). Integrated nutrient management for food security and environmental quality in China. *Advances in Agronomy, 116*, 1–40.

Zhang, C., & Kong, F. (2014). Isolation and identification of potassium-solubilizing bacteria from tobacco rhizospheric soil and their effect on tobacco plants. *Applied Soil Ecology, 82*, 18–25.

CHAPTER THIRTEEN

Use of nano-fertilizers to improve the nutrient use efficiencies in plants

Qurat ul Ain[a,b], Hafiz Athar Hussain[a,b], Qingwen Zhang[a], Ayesha Rasheed[c], Asma Imran[b], Saddam Hussain[d], Namrah Ahmad[e], Huzaima Bibi[f], Komal Shoukat Ali[g]

[a]Institute of Environment and Sustainable Development in Agriculture, Chinese Academy of Agricultural Sciences, Beijing, China
[b]Division of Soil and Environmental Biotechnology, National Institute for Biotechnology and Genetic Engineering (NIBGE), Faisalabad, Pakistan
[c]Center of Agricultural Biochemistry and Biotechnology, University of Agriculture, Faisalabad, Pakistan
[d]Department of Agronomy, University of Agriculture, Faisalabad, Pakistan
[e]Department of Plant Biotechnology, Atta-ur-Rehaman School of Applied Biosciences, National University of Sciences and Technology, Islamabad, Pakistan
[f]Department of Biological Sciences, International Islamic University, Islamabad, Pakistan
[g]Department of Botany, Government College Women University Sialkot, Sialkot, Pakistan

1. Introduction

The application of fertilizer in the crop fields is the common practice for increasing the crop production to fulfill the global demand of food (Chhipa, 2017). However, the nutrient use efficiency is a significant parameter for evaluating the crop production that depends on the fertilizer management. The need of fertilizers is increasing for better yield as Mittal et al. (2020) assessed that the total consumption of fertilizers is expected to rise 263 Mt in 2050. Further, Fatima et al. (2021) stated that the fertilizer application 13 times increased from 15 to 194 million tons between 1950 and 2020. Meanwhile, excess use of mineral fertilizers is unsafe due to increase the pollution issues. However, conventional fertilizers do not provide the required results because most nutrients do not reach at the target areas due to the loss by evaporation, leaching, volatilization, erosion, and runoff or nutrients photolytic deprivation (Manjunatha et al., 2016). Nevertheless, nutrient use efficiency (NUE) has been considered the serious issue of lower soil fertility and reduction of crops yield. So, the improvement of NUE is needed for better cropping systems, supporting the sustainable agricultural production systems and increasing soil quality components (Mahanta et al., 2019).

Sustainable Plant Nutrition
ISBN: 978-0-443-18675-2
https://doi.org/10.1016/B978-0-443-18675-2.00013-4

The use of nanotechnology in the agricultural sectors could reduce the nutrient losses in fertilization and promote yields through the optimized nutrient management. The "nano" is the Greek word which means "dwarf" or one billionth part of a meter and particles with less than 100 nm are considered as nanoparticles (Thakkar et al., 2010). Nano-sized fertilizer has the potential to upraise the agricultural yield by increased nutrients absorption, uptake and translocation in plants due to its unique chemical characteristics. Like, it is stated that the nano-fertilizers are consist on nanoparticles that use nanotechnology for the enhancement of nutrients use efficiency (Khalid et al., 2022). The nano-fertilizers are the nanomaterials which are responsible for providing the one or more type of nutrients to the plants, and support their growth and production (Liu & Lal, 2015). Moreover, Fatima et al. (2021) explain that the nano-fertilizers offers the unique possibility of development in nutrients use efficiency of plants with high-absorption rate, and utilization efficacy and consequently have minimum losses that facilitate the nutrients uptake in plants. Furthermore, nano-fertilizers might be effective alternative to dissolvable inorganic fertilizers due slow release of discharging supplements during crop production (El-Saadony et al., 2021). To absorb the nutrients from the soil, the plants response depends on the morphology and synthesis characteristics of different nanoparticles (NPs). However, the modern nano-sized fertilizer improves the availability of insoluble nutrients and ability to reduce nutrient losses (Prasad et al., 2017). In addition, Zulfiqar et al. (2019) explained that the synthesis of nano-materials with different engineering approaches encapsulated the required nutrients to increase plant uptake efficiency. Compared with conventional water soluble fertilizers, nano-fertilizers showed excellent transport characteristics by plant cells and tissues with controlled mobility (Al-Mamun et al., 2021). So, in light of discussed studies, nano-fertilizer is recommended due to its beneficial effects on nutrient use efficiency (NUE) as well as for crop yield. But the relative economics of nano-fertilizer vis traditional chemical fertilizers and their integrated systems needs the detailed study to pass the meaningful suggestions for the farming communities.

2. Importance of nano-fertilizers in agriculture

The application of nano-sized fertilizer is novel technique toward the modern agriculture as it's novel practice for better crop management and higher production (Mittal et al., 2020). Nano-fertilizers are effective in reduction of the nutrient leaching and volatilization losses, as this formulation may allow selective release linked with time and environmental conditions and may synchronize the release of nutrients with the uptake by crop plants. Further, nano-nutrient fertilizers exhibit high sorption capacity, surface area and control release of nutrients to targeted sites, hence these can be considered as smart nutrient delivery system (Rameshaiah et al., 2015). Therefore, these nutrient formulations may act as possible alternatives to conventional fertilizers ensuring slow and control release of nutrients. Nano-fertilizers increase the nutrient use efficiency, provide better yield and may help in reducing the soil pollution due to overapplication of fertilizers (Naderi & Danesh-Shahraki, 2013). These fertilizer formulations are anticipated to improve the crop growth and development due to its greater absorbance, high reactivity and ability to directly enter the cell through cell wall pores. Previous studies described that the nanofertilizers (NFs) improve the seed germination and the seedling growth because NFs may penetrate into the seed easily and enhance availability of nutrients to the seedling but excessive dose can inhibit the growth of plants (Sharma et al., 2021). Different types of nano-fertilizers can be produced by using different carrier materials e.g., zeolite, hydroxyapatite NPs, mesoporous silica NPs, nitrogen, carbon, zinc, copper, silica, and polymeric NPs (Chaitra et al., 2021). Overall, the nano-fertilizer have been reported in promoted the growth and productivity of agricultural field crops e.g., nano chitosan-based NPK fertilizer have been reported to increase the total sugar contents of wheat grown on clay, sandy and clay—sandy soils (Abdel-Aziz et al., 2016). Furthermore, it is reported that ZnO-NPs promoted the germination and its seedling vigor of peanut plants by improving the leaf chlorophyll contents (Prasad et al., 2012). Similarly (Salama, 2012), stated that the Ag-NPs significantly improve the shoot and root growth, leaf surface area, chlorophyll contents, protein and carbohydrates. Nevertheless, different NPs e.g., Ag-NPs, TiO_2NPs, Si-NPS, Fe_3O_4 have found to increase the plant growth by regulating various responses in wheat, spinach, soybean and maize plants (Hong et al., 2005; Iannone et al., 2016; Suriyaprabha et al., 2012; Yang et al., 2006). Additionally, the foliar spray of nano-fertilizer may lead to increase nutrients use efficiency (NUE) and gives rapid response to the growth of the plants (Mahil & Kumar, 2019). The nano-sized fertilizers could have substantial effects on the soil as toxicity and accumulation and they are extremely resilient to degradation, decomposability in soil and potential to accumulate in soil owing to their biochemical compositions (Rajput et al., 2019). In short, the nano sized fertilizers are getting significance for sustainable agriculture in improvement of crop production, increasing high infiltration rate, nutrient use efficiency (NUE), stress resistance and decrease the wastage of inorganic or chemical fertilizers as well as the cost of cultivation practices.

3. Nano-fertilizers: a better alternative to conventional fertilizers

The development of novel nano-agrochemicals might be interested among the possible nanotechnology applications in agriculture (El-Saadony et al., 2021). However, to increasing the food production with minimum adverse environmental effects, it's essential to developed the new type fertilizers that release the nutrients according to the plant demand. For this, nano-fertilizers could be effective choice that improve the nutrients use efficiency of plants (Mahanta et al., 2019). Moreover, several studies showed that the nano-materials based fertilizer can be more effective than the conventional forms of the fertilizers (as shown in Table 13.1) because controlled release of the nutrients not only enhance the efficiency of plants

Table 13.1 General comparison of Conventional/Chemical fertilizer and nano-fertilizer.

Quality parameters	Nano-fertilizer	Conventional/Chemical fertilizer
Nutrient's solubility	Higher	Lower
Time duration of release	Extended effective duration for release	Used at time of application and remaining changed to insoluble form
Nutrient's losses	Less nutrients losses rate	High nutrients losses rate
Releasing type	Controlled release	Excess release leading to toxicity and soil imbalance
Nutrients use efficiency	Higher	Lower
Nutrient's bioavailability	Higher	Lower
Soil fixation	Less	High
Dispersal of nutrients	Uptake insoluble nutrients as in nano-sized	Low solubility due to bulk sized
Affected surface area	Higher	Lower

nutrients uptake but also potentially decrease the adverse environmental affects related with the loss of nutrients (Al-Mamun et al., 2021). Nevertheless, nano-fertilizers are valuable over the conventional fertilizers as they can improve the soil fertility, yield and the quality parameters of the crop. In addition, optimum dose of nano-fertilizer is beneficial for crops but the extensive usage in agricultural fields may cause toxicity and biocompatibility. Furthermore, Replacement of conventional fertilizers by nano-fertilizers might be influential attempt toward the sustainable agriculture due to its higher infiltration capacity, nano-size, and very high surface area.

In addition, conventional fertilizers are not advantageous in some ways i.e., the organic matters that collected from different plants and animals are not similar that cannot deliver

the same nutrients and organic matter as fertilizer (Toksha et al., 2021). However, N, P and K based chemical fertilizer can reach to groundwater and increase the toxicity, also cause soil and air pollution (Savci, 2012). Moreover, the chemical fertilizers (nitrogen, phosphate and potassium-based) are type of a synthetic compound substance which produced for maximizing the crop yield. Under the effect of continuous use of chemical fertilizers, reduced soil fertility and crops productivity is reported which may be due to lower nutrients use efficiency (Li & Wu, 2008; Sharma et al., 2014). Over dose of P normally fixed in soil and becomes unavailable for plants uptake by forming the chemical bonds with other the nutrients. Previous studies reported that the key mineral nutrients applied to the soil often lost by various ways e.g., N 40%–70%, P 80%–90% and K 50%–90% respectively (Raliya et al., 2017; Zulfiqar et al., 2019).

Furthermore, nano-fertilizers are considered as slow-release fertilizer to overcome the fluctuations i.e., soil acidity, moisture and temperature to enhance plant growth (Vega-Vásquez et al., 2020). Nevertheless, Chhipa and Joshi (2016) stated that the nano-fertilizer has great potential to improve nutrients use efficiency, minimize the cost and reduce environmental deterioration. As compared with Chemical fertilizers, nano-fertilizer have been found to significantly improve the plants growth i.e., plant height, leaf area, number of leaves and biomass by increasing the photosynthetic pigments and translocation photosynthetic products to different parts of the plant (Yomso & Menon, 2021).

Nevertheless, eco-friendly approaches like green or biological synthesis of NPs are good alternatives to the chemical fertilizers. The biological synthesis of NPs can be from fungi, bacteria and the plants which open the new opportunities for the synthesis of inorganic nanoparticles as eco-friendly fertilizers. So, the nano-fertilizers considered as the vital tool in the new age agriculture sector for better crop growth and enhance the nutrients use efficiency.

4. Properties of nano fertilizers

The nano-sized fertilizer is a significant scientific material applied in agricultural fields that is considered as a bridge for gap between molecular structure and bulk material. The advantages of nano-fertilizer are due to nano-sized particles that can easily penetrate to plant cells and reach to the soil (Fatima et al., 2021). The eco-friendly, higher infiltration capacity, slow and controlled release, higher surface area, nutrient use efficiency, stress resistance ability are the unique properties of the nano-fertilizers (as shown in Fig. 13.1) which are helpful for better nutrients management. But some studies revealed that the NPs could cause phytotoxicity through overproduction of reactive oxygen species to lead oxidative stress, protein, cell and DNA damage in the plants (Naderi & Danesh-Shahraki, 2013). So, the optimum level of nano sized fertilizer should be recommended for agricultural sustainability. Nano-fertilizers are effective agricultural inputs that release the nutrients into soils with controlled manner (Singh et al., 2017). The formulation of nano-fertilizer through slow or controlled delivery and conditional release mechanism may have active ingredients to accurately

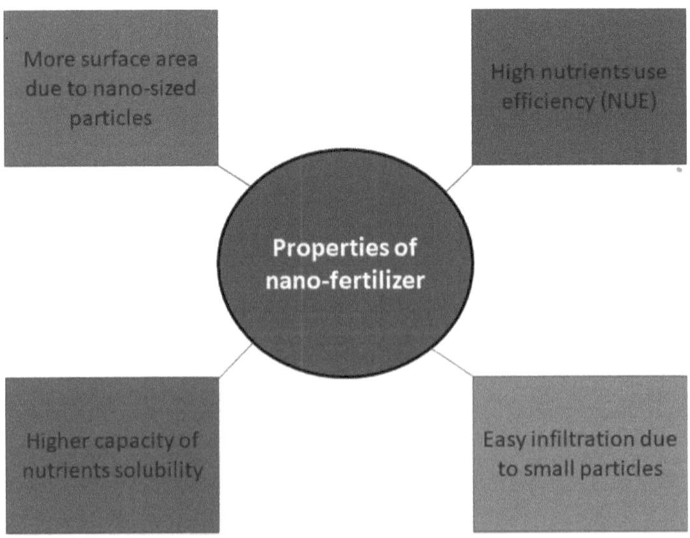

Fig. 13.1 The properties of nano-fertilizer for better nutrients management.

respond the biological demands and environmental triggers (Subramanian et al., 2015). Furthermore, nano-fertilizers act as slow-release to overcome the fluctuations such as temperature, moisture, and soil acidity and to boost the plant growth more efficiently (Vega-Vásquez et al., 2020). Moreover, Cui et al. (2010) stated that the nano-fertilizer could enhance the duration of nutrient release due to its small size particles. The size of the engineered or naturally occurring nano-material range typically between 1 and 100 nm. The nano-fertilizer has property of solubility which supports the insoluble nutrients to increase its dispersion and bioavailability. Another property of nano-fertilizer is high surface area that regulate the various metabolic processes in plants and assist in higher production of photosynthates (Shukla, 2019). So, nano-fertilizer has the unique properties due to minute size which are being explored for new opportunities in the agriculture. Moreover, molecular properties of the nano-fertilizers are totally diverse from the bulk materials due higher surface-to-volume ratio and small size. Thus, nano-fertilizer must be prepared that can retain the essential properties, such as controlled release, higher solubility and stability, increase targeted activities with actual dose for lower eco-toxicity due to safe and easy delivery.

5. Nano-fertilizers for better plant growth and nutrition use efficiency (NUE)

The nano-fertilizer is the vital tools in agriculture to increase the crop growth and production with increasing nutrients use efficiency (NUE) and decrease the nutrients losses and cost of cultivation (Singh, 2017). However, Cui et al. (2010) specified that the nano-fertilizers enhance the efficiency and uptake capacity of the nutrients from soil and reduce the nutrients loss to boost the crop productivity. Furthermore, the higher surface area increases the reactivity of nano-fertilizer with some other compounds, and improve the nutrients uptake and nutrient use efficiency (NUE) in plants (Shukla, 2019). The precision nano fertilizers might have been a very valuable feature for accurate nutrient management with better crop growth as shown in Fig. 13.2. This feature of preciseness may provide nutrient throughout the crop growth period as and when in required quantity. As discussed above, the nano-fertilizer due to 'nano' nature have higher surface area, sophisticated absorption capacity, and controlled-release to the targeted sites resulting in increased uptake in plant cells and by minimizing nutrient loss. Notably, the proper usage to certain optimum concentrations may

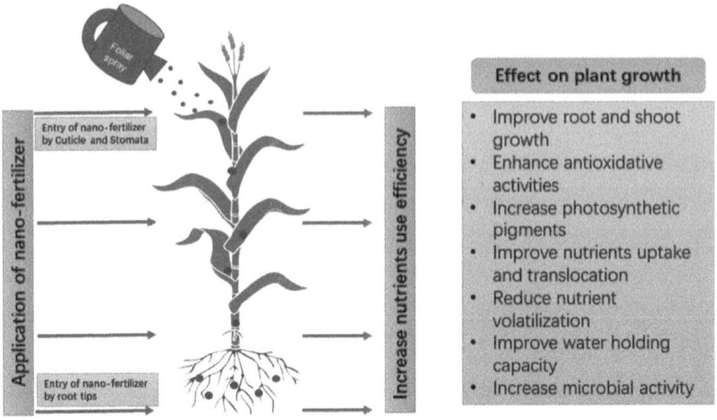

Fig. 13.2 Schematic representation of nano-fertilizer application for nutrients use efficiency and growth improvement in plants.

contribute to the crop growth and bring positive desired effects. Improper concentration of nano-fertilizer i.e., low or over concentration causes the various physiological changes like it can increase various enzyme activities which not needed. Likewise, foliar spray at improper concentration or over dose affect the seedling growth, oxidative stress and exhibited toxicity (Chinnamuthu & Boopathi, 2009). The shoot and root length were regulated with dosage concentrations, and fruit traits like diameter, weight, number of fruits also affected (Fernández-Luqueño et al., 2014).

Nano-fertilizer has greater opportunities to improve inputs use efficiency, minimize costs and also reduce environmental deterioration (Chhipa, 2017). The advantageous effects of various nano-fertilizers on crops are presented in Table 13.2. Nano-fertilizers affect seed germination, chlorophyll content degradation, fresh and dry weight and relative growth rate of plants. Foliar application has ability to improve the efficiency and also rapidity of utilization of nutrients urgently required by the plants and it can increase the growth and yield. Also, this foliar application can cause some effects because of immediate uptake by leaves; it can cause changes in their structure and can cause phytotoxicity which also affects human (Davarpanah et al., 2016). Proper concentration of nano fertilizer can improve the number of flowers and can improve the number of grains which directly can improve yield.

Also, a reduction in nitrogen fixation also can decrease yield. Nano fertilizers applied by foliar can adversely affect seedling growth (Jyothi & Hebsur, 2017). Additionally, nano-fertilizers are mainly synthesized and

Table 13.2 Role of different nano-fertilizer for better nutrients use efficiency (NUE) and growth in plants.

Nano-fertilizer	Dose/concentration	Application method	Plant species	Effect on plants	References
Nano chitosan-NPK	400–500 mg/kg	Spray	Wheat	Increase yield and yield attributes	(Abdel-Aziz et al., 2016)
Synthetic apatite NPs (as P)	100 mg/L	Growing media	Soybean	Increase above ground and below ground biomass and yield	(Liu & Lal, 2014)
Nano-CuO	75–600 mg/kg	Soil	Onion	Increase NUE	(Wang et al., 2020)
Nano-Cu	75, 150, 300 and 600 mg/kg	Soil	Brassica rapa	Improve Cu uptake	(Deng et al., 2020)
ZnO nanomaterial	0–500 mg/kg	Soil	Phaseolus vulgaris	Increase NUE and plant growth	(Medina-Velo et al., 2017)
Fe_2O_3 NPs	2–1000 mg/kg	Soil	Peanut	Enhance Fe uptake, biomass, and SPAD values, plant hormones and antioxidant's enzyme activity	(Rui et al., 2016)
Fe, P, K	N/A	Soil	Crocus sativus L.	Improve flowering traits e.g., dry saffron yield, flower number, stigma length etc.	(Amirnia et al., 2014)
Fe-NFs	35, 75, and 150 $\mu mol\,^{57}Fe\ pot^{-1}$	Soil	Soybean	Increase Fe uptake and plant growth	(Cieschi et al., 2019)
Nano-FeO	0.25, 0.5, 0.75, and 1 g/L	Spray	Glycine max	Yield, leaf, and pod dry weight increased.	(Sheykhbaglou et al., 2010)
IO-NPs	1, 2, 4, and 6 g/L	Seed treatment	Maize		

(Continued)

Table 13.2 Role of different nano-fertilizer for better nutrients use efficiency (NUE) and growth in plants.—cont'd

Nano-fertilizer	Dose/concentration	Application method	Plant species	Effect on plants	References
				Improve growth and chlorophyll content	(Pariona et al., 2017)
Nano-Fe2O3	500 and 4000 ppm	Seed treatment	Peanut	Increase protein content	(Suresh et al., 2016)
NPK	N/A	Ultra sonication method	Vigna raidata L.	Improve enzyme activity and seed vigor index	(Celsia & Mala, 2014)
Nano-size calcite product of $CaCO_3$, SiO_2, MgO and Fe_2O_3	0.5 g/L	Spray	Grapevine	Improve the vine growth, yield, the nutrient contents	(Sabir et al., 2014)
Nano-Fe2O3	2, 20, and 50 mg/L	Seed treatment	Watermelon	Increase germination, seedling growth and activities of antioxidant enzymes	(Li et al., 2013)
γFe2O3	50, 100, 250, 500, 1000 and 2000 mg/L	Spray and soil amendment	Soybean	Increase root and shoot growth, and photosynthetic efficiency	(Alidoust & Isoda, 2013)
Zn-Nps	50–2000 mg/L	Spray	Maize	Enhancing growth, yield, and yield attributes.	(Subbaiah et al., 2016)
ZnO	250–1000 mg/kg	Soil	Green pea	Increase Zn uptake and photosynthetic pigments	(Mukherjee et al., 2016)
	0.25, 0.50, and 0.75 g/L	Seed treatment	Capsicum annuum L.	Enhanced the length of root and shoot, seed	(Afrayeem & Chaurasia, 2017)

				germination, and seedling growth	
	2, 4, 8 & 16 mg/L	Soil	Lycopersicum esculentum	Increased growth, photosynthetic efficiency, antioxidant enzymes, and proline accumulation	(Faizan et al., 2018)
	2 g/L	Spray	Sunflower	Increased leaf area, shoot dry weight, chlorophyll content, net carbon dioxide assimilation rate, and Zn content	(Torabian et al., 2016)
Nano-Cu	80 mg/kg	Soil	Coriandrum sativum	Increased Cu accumulation.	(Zuverza-Mena et al., 2015)
CeO2	400 mg/L	Hydroponic	Cucumber	Increase the scratch and globulins content	(Zhao et al., 2014)
C-NTs	0.05–0.1 mg/L	Soil	Date palm	Increase morphological growth	(Taha et al., 2016)
TiO_2	0.25, 0.5, 1, 2 g/L	Soil	Wheat	Increase seed germination and shoot–root growth	(Ramesh et al., 2014)

incorporated to improve productivity and quality of the crops along with an increase in nutritional value. The macronutrients and micronutrients provided by the nano-fertilizers can improve organic matter and microorganisms (Achari & Kowshik, 2018). The increase in antioxidant activity, concentration difference, inhibited photorespiration, reduce inorganic nitrogen fixation, chlorophyll degradation all these affect the nutritional value (Kumar et al., 2021). This positive result of nano-fertilizers might be due to structure, fixation, concentration, biochemical and physical properties. Furthermore, nano-fertilizers are effective overcome damage and effects of salinity, stress, drought, flooding, temperature fluctuations, heat stress, oxidative stress, and excessive light on crops. Like, biosynthesis of ZnO nanoparticles were reported to helpful for the enhancement of phosphorus mobilization in Mung bean rhizosphere (Raliya et al., 2016). The bio-safety considerations with toxicity at higher concentrations of nano fertilizers and multiple turns of NPK nano fertilizer usage in Pisum sativum plants is examined (Khalifa & Hasaneen, 2018). The development of environmentally friendly biodegradable bio-polymer Chitosan nanostructures supplying salicylic acid and Copper to plants in a slow and sustained manner for an extended period of time is reported (Sharma et al., 2020).

The dosage of Chitosan-based nano-fiber was reportedly resulted in enhancement of plant height, and antioxidant activity (Mirheidari et al., 2022). Sweet basil (Ocimum basilicum L.) is used to regulate blood pressure cholesterol and triglyceride levels. There are reports about improving pharmaceutical results of sweet basil plant with a green route produced zinc nano-complex and iron containing nanoliposome (Farshchi et al., 2021; Tavallali et al., 2018). The usage of these nano materials could effectively replace the traditional fertilizer and increase the oil content in the plant. The combined dose of chemical fertilizer and nano chelated fertilizer when supplemented to peppermint plant, resulted in enhanced essential oil quality and quantity (Ostadi et al., 2020).

6. Formulation of nano-fertilizer for smart nutrients delivery

Nanofertilizers is a nano-sized fertilizer that might be used as slow-release fertilizer to increase the nutrients use efficiency in crop plants and avoid the losses of the nutrients to environment. By absorption or adsorption in a matrix, nanostructured components in nano-fertilizer are frequently included in a carrier complex (Khan et al., 2021). Farmers throughout the world are faced with the onerous job of feeding more mouths each year from agricultural lands that are decreasing as a result of global population growth and rising urbanization (Kumar et al., 2021). Satellite photographs suggest that the world's fertile land is fast running out, whereas the production of food may soon be enough to keep pace along the world's increasing human food consumption. Food grain output has been declining for the past decade, causing concern. Nanotechnology can boost agricultural output by improving plant and animal genetics and delivering molecules to particular places to living cells (Chhipa, 2017). Because there are limited arable lands and water supplies, the agricultural industry can only expand by increasing resource usage efficiency while causing the least amount of harm to the production bed through the proper application of contemporary technology. Nanotechnology, for example, has influenced farming practices, biomedical, geotechnical engineering, safety and wellbeing, water management, thermodynamic efficiency, and a range of certain other sectors (Manjunatha et al., 2016). Nanostructured formulations might release their active components more quickly in response to environmental cues and biological needs, thanks to methods including targeted delivery, sustained and transitive release. According to studies, using nano-fertilizers increases nutrient utilization efficiency decreases soil toxicity, reduces the possible negative consequences of overdosing, and reduces the frequency of application. As a result, nanotechnology holds a lot of promise for attaining sustainable agriculture, particularly in underdeveloped nations (Sabourin, 2015).

7. Macronutrients based nano-fertilizers

Macronutrients are essential for soil fertility and to increase the plant growth e.g., nitrogen, phosphorus, potassium, silicon, calcium have major role in plant metabolic processes. Nitrogen is included in a range of structural, genetic, and metabolic components in plant cells. It's an important part of chlorophyll, the chemical that helps plants to convert solar energy into chemical energy by using water and CO_2 (i.e., photosynthesis). It's also a major component of organic molecules i.e., amino acids, which are the monomers of peptide chains or proteins. Proteins, which function as building elements in plant tissue and cellular structures such as enzymes and others, are responsible for many of the chemical reactions that allow life to exist. Nitrogen is found in energy-transfer compounds such as ATP (adenosine triphosphate), which aids cellular entities in storing and using the energy generated through metabolic reactions (Mejías et al., 2021).

Furthermore, nitrogen is present in deoxyribonucleic acid DNA, the genetic makeup that allows cells (and subsequently entire species e.g., plant) to proliferate and procreate. Because nitrogen molecules of organic nature e.g., ammonium (NH_4^+) ions, and nitrate (NO_3-), and nitrite (NO_2-) ions are all available to plants in three distinct forms. The vast bulk of nitrogen in soil is unavailable to plants. This is because negatively charged nitrate has a lower affinity for soil particle surfaces than positively charged nitrate and hence does not readily sorb on soil (Tamme et al., 2021). To combat the difficulties associated with nitrogen leaching during fertilization, several techniques such as polyolefin resin-coated urea, neem-coated urea, and sulfur-coated urea have been used. Sluggish-release fertilizers, on the other hand, are frequently costly, and the release of nitrogen is slow at times of high N. To limit NH_4^+ release, cation exchangers can be used as fertilizer additions to prevent N loss. Zeolite boosts total crop productivity by retaining and releasing vital nutrients at the right moment (Prasad & Shivay, 2021).

Phosphorus is an essential nutrient in plants because it is a component of several major plant structure components and serves as a catalyst in the conversion of multiple key biochemical processes. Phosphorus is by far the most essential nutrient in plants because of its role in energy transport and storage. P is immobilized in soil due to various edaphic processes, limiting its timely and adequate availability for plant uptake. As a result, present water-soluble P fertilizers have a low usage efficiency, which causes severe environmental and human health issues (Bindraban et al., 2020). Because of its role in energy transport and storage in plants, phosphorus is the most essential nutrient. The most essential types of phosphates involved in energy transfer are adenosine di- and triphosphates (ADP and ATP). Phosphoproteins, phospholipids, sugar phosphates, coenzymes, nucleic acids, and metabolic substrates all contain it. P is required for the formation of reproductive structures in the early stages of crop growth. Plants require greater concentrations of macronutrients than micronutrients (among them nitrogen, phosphorous, and potassium, calcium, Sulfur, and magnesium are included). Similarly, the fertilizer industry creates fertilizer comprising of macronutrients in greater volumes. As a result, it's unexpected that the rate of macronutrient-related Nano-fertilizer basic research and development (R&D) hasn't kept pace with their value in crop productivity (Dimkpa & Bindraban, 2017). The solubility and cation-exchange properties of combinations of rock phosphate and NH_4^+ and K saturated clinoptilolite demonstrated that dissolution and ion-exchange processes in mixtures of zeolite and phosphate rock had the potential to offer slow-release of fertilizers for plants in synthetic soils.

When saturated with monovalent nutritional cations such as NH_4^+ and K+, zeolites (clinoptilolite) have been shown to improve the solubility of phosphate rock (PR). The effectiveness of fertilizer P use by crops in the year it was applied ranged from 18% to 20%. The remaining 78% to 80% is absorbed into the soil P pool, where it is distributed to the crop over months and years (Dimkpa & Bindraban, 2017).

The dissolution and ionic characteristics of rock phosphate and NH_4^+ and K saturated clinoptilolite showed that solubility and ion-exchange mechanisms in composites of zeolite and phosphate rock would have the ability to provide delayed fertilizing in plants and synthetic soils. Zeolites (clinoptilolite) have been demonstrated to enhance the solubility of phosphate rock when swamped with monovalent dietary cations such as NH_4^+ and K+ (PR). The usage of phosphate P by crops during annual application shows 18%–20% effectivity. The remaining 78 to 80% is incorporated into the soil P reservoir, where it is delivered over long periods to the crop (Adhikari et al., 2020).

Some natural zeolites have high levels of exchangeable K+, which can help plants develop faster in potting soil. For example, data on the gradual release of K from K-zeolite was supplied. Because zeolite could adsorb potassium from chemical fertilizer and decrease leakage, the application of 625 kg ha-1 chemical fertilizer combined with 125 kg ha-1 zeolite showed the highest quantity of potassium in the soil (Rajonee et al., 2017). Because of the placement and density of the negative charge in the microstructure of Nano-fertilizer, natural zeolites are extremely selective for K+ rather than sodium or divalent cations like calcium and magnesium (Al-Shahmani & Al-Juthery, 2021). Because of the ion exchangeability of zeolites with chosen nutrient cations, zeolites can become an effective plant development medium for feeding plant roots with additional necessary nutrient cations and anions, according to the researchers. Regardless, that potassium fixation in soil and its dynamic equilibrium mutually maintain potassium levels in the soil to be available for uptake by plants, still applications of nanotechnology in fertilizers and soil sciences can increase nutrient availability and regulation (Dimkpa & Bindraban, 2017).

Nano fertilizer technology has limited insights, and there is a shortage of published research in scholarly publications. Conventional fertilizers have nutrient usage efficiency of just 30%–35%, 18%–20%, and 35%–40% for N, P, and K, respectively. For the previous few decades, the data has remained steady, and study efforts have not yielded productive outcomes. Nano-fertilizers are nutrient transporters made from substrates with nanoscale dimensions ranging from 1 to 100 nm. Nanoparticles have a large surface area and can contain a lot of nutrients and release them slowly and gradually, making it easier for crops to get the nutrients they need without the negative side effects of tailored fertilizer applications (Zulfiqar et al., 2019).

8. Micronutrients based nano-fertilizers

Elements involved essentially in plant development but in considerably fewer quantities than nitrogen, phosphorous, and potassium are micronutrients. Boron (B), copper (Cu), iron (Fe), manganese (Mn), molybdenum (Mo), zinc (Zn), and chloride (Cl). Micronutrients are necessary for healthy plant development and lucrative crop production, even though they are only required in trace amounts. Due to the calcareous character of soils, high pH, poor organic matter, salt stress, persistent drought, high bicarbonate concentration in irrigation water, and uneven use of NPK fertilizers, micronutrient insufficiency is endemic in many Asian nations. Reduced crop yield and quality, inadequate plant morphological structure (such as fewer small xylem vessels), pervasive infection of diverse insect pests and diseases, minimal stimulation of phytosiderophores, and fairly low fertilizer utilization are among the adverse effects of micronutrient deficiency-induced plant tolerance. The capacity of five natural zeolites and bentonite minerals to adsorb zinc and iron and then release it was reported upon investigation. The Langmuir and Freundlich equations are being used to calculate the sorption potential of such ions (Tarafder et al., 2020). Natural zeolites, notably chabazite and bentonite minerals, appear to have a significant potential for Zn and Fe sorption, as well as a high capacity for slow-release fertilizers, according to the findings. The delayed delivery of Zn is due to the mineral's parsimonious dissolution rate and the concentration impact of the exchange, which releases trace nutrients to zeolite transfer sites where they are more available for plant uptake. For 3 years, researchers experimented with foliar applications of ion-exchanged zeolite on winter wheat. The increase in raw protein as a consequence of zinc zeolite treatment was better than the rise in raw protein as a result of copper-zeolite treatment (Ekanayake & Godakumbura, 2019). According to Pandey et al. (2010), zinc-rich ZnO NPs enhanced the amount of IAA in roots (sprouts), indicating an increase in plant development rate since zinc is an important nutrient for plants.

Boron is a necessary mineral for plants, but when it accumulates in excessive amounts, it may be hazardous to organisms. Several researchers have looked at the boron adsorption by clays, soils, and other minerals. Boron boosted root yield, shot yield, and biological yield considerably, and its treatments resulted in the highest root yield, shoot yield, biological yield, and sugar beet quality (Kandil et al., 2020). Furthermore, Mo is a major component of the nitrogenase enzyme found in nitrogen-fixing bacteria, which is critical for legume crops. Mo is vital for N fixation in legumes and nitrate reduction control in other plants, as well as amino acid and protein production. The plant enzyme nitrate reductase requires molybdenum to function. Excess Mo NP exposure, on the other hand, might impede root growth/elongation, delay seed germination, raise nitrate reductase levels, and result in oxidative imbalance (Huang et al., 2021).

Pollution and major health hazards have resulted from the overuse of mineral fertilizers and dangerous pesticides. Nanoscience may be able to address these challenges by developing higher-performing nanomaterials. Insights at the evolution of nano-fertilizers and nano-pesticides, as well as their use in agricultural systems, has been reported (Butt & Naseer, 2020). N, P, K, Fe, Mn, Zn, Cu, Mo, and carbon nanotubes are examples of nano-fertilizers that have greater release and targeted delivery efficiency. In comparison to traditional pesticides, nano-pesticides such as Ag, Cu, SiO2, ZnO, and nano-formulations provide improved broad-spectrum insect control (Chhipa, 2017).

9. Conclusion

The agriculture field are facing the challenges of nutrient deficiency, crop yield reduction, weakening the soil organic matter, low water availability as the result of poor nutrients use efficiencies. However, the application of different nano-fertilizers has greater role in improving nutrients use efficiency (NUE), enhancing the crop yield, reducing the environmental pollution hazard and the fertilization cost for crop production. Because, the nano-sized fertigation is more soluble, reactive, and could increase the penetration through the cuticle for targeted delivery. Meanwhile, plant root entry also increases the passage of nutrient uptake as it's highly porous to nano-fertilizer compared to inorganic fertilizer. Furthermore, the optimum concentrations of nano-fertilizers could improve crop growth but excessive dose may lead to reduced crop growth. Thus, the optimization of nano-fertilizer dose is prime important to improve nutrients use efficiency in different crops. In addition, the behavior and properties of different nano-sized fertilizer to different plant species and environmental conditions are not clear, so the field based detailed studies are required to provide exact fertigation input approach to enhance nutrients use efficiency as ecofriendly way in near future.

References

Abdel-Aziz, H. M. M., Hasaneen, M. N., & Omer, A. M. (2016). Nano chitosan-NPK fertilizer enhances the growth and productivity of wheat plants grown in sandy soil. *Spanish Journal of Agricultural Research, 14*(1), 17.

Achari, G. A., & Kowshik, M. (2018). Recent developments on nanotechnology in agriculture: Plant mineral nutrition, health, and interactions with soil microflora. *Journal of Agricultural and Food Chemistry, 66*(33), 8647–8661.

Adhikari, B., Dhungana, S. K., Kim, I. D., & Shin, D. H. (2020). Effect of foliar application of potassium fertilizers on soybean plants under salinity stress. *Journal of the Saudi Society of Agricultural Sciences, 19*(4), 261–269.

Afrayeem, S. M., & Chaurasia, A. K. (2017). Effect of zinc oxide nanoparticles on seed germination and seed vigour in chilli (Capsicum annuum L.). *Journal of Pharmacognosy and Phytochemistry, 6*(5), 1564–1566.

Al-Mamun, M. R., Hasan, M. R., Ahommed, M. S., Bacchu, M. S., Ali, M. R., & Khan, M. Z. H. (2021). *Nanofertilizers toward sustainable agriculture and environment* (p. 101658). Environmental Technology & Innovation.

Al-Shahmani, A. M. K., & Al-Juthery, H. W. A. (April 2021). Response of rice (Oryza sativa L.) to silica fertilization and spraying with nano-potassium and calcium. In *IOP conference series: Earth and environmental science* (p. 012068). IOP Publishing. Vol. 735, No. 1.

Alidoust, D., & Isoda, A. (2013). Effect of γFe 2 O 3 nanoparticles on photosynthetic characteristic of soybean (Glycine max (L.) Merr.): Foliar spray versus soil amendment. *Acta Physiologiae Plantarum, 35*(12), 3365–3375.

Amirnia, R., Bayat, M., & Tajbakhsh, M. (2014). Effects of nano fertilizer application and maternal corm weight on flowering at some saffron (Crocus sativus L.) ecotypes. *Turkish Journal of Field Crops, 19*(2), 158–168.

Bindraban, P. S., Dimkpa, C. O., & Pandey, R. (2020). Exploring phosphorus fertilizers and fertilization strategies for improved human and environmental health. *Biology and Fertility of Soils, 56*(3), 299–317.

Butt, B. Z., & Naseer, I. (2020). "Nanofertilizers." *Nanoagronomy. Springer, Cham,*, 125–152.

Chaitra, P., Kalia, A., Ahuja, R., SPK, S., & Sikka, R. (2021). Importance of nano fertilizers in sustainable agriculture. *Environmental Science Ecology: Current Research, 2*, 1029.

Chhipa, H. (2017). Nanofertilizers and nanopesticides for agriculture. *Environmental Chemistry Letters, 15*(1), 15–22.

Chhipa, H., & Joshi, P. (2016). Nanofertilisers, nanopesticides and nanosensors in agriculture. *Nanoscience in Food and Agriculture, 1*, 247–282. Springer, Cham.

Chinnamuthu, C. R., & Boopathi, P. M. (2009). Nanotechnology and agroecosystem. *Madras Agricultural Journal, 96*(1/6), 17–31.

Cieschi, M. T., Polyakov, A. Y., Lebedev, V. A., Volkov, D. S., Pankratov, D. A., & Veligzhanin, A. A. (2019). Eco-friendly iron-humic nanofertilizers synthesis for the prevention of iron chlorosis in soybean (Glycine max) grown in calcareous soil. *Frontiers in Plant Science, 10*, 413.

Cui, H. X., Sun, C. J., Liu, Q., Jiang, J., & Gu, W. (June 2010). Applications of nanotechnology in agrochemical formulation, perspectives, challenges and strategies. In *International conference* on Nanoagri (pp. 28–33). Sao pedro, Brazil.

Davarpanah, S., Tehranifar, A., Davarynejad, G., Abadía, J., & Khorasani, R. (2016). Effects of foliar applications of zinc and boron nano-fertilizers on pomegranate (Punica granatum cv. Ardestani) fruit yield and quality. *Scientia Horticulturae, 210*, 57–64.

Deng, C., Wang, Y., Cota-Ruiz, K., Reyes, A., Sun, Y., et al. (2020). Bok choy (Brassica rapa) grown in copper oxide nanoparticles-amended soils exhibits toxicity in a phenotype-dependent manner: Translocation, biodistribution and nutritional disturbance. *Journal of Hazardous Materials, 398*, 122978.

Dimkpa, C. O., & Bindraban, P. S. (2017). Nanofertilizers: New products for the industry? *Journal of Agricultural and Food Chemistry, 66*(26), 6462–6473.

Ekanayake, S. A., & Godakumbura, P. I. (2019). *Synthesis of a micronutrient nanofertilizer by incorporating zinc oxide and copper oxide nanoparticles to alginic acid.*

El-Saadony, M. T., ALmoshadak, A. S., Shafi, M. E., Albaqami, N. M., Saad, A. M., et al. (2021). Vital roles of sustainable nano-fertilizers in improving plant quality and quantity-an updated review. *Saudi Journal of Biological Sciences, 28*(12), 7349–7359.

Faizan, M., Faraz, A., Yusuf, M., Khan, S. T., & Hayat, S. (2018). Zinc oxide nanoparticle-mediated changes in photosynthetic efficiency and antioxidant system of tomato plants. *Photosynthetica, 56*, 678–686.

Farshchi, H. K., Azizi, M., Teymouri, M., Nikpoor, A. R., & Jaafari, M. R. (2021). Synthesis and characterization of nanoliposome containing Fe2+ element: A superior nano-fertilizer for ferrous iron delivery to sweet basil. *Scientia Horticulturae, 283*, 110110.

Fatima, F., Hashim, A., & Anees, S. (2021). Efficacy of nanoparticles as nanofertilizer production: A review. *Environmental Science and Pollution Research, 28*(2), 1292–1303.

Fernández-Luqueño, F., López-Valdez, F., ValerRodríguez, M. F., Pariona, N., Hernández-López, J. L., et al. (2014). Effects of nanofertilizers on plant growth and development, and their interrelationship with the environment. In *Fertilizers: components, uses in agriculture and environmental impacts. Nova Science Publishers, Inc.*., 211–224.

Hong, F., Yang, F., Liu, C., Gao, Q., Wan, Z., Gu, F., ... Yang, P. (2005). Influences of nano-TiO_2 on the chloroplast aging of spinach under light. *Biological Trace Element Research, 104*(3), 249–260.

Huang, X., Cervantes-Avilés, P., Li, W., & Keller, A. A. (2021). *Drilling into the metabolomics to enhance insight on corn and wheat responses to molybdenum trioxide nanoparticles.* Environmental Science & Technology.

Iannone, M. F., Groppa, M. D., de Sousa, M. E., van Raap, M. B. F., & Benavides, M. P. (2016). Impact of magnetite iron oxide nanoparticles on wheat (*Triticum aestivum* L.) development: Evaluation of oxidative damage. *Environmental and Experimental Botany, 131*, 77–88.

Jyothi, T. V., & Hebsur, N. S. (2017). Effect of nanofertilizers on growth and yield of selected cereals-A review. *Agricultural Reviews, 38*(2), 112–120.

Kandil, E. E., Abdelsalam, N. R., Aziz, A. A. A. E., Ali, H. M., & Siddiqui, M. H. (2020). Efficacy of nanofertilizer, fulvic acid and boron fertilizer on sugar beet (Beta vulgaris L.) yield and quality. *Sugar Tech, 22*(5), 782–791.

Khalid, U., Sher, F., Noreen, S., Lima, E. C., Rasheed, T., Sehar, S., & Amami, R. (2022). Comparative effects of conventional and nano-enabled fertilizers on morphological and physiological attributes of Caesalpinia bonducella plants. *Journal of the Saudi Society of Agricultural Sciences, 21*(1), 61–72.

Khalifa, N. S., & Hasaneen, M. N. (2018). The effect of chitosan–PMAA–NPK nanofertilizer on Pisum sativum plants. *3 Biotech, 8*(4), 1–12.

Khan, M. Z. H., Islam, M. R., Nahar, N., Al-Mamun, M. R., Khan, M. A. S., & Matin, M. A. (2021). Synthesis and characterization of nanozeolite based composite fertilizer for sustainable release and use efficiency of nutrients. *Heliyon, 7*(1), e06091.

KUMAR, Y., SINGH, K. T. T., & RALIYA, R. (2021). Nanofertilizers and their role in sustainable agriculture. *Annals of Plant and Soil Research, 23*(3), 238–255.

Li, J., Chang, P. R., Huang, J., Wang, Y., Yuan, H., & Ren, H. (2013). Physiological effects of magnetic iron oxide nanoparticles towards watermelon. *Journal of Nanoscience and Nanotechnology, 13*(8), 5561–5567.

Liu, R., & Lal, R. (2014). Synthetic apatite nanoparticles as a phosphorus fertilizer for soybean (Glycine max). *Scientific Reports, 4*(1), 1–6.

Li, D. P., & Wu, Z. J. (2008). Impact of chemical fertilizers application on soil ecological environment. *Ying Yong Sheng Tai Xue Bao=The Journal of Applied Ecology, 19*(5), 1158–1165.

Liu, R., & Lal, R. (2015). Potentials of engineered nanoparticles as fertilizers for increasing agronomic productions. *Science of the Total Environment, 514*, 131–139.

Mahanta, N., Ashok, D., & Montrishna, R. (2019). Nutrient use efficiency through nano fertilizers. *International Journal of Chemical Studies,* 7(3), 2839–2842.

Mahil, E., & Kumar, B. (2019). Foliar application of nanofertilizers in agricultural crops—A review. *Journal of Farm Sciences, 32*(3), 239–249.

Manjunatha, S. B., Biradar, D. P., & Aladakatti, Y. R. (2016). Nanotechnology and its applications in agriculture: A review. *Journal of Farm Science, 29*(1), 1–13.

Medina-Velo, I. A., Dominguez, O. E., Ochoa, L., Barrios, A. C., Hernández-Viezcas, J. A., White, J. C., et al. (2017). Nutritional quality of bean seeds harvested from plants grown in different soils amended with coated and uncoated zinc oxide nanomaterials. *Environmental Science: Nano, 4*(12), 2336–2347.

Mejías, J. H., Salazar, F. J., Pérez, L., Hube, S., Rodriguez, M., & Alfaro, M. A. (2021). Nanofertilizers: A cutting-edge approach to increase nitrogen use efficiency in grasslands. *Frontiers in Environmental Science, 9*, 52.

Mirheidari, F., Hatami, M., & Ghorbanpour, M. (2022). Effect of different concentrations of IAA, GA3 and chitosan nano-fiber on physio-morphological characteristics and metabolite contents in roselle (Hibiscus sabdariffa L.). *South African Journal of Botany, 145*, 323–333.

Mittal, D., Kaur, G., Singh, P., Yadav, K., & Ali, S. A. (2020). Nanoparticle-based sustainable agriculture and food science: Recent advances and future outlook. *Frontiers in Nanotechnology, 2*, 10.

Mukherjee, A., Sun, Y., Morelius, E., Tamez, C., Bandyopadhyay, S., Niu, G., … Gardea-Torresdey, J. L. (2016). Differential toxicity of bare and hybrid ZnO nanoparticles in green pea (Pisum sativum L.): A life cycle study. *Frontiers in Plant Science, 6*, 1242.

Naderi, M. R., & Danesh-Shahraki, A. (2013). Nanofertilizers and their roles in sustainable agriculture. *International Journal of Agriculture and Crop Sciences (IJACS), 5*(19), 2229–2232.

Ostadi, A., Javanmard, A., Machiani, M. A., Morshedloo, M. R., Nouraein, M., Rasouli, F., & Maggi, F. (2020). Effect of different fertilizer sources and harvesting time on the growth characteristics, nutrient uptakes, essential oil productivity and composition of Mentha x piperita L. *Industrial Crops and Products, 148*, 112290.

Pandey, A. C., Sanjay, S., & Yadav, S. (2010). Application of ZnO nanoparticles in influencing the growth rate of Cicer arietinum. *Journal of Experimental Nanoscience, 5*(6), 488–497.

Pariona, N., Martinez, A. I., Hdz-García, H. M., Cruz, L. A., & Hernandez-Valdes, A. (2017). Effects of hematite and ferrihydrite nanoparticles on germination and growth of maize seedlings. *Saudi Journal of Biological Sciences, 24*(7), 1547–1554.

Prasad, R., Bhattacharyya, A., & Nguyen, Q. D. (2017). Nanotechnology in sustainable agriculture: Recent developments, challenges, and perspectives. *Frontiers in Microbiology, 8*, 1014.

Prasad, R., & Shivay, Y. S. (2021). Enhanced efficiency fertilizers or slow-release and NI/UI blended nitrogen fertilizers. *Indian Journal of Fertilisers, 17*(4), 316–321.

Prasad, T. N. V. K. V., Sudhakar, P., Sreenivasulu, Y., et al. (2012). Effect of nanoscale zinc oxide particles on the germination, growth and yield of peanut. *Journal of Plant Nutrition, 35*(6), 905–927.

Rajonee, A. A., Zaman, S., & Huq, S. M. I. (2017). Preparation, characterization and evaluation of efficacy of phosphorus and potassium incorporated nano fertilizer. *Advances in Nanoparticles, 6*(02), 62.

Rajput, R., Pokhriya, P., Panwar, P., Arunachalam, A., & Arunachalam, K. (2019). Soil nutrients, microbial biomass, and crop response to organic amendments in rice cropping system in the Shiwaliks of Indian Himalayas. *International Journal of Recycling of Organic Waste in Agriculture, 8*(1), 73–85.

Raliya, R., Franke, C., Chavalmane, S., Nair, R., Reed, N., & Biswas, P. (2016). Quantitative understanding of nanoparticle uptake in watermelon plants. *Frontiers in Plant Science, 7*, 1288.

Raliya, R., Saharan, V., Dimkpa, C., & Biswas, P. (2017). Nanofertilizer for precision and sustainable agriculture: Current state and future perspectives. *Journal of Agricultural and Food Chemistry, 66*(26), 6487–6503.

Rameshaiah, G. N., Pallavi, J., & Shabnam, S. (2015). Nano fertilizers and nano sensors—an attempt for developing smart agriculture. *International Journal of Engineering Research and General Sciences, 3*(1), 314–320.

Ramesh, M., Palanisamy, K., Babu, K., & Sharma, N. K. (2014). Effects of bulk & nano-titanium dioxide and zinc oxide on physio-morphological changes in *Triticum aestivum* Linn. *Journal of Global Biosciences, 3*(2), 415–422.

Rui, M., Ma, C., Hao, Y., Guo, J., Rui, Y., Tang, X., ... Zhu, S. (2016). Iron oxide nanoparticles as a potential iron fertilizer for peanut (*Arachis hypogaea*). *Frontiers in Plant Science, 7*, 815.

Sabir, A., Yazar, K., Sabir, F., Kara, Z., Yazici, M. A., & Goksu, N. (2014). Vine growth, yield, berry quality attributes and leaf nutrient content of grapevines as influenced by seaweed extract (Ascophyllum nodosum) and nanosize fertilizer pulverizations. *Scientia Horticulturae, 175*, 1–8.

Sabourin, V. (2015). Commercial opportunities and market demand for nanotechnologies in agribusiness sector. *Journal of Technology Management & Innovation, 10*(1), 40–51.

Salama, H. M. (2012). Effects of silver nanoparticles in some crop plants, common bean (Phaseolus vulgaris L.) and corn (Zea mays L.). *International Research Journal of Biotechnology, 3*(10), 190–197.

Savci, S. (2012). An agricultural pollutant: Chemical fertilizer. *International Journal of Environmental Science and Development, 3*(1), 73.

Sharma, G., Kumar, A., Devi, K. A., Prajapati, D., Bhagat, D., Pal, A., ... Saharan, V. (2020). Chitosan nanofertilizer to foster source activity in maize. *International Journal of Biological Macromolecules, 145*, 226–234.

Sharma, U., Paliyal, S. S., Sharma, S. P., & Sharma, G. D. (2014). Effects of continuous use of chemical fertilizers and manure on soil fertility and productivity of maize—wheat under rainfed conditions of the Western Himalayas. *Communications in Soil Science and Plant Analysis, 45*(20), 2647–2659.

Sharma, S., Rana, V. S., Pawar, R., Lakra, J., & Racchapannavar, V. (2021). Nanofertilizers for sustainable fruit production: A review. *Environmental Chemistry Letters, 19*(2), 1693–1714.

Sheykhbaglou, R., Sedghi, M., Shishevan, M. T., & Sharifi, R. S. (2010). Effects of nano-iron oxide particles on agronomic traits of soybean. *Notulae Scientia Biologicae, 2*(2), 112–113.

Shukla, Y. M. (2019). Nanofertilizers: A recent approach in crop production. In *Nanotechnology for agriculture: Crop production & protection* (pp. 25–58). Singapore: Springer.

Singh, M. D. (2017). Nano-fertilizers is a new way to increase nutrients use efficiency in crop production. *International Journal of Agriculture Sciences, ISSN*, 0975–3710.

Subbaiah, L. V., Prasad, T. N. V. K. V., Krishna, T. G., Sudhakar, P., Reddy, B. R., & Pradeep, T. (2016). Novel effects of nanoparticulate delivery of zinc on growth, productivity, and zinc biofortification in maize (Zea mays L.). *Journal of Agricultural and Food Chemistry, 64*(19), 3778–3788.

Subramanian, K. S., Manikandan, A., Thirunavukkarasu, M., & Rahale, C. S. (2015). Nanofertilizers for balanced crop nutrition. In *Nanotechnologies in food and agriculture* (pp. 69–80). Cham: Springer.

Suresh, S., Karthikeyan, S., & Jayamoorthy, K. (2016). Effect of bulk and nano-Fe_2O_3 particles on peanut plant leaves studied by Fourier transform infrared spectral studies. *Journal of Advanced Research,* 7(5), 739–747.

Suriyaprabha, R., Karunakaran, G., Yuvakkumar, R., Rajendran, V., & Kannan, N. (2012). Silica nanoparticles for increased silica availability in maize (Zea mays. L) seeds under hydroponic conditions. *Current Nanoscience, 8*(6), 902–908.

Taha, R. A., Hassan, M. M., Ibrahim, E. A., Abou Baker, N. H., & Shaaban, E. A. (2016). Carbon nanotubes impact on date palm in vitro cultures. *Plant Cell, Tissue and Organ Culture (PCTOC), 127*(2), 525–534.

Tamme, R., Pärtel, M., Kõljalg, U., Laanisto, L., Liira, J., Mander, Ü., … Zobel, M. (2021). Global macroecology of nitrogen-fixing plants. *Global Ecology and Biogeography, 30*(2), 514–526.

Tarafder, C., Daizy, M., Alam, M. M., Ali, M. R., Islam, M. J., Islam, R., … Khan, M. Z. H. (2020). Formulation of a hybrid nanofertilizer for slow and sustainable release of micronutrients. *ACS Omega, 5*(37), 23960–23966.

Tavallali, V., Rowshan, V., & Bahmanzadegan, A. (2018). Variations in sweet basil in response to Green synthesized Zinc-Amino nano complexes. *Journal of Cleaner Production, 196*, 452–459.

Thakkar, K. N., Mhatre, S. S., & Parikh, R. Y. (2010). Biological synthesis of metallic nanoparticles. *Nanomedicine: Nanotechnology, Biology and Medicine, 6*(2), 257–262.

Toksha, B., Sonawale, V. A. M., Vanarase, A., Bornare, D., Tonde, S., Hazra, C., … Chatterjee, A. (2021). *Nanofertilizers: A review on synthesis and impact of their use on crop yield and environment* (p. 101986). Environmental Technology & Innovation.

Torabian, S., Zahedi, M, & Khoshgoftar, A. H. (2016). Effects of foliar spray of two kinds of zinc oxide on the growth and ion concentration of sunflower cultivars under salt stress. *Journal of Plant Nutrition, 39*(2), 172–180.

Vega-Vásquez, P., Mosier, N. S., & Irudayaraj, J. (2020). Nanoscale drug delivery systems: From medicine to agriculture. *Frontiers in Bioengineering and Biotechnology, 8*, 79.

Wang, Y., Deng, C., Cota-Ruiz, K., Peralta-Videa, J. R., Sun, Y., Rawat, S., … Gardea-Torresdey, J. L. (2020). Improvement of nutrient elements and allicin content in green onion (Allium fistulosum) plants exposed to CuO nanoparticles. *Science of The Total Environment, 725*, 138387.

Yang, F., Hong, F., You, W., Liu, C., Gao, F., Wu, C., & Yang, P. (2006). Influence of nano-anatase TiO_2 on the nitrogen metabolism of growing spinach. *Biological Trace Element Research, 110*(2), 179–190.

Yomso, J., & Menon, S. (2021). Impact of nanofertilizers on growth and yield parameters of rice crop; A Review. *The Pharma Innovation Journal, 10*(6), 249–253.

Zhao, L., Peralta-Videa, J. R., Rico, C. M., Hernandez-Viezcas, J. A., Sun, Y., Niu, G., … Gardea-Torresdey, J. L. (2014). CeO_2 and ZnO nanoparticles change the nutritional qualities of cucumber (Cucumis sativus). *Journal of Agricultural and Food Chemistry, 62*(13), 2752–2759.

Zulfiqar, F., Navarro, M., Ashraf, M., Akram, N. A., & Munné-Bosch, S. (2019). Nanofertilizer use for sustainable agriculture: Advantages and limitations. *Plant Science, 289*, 110270.

Zuverza-Mena, N., Medina-Velo, I. A., Barrios, A. C., Tan, W., Peralta-Videa, J. R., & Gardea-Torresdey, J. L. (2015). Copper nanoparticles/compounds impact agronomic and physiological parameters in cilantro (Coriandrum sativum). *Environmental Science: Processes & Impacts, 17*(10), 1783−1793.

Celsia, A. R., & Mala, R. (2014). Fabrication of nano structured slow release fertilizer system and its influence on germination and biochemical characteristics of vigna raidata. *Journal of ChemTech Research, 6*(10), 4497−4503.

CHAPTER FOURTEEN

Mulching with organic matters has potential of remediation in ecosystem

Misbah Naz[a], Sania Zaib[b], Muhammad Tariq[c], Muhammad Ammar Raza[d], Zhicong Dai[a,e], Daolin Du[a], Muhammad Jafar Jaskani[f], Sohail Ahmed Rajper[g], Muhammad Ahsan[h]

[a]Institute of Environment and Ecology, School of The Environment and Safety Engineering, Jiangsu University, Zhenjiang, China
[b]Department of Biochemistry, Biological Sciences, Quaid-i-Azam University, Islamabad, Pakistan
[c]Department of Pharmacology, Lahore Pharmacy Collage, Lahore, Pakistan
[d]School of Food Science and Biotechnology, Key Laboratory of Fruits and Vegetables Postharvest and Processing Technology Research, Zhejiang Gongshang University, Hangzhou, China
[e]Jiangsu Collaborative Innovation Center of Technology and Material of Water Treatment, Suzhou University of Science and Technology Suzhou, Suzhou, China
[f]Faculty of Agriculture, Institute of Horticultural Sciences, University of Agriculture, Faisalabad, Pakistan
[g]Department of Biotechnology, Sindh Agricultural University, Tandojam, Pakistan
[h]Department of Horticultural Sciences, The Islamia University of Bahawalpur, Bahawalpur, Pakistan

1. Introduction

Soil is an essential part of the agriculture system. It is the medium where various chemical, physical and biological processes take place that ultimately helps in improving agricultural productivity. Regeneration of soil usually occurs naturally in an undisturbed ecosystem, however continuous planting of crops in arable lands leads to degradation and depletion of nutrients. Continuous farming leads to low and declining soil fertility, which is the main restriction factor for crop production and sustainable development (Ngosong et al., 2019). Nevertheless, considering the dynamic nature of cultivated soil affected by farm management practices, there is a need of regular monitoring of the health of the soil. Soil an important natural source must be managed properly for sustainable development. Achievement of sustainability is possible under the conditions of system resource conservation, social and cultural support, commercial competition and environmental friendliness. For improving the fertility and productivity of soil in cultivated land systems mulch is used as a sustainable strategy (Shah & Wu, 2019). Mulching is usually the spreading of covering materials on the

Sustainable Plant Nutrition
ISBN: 978-0-443-18675-2
https://doi.org/10.1016/B978-0-443-18675-2.00018-3

soil surface which helps in minimizing moisture loss and weed population and in enhancing crop yield (Nalayini, 2007). Covering has several positive effects on plant soil and environment (Indoria et al., 2017).

In the late 1930s mulch received attention because it can change the adjoining conditions of agricultural and horticultural lands and of forest areas. By-products from agriculture and forestry were used as mulch in the mid-1900s (Pirone, 1941). Though, some other materials could also be used, such as the pruning of trees and shrubs, animal excrement, stubble, and crop residues. The landscape cover was also seen in 1957, but scientific research was not conducted on it (Haque et al., 1995). Some early studies have shown that coverings can cause many damaging effects (Gomiero et al., 2011). In 1941, deep mulching was also practiced for shrub and trees planting because deep mulching could protect against the dehydration stresses such as drought and frost stress. When the same amount of mulch material is used as the mulch, more water saving can be achieved compared with the material mixed into the soil (Watschke & Schmidt, 1992).

Mulch improves soil health through enhancing water retention, reducing evaporation, regulating temperature, increasing nutrient absorption by root, reducing salinity, controlling weeds, pests and crop diseases (Han et al., 2020). For improving the soil health although organic mulch materials are usually used in arable land system, however recently the utilization of inorganic plastic materials has gained global attention. Nevertheless, the widespread use of plastic mulch will have severe impacts on soil and environment affecting agricultural productivity and endangering sustainable development (Rosemeyer, 2001). Hence for the assurance of sustainable development, continuous monitoring of soil health is necessary based on various covering materials and local environmental conditions. In general, keeping in mind the precise needs of farmers and the local conditions of environment, the careful selection of plastic films the inorganic mulching material should be done. However, for improving the productivity and soil health, organic mulching is a more viable option (Iqbal et al., 2020).

Coverage has different modes of action on pests, including interference with visual and olfactory host discovery capabilities. It also induces suicidal attraction to sun-heated coverings, especially inorganic coverings, while organic coverings promote the accumulation of natural enemies of pests. For example, ground predatory beetles prefer high coverage areas, while ground predators and parasitic wasps prefer living mulch with a rich supply of nectar to attract more replacement prey and provide necessary refuge (Reddy, 2016). For sustainable soil management, there are large numbers of views, concepts and paradigms including integrated management of nutrients, fertility of soil and of natural resources, organic agriculture and agricultural conservation (Ngosong et al., 2019). These basically aim to improve fertility of soil and productivity while protecting natural resource base being fragile in nature. Coverage is an important part of these views, however organic coverage is limited owing to huge amount and labor costs needed for collection, transport and application (Erenstein, 2002). Therefore, the use of high value residue mulch must be adapted according to local environmental conditions and based on specific needs of farmers. Even though organic mulch is usually used in agriculture, inorganic mulch is used more and more, but its all-round impact on the environment remains to be determined, particularly the long-term impact of plastics. In general organic and inorganic mulches found to be effective in improving soil health and productivity, and the benefits usually exceed the linked costs (Kasirajan & Ngouajio, 2012). The interaction between coverage practices and conservation agriculture and the global climate environment is shown in Fig. 14.1.

2. Potential agricultural and environmental benefits of mulches

Over the years, global temperature is rising owing to rapid industrialization and urbanization, thereby disrupting the balance of global agricultural ecosystems (Appannagari, 2017). So, there is need of new

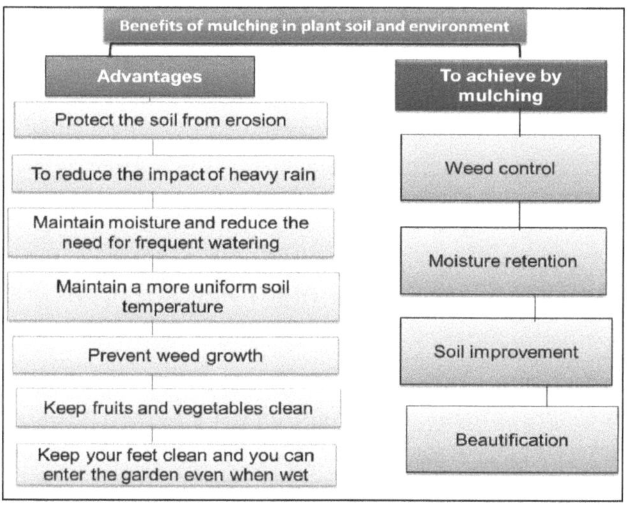

Fig. 14.1 Benefits of mulching in the ecosystem.

environmental-friendly agricultural practices to achieve sustainable food production. Mulch usually helps in lowering evaporation rate of soil, maintaining moisture content, and temperature of soil, and controlling the growth of weed, and improving the microbial activity (Bahadur et al., 2015). In addition, mulch can provide environmental, economic and aesthetic benefits. Besides, mulch is broadly used in restoration sites to plant trees that do not require special care. Mulch is also helpful in coping various environmental stress conditions. This chapter focuses on the significant effects of mulch on the production and establishment of various crops in nature. The mulch retains soil moisture, improves soil nutrient status, controls erosion loss, suppresses crop weeds, and removes the residual effects of pesticides, fertilizers and heavy metals (Iqbal et al., 2020). The mulch enhances the aesthetic value of the landscape and the economic value of the crop. This chapter also describes some of the issues related to various covering materials. Covering materials are contradictory, because some researchers like to cover, while others have expressed some concerns (Kader et al., 2017). The choice of mulch material is important in terms of types of crops, management practices and the conditions of climate. Suitable mulching technology may help in providing the above-mentioned benefits for agro-ecosystems. Therefore, in the future, we should explore the effects of low-cost, biodegradable and environmentally friendly covering materials on soil microbes, soil erosion, nutrient balance and plant growth (Briassoulis & Degli Innocenti, 2017).

Sometimes both bark mulch and wood chips are used along with fabrics and plastics of landscape. Faric or plastic are usually placed on the soil and covered with a layer of fragments of bark. A caveat to this approach: Although, at first fabric or plastic can provide protection against weeds, but with the passage of time as the mulch decomposes, weeds start to grow in the mulch. In addition, the barrier between mulch and soil also prevents improving soil conditions and hence making it more difficult to grow more plants.

3. Mulch can potentially maintain soil moisture

Strong winds, high temperature, severe weather conditions and competing plantations such as weeds are the factors which play a role in losing soil moisture content and transforming it into barren land. According to an estimation the presence of weeds may cause water loss as much as 25% due to evapotranspiration (Chakraborty & Newton, 2011). In this regard mulch plays a significant role in reducing weed infestation and evaporation loss, and in increasing soil infiltration and retention. According to reports, straw mulching can reduce the evaporation rate by 35% (Russel, 1940). Similarly, in accordance to (Kacinski, 1951) nonliving mulch materials have the strongest ability to retain water in the soil compared to uncovered soil (Abouziena et al., 2014). Mulches usually help in reducing the irrigation demands of crops, and sometimes they can fully meet irrigation needs (Singh et al., 2010). Some organic mulches acting as sponge help in retaining rainfall and irrigation water, thereby protecting runoff and providing water when the crop needs it. By using straw as mulch, runoff was reduced by 43% (Borst & Woodburn, 1942). Due to its water retention capacity, the mulch reduces supplemental irrigation and at the same time reduces runoff in the soil profile (Döring et al., 2005).

4. Mulch improves soil fertility

Organic mulch and live mulch have many beneficial effects on soil quality in improving nutrient uptake. Though, the type of material, characteristics of soil and environmental conditions determine the increase, decrease or no effect on nutrients of soil. The use of organic mulches is more effective because they are decomposed in the same environment and provide nutrients as described in Fig. 14.2. Wood chips, bark, straw and green manure mulches provide more nutrients than inorganic mulches

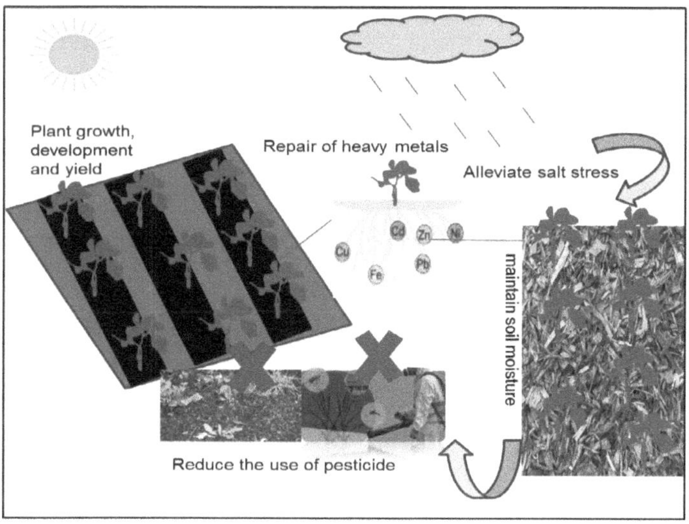

Fig. 14.2 Environmental assistances of mulching in agro-ecosystem.

(Qu et al., 2019). The study of Chalker-Scott (2007) showed that uncomposted bark or straw (low-nutrient mulch) reduces the nitrogen content of the soil without having any effect on plant nutrition, thereby reducing watershed pollution. Mulches with high nitrogen content can assist in increasing the crop yield. Mulch with low nitrogen content can increase plant nutrition and soil fertility such as outdoor tree mulch sawdust mulch and bark mulch and soil. Compared with nitrogen-rich plants or grass mulch, husk mulch is the most effective for enhancing the nutrient contents (Pinamonti, 1998).

5. Mulch helps in alleviating salt stress

Healthy soil can ensure food security and sustainable crop production. However, owing to human disturbance and several other variables, the problem of soil salinity has become increasingly serious. Poor management and misuse of these soils can lead to soil erosion, reduced crop productivity, reduced water retention ability of soil, reduced biodiversity of soil, which ultimately results in carbon loss and desertification (Rojas et al., 2016). Irrigation of plants with untreated municipal wastewater having high salt content badly affects the growth and development of crops. In addition, the wide use of fertilizers, pesticides, detergents and other chemicals may add a lot of salt to the soil (Larney & Angers, 2012). Mulch can effectively solve the salt problem through increasing the water holding capacity of soil and by reducing evaporation. Several studies have shown that coverings can reduce the toxicity of salts (Meena et al., 2019). It has been observed that organic mulch is more helpful for soil reclamation and leads to soil desalination. Organic coverings can be decomposed by microorganisms, thereby degrading many harmful residues and reducing salt pollution (Xie et al., 2019).

6. Role of mulching in growth, development and yield of plant

There is extensive use of covering materials for the establishment of various herbs and tree species. Several studies have shown that, compared with uncovered treatments, mulch has a positive effect on the germination and survival of new plants, as well as seedling transplantation and overall performance of crop plants (Table 14.1). This way, coverage is conducive to maximize yield using minimum input resources (Iqbal et al., 2020). Straw mulch combined with large-scale precision planting is a suitable measure

Table 14.1 Covering can enrich and protect the soil and help provide a better growth environment.

Material	Amount to apply	Important	References
Bark mulch	2–4 inches	Smaller chips can easily be spread around small plants. It is very appropriate for utilization around shrubs, trees and perennial gardens. Mulches must be kept one or two inches away from the trunk, when spreading around the tree. Covering of few inches is sufficient. Application of 6 or 8 inch tall covering as seen often is not needed.	(Tripathi & Kamal Tiwari, 2020)
Wood chips	2–4 inches	Similar to bark mulch. Composting may be advantageous for the mixture of fresh wood chips with a large number of leaves.	(Hoover-Litty & Hanlin, 1985)
Leaves	3–4 inches	Chopping and composting before spreading is best. Dry leaves should be applied about 6 inches.	(Hong-hai et al., 2013)
Grass clippings	2–3 inches	Due to compacting and rotting, thick layers become very sticky and smelly. Add other layers when the clipping is broken down. Do not use newspaper clippings on lawns that have been treated with herbicides.	(Donjadee & Tingsanchali, 2016)
Newspaper	1/4 inch	Spread some newspapers and lightly cover with grass clippings or other covering materials to secure it. If there is no other covering material, cover the paper edges with soil. Application on windy day may be problematic.	(Frochot & Levy, 1986)
Compost	3–4 inches	Excellent material for fertile soil.	(Pinamonti, 1998)

to compensate for the reduction in winter wheat grain yield and improve grain quality (Ahmad et al., 2020; Iqbal et al., 2020). The use of sorghum straw mulching planting pot cultivation technology, which is a technology designed for small farmers with limited livestock capacity, can significantly increase the yield of dryland sorghum and the weight of 1000 grains (Masaka et al., 2020). Based on past research results, coverage can increase agricultural output from 7% to 47%, as described in Table 14.1. Corn has the highest growth rate, while wheat has the lowest growth rate.

6.1 Mulch helps in reducing diseases

Covering materials can reduce evaporation losses, thereby retaining soil moisture and hence reducing irrigation water demand. The mulch can be used as a barrier to the beating of irrigation water or raindrops carrying spores of various diseases, which attach to the leaves and branches of fragile plants (Wang et al., 2019). The mulch supports the growth of beneficial microbes by providing nutrition, these beneficial microbes than compete with pathogens or sometime release chemicals to inhibit the pathogens; this way they help in reducing diseases. It is necessary that before using the mulch material it must be composted fully because the covering material of diseased plants contains pathogens inside, which might be transferred to healthy plants (Cook, 1986). Sometimes, the covering material is subjected to temperature treatment to destroy all beneficial and harmful organisms contained in the covering. Commercially available organic coverings are usually sterile in nature.

6.2 Mulch helps in weed reduction

Mulch is a useful tool for controlling weed populations in nurseries and fields, as shown in Fig.14.1 however, until now, people have not fully understood the phenomenon of weed reduction (Case et al., 2005) the number of weeds was reduced by 92% compared to the nonmulch treatment. Spread of the mulches on the surface of soil make it a barrier for the light passing through it, consequently it reduces the germination of small seed weeds. Almost various types (15 types) of mulches are used. According to one study there is no difference between all mulch types, but the bare soil treatment has significant differences in reducing weeds (Humphreys et al., 2011). Although, mulches act as a physical barrier to weeds, however, when organic mulches decompose, then these weeds quickly emerge out of the soil surface. Some organic mulch also plays an allelopathic effect and releases some toxic chemicals that help to reduce weeds. In addition, the environment created by the mulch is very helpful to the beneficial microorganisms that feed on weeds or weed (Grundy & Bond, 2007).

Similarly, living mulch also help in reducing weed species by competing for basic resources such as nutrients, water, oxygen and light etc. and also through allelopathic effects on weeds. Some cover crops and ground cover plants also help reduce the germination and growth of weed seeds. The number of mulches should be limited under field conditions, if too many live types of mulch are used, they will compete for basic resources. Although plastic films and landscape fabrics have the ability to successfully reduce the number of weeds, they have not been used for long. Plastic coverings, such as white and green plastic coverings, allow photo-synthetically active radiation (PAR) to pass through, so that weeds will grow under these coverings. While, dark coverings do not allow these radiations to pass through, as a result weeds amount reduces. Some tenacious weeds can grow above the mulch surface and sometimes they can even penetrate the plastic mulch (Chalker-Scott, 2007). Replacing fabric and plastic mulch in addition to having a time-consuming process is also a costly method of covering the soil surface. Replacement of these mulches can even damage the crops we are interested in (Weston & Duke, 2003).

6.3 Mulch helps in repairing of heavy metals

Heavy metals found in the soil are usually harmful for animals and human health. Mulch plays an important role in removing these heavy metals and making the soil beneficial (Chalker-Scott, 2007). The leaves of arborvitae, eucalyptus, pine and poplar are mainly used to remove these heavy metals from soil solutions (Schwan et al., 1949). Wood chips and compost in forest areas can be compounded with copper metal and converted into a form that is nontoxic to normal crop growth (Kuo et al., 2004).

6.4 Mulches assist in reducing the use of pesticides

Covering materials can reduce the biotic stress level by dealing with various pathogens affecting the plant. Plants are resistant to weeds and other harmful pests, so that no fungicides, insecticides and herbicides of any kind are used (Altman & Campbell, 1977). Reducing the use of such chemicals will benefit farmers, because there is no need to spend money on such chemicals, and not using chemicals will be beneficial for useful soil microbes as well as for the environment (Wilson & Tisdell, 2001).

6.5 Mulching as water-saving technique in dryland agriculture

Irregular rainfall pattern in arid and semiarid regions together with global warming system has resulted in shortage of agriculture water resources. In order to alleviate the pressure of water shortage in agriculture, mulching as a water-saving technology in the cultivation of rain-fed crops has a vital impact (Deng et al., 2006). It is mainly used for maintaining moisture content of soil, reducing temperature and evaporation of soil which are the major factors of crop yield reduction. There are several strategic effects of coverage on soil ecosystem, climate and on crop growth. The mulch by insulating the soil assists in providing a buffer against the effects of hot and cold temperatures (Blanco-Canqui & Lal, 2009). In dryland areas, mulch acts as a water saving technology through maintaining moisture content, regulating temperature and by reducing evaporation of soil (Reicosky, 2007). In rain-fed agricultural systems, land cover is widely used as a water-saving technology (Deng et al., 2006). Plastic film mulching is more effective in retaining soil moisture than wheat straw mulching (Wu et al., 2016). Mulch usually maintains the moisture content of soil through reducing surface evaporation, regulating temperature as well as by controlling soil erosion, thereby reducing the need for irrigation during crop planting The soil water and heat transfer mechanism under the mulch is very important to improve the availability of the mulch system. It is still unknown that how much water can be saved by mulch due to interaction between microclimate, soil environment and plant growth (McMillen, 2013).

6.6 Mulch improves input efficiency

By improving the soil environment, the beneficial effects of mulching on crop growth and yield increase the efficiency of water and nutrient use. Due to reduced evaporation and runoff loss under the mulch, the increase in the amount of available water in the upper layer is one of the reasons for improving the efficiency of input use (Li et al., 2013). In addition, the change of the soil hydrothermal state under the mulch is conducive to root proliferation, and water and nutrients can be extracted from a larger soil volume, thereby improving the efficiency of water and nutrient use. Coverage can better coordinate water demand and supply, thereby completely extracting the water stored in the profile, thereby improving water use efficiency (Gan et al., 2013). Since the grain production-water use relationship is linear within the limited water supply range, any increase in water supply exceeding the minimum threshold will be directly reflected in the increase in grain production and water use efficiency (WUE) (Lindroth et al., 1994). Nowadays, the major challenges of intensive horticulture are increasing crop yield, enhancing nutrient use efficiency, producing high quality products and protecting environment. But high-yield sustainability and reduced environmental impact are in conflict. Nitrogen is the most vital nutrient required by horticultural crops. Although the toxicity of NO_3 in groundwater increases costs and conceivable adverse effects, sources of inorganic nitrogen fertilizers are being widely used (Reddy, 2016). The main part of this reactive nitrogen is used as fertilizer in horticultural crops/planting systems, escapes to the environment, has a chain negative impact on ecosystem functions, and causes global warming. In this regard, green manure crops and organic fertilizers have great potential to partially replace inorganic nitrogen (Galloway et al., 2002).

In addition to nitrogen used as fertilizer, plant available nitrogen is extracted from SOM through mineralization. The available nitrogen status of the soil varies depending on the planting system and management practices. The amount of nitrogen fertilizer to be applied on horticultural crops is usually based upon the yield potential of the cultivar, planting system, nutritional status of soil, and growth conditions. Since the nutrition and reproduction stages of these crops overlap, plants need nutrition even until the fruit is ripe (Olfs et al., 2005). These practices also help in reducing nitrogen loss, thereby making them environmentally friendly (Jat et al., 2015).

7. Future research and gaps

Farmers in developing countries do not know the environmental pollution of plastic coverings. Plastic film negatively affects the soil ecosystem and environment, so it is important to find new covering materials. Mulch from plant sources can be used as the beginning of the life cycle of renewable resources. For the newly developed covering based on biodegradable, petroleum and textile materials used to protect water resources of agriculture still have limitations in the research of economic analysis. In addition, recycled paper-based mulch (such as newspapers) will release ink to the soil surface (Blaikie, 1995), which can lead to physical and chemical interactions with the soil, requiring in-depth investigation in the future. Further research is needed, including the appropriate coverage rate and plant response. Future research work may also focus on the cost-effectiveness of hydraulic coverage, spray coverage, biodegradability, biofilm, and petroleum-based coverage in agriculture.

Modeling methods for different covering materials need to consider the effective usability of the system. Organic mulch protects direct rainfall infiltration, and plastic mulch prevents the flow of water to the soil profile. Therefore, the water flow mechanism of the organic covering and the plastic covering exhibits different characteristics, and a modeling method is required. For instance, the modeling effects of the water vapor flow and heat transfer process of crop growth in different climatic regions through different thicknesses of organic cover soil may receive attention in future research. In addition, mulches of different plastic colors exhibit different optical properties, so the color of the mulch may affect the temperature of soil and canopy distribution of plant. Hence, the numerical model needs to pay attention to the interaction between mulch, soil and plant canopy interface. It may help in creating a new window of opportunities for the effective utilization of mulch systems in agricultural soils in the future. In the future, it is necessary to simulate the interaction of soil water under cover soil in terms of soil heat capacity and thermal conductivity.

8. Summary remarks

In modern agricultural production, mulching has become a vital water saving technique in arid and semiarid areas. Covering materials protect the surface of soil from sunlight, thereby reducing evaporation through maintaining soil moisture and regulating the temperature of soil. The water use in the root zone of soil is a key use of water in the soil root zone is a key phenomenon to effectively enhance water utilization and saving water resources through mulching. In addition, to having role as water saving technology, mulching is also responsible for beautifying farmland. The choice of covering material chiefly depends on the availability of material, environment, durability and cost-effectiveness. For sustainable use, there is also need that it must be environmentally feasible. Hence, it is concluded that the use of different covering materials may help in saving agricultural water resources, thereby increasing the yield of rain-fed cultivated crops.

9. Conclusion

Efficient utilization of water is a key factor in the growth period of crops, which can greatly increase yields. Therefore, the use of mulch to maintain soil moisture may be an effective option for water saving and increasing dryland agricultural production. Mulch technology establishes a connection between soil and agro meteorology, which can change the growth environment of crops. By improving the soil environment, the beneficial effects of mulching on crop growth and yield increase the efficiency of water and nutrient use efficiency. Due to reduced evaporation and runoff loss under the mulch, the increase in the amount of available water in the upper layer is one of the reasons for improving the efficiency of input use.

References

Abouziena, H. F., El-Saeid, H. M., & Amin, A. (2014). Water loss by weeds: A review. *International Journal of ChemTech Research, 7*(01), 323–336.

Ahmad, S., Raza, M., Saleem, M., Zaheer, M., Iqbal, R., Haider, I., Aslam, M., Ali, M., & Khan, I. (2020). Significance of partial root zone drying and mulches for water saving and weed suppression in wheat. *Journal of Animal Plant Science, 30*, 154–162.

Altman, J., & Campbell, C. L. (1977). Effect of herbicides on plant diseases. *Annual Review of Phytopathology, 15*(1), 361–385.

Appannagari, R. R. (2017). Environmental pollution causes and consequences: A study. *North Asian International Research Journal of Social Science and Humanities, 3*(8), 151–161.

Bahadur, S., Verma, S., Prasad, S., Madane, A., Maurya, S., Gaurav, V. V., & Sihag, S. (2015). Eco-friendly weed management for sustainable crop production-A review. *Journal of Crop Weed, 11*(1), 181–189.

Blaikie, P. (1995). Changing environments or changing views? A political ecology for developing countries. *Geography*, 203–214.

Blanco-Canqui, H., & Lal, R. (2009). Crop residue removal impacts on soil productivity and environmental quality. *Critical Reviews in Plant Science, 28*(3), 139–163.

Borst, H., & Woodburn, R. (1942). The effect of mulching and methods of cultivation on runoff and erosion from Muskingum silt loam. *Agricultural Engineering, 23*(1), 19–22.

Briassoulis, D., & Degli Innocenti, F. (2017). Standards for soil biodegradable plastics. In *Soil degradable bioplastics for a sustainable modern agriculture* (pp. 139–168). Springer.

Case, L., Mathers, H., & Senesac, A. (2005). A review of weed control practices in container nurseries. *HortTechnology, 15*(3), 535–545.

Chakraborty, S., & Newton, A. C. (2011). Climate change, plant diseases and food security: An overview. *Plant Pathology, 60*(1), 2–14.

Chalker-Scott, L. (2007). Impact of mulches on landscape plants and the environment—a review. *Journal of Environmental Horticulture, 25*(4), 239–249.

Cook, R. J. (1986). Plant health and the sustainability of agriculture, with special reference to disease control by beneficial microorganisms. *Biological Agriculture & Horticulture, 3*(2–3), 211–232.

Deng, X.-P., Shan, L., Zhang, H., & Turner, N. C. (2006). Improving agricultural water use efficiency in arid and semiarid areas of China. *Agricultural Water Management, 80*(1–3), 23–40.

Donjadee, S., & Tingsanchali, T. (2016). Soil and water conservation on steep slopes by mulching using rice straw and vetiver grass clippings. *Agriculture and Natural Resources, 50*(1), 75–79.

Döring, T. F., Brandt, M., Heß, J., Finckh, M. R., & Saucke, H. (2005). Effects of straw mulch on soil nitrate dynamics, weeds, yield and soil erosion in organically grown potatoes. *Field Crops Research, 94*(2–3), 238–249.

Erenstein, O. (2002). Crop residue mulching in tropical and semi-tropical countries: An evaluation of residue availability and other technological implications. *Soil and Tillage Research, 67*(2), 115–133.

Frochot, H., & Levy, G. (1986). Effectiveness of a newspaper mulch on juvenile growth of wild cherry (Prunus avium). In *Annales des Sciences Forestières* (pp. 263–267).

Galloway, J. N., Cowling, E. B., Seitzinger, S. P., & Socolow, R. H. (2002). Reactive nitrogen: Too much of a good thing? *Ambio, 31*(2), 60–63.

Gan, Y., Siddique, K. H., Turner, N. C., Li, X.-G., Niu, J.-Y., Yang, C., Liu, L., & Chai, Q. (2013). Ridge-furrow mulching systems—an innovative technique for boosting crop productivity in semiarid rain-fed environments. *Advances in Agronomy, 118*, 429–476.

Gomiero, T., Pimentel, D., & Paoletti, M. G. (2011). Environmental impact of different agricultural management practices: Conventional vs. organic agriculture. *Critical Reviews in Plant Sciences, 30*(1–2), 95–124.

Grundy, A., & Bond, B. (2007). *Use of non-living mulches for weed control* (pp. 135–153). Non-Chemical Weed Management.

Han, Q., Harayama, H., Uemura, A., Ito, E., Utsugi, H., Kitao, M., & Maruyama, Y. (2020). High biomass productivity of short-rotation willow plantation in boreal Hokkaido achieved by mulching and cutback. *Forests, 11*(5), 505.

Haque, I., Powell, J., & Ehui, S. K. (1995). Improved crop-livestock production strategies for sustainable soil management in tropical Africa. *Soil Management: Experimental Basis for Sustainability and Environmental Quality, 4*, 293.

Hong-hai, L., Huan-yong, H., Ya-li, Z., & Wang-feng, Z. (2013). Effects of drought and re-watering on endogenous hormone contents of cotton roots and leaves under drip irrigation with mulch. *Yingyong Shengtai Xuebao, 24*(4).

Hoover-Litty, H., & Hanlin, R. T. (1985). The mycoflora of wood chips to be used as mulch. *Mycologia, 77*(5), 721–731.

Humphreys, E., Eberbach, P., Katupitiya, A., & Kukal, S. (2011). Growth, yield and water productivity of zero till wheat as affected by rice straw mulch and irrigation schedule. *Field Crops Research, 121*(2), 209–225.

Indoria, A., Rao, C. S., Sharma, K., & Reddy, K. S. (2017). Conservation agriculture—a panacea to improve soil physical health. *Current Science*, 52–61.

Iqbal, R., Raza, M. A. S., Valipour, M., Saleem, M. F., Zaheer, M. S., Ahmad, S., Toleikiene, M., Haider, I., Aslam, M. U., & Nazar, M. A. (2020). Potential agricultural and environmental benefits of mulches—a review. *Bulletin of the National Research Centre, 44*, 1–16.

Jat, L. K., Singh, Y., Meena, S. K., Meena, S. K., Parihar, M., Jatav, H., Meena, R. K., & Meena, V. S. (2015). Does integrated nutrient management enhance agricultural productivity. *Journal of Pure Applied Microbiology, 9*(2), 1211–1221.

Kacinski, N. (1951). Sowing oak in micro depressions as a means of combating drought on light chestnut soils. *Pocvoved, 10*, 585–603.

Kader, M., Senge, M., Mojid, M., & Ito, K. (2017). Recent advances in mulching materials and methods for modifying soil environment. *Soil and Tillage Research, 168*, 155–166.

Kasirajan, S., & Ngouajio, M. (2012). Polyethylene and biodegradable mulches for agricultural applications: A review. *Agronomy for Sustainable Development, 32*(2), 501–529.

Kuo, S., Ortiz-Escobar, M., Hue, N., Hummel, R., & Pandalai, S. (2004). Composting and compost utilization for agronomic and container crops. *Recent Research Development and Environment Biology, 1*(Part II), 451–513.

Larney, F. J., & Angers, D. A. (2012). The role of organic amendments in soil reclamation: A review. *Canadian Journal of Soil Science, 92*(1), 19–38.

Lindroth, A., Verwijst, T., & Halldin, S. (1994). Water-use efficiency of willow: Variation with season, humidity and biomass allocation. *Journal of Hydrology, 156*(1–4), 1–19.

Li, S., Wang, Z., Li, S., Gao, Y., & Tian, X. (2013). Effect of plastic sheet mulch, wheat straw mulch, and maize growth on water loss by evaporation in dryland areas of China. *Agricultural Water Management, 116*, 39–49.
Masaka, J., Dera, J., & Muringaniza, K. (2020). Dryland grain sorghum (sorghum bicolor) yield and yield component responses to tillage and mulch practices under subtropical african conditions. *Agricultural Research, 9*(3), 349–357.
McMillen, M. (2013). *The effect of mulch type and thickness on the soil surface evaporation rate.*
Meena, M. D., Yadav, R. K., Narjary, B., Yadav, G., Jat, H., Sheoran, P., Meena, M. K., Antil, R., Meena, B., & Singh, H. (2019). Municipal solid waste (msw): Strategies to improve salt affected soil sustainability: A review. *Waste Management, 84*, 38–53.
Nalayini, P. (2007). *Poly mulching-a case study to increase cotton productivity, Model training course on Cultivation of Long Staple Cotton (ELS).* Regional Station, Coimbatore, India: Central Institute for Cotton Research, 2007.
Ngosong, C., Okolle, J. N., & Tening, A. S. (2019). *Mulching: A sustainable option to improve soil health* (pp. 231–249). Soil fertility management for sustainable development.
Olfs, H. W., Blankenau, K., Brentrup, F., Jasper, J., Link, A., & Lammel, J. (2005). Soil-and plant-based nitrogen-fertilizer recommendations in arable farming. *Journal of Plant Nutrition and Soil Science, 168*(4), 414–431.
Pinamonti, F. (1998). Compost mulch effects on soil fertility, nutritional status and performance of grapevine. *Nutrient Cycling in Agroecosystems, 51*(3), 239–248.
Pirone, P. (1941). Freak weather damages trees and shrubs. *New Jersey Agriculture, 23*(3).
Qu, B., Liu, Y., Sun, X., Li, S., Wang, X., Xiong, K., Yun, B., & Zhang, H. (2019). Effect of various mulches on soil physico—chemical properties and tree growth (Sophora japonica) in urban tree pits. *Plos One, 14*(2), e0210777.
Reddy, P. P. (2016). Agro-ecological pest management. In *Sustainable intensification of crop production* (pp. 269–287). Springer.
Reicosky, D. (2007). Conservation agriculture: Environmental benefits of reduced tillage and soil carbon management in water-limited areas of Central Asia. In *Climate change and terrestrial carbon sequestration in central Asia* (pp. 217–228). CRC Press.
Rojas, R. V., Achouri, M., Maroulis, J., & Caon, L. (2016). Healthy soils: A prerequisite for sustainable food security. *Environmental Earth Sciences, 75*(3), 180.
Rosemeyer, M. E. (2001). Improving agroecosystem sustainability using organic (plant-. based) mulch. *Agroecosystem Sustainability: Developing Practical Strategies*, 67–91.
Russel, J. (1940). The effect of surface cover on soil moisture losses by evaporation. *Proceedings Soil Science Society of America, 4*, 65–70.
Schwan, H., Hodges, D. J., & Weaver, C. N. (1949). Influence of grazing and mulch on forage growth. *Rangeland Ecology & Management/Journal of Range Management Archives, 2*(3), 142–148.
Shah, F., & Wu, W. (2019). Soil and crop management strategies to ensure higher crop productivity within sustainable environments. *Sustainability, 11*(5), 1485.
Singh, R., Kundu, D., & Bandyopadhyay, K. (2010). Enhancing agricultural productivity through enhanced water use efficiency. *Journal of Agricultural Physics, 10*(2), 1–15.
Tripathi, A., & Kamal Tiwari, R. (2020). *A Simplified Sub-Surface Soil Salinity estimation using Synergy of Sentinel-1 SAR and Sentinel-2 multispectral satellite data, for early stages of wheat crop growth in Rupnagar.* Punjab, India: Land Degradation & Development.
Wang, X., Fan, J., Xing, Y., Xu, G., Wang, H., Deng, J., Wang, Y., Zhang, F., Li, P., & Li, Z. (2019). The effects of mulch and nitrogen fertilizer on the soil environment of crop plants. *Advances in Agronomy, 153*, 121–173.
Watschke, T., & Schmidt, R. (1992). Ecological aspects of turf communities. *Turfgrass, 32*, 129–174.
Weston, L. A., & Duke, S. O. (2003). Weed and crop allelopathy. *Critical Reviews in Plant Sciences, 22*(3–4), 367–389.

Wilson, C., & Tisdell, C. (2001). Why farmers continue to use pesticides despite environmental, health and sustainability costs. *Ecological Economics, 39*(3), 449–462.
Wu, Y., Huang, F., Zhang, C., & Jia, Z. (2016). Effects of different mulching patterns on soil moisture, temperature, and maize yield in a semi-arid region of the Loess Plateau, China. *Arid Land Research and Management, 30*(4), 490–504.
Xie, X., Pu, L., Shen, H., Wang, X., Zhu, M., Ge, Y., & Sun, L. (2019). Effects of soil reclamation on the oat cultivation in the newly reclaimed coastal land, eastern China. *Ecological Engineering, 129*, 115–122.

CHAPTER FIFTEEN

How and why to prevent over fertilization to get sustainable crop production

Ghulam Mustafa[a], Naveed Hayat[a], Bader Alhafi Alotaibi[b]

[a]Department of Economics, Division of Management and Administrative Science, University of Education, Lahore, Pakistan

[b]Department of Agricultural Extension and Rural Society, King Saud University, Riyadh, Saudi Arabia

1. Introduction

21st century marked with many challenges such as Covid-19, food insecurity, environmental problems and pest (e.g., locust) attacks. In the presence of such challenges, increasing the agriculture production while keeping environment safe is one of the major challenges (Berg, 2009). World population is increasing day by day, and, in such circumstances, demand for agricultural products increases. As per FAO's estimates, the demand of food will increase to 60% by 2030 (Mia & Shamsuddin, 2010). To fulfill the ever increasing world population food demand, farmers explored different crop nourishment strategies. Among these strategies, chemical fertilizer is one of accessible and cheap techniques. Chemicals pasted an adverse impact both on the environment and living organisms although such fertilizations may make a country self-sufficient in food production at the cost heavy health budgets.

Overfertilizer (higher-than-recommended) application is considered as an insurance against unfavourable circumstances such as floods, hurricanes or dry period. It maintains sufficient levels of plant nutrients during periods of excessive rainfall. Overfertilization facilitates an extended harvest period and it may capitalizes on favorable market conditions. Excess fertilizer application may exacerbate soil salinity that reduces agricultural yields and economic benefits after extreme weather events such as hurricanes in addition to increasing the likelihood of eutrophic waters (Nosheen et al., 2021). The chemical fertilizers cause imbalance in soil nutrients, reduce soil fertility and its water-holding capacity, and result in deterioration of soil and unacceptable level of water pollution (Sprent & Sprent, 1990). These are most of

Sustainable Plant Nutrition
ISBN: 978-0-443-18675-2
https://doi.org/10.1016/B978-0-443-18675-2.00019-5

time expensive and casted a heavy budget on farmers' limited income. Their management strategies are necessary. Those management strategies (discussed in this chapter) are nontoxic, cost-effective, easy to apply, and eco-friendly. They (e.g., biofertilzers and exiting chemical fertilizer techniques) serve as a good substitutes for harmful chemical fertilizers and help to maintain soil structure and biodiversity of the agricultural land (Deepali & Gangwar, 2010; Thomas & Singh, 2019).

2. Losses of over-fertilization

2.1 Economic losses of over fertilization

Economic losses of over fertilization can be seen from agronomy, eco-environmentally and profit maximization prospective. The agronomy perspective emphasizes on the maximization of agricultural output. For crop production, if the chemical fertilizer is the only variable input, then crop yield curve is shown in Fig. 15.1. Point "A" in Fig. 15.1 depicts the input level for maximizing output, up to point "A" the farmer can use the chemical fertilizer but beyond point "A" if the farmer increases the level of chemical fertilizer then his/her output decreases. Most of the farmers did not adopted the optimal level of chemical fertilization because they did not haw proper knowledge about the optimum use of chemical fertilizer and resultantly they get lower amount of output from a given crop (Xue et al., 2020).

The eco-environment fertilization standard emphasizes on the use of chemical fertilizers to an extent which do not causes ecological and environmental imbalances. The over fertilization to the field causes point-source

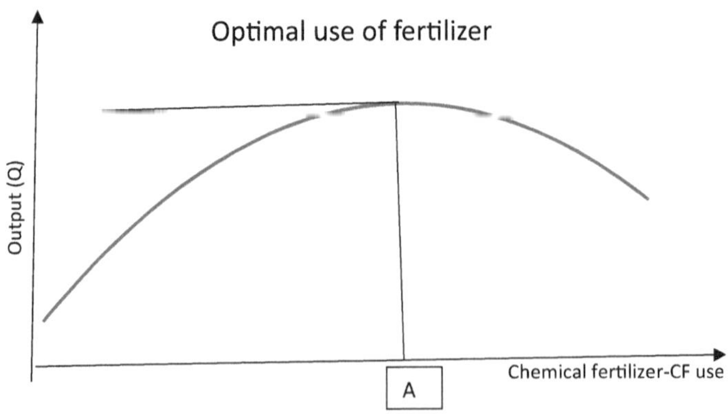

Fig. 15.1 Optimal application of chemical fertilizers in agronomy.

pollution and destroy the ecological environment. The farmers use heavy amount of chemical fertilizers which causes ecological and environmental destructions but they take little responsibility for this ecological destruction. The eco-environment perspective of over fertilization focuses on the maximization of profit, taking consideration of the production cost and the environmental cost caused by the use of chemical fertilizers (Chen et al., 2014). The optimum chemical fertilizer use occurs at the point where the marginal value of the agriculture product (VMP) equals to the marginal cost of the product (SMC), which is given as point "S" in Fig. 15.2 and usually called the social optimal point. Such point is indicated by previous study (Xue et al., 2020).

The economics perspective emphasizes the maximization of profits, merely taking into consideration the production cost (Qiu et al., 2014). When the produce value is high, it is suggested that some environmental value can be abandoned appropriately. Because the negative externality caused by the application of chemical fertilizers on the environment is not taken into consideration, the marginal cost line (PMC) moves down. In Fig. 15.2, the optimal chemical fertilizers use is denoted at point "P", which is also known as the private optimal point. Since rational smallholder farmers (Balogh, 1964), whose goal is to pursue maximized profit, it is reasonable to measure the optimal amount of chemical fertilizers application by farmers from the perspective of economics. In addition, economics perspective is considered appropriate for examining farmers' optimal fertilization use (Xue et al., 2020).

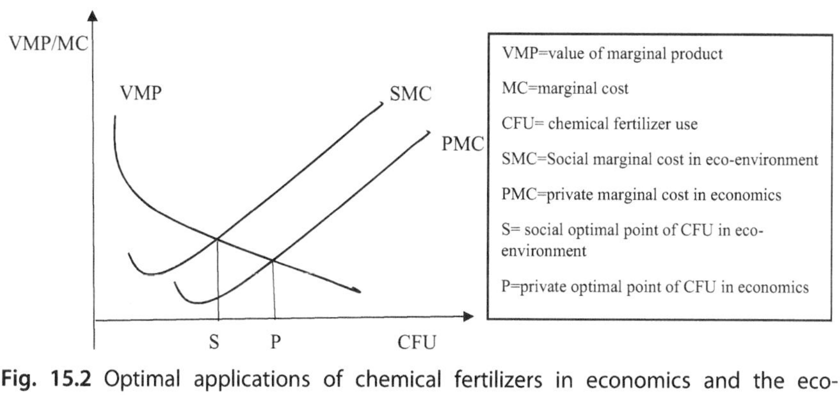

Fig. 15.2 Optimal applications of chemical fertilizers in economics and the eco-environment.

2.2 Losses to plants and soils

Too high salt concentrations which alter the soil are the result of using excess fertilizers, and this hurts beneficial soil microorganisms. Over fertilization can lead to sudden plant growth, whose root system is weak and insufficient to supply adequate water and other nutrients to the plant. Thus, plant does not get proper food production and its growth cannot be sustained. Minimum number of flowers is also an indicator of improper growth and lack of nutrients. Plants need soluble salts, which can also ne inculcated by the use of fertilizers. These fertilizers are mainly minerals which can easily dissolve in water and can be transported easily also. So, the fertilizers which are given to plants should be easily dissolved in water to enable roots to uptake them to all parts of the plant. But the use of these fertilizers should be limited. When these are used in excessive quantity it becomes a trouble for roots to carry and exchange them. Due to excessive quantity, these fertilizers become unable to dissolve naturally and thus they raise soil salinity and alter the ph. Due to lower pH nutrients are less available to plants. High pH soil is alkaline while low pH soil is acidic. Most of the plant nutrients are present in that soil with neutral pH, which is between 5.5 and 7.5, so nutritionally balanced.

Another loss is that the overuses of chemical fertilizers increase soil acidification. Though these chemical fertilizers help plants grow faster but these plants are not strong and healthy, as they do not get enough time to establish strong roots and stem, or better quality fruits and vegetables. Weak roots and stem lack good immune system and the weak plant will be more susceptible to pests and diseases. The use of chemical fertilizers also cause root and fertilizer burn.

In addition, some fertilizers merely do not dissolve in water as they are not water soluble; like phosphorus, and its overuse may cause hardening of soil. On the other hand fertilizers containing nitrogenous contents damage topsoil when applied to the fields in large amount for a long period of time and deteriorates the crop yield. Furthermore these nitrogen salts are water soluble and absorb in water too quickly, causing dehydration and dryness in plants. It's also contaminates in soil and groundwater and remains here for decades, causing accumulative effect on soil and ultimately on plants.

2.3 Losses to human and animal life

Excessive fertilization is harmful for human health. In this case those effects of chemical fertilizers are lethal as it contains a direct toxic effect. The chemical reaction between nitrites and secondary amines; present in chemical fertilizers creates a substantial role of nitrosamine exposure in Alzheimer's disease, diabetes mellitus and such diseases. These chemical fertilizers also cause damage in deoxyribonucleic acid and lipids peroxidation. Proinflammatory cytokines gets activated; which together with lipids preoxidation increases cellular degeneration and eventually leads to death. Nitrate which is used in these fertilizers is a major component of polluted water which is causing major and lethal diseases. The nitrogen contents get dissolved in underground water. Drinking nitrate and nitrite contaminated water causes methemoglobinemia in children. The blue coloration of skin and even cancer can be result of drinking this polluted water. It can also damage respiratory and vascular system. This water is also causing eutrophication and mass death of fish and other aerobic organisms. In addition Cadmium is another chemical content which is adversely affecting human health. Normally only 5% of the cadmium in food is absorbed in gastrointestinal tract. This ultimately enters the human tissues, results in diseases like tracheobronchitis, pneumonitis, pulmonary edema and osteoporosis. This can also lead to kidney and bone damage. For wildlife it also has adverse effects. The water with excess fertilizers causes low quantity of dissolved oxygen in surface water, due to which fishes and other aquatic species suffocate.

2.4 Losses to the environment

The deleterious effect of excess chemical fertilizers starts from their manufacturing process, whose products and byproducts pollute the air in the form of some toxic gases like NH_4, CO_2, and CH_4 etc. CO_2 one of the major causes of global warming and depletion of Ozone layer, hence exposing the humans to harmful ultraviolet rays. In presence of soil bacteria, nitrogen fertilizers can be converted into nitrous oxide which is a Greenhouse gas. These nitrous oxides (NO, N_2O, NO_2) are responsible for severe air pollution. The global warming potential of these gases is 310 times more than that of carbon dioxide. Another lethal gas ammonia emitted from fertilizers, forms nitric acid and sulfuric acid after deoxidation and causes acid rain. Methane is also important Greenhouse gas as its concentration is increased by applying ammonium based fertilizers up to the safe limit. When the application of fertilizers exceeds the recommended limit, these get lost in the environment in different ways. This can be drainage, leaching or surface flow; such that nitrogen gets leached down the ground water and also affects the roots of the plant. Even if these fertilizers are used under certain limit, plants only use up to 50% of the N fertilizers to 2%–20% is that which get volatilized, 15%–25% is that which reacts to organic compounds and the 2%–10% which remains, interferes surface and groundwater. This nitrogen can be a factor responsible for eutrophication when biomass growth increases.

3. Over fertilization management

Fertilizer management significantly affect the crop production (agriculture production). This show that by using optimum level of chemical fertilizers the farmers can increase the crop yield (Zhang et al., 2016). A proper management of controlling over fertilization include: the timing of fertilization, the consistency between the fertilizer types and nutrient requirements as well as the technologies used for fertilization (Xue et al., 2020).

3.1 The timing of fertilization

The various nutrients released by the soil organic matter cannot fully meet the normal needs of crop growth. Therefore, it is necessary to continuously supplement nutrients through artificial fertilization to ensure the healthy growth of the crop (Cai et al., 2002). When applied according to the growth cycle of a crop and its nutrient requirements at each stage, artificial fertilization could play an important role in improving the crop yield during the critical periods of crop nutrient demand (Grahmann et al., 2016; Melaku et al., 2018). The appropriate timing of top-dressing, for example, is found to contribute to an increase in crop yield (Zhang et al., 2016), because fertilizing at appropriate times is helpful for the growth of the crop and its competition with weeds for nutrients (Blackshaw et al., 2004). On the other hand, delaying or advancing the fertilization time would reduce crop yield (Fischer et al., 1993; Scharf et al., 2002; Xue et al., 2020).

3.2 The consistency between the types of fertilizers applied and the nutrient requirements of crop

At various stages of growth, crops vary in the quantity and type of nutrients needed. Farmers need to apply fertilizers in accordance with the nutrient requirements of crops, the soil environment and climatic conditions to increase the efficiency of their fertilizer use. In order to reduce the frequency of fertilization and improve fertilizer utilization efficiency, slow-release fertilizers and inhibitors can be applied. Slow-release fertilizers and inhibitors can slow down the release rate and conversion rate of fertilizer nutrients, respectively, but they have not yet been widely applied to the agriculture. Fertilization application should consider a combination of both organic fertilizers and chemical fertilizers. The application of excessive chemical fertilizers is more likely to cause overfertilization (Pan et al., 2017). If Farmer fail to apply an appropriate type of fertilizer at an appropriate time, it may lead to the nutrient deficiency of crops. In this circumstance, the accurate identification of a deficiency in nutrients is the premise for farmers to prescribe the right medicine. If the farmers cannot correctly identify the disease, they may mistakenly use prescription, missing the opportunity for appropriate treatment. As a remedy, farmers tend to apply chemical fertilizers for recovering some losses, possibly leading to overfertilization (Xue et al., 2020).

3.3 The technologies used for fertilization

The optimal fertilization largely on the technologies of fertilization (Adams et al., 2018). Advanced fertilization technologies can increase the fertilizer-use efficiency and can reduce costs of the farmers (Pan et al., 2017). The fertilization technologies include the technologies for determining the fertilizer type to be applied and helping the crop absorb nutrients. Commonly, the farmers applied two fertilization technologies in the field namely short-time fast Fourier transform (STFFT) and WFIT. STFFT is a fertilization technology that simultaneously considers the fertilizers demanded by the crop and supplied by the soil (Marenya & Barrett, 2009). Its purpose is to increase crop yield and protect ecosystems, by means of balancing the supply of various nutrients, increasing the fertilizer-use efficiency and reducing chemical fertilizers application. The field trial results and farmers' surveys results have both proved that using STFFT can achieve these goals (Ma et al., 2014; Xu et al., 2014). The WFIT is a technology that integrates fertilization with irrigation. It enhances effectiveness, mobility and solubility of nutrients in the soil by means of small amount, fractionated, dripped, which helps promote the efficiency of the absorption of nutrients by the crop (McBeath et al., 2007). In addition, WFIT has the advantages of reducing the chemical fertilizers amount, increasing crop yields and improving the quality of crops (Bhat et al., 2007). The adoption of WFIT, therefore, is expected to help reduce overfertilization (Xue et al., 2020).

3.4 Fertilizer best management practices (FBMPs)

Fertilizer best management practices (FBMPs) are a part of an integrated farming system that includes crop management and all of the soil and plant nutrient management components of a complete farming system. There are integrated crop management, integrated soil fertility management, integrated plant nutrient management and fertilizer best management practices. Based on nutrient stewardship principles, FBMPs not only fulfill the four management objectives of productivity, profitability, cropping system sustainability, and a favorable biophysical and social environment. Specific and universal scientific principles in the development and implementation of FBMPs have been described and discussed to enhance the efficiency of nutrient use through a variety of fertilizer materials and new technologies not only to enhance crop production but also to reduce the negative impacts of fertilizer use on air and water resources (Reetz, 2016; Reetz et al., 2015).

3.5 Adaptive management

In practice, nutrient management, especially nitrogen management, is integrated with other crop management practices in developing a complete production system. To select fertilizer BMPs for a given farm, it is important to use the best science available for optimizing the components of the system including their interactions. Nutrient stewardship wants continuous adaptation to the varying agricultural system in which it is implemented. An adaptive management process for development and adoption of fertilizer BMPs is required to respond to changing conditions in the production system. Adaptive management is a continuous loop that responds to the experience gained from implementing and evaluating practices. The farmer's experience, aided by research from universities and industry sources, guides the decision process. Site factors and stakeholder inputs supply additional information to be considered. The farmer ultimately makes the decisions on which practices and inputs are adopted or modified in the production system. He or she is also ultimately responsible for economic, environmental, and regulatory benefits and costs. Analysis of the outcomes of these decisions provides feedback to adjust the management decisions for future action. The nutrient source, rate, timing and placement decisions are mutually dependent, and the management objectives will change according to local situations, the farmer's objectives, and stakeholder input regarding the comparative priority among system performance indicators. The relative importance of these objectives will significantly affect what is defined as right in terms of source, rate, timing, and placement. Sound science is essential to ensure that the practices chosen have the highest likelihood of attaining the management objectives. For each practice, it is important to consider the source, rate, timing and placement of nutrients. In most cases there are multiple options for each of the 4Rs in relation to each practice (Reetz et al., 2015).

3.6 Smart nitrogen fertilizer management by remote sensing

Currently, for the farm decision-making process, GIS data and plant/soil parameters obtained through remote sensing and other analyses are used. Several developed countries, like Sweden, USA, Australia, and Canada, have already implemented national projects to map the soil properties, like organic matter content and other physical and chemical soil properties of soil, in different regions using GIS, for the development of soil management practices (Say et al., 2018; Söderström et al., 2016). On the other hand, vegetation indices (VIs), which are generated from multispectral data, allow the evaluation of crop growth and nitrogen use level at the farm (Osco et al., 2019). The mixture of these information allows the control of a variable-rate applicator (VRA) equipped with a GPS-positioned tractor, which accurately conducts nitrogen fertilization at the right positions and time. This can be accomplished because of the increased resolution of GPS and satellite images, while it is also related to the development of cameras, sensors, and imaging methods (Maes & Steppe, 2019). At the early stage of satellite imaging, natural plant habitation was monitored by VIs, which were transformed from the spectral reflectance of red and near-infrared (NIR) (Lloyd, 1990). The infrared ratio is based on the fact that healthy vegetation reflects more NIR than red light, which is partially explained by the difference in the amount of chlorophyll between healthy vegetation and stressed vegetation (Gitelson et al., 2003; Herrero et al., 2020). Several VIs have been suggested for modeling plant growth capacity, yield, and nitrogen status. Among the various VIs, the normalized difference vegetation index (NDVI) is the most popular VI which is widely used to define the nitrogen status of crops. NDVI is calculated as NDVI = (NIR-Red)/(NIR + Red), and it can well describe the variation in chlorophyll content and thus in the crop nitrogen status (Vizzari et al., 2019). Chlorophyll fluorescence spectral emission ranges approximately 650–800 nm in intact leaves with two maxima: one in the red spectral region around 685–690 nm (F685) and the other in the far-red spectral region around 730–740 nm (F740) driven by solar or laser radiation, which allows the assessment of chlorophyll contents more precisely than VIs, has recently been used for the determination of crop nitrogen status. Uninterrupted development of imaging technologies, which include unmanned aerial vehicles, hyperspectral data (camera/sensor; see next section for details on hyperspectral imaging, HSI), and data analysis (e.g., machine learning, ML), allow more precise nitrogen fertilizer management. Several new VIs have been offered by the data mining of HSI, which allow the prediction of the nitrogen content in crops other than well-studied grain crops, for example in spinach, citrus species, and cucumber. The nitrogen status of crops, which is linked with plant growth and yield, is more valuable than tissue nitrogen content for efficient nitrogen management. In order to enhance the accurateness of nitrogen status prediction, multispectral imaging (MSI) is combined with data obtained by other methods, such as HSI, visible red, green, and blue (RGB) images, and data obtained by sensors Portable sensors that calculate the volume of chlorophyll and nitrogen status are currently used at farming sites. Two nondestructive sensors have been widely adopted for calculating the plant nitrogen status, namely soil plant analysis development and GreenSeeker. These sensors can easily find proximal data at farming sites and can be combined with remote sensing data. This integration sometimes enhances the effectiveness of the nitrogen fertilization model. For phenotyping major cereal crops in the field, it is essential to develop airborne images using UAVs (e.g., drones) with various prediction models (Agrahari et al., 2021; Maes & Steppe, 2019).

4. Integrating sustainability goals and cropping system management objectives

At the practical level, the farmers manage cropping systems for multiple purposes. Fertilizer best management practices (BMPs) should be design in such a way that it meets the agronomic and economic needs of farmers, but at the same time should also restrict nutrient losses that impact the environment and ecological services the other stakeholders want to have protected. The appropriate BMPs are those practices that support various stakeholder objectives. At the field level, it can be difficult to adopt a specific crop management practices that achieve the economic, environmental and social aspects of sustainability. Therefore, it is useful to imagine cropping system objectives as the vehicle for connecting practices to sustainability. System objectives vary with the region, sector and, often, over time, and they depend on the inputs of stakeholders, including farmers, consumers, rural residents, and other citizens. However, four common practical management objectives at the farm level are: productivity, profitability, durability of the cropping system, and environmental health. These practices also affect the wider concept of soil health that relates to the long-term resilience and durability of the cropping system. The overall farm management system should contribute to the preservation and improvement of soil health, including nutrient availability, water-holding capacity, structure, biological activity and other measures. Fertilizer BMPs are part of a larger, interlinked set of nutrient, crop, soil, water and farm management practices. For a best fertilizer management practice, it must harmonize with the other agronomic practices to address the full range of farm-level management objectives. The development, assessment and improvement of fertilizer BMPs at the farm level must consider these multiple objectives, as must the selection of indicators reflecting their combined impact at different scales, from the field to the global level (Reetz et al., 2015). The set of cropping system management objectives at the farm level mentioned above can be defined and measured by Reetz et al. (2015) as follows:

4.1 Productivity

For cropping systems, the primary measure of productivity is yield per unit area of cropland per unit of time and per unit of total inputs. The quality of the yield is part of the productivity measure. Both can effect profitability. Productivity should be considered in terms of all resources such that land, water, energy, labor employed at the firm. Multiple productivities should be calculated to evaluate productivity accurately.

4.2 Profitability

Profitability is determined by the difference between the revenue and the cost of production. Primary measure of profitability at farm level is net profit per unit of cropland per unit of time. The profitability impact of a specific management practice is related to its economic efficiency, which is the increase in yield value compared to the cost of the practice.

4.3 Durability of the cropping system

Durability refers to the ability of the cropping system to maintain resource quality over time. A durable production system is one where the quality and efficiency of the resources used do not diminish over time. Therefore, for a cropping system durability, outputs can be sustained or increased over time, without a need for increased inputs. With good farm management practices, the cropping system durability may increase, particularly on degraded soils, if increased crop productivity and photosynthesis lead to increased return of crop residues to the soil. Greater residue return can increase soil organic matter content, contributing to better soil health and improved productivity.

4.4 Environmental health

Crop production systems have several undesirable effects on the ecosystems because the crop production release environmental harmful material to water and air. These environmental impacts can be felt at local, national, continental or global levels. Specific environmental effects from crop production can be controlled by practices designed to optimize resource use efficiency. However, not all environmental effects can be controlled to the same level. Some environmentally important losses, such as those of phosphorous or nitrous oxide (N_2O), involve only a small fraction of the input applied. Others, such as ammonia (NH_3) volatilization or dinitrogen (N_2) emission from denitrification, may involve large losses, but are largely controlled by consideration of impacts on profitability.

5. Possibilities to strengthen fertilizer management

5.1 Enact comprehensive national policies for fertilizers

Many countries of the world, experiences shortcomings in fertilizer quality control. The national polices of these countries may have gaps. In some countries policies regarding pollutants that come from fertilizers, in food, feed and drinking water are not exists. On the other hand, some countries and regions are doing fairly well in controlling the pollution causes by over fertilization. Harmonizing policies at regional levels could lower the costs of implementing these policies (Zhongming et al., 2021).

5.2 Fill information and knowledge gaps for effective fertilizer and nutrient management

A large information gaps exists in the use, efficacy and impacts of fertilizers. In the case of some of the information at this time available there are concerns about its consistency and the comparability of data from different research studies. In some countries efforts to make technical information available to the public is ongoing. In this regard, strengthening partnerships between research programs at national, regional and international levels will be helpful as noted by Zhongming et al. (2021).

5.3 Strengthen global policies on sustainable and safe use of fertilizers

Partnerships that encourage information sharing between scientists and global policymakers can contributes to filling information gaps at the global level. Examples of such partnerships consist of the Intergovernmental Technical Panel on Soils and the Intergovernmental Panel on Climate Change, which provide scientific and technical assistance and guidance at the global level. Contribution by scientists in such partnerships improves chances that the global policies, decisions and suggestions arising from them will be implemented.

5.4 Scale up training of all relevant stakeholders in fertilizer and nutrient management

There is need to scale up regular training of knowledge programs to equip them with up-to-date information, which they can then convey to farmers. Development of a joint platform including knowledge generators, knowledge disseminators, and other important stakeholders could help ensure that messages are harmonized.

5.5 Ensure that suitable and affordable fertilizers are accessible

Use of fertilizer in some least developed and developing countries is very limited. In subSaharan Africa this amount is just 50 kg/ha/year. Resultantly, in the region crop yields are low and many people do not have access to sufficient food full of nutrition. A main reason for lack of access to affordable fertilizer is that fertilizer prices in the region are higher as compared to other regions. Provision of affordable fertilizer in all over the world should be accompanied by controls such that on quality and the use of suitable practices and technologies to minimize adverse impacts.

6. Conclusion

Overfertilization may make a country self-reliant on food at the cost of extreme economics, environment, health and soil fertility losses. Hence, proper fertilizer management strategies are recommended. The study in hand tried to highlight the losses created by over fertilization and their fertilization management strategies. The current chapter found that a threshold level for fertilizer use. Up to that threshold, productivities of crop increase and after that output declines. Thus, productivity response to chemical fertilizer for various crops is inverted U-shaped. Upper threshold of inverted U-shape is the balance use of fertilizer. All-embracing efforts have to be made to train farmers to exercise balanced use of fertilizers. A few fertilizer organizations and institutions have come ahead to train the villagers, booklet of literature in local languages associated with balanced use of fertilizers for better crop yields in a sustainable way. The real time has come; the farmers, researchers and different associated groups have to come ahead and act on this respect. The study found the threshold for overall agriculture however future studies required to find the fertilizer threshold for each crop. By this way, a research-based fertilizer recommendations can be found to address global eutrophication of fresh water systems.

References

Adams, C. B., Thapa, S. B., Fan, Y., & Park, S. (2018). Agronomic and economic effects of two enhanced-efficiency urea fertilizer technologies on Southern Great Plains winter wheat. *Agronomy Journal, 110*(3), 1097–1102.

Agrahari, R. K., Kobayashi, Y., Tanaka, T. S. T., Panda, S. K., & Koyama, H. (2021). Smart fertilizer management: The progress of imaging technologies and possible implementation of plant biomarkers in agriculture. *Soil Science and Plant Nutrition, 67*(3), 248–258.

Balogh, T. (1964). *Transforming traditional agriculture: Jstor.*

Berg, G. (2009). Plant–microbe interactions promoting plant growth and health: Perspectives for controlled use of microorganisms in agriculture. *Applied Microbiology and Biotechnology, 84*(1), 11–18.

Bhat, R., Sujatha, S., & Balasimha, D. (2007). Impact of drip fertigation on productivity of arecanut (Areca catechu L.). *Agricultural Water Management, 90*(1–2), 101–111.

Blackshaw, R. E., Molnar, L. J., & Janzen, H. H. (2004). Nitrogen fertilizer timing and application method affect weed growth and competition with spring wheat. *Weed Science, 52*(4), 614–622.

Cai, G., Chen, D., Ding, H., Pacholski, A., Fan, X., & Zhu, Z. (2002). Nitrogen losses from fertilizers applied to maize, wheat and rice in the North China Plain. *Nutrient Cycling in Agroecosystems, 63*(2), 187–195.

Chen, X., Cui, Z., Fan, M., Vitousek, P., Zhao, M., Ma, W., … Yang, J. (2014). Producing more grain with lower environmental costs. *Nature, 514*(7523), 486–489.

Deepali, G. K., & Gangwar, K. (2010). *Biofertilizers: An ecofriendly way to replace chemical fertilizers.* Available online: https://www.krishisewa.com/articles/organic-agriculture/237-biofertilizer-substitute-chemical.html

Fischer, R., Howe, G., & Ibrahim, Z. (1993). Irrigated spring wheat and timing and amount of nitrogen fertilizer. I. Grain yield and protein content. *Field Crops Research, 33*(1–2), 37–56.

Gitelson, A. A., Gritz, Y., & Merzlyak, M. N. (2003). Relationships between leaf chlorophyll content and spectral reflectance and algorithms for non-destructive chlorophyll assessment in higher plant leaves. *Journal of Plant Physiology, 160*(3), 271–282.

Grahmann, K., Govaerts, B., Fonteyne, S., Guzmán, C., Soto, A. P. G., Buerkert, A., & Verhulst, N. (2016). Nitrogen fertilizer placement and timing affects bread wheat (*Triticum aestivum*) quality and yield in an irrigated bed planting system. *Nutrient Cycling in Agroecosystems, 106*(2), 185–199.

Herrero, H., Waylen, P., Southworth, J., Khatami, R., Yang, D., & Child, B. (2020). A healthy park needs healthy vegetation: The story of gorongosa national park in the 21st century. *Remote Sensing, 12*(3), 476.

Lloyd, D. (1990). A phenological classification of terrestrial vegetation cover using shortwave vegetation index imagery. *Remote Sensing, 11*(12), 2269–2279.

Maes, W. H., & Steppe, K. (2019). Perspectives for remote sensing with unmanned aerial vehicles in precision agriculture. *Trends in Plant Science, 24*(2), 152–164.

Ma, L., Feng, S., Reidsma, P., Qu, F., & Heerink, N. (2014). Identifying entry points to improve fertilizer use efficiency in Taihu Basin, China. *Land Use Policy, 37*, 52–59.

Marenya, P. P., & Barrett, C. B. (2009). Soil quality and fertilizer use rates among smallholder farmers in western Kenya. *Agricultural Economics, 40*(5), 561–572.

McBeath, T., McLaughlin, M., Armstrong, R., Bell, M., Bolland, M., Conyers, M., … Mason, S. (2007). Predicting the response of wheat (*Triticum aestivum* L.) to liquid and granular phosphorus fertilisers in Australian soils. *Soil Research, 45*(6), 448–458.

Melaku, N. D., Bayu, W., Ziadat, F., Strohmeier, S., Zucca, C., Tefera, M. L., … Klik, A. (2018). Effect of nitrogen fertilizer rate and timing on sorghum productivity in Ethiopian highland Vertisols. *Archives of Agronomy and Soil Science, 64*(4), 480–491.

Mia, M. B., & Shamsuddin, Z. H. (2010). Rhizobium as a crop enhancer and biofertilizer for increased cereal production. *African Journal of Biotechnology, 9*(37), 6001–6009.

Nosheen, S., Ajmal, I., & Song, Y. (2021). Microbes as biofertilizers, a potential approach for sustainable crop production. *Sustainability, 13*(4), 1868.

Osco, L. P., Ramos, A. P. M., Moriya, É. A. S., de Souza, M., Junior, J. M., Matsubara, E. T., … Creste, J. E. (2019). Improvement of leaf nitrogen content inference in Valencia-orange trees applying spectral analysis algorithms in UAV mounted-sensor images. *International Journal of Applied Earth Observation and Geoinformation, 83*, 101907.

Pan, D., Kong, F., Zhang, N., & Ying, R. (2017). Knowledge training and the change of fertilizer use intensity: Evidence from wheat farmers in China. *Journal of Environmental Management, 197*, 130–139.

Qiu, H., Luan, H., Li, J., & Wang, Y. (2014). Effects of risk aversion on households' excessive use of chemical fertilizers. *China Rural Economy, 3*, 85–96.

Reetz, H. F. (2016). *Fertilizers and their efficient use.* IFA: International Fertilizer industry Association.

Reetz, H., Heffer, P., & Bruulsema, T. W. (2015). 4R nutrient stewardship: A global framework for sustainable fertilizer management. In *Managing water and fertilizer for sustainable agricultural intensification* (pp. 65–86), 2015.

Say, S. M., Keskin, M., Sehri, M., & Sekerli, Y. E. (2018). Adoption of precision agriculture technologies in developed and developing countries. *Online Journal of Science and Technology, 8*(1), 7–15.

Scharf, P. C., Wiebold, W. J., & Lory, J. A. (2002). Corn yield response to nitrogen fertilizer timing and deficiency level. *Agronomy Journal, 94*(3), 435–441.

Söderström, M., Sohlenius, G., Rodhe, L., & Piikki, K. (2016). Adaptation of regional digital soil mapping for precision agriculture. *Precision Agriculture, 17*(5), 588–607.

Sprent, J. I., & Sprent, P. (1990). Nitrogen fixing organisms: Pure and applied aspects. *Nitrogen Fixing Organisms: Pure and Applied Aspects, 19*, 288.

Thomas, L., & Singh, I. (2019). Microbial biofertilizers: Types and applications. In *Biofertilizers for sustainable agriculture and environment* (pp. 1–19). Berlin/Heidelberg, Germany: Springer.

Vizzari, M., Santaga, F., & Benincasa, P. (2019). Sentinel 2-based nitrogen VRT fertilization in wheat: Comparison between traditional and simple precision practices. *Agronomy, 9*(6), 278.

Xue, C., Zhang, T., Yao, S., & Guo, Y. (2020). Effects of households' fertilization knowledge and technologies on over-fertilization: A case study of grape growers in shaanxi, China. *Land, 9*(9), 321.

Xu, X., He, P., Qiu, S., Pampolino, M. F., Zhao, S., Johnston, A. M., & Zhou, W. (2014). Estimating a new approach of fertilizer recommendation across small-holder farms in China. *Field Crops Research, 163*, 10–17.

Zhang, W., Cao, G., Li, X., Zhang, H., Wang, C., Liu, Q., ... Jiang, R. (2016). Closing yield gaps in China by empowering smallholder farmers. *Nature, 537*(7622), 671–674.

Zhongming, Z., Linong, L., Xiaona, Y., Wangqiang, Z., & Wei, L. (2021). *Environmental and health impacts of pesticides and fertilizers and ways of minimizing them.*

CHAPTER SIXTEEN

Molecular basis of abiotic stress alleviation by nanoparticles

Asad Jamil, Muhammad Umair, Ali Rizwan and Rubaz Ahmed
Institute of Soil and Environmental Sciences, University of Agriculture, Faisalabad, Pakistan

1. Introduction

Abiotic stress refers to any environmental factor that might impact plant growth and productivity (Yadav et al., 2020). Abiotic stress has become a focus of plant scientists and agronomists in the last half-century as a result of high population expansion (Raza et al., 2019). Throughout their life cycle, plants are subjected to numerous forms of abiotic stress and have evolved diverse defense systems to cope with them via distinct physiological pathways (Gull et al., 2019). Abiotic stressors like heavy metals, salinity, temperature, drought and UV radiations have detrimental effects on plant physiology and growth (Ayub et al., 2019, 2020; Rehman et al., 2021; Zaidi et al., 2014). About 90% of the world's arable land is vulnerable to one or more of the above-mentioned stresses, resulting in yearly output losses of up to 70% for important food crops (Mantri et al., 2012). These stressors cause the generation of reactive oxygen species (ROS), which triggers oxidative impairment in the plants (Rehman et al., 2021). The agricultural productivity has been retried since reduced agriculture area, and abiotic factors (Ayub et al., 2020).

Nanoparticles (NPs) are substances with a diameter of 1−100 nm, having tremendous applications in field of agriculture and plant science, to reduce the abiotic stresses (Khan et al., 2019). Unique physico-chemical characteristics of NPs including small size, porosity, catalytic activities, crystallinity, and aggregating abilities have widen the agricultural application of NPs (Rajput et al., 2018; Usman et al., 2020). These particles are part of a smart technology aimed at increasing agro-chemicals efficiency and controlling release and reducing ecological losses (Shang et al., 2019). They can be made in a variety of ways, including physical, chemical, and biological techniques (Jamkhande et al., 2019; Singh et al., 2016), and they have variety of effects when they interact with plants (Ayub et al., 2019; Tripathi, Shweta

Sustainable Plant Nutrition
ISBN: 978-0-443-18675-2
https://doi.org/10.1016/B978-0-443-18675-2.00017-1

Singh, et al., 2017). Nano-agrochemicals can be applied to plant via two ways by soil or foliar application. When they encounter with plants, they tend to enter either through root epidermis or via aerial surfaces, using both apoplastic and symplastic pathways in accordance with the prevailing conditions (Arif et al., 2018; Tripathi, Shweta Singh, et al., 2017; Zahedi, Moharrami, Sarikhani, & Padervand, 2020a,b). The movement (absorption and transportation) of NPs inside the plant matrices is both active and passive depending on their size.

The use of nano-agrochemicals not only enhance the efficiency of inputs but also improve the morpho-physiological attributes of plants including molecular and biochemical aspects (Rizwan et al., 2019). At molecular level, the effect of NPs has been evaluated in different crop plants and the results revealed that NPs influence the activity of antioxidants enzymes by regulating their production at transcription level (Abdoli et al., 2020; Husen & Siddiqi, 2014; Lee et al., 2013). Moreover, the incorporation of CeO, Se, TiO_2 and Ag NPs improves the gene activities that are involved in biotic and abiotic stress management (Ayub et al., 2019; Kaveh et al., 2013; Landa, 2021). During abiotic stress, NPs regulate the accumulation process of osmolytes and the activity of antioxidants (Abdoli et al., 2020; Moradbeygi et al., 2020; Tariq et al., 2021). Furthermore, nanofertilizers promote nitrogen metabolism and increase chlorophyll and protein levels in crop plants (Alsaeedi et al., 2019; Beig et al., 2022) which improve a series of biochemical parameters, resulting in higher agricultural yields (Alharby et al., 2021; Ayub et al., 2019; Bashir et al., 2021; Gruyer et al., 2014).

The purpose of this chapter is to discuss the impacts of different abiotic stresses on plants growth that cause significant reduction in morphological and physiological attributes and the possible roles and mechanisms of NPs for mitigating the negative impacts caused by abiotic stresses, and to improve the agricultural productivity using NPs.

2. Nanoparticles uptake and translocation mechanism in plants

The use of NPs to increase crop productivity can be beneficial, however, large-scale application of NPs in agricultural areas is still in its infancy (Rajput, Minkina, et al., 2021). NPs availability, uptake, translocation, and their adsorption are greatly influenced by their size. As a result, effective strategies for introducing beneficial NPs and making them accessible to plants are necessary (Khan et al., 2019). NPs can be introduced into soil, or by foliar application, on plants (Mittal et al., 2020). A major hurdle to the entry of NPs in plant is the existence of cell wall in plant cells (Zhang et al., 2019). Additionally, cell wall degrading enzymes have been employed to break down this barrier, and protoplasts have been used to introduce NPs for improved transportation (Ahmar et al., 2021). Biolistic techniques have been explored as well, but they are pricey and have limited uses (Nafady et al., 2019). Another way to use NPs is to incorporate them in hydroponics, where an ideal environment with precise conditions is maintained for greater NP absorption (Mittal et al., 2020).

The NPs are absorbed through the root epidermis and/or the aerial surfaces, using both apoplastic and symplastic pathways (Ali and Khan, 2021; Shukla et al., 2016). When NPs are sprayed on aerial parts, they make their entry using pathways from hydathodes or lenticles, trichomes, stomata, stigma, and carried via phloem (Su et al., 2019). When applied in soil, they enter the vascular bundles through the root tip meristem, where casparian strips are underdeveloped and reach to the shoot (Shukla et al., 2018). NPs can penetrate the cell via endocytosis, despite the fact that the casparian strips, cell membrane, and cell wall prevent their absorption and transport to the cells (Shukla et al., 2016). Aquaporins, carrier proteins, and other cell transporters also facilitate their entrance into cells (Kapilan et al., 2018). The NPs with a diameter of 4–100 nm can rupture the layer, overcome the cuticle barrier, and gather in the substomatal cavity (Banerjee et al., 2022). NPs can also penetrate the seed coat via parenchymatous intercellular gaps which are mediated by aquaporins (Liu et al., 2016). The amount of NPs taken in and accumulated by plant cells is determined by their charge and some other characteristics (Khan et al., 2019). The outermost layers of a plant cell contain both hydrophobic and hydrophilic components along with negatively charged membranes (Alberts et al., 2002). As a result, a negatively charged layer (cell wall) promotes the absorption of positively charged NPs. For example, Li et al. (2019) reported that CeO_2 NPs have a positive surface charge and tend to absorb at considerably larger concentrations, indicating that transport efficiency is charge dependent (Li et al., 2019). Capillary action and osmotic pressure allow NPs to enter the cell (Shukla et al., 2016). The receptors, transporters, and other membrane proteins of NPs are changed when they contact with the outermost layer (Ji et al., 2017).

The NPs are selectively absorbed by the plants, if NPs are larger in size then they absorb via cellular pores or may not absorbed and if NPs are smaller in size, they can be easily absorbed, transported, and eventually accumulated in various areas of the plant (Khan et al., 2019). The plasmodesmata, cell wall, casparian strip, cell membrane, hydathodes, vasculature, stomata, and cuticle all have size exclusion limits (SELs) that influence NP uptake and transport (Schwab et al., 2015). NPs between 36 and 50 nm have been proven to penetrate and translocate, however this value exceeds the SELs of barriers (Barua & Mitragotri, 2014). Many studies suggested that NPs transmit from 1 cell to another via both apoplastic and symplastic mechanisms. NPs reach the plant shoot through a series of complex processes (Etesami et al., 2021). Pore size determines NP absorption and transport, which involves both active and passive processes (Shang et al., 2019).

Despite the fact that cell wall has small pores (plasmodesmata) through which NPs can pass and move in the space between the cell membrane and cell wall via the apoplastic route, the cell wall is the initial point of contact for NPs that countertheir entry in plant (Sanzari et al., 2019). However, various cationic and anionic cell membrane transporters further allow NPs to enter the cell which promote NPs absorption and translocation (Islam et al., 2021). Sulfur transporters (STR), aquaporins, H^+ ATPase, and potassium transporters (KTR) all have been identified in the transport of NPs (Noori et al., 2020). Nanoparticles applied on foliar parts tend to enter the plant through trichomes or stomata, where they are carried by phloem to other parts of the plant (Khan et al., 2019). Once the NPs are absorbed, they make nanocrystals either inside or outside of the cell and these specific forms of NPs are then transported to other locations via phloem and xylem (Khan et al., 2021).

3. Effect of nanoparticles on growth and physiology of plants under stress conditions

Application of NPs is ubiquitous in every discipline of science because of their unique physical and chemical properties. The nature of impact of NPs, either positive or negative, depends upon shape, concentration, nature of coatings, size and stability (Rastogi et al., 2018). It is fact that plants are quite sensible multicellular organisms, having capability to develop strategies to survive under stressed environment. Currently scientists are working to develop strategies to enhance the ability of plants to survive under stressed environment. Nanoparticles can improve plant metabolism because of their potential and unique physiochemical properties (Noori et al., 2019; Siddiqui et al., 2015). Engineered NPs can enter into plant cells and translocate DNA and materials into plant cells (Buriak et al., 2022). Researchers in plant science and other sectors can use nanotechnology to create new approaches for incorporating nanoparticles into plants, which can enhance existing functions and introduce new ones. Low doses of NPs are reported to improve seed germination in many plant species (Iqbal et al., 2021; Yin et al., 2012). Nanoparticles improve photosynthetic efficiency as well as chlorophyll content in various plant species when applied at moderate doses. Furthermore, their application improves the sugar contents in plants (Mohamed et al. 2017). They have ability to improve germination, growth (biomass of root and shoot) and photosynthetic process. Various NPs have ability to cause morphological and physiological changes in plant. Different doses of NPs, their composition, size and properties have both positive as well as negative affects on different species of plants (Ma et al., 2010). Effectiveness of NPs in improving crop growth depends upon their concentration as well as differ from plant to plant.

3.1 Improvement in seed germination

Germination of seed is first and most important step which help to determine the success of plant growth in soils deteriorated salinity and metals (Uçarlı et al., 2020). The utilization of nanomaterials to improve plant growth is novel and effective approach to improve plant growth under stressed conditions (Etesami et al., 2021). NPs affect the plant species in different way, including morphological, physiological, and biochemical aspects (Mohamed et al., 2017; Noori et al., 2019; Siddiqui et al., 2015). Through an eco-friendly and sustainable nanotechnological technique, seed priming with NPs can improve seed germination, growth, and yield while retaining fruit quality and appearance (Acharya et al., 2020). Nanopriming has extrabenefit of being able to stimulate metabolic systems that are ordinarily active during the early stages of germination.

3.2 Nano-fertilizers as plant growth promotors

Plants are susceptible due to variable micronutrient availability throughout their growth period, and their deficiency have detrimental effects on the plant growth (Siddiqui et al., 2022). It is not easy to control the distribution of micronutrients to a given crop in a traditional nutrient management system, but nano fertilizers allow farmers to supply proper levels of nutrients (Verma et al., 2022). Nano fertilizers increase the phyto-availability of nutrients because of their highest surface area, small size and high reactivity, which results in better growth of plants under abiotic stresses (Buriak et al., 2022; Verma et al., 2022). Micronutrients have wide range of functions, and their appropriate supply enhance the plant growth and yield, consequently protecting plants against the negative impacts of numerous biotic and abiotic stresses (Siddiqui et al., 2022). Nanomaterials also have various other roles in agriculture including nano-pesticides, nano-fungicides, and nano-insecticides for better crop production under stress conditions (Abdollah-dokht et al., 2022). The combination of nanomaterials and fertilizer results in enhanced and efficient absorption of nutrients and important chemicals for plants due to the high reactivity of nanomaterials (Babu et al., 2022). Uptake, dispersion, and accumulation of nutrients applied through nano fertilizers in plants are all controlled by various intrinsic and extrinsic factors, as well as the path of exposure (Chugh et al., 2021). Among most crucial inherent elements that determine the efficiency of nanoparticle are particle size and surface coating of NPs (Khan et al., 2019).

3.3 Increase in photosynthetic activity

Photosynthesis is a process which convert light energy into chemical energy, provide carbohydrate for the growth and development of plants (Goswami & Mathur, 2022). The utilization efficiency of plants for growth and development is only 2%–4% for the sun light (Amthor et al., 2010). Abiotic stresses severely affect the photosynthetic activities of the crops resulting in reduced growth. NPs have been reported to upregulate the photosynthetic processes. Under Co stress, ZnO NPs greatly increase the plant growth, biomass, and photosynthetic apparatus in maize (Salam et al., 2022). Similarly, Si-NPs act as catalysts in redox reactions, affecting photosynthesis and respiration (Tripathi, Singh, et al., 2017). In *Brassica juncea* the application of Ag-NPs at the rate of 25, 50 and 100 ppm enhanced plant growth and development (Mehmood, 2018). In another study, foliar application of ZnO-NPs in rice improved biomass, photosynthesis, protein content, antioxidant enzyme activity, mineral nutrient content, and reduced Cd levels (Faizan et al., 2021). The use of TiO_2 NPs increases the activity of rubisco and antioxidant enzymes, as well as photosynthetic rate and chlorophyll contents, resulting in higher agricultural yields (Gohari et al., 2020). Phytotoxicity of Cu^{2+} in corn seedlings reduced by increase in antioxidative enzyme activities (SOD, POD, CAT), which are responsible for alleviating oxidative stress (Xin et al., 2022).

3.4 Role of NPs in antioxidant enzymes activities

Nanoparticles play a key role to enhance the performance of enzymatic activities against exogenous stresses (Gohari et al., 2020). The formation of proline and methylation of cytosine is controlled by ZnO NPs, which enhances the resistance of crops against salts stress (Noohpisheh et al., 2020). The application of copper oxide NPs when applied at the rate of 52–86 μM, also have ability to enhance the activity of superoxide dismutase enzyme and peroxidase enzyme (Van Nguyen et al., 2021). Correspondingly, application of CeO_2 NPs at the rate of 50 ppm on cucumber crop significantly improved the catalase enzymatic activity. Similarly, TiO_2 when applied at the rate of 50–200 ppm, enhanced the performance of many antioxidants such as superoxide dismutase, APX, CAT, and guaiacol peroxidase (GP) under water stress (Aghdam et al., 2015).

NPs have capability to improve the activity of various enzymes like glutamic–pyruvic transaminase, glutamine synthase, glutamate dehydrogenase as well as nitrate reductase and the uptake of nitrates and transformation of inorganic form of nitrogen to organic form of nitrogen in crops (Lei et al., 2008; Yang et al., 2006). CeO_2 nanoparticles have the ability to neutralize ROS as well as to increase the photosynthesis rate (Ayub et al., 2019). Ag NPs when applied at the rate of 25 mg kg^{-1}, play crucial role in formation of proteins as well as carbohydrates and significantly improve several physical and chemical parameters of plants (Jaskulak et al., 2019). several other researches have demonstrated the beneficial effects of nanoparticles on antioxidant attributes (Rico et al., 2015; Tripathi, Singh, et al., 2017).

4. Role of NPs in various abiotic stress management

Abiotic stressors like drought, salinity, mineral deficiency, temperature and heavy metals have direct impact on crop development and yield (Imran et al., 2021). Because of their sessile nature, plants have no option but to endure numerous environmental stresses throughout their life cycle. NP-mediated impacts on plant growth and development in terms of stress tolerance have been observed in various studies. Under oxidative stress, NPs are involved in upregulating the activity of antioxidative enzymes such as catalase, superoxide dismutase, and peroxidase (Torabian, 2018).

Plants have the potential to adapt or to cope with environmental stresses like cold, salinity, heavy metals, drought, and temperature stress etc. The cellular as well as molecular responses of plants to numerous abiotic stresses have been examined in depth (Duque et al., 2013). The primary response of plants in stress situation is comprise of a transient increase in cytoplasmic Ca^{2+}, rise in mitogen-activated protein kinase (MAPK) pathways, increased secondary messengers (polyphosphate, inositol), abscisic acid, and ROS (Jalil et al., 2019). In response to oxidative stress, the regulation of proteins involve cellular protection of plants as well as expressions of specific genes (Mahalingam & Fedoroff, 2003). Secondary metabolites help plants cope with abiotic stress by stabilizing cell structure, protecting photosystems from ROS, signal transduction, and polyamine production (Rehman et al., 2021). Plant cell walls serve as a physical barrier to stress sensitivity and play an important role in plant adaptation during abiotic and biotic stress environments (Degenhardt and Gimmler, 2000). Extracellular peroxidases are active in cell wall modification and accumulate reactive oxygen species (ROS) when exposed to stress (Jalil et al., 2021). In the plant defense response, oxidative stress promotes the production of reactive oxygen species (ROS), the buildup of phenylpropanoid, enzyme biosynthesis, and gene expression (Daudi et al., 2012). Various abiotic stresses faced by plant during their life cycle have been shown in Fig. 16.1. Because NPs are small and have a large surface area, they quickly permeate plant cells, allowing for faster absorption and targeted delivery to specific locations and substances than conventional or bulk chemical forms (Khan et al., 2019). Several NPs have been employed to protect fruits, flowers, and vegetables against the detrimental effects of diverse abiotic stressors.

4.1 Drought stress

Among most frequent abiotic stresses, drought stress in most common that considerably reduces the crop production in arid and semi-arid regions

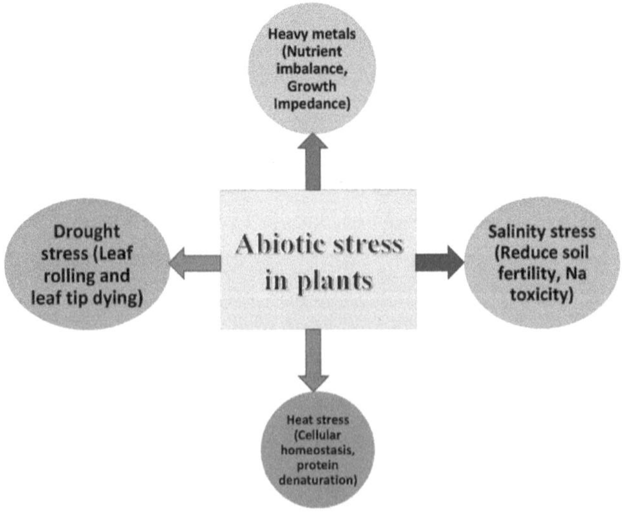

Fig. 16.1 Effect of different abiotic stress on plant growth and physiology.

(Yadav et al., 2020). Drought stress has an impact on plants at all phases of their growth, from morphological to molecular levels (Farooq et al., 2009). Drought can reduce yields in many plant species depending on the degree and duration of the drought, but drought stress after anthesis is always deleterious to grain output regardless of severity (Khatun et al., 2021). Overall, nutrient accessibility to root zone reduces leading to poor absorption at root hair and translocation in xylem and phloem channels, resulting in poor nutrient metabolism in cells and tissues. Due to their small size and large surface area, the use of NPs is a potential and cost-effective approach for mitigating drought stress (Seleiman et al., 2021). Plant resilience to drought stress can be improved by using nanoparticles in different fractions (Rajput, Singh, et al., 2021). Due to nanoparticles' unique physical and chemical qualities, they can easily enter plant cells and alter plant growth and development by affecting their metabolism through a variety of interactions and activating the metabolism to mitigate stress conditions (Salam et al., 2022). The addition of silicon to two sorghum cultivars with variable drought resistance enhanced drought tolerance by reducing the shoot to root ratio, which might be due to improved root growth and photosynthesis rate stability. This root structure might led to improved sorghum drought tolerance by increasing plant water absorption efficiency (Hattori et al., 2005). Semida et al. (2021) found that ZnO nanoparticles can increase the percentage of seed germination in soybeans that have been exposed to water stress. Under drought stress, using ZnO nanoparticles enhanced the fresh and dry weight of seeds, demonstrating that ZnO nanoparticles were effective for seed germination and seedling growth (Sedghi et al., 2013). Iron is an essential micronutrient for plant growth and development and its deficiency causes substantial abnormalities in plant metabolism and chlorosis (Rout & Sahoo, 2015). Iron nanoparticles have a significant influence on plant parameters such as seeds per boll, bolls per branch, grain weight, and yield under drought conditions. The foliar application of Zn NPs significantly reduced the drought stress in cucumber cultivars resulted in better yield (Ghani et al., 2022). Similarly, Dayar et al. (2014) reported that Fe nanoparticles improved the biomass of plant at granulation and flowering stage, further it found better at the blooming stage as compared to the seed production stage. Foliar application of Ti NPs showed ameliorating affects on wheat growth under drought stress. Application of Ti NPs @ 0.02% improved biomass, spike number, 1000 grains weight, spike weight, seed number, harvest index, yield as well as gluten and starch content (Jaberzadeh et al., 2013) (Table 16.1).

Table 16.1 Role of nanoparticles in mitigating the drought stress.

Nanomaterial	Concentration	Plant species	Plant response	References
TiO_2	0.01%, 0.02%, and 0.03%	Wheat	Increased growth and yield of wheat as well as starch and gluten content	(Jaberzadeh et al., 2013)
TiO_2	0, 10, 100, and 500 mg L^{-1}	Wheat	Increased carotenoids and chlorophyll, improved growth and biomass attributes, decreased malondialdehyde and H_2O_2	(Faraji and Sepehri, 2020)
SiO_2	30, 60 mg kg^{-1}	Wheat	All the growth and physiological parameters were enhanced.	(Behboudi et al., 2018)
ZnO	25–100 mg kg^{-1}	Cucumber	Improved stomatal conductance, biomass, and photosynthetic rate.	(Ghani et al., 2022)
ZnO, PGPR	10 mg kg^{-1}	Wheat	Improved the biochemical metabolism in wheat under drought and temperature stress	(Azmat et al., 2022)

SiO_2	0, 10, 50 and 100 mg L^{-1}	Hawthorns	Increased xylem water potential, plant biomass, and malondialdehyde content.	(Ashkavand et al., 2015)
SiO_2, Se	100 mg kg^{-1}	Strawberry	Increased proline content, phenolics, free amino acids, soluble proteins, antioxidant enzymes (superoxide dismutase, peroxidase and catalase)	(Zahedi et al., 2020a, b)

4.2 Heavy metals

The excessive accumulation of non-essential ions like toxic heavy metals in plant cells considerably affect the plant metabolism under normal growth conditions by producing oxidative stress (Sachdev et al., 2021). The accumulation of harmful heavy metal ions causes a considerable loss in plant development and yield under various HMs sress (Rehman et al., 2021). These toxic ions are largely cytotoxic, disrupting a variety of physiological functions such as nutrition transport, signaling, and photosynthesis (Shahzad et al., 2018). Furthermore, HMs stress affect the ionic homeostasis in plant cells by lowering or substituting ionic absorption. Cd competes with Zn for absorption by plants under HMs pollution, for example, Cd employs Zn transporters to enter plants. As a result, it's critical to either limit hazardous ion absorption or increase the plant's ability to escape toxicity.

Nanotechnology, or the use of nanoparticles, has been found to be particularly successful at reducing ionic stress caused by a variety of abiotic factors. NPs can prevent deleterious ion accumulation in cells, protect plants from ionic stress, and reduce the toxic effects of heavy metals in plants (Tanveer et al., 2020). These NPs bind the metal ions quickly due to large surface area, and reduce their bioavailability. Quantum dots, for example, lowered the availability of lead and copper and also reduced their mobility (Bade et al., 2012). When heavy metals penetrate plant cells and cause oxidative damage or ionic toxicity, these NPs activate the plant's defense mechanism and induce an increase in the concentration of essential ions like K^+ and Ca^{2+}. In an investigation conducted by (Tripathi et al., 2015) showed that the application of Si-NPs was found to have beneficial effects on plants under Cr toxicity. The use of zero-valent iron NPs has also been proposed as a potential strategy for reducing chromium contamination (Sadeghi et al., 2021). Furthermore, Aqeel et al. (2022) discovered that using nanoparticles lowers HMs toxicity in plants. It is concluded from above discussion, that using NPs in minimizing HMs toxicity, and enhancing plant physiology under HMs stress is a highly promising technique. In Table 16.2 the effects of various NPs in combating the HMs stress in plants have been summarized.

4.3 Salinity stress

Salinity is the most important abiotic stress, causing a 20% reduction in food production and an imbalance in plant mechanisms associated with growth and yield (Ma et al., 2020). Salt stress causes the formation of reactive oxygen species (ROS) and harmful ions, which disrupts important physiological

Table 16.2 Remediation of heavy metals stress by nanoparticles.

Type of NPS	Experiment type	Dose	Test crop	Results	References
Fe + ZnO + Si (NPs)	Field experiment	Fe (5 mg L^{-1}), ZnO (25 mg L^{-1}), Si (300 mg L^{-1})	Wheat	All the growth and physiological parameters were enhanced under Cd stress by incorporation of NPs simultaneously.	(Hussain et al., 2021)
ZnO NPs	Pot experiment	100 mg L^{-1}	Wheat	Under Cd and drought, foliar application of ZnO NPs enhanced the chlorophyll contents and reduced the oxidative stress in wheat.	(Rai-Kalal & Jajoo., 2021)
FeO	Pot experiment	100 mg L^{-1}	Wheat	FeO NPs improved the plant growth antioxidant concentrations under Cd and salinity stress	(Manzoor et al., 2021)
Fe_2O_3	Pot experiment	100 mg L^{-1}	Wheat	Fe_2O_3 enhanced the wheat growth, yield and reduced the oxidative stress.	(Shen et al., 2020)
ZnO + FeO	Pot experiment	ZnO (100 mg L^{-1}), FeO (20 mg L^{-1})	Wheat	Seed priming with ZnO and FeO NPs reduced the oxidative stress and enhance the photosynthetic activity in wheat	(Rizwan et al., 2019)

(Continued)

Table 16.2 Remediation of heavy metals stress by nanoparticles.—cont'd

Type of NPS	Experiment type	Dose	Test crop	Results	References
Iron (Fe), silicon (si) and zinc (Zn)	Pot experiment	60 $mg\ kg^{-1}$	Wheat	Incorporation of NPs enhanced the nutrient availability, resulting in improved plant growth under Cd stress.	(Khalid et al., 2021)
Si NPS	Pot experiment	1200 $mg\ kg^{-1}$	Wheat	Increased yield, antioxidant enzymes, and nutrients acquisition	(Ali et al., 2019)
ZnO	Hydroponic experiment	20 $mg\ L^{-1}$	Water pepper (*Persicaria hydropiper* L.)	Reduced Pb concentration in plant and enhance the growth.	(Hussain et al., 2021)
Silver (Ag)	Pot experiment	25 $mg\ kg^{-1}$	Lupinus luteus L.	Improved GPX activity and metal-lothioneins expression	(Jaskulak et al., 2019)
Zinc oxide (ZnO)	Pot experiment	100 $mg\ L^{-1}$	*Oryza sativa* L.	Improved photosynthesis and yield with reduction in Cd accumulation	(Ali et al., 2019)

processes in plants (Sharma et al., 2012). Salt stress has a deleterious impact on a variety of physiological and biochemical activities in plants, including photosynthesis, protein synthesis, and lipid synthesis (Tanveer & Shabala, 2018). By 2050, soil salinization is expected to affect over half of all cultivated land, and this number is expected to rise further as a result of global change in climatic patterns and other environmental issues. Several defense techniques including marker-assisted breeding, conventional breeding, and genetically engineered crops employing transgenic technologies are still in infancy to produce salt resistant germplasms (Babar et al., 2020). Soils with high salt contents lead to a low osmotic potential which reduce the uptake of plant nutrients and increases stress on plant (Evelin et al., 2019). Cellular ionic toxicity is caused by a significant accumulation of Na^+ or Cl^- ions in plant cells when they are stressed by salt (Tanveer & Shah, 2017; Wu et al., 2015). Salt stress causes considerable K^+ transport from leaf mesophyll cells and increases Na^+ concentration in the cytosol in plants (Shabala & Cuin, 2008). The use of NPs helps to alleviate salt stress by lowering the harmful consequences of salinization. For example, Si improved seed germination, carbon-assimilation, antioxidants system, leaf turgor and reduced the salt stress (Haghighi & Pessarakli, 2013; Qados, 2015). Further, nano-Si application considerably reduced salt stress in tomatoes by lowering Na^+ toxicity and ROS generation (Haghighi et al., 2013). On agricultural plants, such as basil, lentils, and broad beans, similar effects have been studied for similar effects of n-Si (Qados, 2015) (Table 16.3).

5. Conclusion

Nanoparticles due to their small size can easily penetrate into the plant cells, and act as potential stimulant for plant growth and physiology. Furthermore, because NPs have a relatively broad spectrum of activity, they can protect plants from the negative effects of a variety of abiotic stressors (increase in water uptake into seeds, metabolism of starch reserves, stimulation of photosynthesis, changes in phytohormone levels, modulation of oxidative stress, or influenced uptake of nutrients). The bulk of research, on the other hand, have been undertaken to better understand one or more forms of stress situations. Future research should concentrate on more realistic stress conditions in a real-world setting. As a result, additional research is needed to understand how NPs affect the plant cell's antioxidant system, resulting in enhanced tolerance to various types of stress. Such knowledge

Table 16.3 Remediation of salinity stress by nanoparticles.

Type of NPS	Dose	Test crop	Results	References
Mn	0.05, 0.1, 0.5 and 1 mg L^{-1}	Mung bean (vigna radiata)	Reduced the ROS production, enhanced the photosynthetic activity	(Pradhan et al., 2013)
Ag	60 mg kg^{-1}	*Phaseolus vulgaris* L., *Zea mays* L.	Positively affected the fresh root and shoot dry weight	(Salama, 2012)
TiO_2	1000 mg L^{-1}	Wheat (triticum aestivum L.)	Improved the no. of tillers and yield	(Yang et al., 2007)
ZnS	10 and 40 mg L^{-1}	Rice and wheat	ZnO NPs and humic acid enhanced the growth and yield	(Mirzadeh et al., 2013)
TiO_2 and Fe_3O_4	0, 50, 100, 150, 200 and 250 mg kg^{-1}	Lactuca sativa	Improved P availability and uptake	(Zahra et al., 2015)
NiO nanoparticles	120 mg kg^{-1}	*Triticum aestivum*	Severe growth retardation and oxidative damage in wheat	(Saleh et al., 2019)
CeO_2 and CuO	200 and 50 mg kg^{-1} respectively	Cucumis sativus	CeO_2 decreased firmness of fruit. Moreover, fruit Zn concentration was decreased by 25% than control @ 200 mg kg^{-1} concentrations.	(Hong et al., 2016)
TiO_2 and ZnO	0–500 mg L^{-1}	Tomato	Enhanced the root growth, No significant effect on rate of germination	(Raliya et al., 2015)

might aid in the development of smart NPs in the future that can assist mitigate stress and maintain long-term agricultural yield.

Future research should concentrate on the interaction of NPs with signaling systems that result in a shift in defense enzyme gene expression. Furthermore, there are currently certain limits to employing pricey NPs in agriculture. Treatment using gold and silver nanoparticles is not cost-effective. Understanding the mechanism by which these NPs modulate plant immunity may thus aid in the development of a new technique for enhancing plant tolerance. Because of changing climatic circumstances, soil types, plants to be treated, and, most significantly, the physicochemical character of metallic/non-metallic NPs, many of the produced NPs are still limited in their field applicability.

However, further research into the synergistic effects of the various therapies is required before definitive conclusions can be drawn. Long-term research on cereal crops and other important crops are also recommended to show the link between NP dosage, soil types, and ecological effects. When assessing the toxicological impact of NPs on various creatures and habitats before using their potential advantages, caution is required.

References

Abdoli, S., Ghassemi-Golezani, K., & Alizadeh-Salteh, S. (2020). Responses of ajowan (*Trachyspermum ammi* L.) to exogenous salicylic acid and iron oxide nanoparticles under salt stress. *Environmental Science and Pollution Research, 27*(29), 36939–36953.

Abdollahdokht, D., Gao, Y., Faramarz, S., Poustforoosh, A., Abbasi, M., Asadikaram, G., & Nematollahi, M. H. (2022). Conventional agrochemicals towards nano-biopesticides: An overview on recent advances. *Chemical and Biological Technologies in Agriculture, 9*(1), 1–19.

Acharya, P., Jayaprakasha, G. K., Crosby, K. M., Jifon, J. L., & Patil, B. S. (2020). Nanoparticle-mediated seed priming improves germination, growth, yield, and quality of watermelons (*Citrullus lanatus*) at multi-locations in Texas. *Scientific Reports, 10*(1). https://doi.org/10.1038/s41598-020-61696-7, 5037–5037.

Aghdam, M. T. B., Mohammadi, H., & Ghorbanpour, M. (2015). Effects of nanoparticulate anatase titanium dioxide on physiological and biochemical performance of Linum usitatissimum (Linaceae) under well-watered and drought stress conditions. *Brazilian Journal of Botany, 39*(1), 139–146. https://doi.org/10.1007/s40415-015-0227-x

Ahmar, S., Mahmood, T., Fiaz, S., Mora-Poblete, F., Shafique, M. S., Chattha, M. S., & Jung, K.-H. (2021). Advantage of nanotechnology-based genome editing system and its application in crop improvement. *Frontiers in Plant Science, 12*. https://doi.org/10.3389/fpls.2021.663849, 663849–663849.

Alberts, B., Bray, D., Lewis, J., Raff, M., Roberts, K., & Watson, J. D. (2002). *Molecular biology of the cell*. London: Garland Pub. Inc.

Alharby, H. F., Rizwan, M., Iftikhar, A., Hussaini, K. M., Rehman, M. Z., Bamagoos, A. A., Alharbi, B. M., Asrar, M., Yasmeen, T., & Ali, S. (2021). Effect of gibberellic acid and titanium dioxide nanoparticles on growth, antioxidant defense system and mineral nutrient uptake in wheat. *Ecotoxicology and Environmental Safety, 221*, Article 112436. 34171689.

Ali, S., Rizwan, M., Hussain, A., Zia ur Rehman, M., Ali, B., Yousaf, B., Wijaya, L., Alyemeni, M. N., & Ahmad, P. (2019). Silicon nanoparticles enhanced the growth and reduced the cadmium accumulation in grains of wheat (*Triticum aestivum* L.). *Plant Physiology and Biochemistry, 140*, 1–8. https://doi.org/10.1016/j.plaphy.2019.04.041

Ali, S., Mehmood, A., & Khan, N. (2021). Uptake, translocation, and consequences of nanomaterials on plant growth and stress adaptation. *Journal of Nanomaterials, 2021*, Article 6677616. https://doi.org/10.1155/2021/6677616, 2021.

Alsaeedi, A., El-Ramady, H., Alshaal, T., El-Garawany, M., Elhawat, N., & Al-Otaibi, A. (2019). Silica nanoparticles boost growth and productivity of cucumber under water deficit and salinity stresses by balancing nutrients uptake. *Plant Physiology and Biochemistry, 139*, 1–10. https://doi.org/10.1016/j.plaphy.2019.03.008

Amthor, J. S. (2010). From sunlight to phytomass: On the potential efficiency of converting solar radiation to phyto-energy. *New Phytologist, 188*(4), 939–959. https://doi.org/10.1111/j.1469-8137.2010.03505.x

Aqeel, U., Aftab, T., Khan, M. M. A., Naeem, M., & Khan, M. N. (2022). A comprehensive review of impacts of diverse nanoparticles on growth, development and physiological adjustments in plants under changing environment. *Chemosphere, 291*, 132672. https://doi.org/10.1016/j.chemosphere.2021.132672

Arif, N., Yadav, V., Singh, S., Tripathi, D. K., Dubey, N. K., Chauhan, D. K., & Giorgetti, L. (2018). Interaction of copper oxide nanoparticles with plants. In *Nanomaterials in plants, algae, and microorganisms* (pp. 297–310). Elsevier.

Ashkavand, P., Tabari, M., Zarafshar, M., Tomásková, I., & Struve, D. (2015). Effect of SiO_2 nanoparticles on drought resistance in hawthorn seedlings. *Leśne Prace Badawcze., 76*.

Ayub, M. A., Sohail, M. I., Umair, M., Rehman, M. Z., Usman, M., Sabir, M., Rizwan, M., Ali, S., & Ahmad, Z. (2019). Cerium oxide nanoparticles: Advances in synthesis, prospects and application in agro-ecosystem. *Comprehensive Analytical Chemistry*, 209–250.

Ayub, M. A., Usman, M., Faiz, T., Umair, M., Haq, A., Rizwan, M., Ali, S., & Rehman, M. Z. (2020). Restoration of degraded soil for sustainable agriculture. In *Soil health restoration and management* (pp. 31–81).

Azmat, A., Tanveer, Y., Yasmin, H., Hassan, M. N., Shahzad, A., Reddy, M., & Ahmad, A. (2022). Coactive role of zinc oxide nanoparticles and plant growth promoting rhizobacteria for mitigation of Synchronized effects of heat and drought stress in wheat plants. *Chemosphere*, 133982.

Babar, U., Nawaz, M. A., Arshad, U., Azhar, M. T., Atif, R. M., Golokhvast, K. S., Tsatsakis, A. M., Shcerbakova, K., Chung, G., & Rana, I. A. (2020). Transgenic crops for the agricultural improvement in Pakistan: A perspective of environmental stresses and the current status of genetically modified crops. *GM Crops & Food, 11*(1), 1–29.

Babu, S., Singh, R., Yadav, D., Rathore, S. S., Raj, R., Avasthe, R., Yadav, S. K., Das, A., Yadav, V., Yadav, B., & Shekhawat, K. (2022). Nanofertilizers for agricultural and environmental sustainability. *Chemosphere, 292*, 133451.

Bade, R., Oh, S., & Shin, W. S. (2012). Diffusive gradients in thin films (DGT) for the prediction of bioavailability of heavy metals in contaminated soils to earthworm (Eisenia foetida) and oral bioavailable concentrations. *Science of the Total Environment, 416*, 127–136. https://doi.org/10.1016/j.scitotenv.2011.11.007

Banerjee, A., & Roychoudhury, A. (2022). Explicating the cross-talks between nanoparticles, signaling pathways and nutrient homeostasis during environmental stresses and xenobiotic toxicity for sustainable cultivation of cereals. *Chemosphere, 286*, 131827. https://doi.org/10.1016/j.chemosphere.2021.131827

Barua, S., & Mitragotri, S. (2014). Challenges associated with penetration of nanoparticles across cell and tissue barriers: A review of current status and future prospects. *Nano Today, 9*(2), 223–243.

Bashir, A., Rehman, M. Z., Hussaini, K. M., Adrees, M., Qayyum, M. F., Sayal, A. U., Rizwan, M., Ali, S., Alsahli, A. A., & Alyemeni, M. N. (2021). Combined use of zinc nanoparticles and co-composted biochar enhanced wheat growth and decreased Cd concentration in grains under Cd and drought stress: A field study. *Environmental Technology & Innovation, 23*, 101518.

Behboudi, F., Tahmasebi Sarvestani, Z., Kassaee, M. Z., Modares Sanavi, S. A. M., & Sorooshzadeh, A. (2018). Improving growth and yield of wheat under drought stress via application of SiO_2 nanoparticles. *Journal of Agricultural Science and Technology, 20*(7), 1479–1492.

Beig, B., Niazi, M. B. K., Sher, F., Jahan, Z., Malik, U. S., Khan, M. D., Américo-Pinheiro, J. H. P., & Vo, D. V. N. (2022). Nanotechnology-based controlled release of sustainable fertilizers. A review. *Environmental Chemistry Letters*, 1–18.

Buriak, J. M., Liz-Marzán, L. M., Parak, W. J., & Chen, X. (2022). Nano and plants. *ACS Nano, 16*(2), 1681–1684.

Chugh, G., Siddique, K. H., & Solaiman, Z. M. (2021). Nanobiotechnology for agriculture: Smart technology for combating nutrient deficiencies with nanotoxicity challenges. *Sustainability, 13*(4), 1781.

Daudi, A., & O'Brien, J. (2012). Detection of hydrogen peroxide by DAB staining in arabidopsis leaves. *Bio-Protocol, 2*(18). https://doi.org/10.21769/bioprotoc.263

Davar, Z. F., Roozbahani, A., & Hosnamidi, A. (2014). *Evaluation the effect of water stress and foliar application of Fe nanoparticles on yield, yield components and oil percentage of safflower*. Carthamus tinctorious L.

Degenhardt, B. (2000). Cell wall adaptations to multiple environmental stresses in maize roots. *Journal of Experimental Botany, 51*(344), 595–603. https://doi.org/10.1093/jexbot/51.344.595

Duque, A. S., de Almeida, A. M., da Silva, A. B., da Silva, J. M., Farinha, A. P., Santos, D., Fevereiro, P., & de Sousa Araújo, S. (2013). *Abiotic stress responses in plants: Unraveling the complexity of genes and networks to survive. Abiotic stress-plant responses and applications in agriculture* (pp. 49–101).

Etesami, H., Fatemi, H., & Rizwan, M. (2021). Interactions of nanoparticles and salinity stress at physiological, biochemical and molecular levels in plants: A review. *Ecotoxicology and Environmental Safety, 225*, 112769. https://doi.org/10.1016/j.ecoenv.2021.112769

Evelin, H., Devi, T. S., Gupta, S., & Kapoor, R. (2019). Mitigation of salinity stress in plants by arbuscular mycorrhizal symbiosis: Current understanding and new challenges. *Frontiers in Plant Science, 10*, 470. https://doi.org/10.3389/fpls.2019.00470

Faizan, M., Bhat, J. A., Hessini, K., Yu, F., & Ahmad, P. (2021). Zinc oxide nanoparticles alleviates the adverse effects of cadmium stress on Oryza sativa via modulation of the photosynthesis and antioxidant defense system. *Ecotoxicology and Environmental Safety, 220*, 112401. https://doi.org/10.1016/j.ecoenv.2021.112401

Faraji, J., & Sepehri, A. (2020). Exogenous nitric oxide improves the protective effects of TiO2 nanoparticles on growth, antioxidant system, and photosynthetic performance of wheat seedlings under drought stress. *Journal of Soil Science and Plant Nutrition, 20*(2), 703–714.

Farooq, M., Wahid, A., Kobayashi, N., Fujita, D., & Basra, S. M. A. (2009). Plant drought stress: Effects, mechanisms and management. *Agronomy for Sustainable Development, 29*(1), 185–212. https://doi.org/10.1051/agro:2008021

Ghani, M. I., Saleem, S., Rather, S. A., Rehmani, M. S., Alamri, S., Rajput, V. D., & Liu, M. (2022). Foliar application of zinc oxide nanoparticles: An effective strategy to mitigate drought stress in cucumber seedling by modulating antioxidant defense system and osmolytes accumulation. *Chemosphere, 289*, 133202.

Gohari, G., Mohammadi, A., Akbari, A., Panahirad, S., Dadpour, M. R., Fotopoulos, V., & Kimura, S. (2020). Titanium dioxide nanoparticles (TiO_2 NPs) promote growth and ameliorate salinity stress effects on essential oil profile and biochemical attributes of Dracocephalum moldavica. *Scientific Reports, 10*(1), 1–14.

Goswami, P., & Mathur, J. (2022). Application of agro-waste-mediated silica nanoparticles to sustainable agriculture. *Bioresources and Bioprocessing, 9*(1). https://doi.org/10.1186/s40643-022-00496-5

Gruyer, N., Dorais, M., Bastien, C., Dassylva, N., & Triffault-Bouchet, G. (2014). Interaction between silver nanoparticles and plant growth. *Acta Horticulturae, 1037*, 795–800. https://doi.org/10.17660/actahortic.2014.1037.105

Gull, A., Ahmad Lone, A., & Islam Wani, N. (2019). Biotic and abiotic stresses in plants. In *Abiotic and biotic stress in plants*. IntechOpen.

Haghighi, M., & Pessarakli, M. (2013). Influence of silicon and nano-silicon on salinity tolerance of cherry tomatoes (Solanum lycopersicum L.) at early growth stage. *Scientia Horticulturae, 161*, 111–117. https://doi.org/10.1016/j.scienta.2013.06.034

Hattori, T., Inanaga, S., Araki, H., An, P., Morita, S., Luxova, M., & Lux, A. (2005). Application of silicon enhanced drought tolerance in Sorghum bicolor. *Physiologia Plantarum, 123*(4), 459–466. https://doi.org/10.1111/j.1399-3054.2005.00481.x

Hong, J., Wang, L., Sun, Y., Zhao, L., Niu, G., Tan, W., … Gardea-Torresdey, J. L. (2016). Foliar applied nanoscale and microscale CeO2 and CuO alter cucumber (Cucumis sativus) fruit quality. *Science of the Total Environment, 563*, 904–911.

Husen, A., & Siddiqi, K. S. (2014). Plants and microbes assisted selenium nanoparticles: Characterization and application. *Journal of Nanobiotechnology, 12*(1), 1–10.

Hussain, A., Rizwan, M., Ali, S., Rehman, M. Z.u., Qayyum, M. F., Nawaz, R., Ahmad, A., Asrar, M., Ahmad, S. R., Alsahli, A. A., & Alyemeni, M. N. (2021). Combined use of different nanoparticles effectively decreased cadmium (Cd) concentration in grains of wheat grown in a field contaminated with Cd. *Ecotoxicology and Environmental Safety, 215*, 112139. https://doi.org/10.1016/j.ecoenv.2021.112139

Imran, Q. M., Falak, N., Hussain, A., Mun, B. G., & Yun, B. W. (2021). Abiotic stress in plants; stress perception to molecular response and role of biotechnological tools in stress resistance. *Agronomy, 11*(8), 1579.

Iqbal, T., Irfan, F., Afsheen, S., Zafar, M., Naeem, S., & Raza, A. (2021). Synthesis and characterization of Ag–TiO_2 nano-composites to study their effect on seed germination. *Applied Nanoscience, 11*(7), 2043–2057. https://doi.org/10.1007/s13204-021-01912-6

Islam, A., Mandal, C., & Habib, A. (2021). Antibacterial potential of synthesized silver nanoparticles from leaf extract of Moringa oleifera. *Journal of Advanced Biotechnology and Experimental Therapeutics, 4*(1), 67. https://doi.org/10.5455/jabet.2021.d108

Jaberzadeh, A., Moaveni, P., Tohidi Moghadam, H. R., & Zahedi, H. (2013). Influence of bulk and nanoparticles titanium foliar application on some agronomic traits, seed gluten and starch contents of wheat subjected to water deficit stress. *Notulae Botanicae Horti Agrobotanici Cluj-Napoca, 41*(1), 201. https://doi.org/10.15835/nbha4119093

Jalil, S. U., & Ansari, M. I. (2019). Nanoparticles and abiotic stress tolerance in plants: Synthesis, action, and signaling mechanisms. *Plant Signaling Molecules*, 549–561.

Jalil, S. U., & Ansari, M. I. (2021). Role of nanomaterials in regulating reactive species as a signaling molecule of abiotic stress in plants. In *Nanobiotechnology* (pp. 291–304). Springer International Publishing.

Jamkhande, P. G., Ghule, N. W., Bamer, A. H., & Kalaskar, M. G. (2019). Metal nanoparticles synthesis: An overview on methods of preparation, advantages and disadvantages, and applications. *Journal of Drug Delivery Science and Technology, 53*, 101174.

Jaskulak, M., Rorat, A., Grobelak, A., Chaabene, Z., Kacprzak, M., & Vandenbulcke, F. (2019). Bioaccumulation, antioxidative response, and metallothionein expression in *Lupinus luteus* L. exposed to heavy metals and silver nanoparticles. *Environmental Science and Pollution Research, 26*, 16040–16052. https://doi.org/10.1007/s11356-019-04972-y

Ji, Y., Zhou, Y., Ma, C., Feng, Y., Hao, Y., Rui, Y., Wu, W., Gui, X., Le, V. N., Han, Y., Wang, Y., Xing, B., Liu, L., & Cao, W. (2017). Jointed toxicity of TiO_2 NPs and Cd to rice seedlings: NPs alleviated Cd toxicity and Cd promoted NPs uptake. *Plant Physiology and Biochemistry, 110*, 82–93. https://doi.org/10.1016/j.plaphy.2016.05.010

Kapilan, R., Vaziri, M., & Zwiazek, J. J. (2018). Regulation of aquaporins in plants under stress. *Biological Research, 51*(1), 1–11.

Kaveh, R., Li, Y.-S., Ranjbar, S., Tehrani, R., Brueck, C. L., & Van Aken, B. (2013). Changes in *Arabidopsis thaliana* gene expression in response to silver nanoparticles and silver ions. *Environmental Science & Technology, 47*(18), 10637–10644. https://doi.org/10.1021/es402209w

Khalid, H. I., Murtaza, G., Ullah, S., & Waraich, E. A. (2021). Nanoparticles induced restricted cadmium mobility and accumulation in wheat from contaminated soil. In *Research square platform LLC*.

Khan, I., Saeed, K., & Khan, I. (2019). Nanoparticles: Properties, applications and toxicities. *Arabian Journal of Chemistry, 12*(7), 908–931. https://doi.org/10.1016/j.arabjc.2017.05.011

Khan, M., Ware, P., & Shimpi, N. (2021). Synthesis of ZnO nanoparticles using peels of Passiflora foetida and study of its activity as an efficient catalyst for the degradation of hazardous organic dye. *SN Applied Sciences, 3*(5). https://doi.org/10.1007/s42452-021-04436-4

Khatun, M., Sarkar, S., Era, F. M., Islam, A. K. M. M., Anwar, M. P., Fahad, S., Datta, R., & Islam, A. K. M. A. (2021). Drought stress in grain legumes: Effects, tolerance mechanisms and management. *Agronomy, 11*(12), 2374. https://doi.org/10.3390/agronomy11122374

Landa, P. (2021). Positive effects of metallic nanoparticles on plants: overview of involved mechanisms. *Plant Physiology and Biochemistry, 161*, 12–24.

Lee, S., Kim, S., Kim, S., & Lee, I. (2013). Assessment of phytotoxicity of ZnO NPs on a medicinal plant, Fagopyrum esculentum. *Environmental Science and Pollution Research, 20*(2), 848–854.

Lei, Z., Mingyu, S., Xiao, W., Chao, L., Chunxiang, Q., Liang, C., … Fashui, H. (2008). Antioxidant stress is promoted by nano-anatase in spinach chloroplasts under UV-B radiation. *Biological Trace Element Research, 121*(1), 69–79.

Li, J., Tappero, R. V., Acerbo, A. S., Yan, H., Chu, Y., Lowry, G. V., & Unrine, J. M. (2019). Effect of CeO_2 nanomaterial surface functional groups on tissue and subcellular distribution of Ce in tomato (Solanum lycopersicum). *Environmental Science: Nano, 6*(1), 273–285.

Liu, R., Zhang, H., & Lal, R. (2016). Effects of stabilized nanoparticles of copper, zinc, manganese, and iron oxides in low concentrations on lettuce (*Lactuca sativa*) seed germination: Nanotoxicants or nanonutrients? *Water, Air, & Soil Pollution, 227*(1), 1–14.

Ma, Y., Kuang, L., He, X., Bai, W., Ding, Y., Zhang, Z., & Chai, Z. (2010). Effects of rare earth oxide nanoparticles on root elongation of plants. *Chemosphere, 78*(3), 273–279.

Ma, Y., Dias, M. C., & Freitas, H. (2020). Drought and salinity stress responses and microbe-induced tolerance in plants. *Frontiers in Plant Science, 11*. https://doi.org/10.3389/fpls.2020.591911, 591911-591911.

Mahalingam, R., & Fedoroff, N. (2003). Stress response, cell death and signalling: The many faces of reactive oxygen species. *Physiologia Plantarum, 119*(1), 56–68. https://doi.org/10.1034/j.1399-3054.2003.00156.x

Mantri, N., Patade, V., Penna, S., Ford, R., & Pang, E. (2012). Abiotic stress responses in plants: Present and future. In *Abiotic stress responses in plants* (pp. 1–19). https://doi.org/10.1007/978-1-4614-0634-1_1

Manzoor, N., Ahmed, T., Noman, M., Shahid, M., Nazir, M. M., Ali, L., Alnusaire, T. S., Li, B., Schulin, R., & Wang, G. (2021). Iron oxide nanoparticles ameliorated the cadmium and salinity stresses in wheat plants, facilitating photosynthetic pigments and restricting cadmium uptake. *Science of the Total Environment, 769*, 145221. https://doi.org/10.1016/j.scitotenv.2021.145221

Mehmood, A. (2018). Brief overview of the application of silver nanoparticles to improve growth of crop plants. *IET Nanobiotechnology, 12*(6), 701–705.

Mirzadeh, S., Darezereshki, E., Bakhtiari, F., Fazaelipoor, M. H., & Hosseini, M. R. (2013). Characterization of zinc sulfide (ZnS) nanoparticles biosynthesized by Fusarium oxysporum. *Materials Science in Semiconductor Processing, 16*(2), 374–378.

Mittal, D., Kaur, G., Singh, P., Yadav, K., & Ali, S. A. (2020). Nanoparticle-Based sustainable agriculture and food science: Recent advances and future outlook. *Frontiers in Nanotechnology, 2*. https://doi.org/10.3389/fnano.2020.579954

Mohamed, A. K. S. H., Qayyum, M. F., Abdel-Hadi, A. M., Rehman, R. A., Ali, S., & Rizwan, M. (2017). Interactive effect of salinity and silver nanoparticles on photosynthetic and biochemical parameters of wheat. *Archives of Agronomy and Soil Science, 63*(12), 1736–1747. https://doi.org/10.1080/03650340.2017.1300256

Moradbeygi, H., Jamei, R., Heidari, R., & Darvishzadeh, R. (2020). Fe_2O_3 nanoparticles induced biochemical responses and expression of genes involved in rosmarinic acid biosynthesis pathway in Moldavian balm under salinity stress. *Physiologia Plantarum, 169*(4), 555–570. https://doi.org/10.1111/ppl.13077

Nafady, N. A., Alamri, S. A. M., Hassan, E. A., Hashem, M., Mostafa, Y. S., & Abo-Elyousr, K. A. M. (2019). Application of ZnO-nanoparticles to manage Rhizopus soft rot of sweet potato and prolong shelf-life. *Folia Horticulturae, 31*(2), 319–329. https://doi.org/10.2478/fhort-2019-0025

Noohpisheh, Z., Amiri, H., Mohammadi, A., & Farhadi, S. (2020). Effect of the foliar application of zinc oxide nanoparticles on some biochemical and physiological parameters of Trigonella foenum-graecum under salinity stress. *Plant Biosystem - An International Journal of Dealing with All Aspiring Plant Biology, 155*, 267–280. https://doi.org/10.1080/11263504.2020.1739160

Noori, A., Donnelly, T., Colbert, J., Cai, W., Newman, L. A., & White, J. C. (2019). Exposure of tomato (Lycopersicon esculentum) to silver nanoparticles and silver nitrate: Physiological and molecular response. *International Journal of Phytoremediation, 22*(1), 40–51. https://doi.org/10.1080/15226514.2019.1634000

Noori, A., Ngo, A., Gutierrez, P., Theberge, S., & White, J. C. (2020). Silver nanoparticle detection and accumulation in tomato (Lycopersicon esculentum). *Journal of Nanoparticle Research, 22*, 131.

Pradhan, S., Patra, P., Das, S., Chandra, S., Mitra, S., Dey, K. K., Akbar, S., Palit, P., & Goswami, A. (2013). Photochemical modulation of biosafe manganese nanoparticles on vigna radiata: A detailed molecular, biochemical, and biophysical study. *Environmental Science & Technology, 47*(22), 13122–13131.

Qados, A. (2015). Mechanism of nanosilicon-mediated alleviation of salinity stress in faba bean (*Vicia faba* L.) plants. *American Journal of Experimental Agriculture*, 7(2), 78–95. https://doi.org/10.9734/ajea/2015/15110

Rai-Kalal, P., Tomar, R. S., & Jajoo, A. (2021). H_2O_2 signaling regulates seed germination in ZnO nanoprimed wheat (*Triticum aestivum* L.) seeds for improving plant performance under drought stress. *Environmental and Experimental Botany, 189*, 104561. https://doi.org/10.1016/j.envexpbot.2021.104561

Rajput, V. D., Minkina, T. M., Behal, A., Sushkova, S. N., Mandzhieva, S., Singh, R., Gorovtsov, A., Tsitsuashvili, V. S., Purvis, W. O., Ghazaryan, K. A., & Movsesyan, H. S. (2018). Effects of zinc-oxide nanoparticles on soil, plants, animals and soil organisms: A review. *Environmental Nanotechnology, Monitoring & Management, 9*, 76–84. https://doi.org/10.1016/j.enmm.2017.12.006

Rajput, V. D., Minkina, T., Kumari, A., Harish, Singh, V. K., Verma, K. K., Mandzhieva, S., Sushkova, S., Srivastava, S., & Keswani, C. (2021). Coping with the challenges of abiotic stress in plants: New dimensions in the field application of nanoparticles. *Plants, 10*(6), 1221. https://doi.org/10.3390/plants10061221

Rajput, V. D., Singh, A., Minkina, T., Rawat, S., Mandzhieva, S., Sushkova, S., Shuvaeva, V., Nazarenko, O., Rajput, P., Komariah, Verma, K. K., Singh, A. K., Rao, M., & Upadhyay, S. K. (2021). Nano-enabled products: Challenges and opportunities for sustainable agriculture. *Plants, 10*(12), 2727. https://doi.org/10.3390/plants10122727

Raliya, R., Nair, R., Chavalmane, S., Wang, W.-N., & Biswas, P. (2015). Mechanistic evaluation of translocation and physiological impact of titanium dioxide and zinc oxide nanoparticles on the tomato (Solanum lycopersicum L.) plant. *Metallomics,* 7(12), 1584–1594. https://doi.org/10.1039/c5mt00168d

Rastogi, A., Singh, P., Haraz, F. A., & Barhoum, A. (2018). Biological synthesis of nanoparticles: An environmentally benign approach. In *Fundamentals of nanoparticles* (pp. 571–604). Elsevier.

Raza, A., Mehmood, S. S., Tabassum, J., & Batool, R. (2019). Targeting plant hormones to develop abiotic stress resistance in wheat. In *Wheat production in changing environments* (pp. 557–577). Springer Singapore.

Rehman, M. Z., Ayub, M. A., Umair, M., Ahmad, Z., Khalid, H., Ali, S., & Rizwan, M. (2021). Synthesis and regulation of secondary metabolites in plants in conferring tolerance against pollutant stresses. In *Approaches to the remediation of inorganic pollutants* (pp. 57–76).

Rico, C. M., Peralta-Videa, J. R., & Gardea-Torresdey, J. L. (2015). Chemistry, biochemistry of nanoparticles, and their role in antioxidant defense system in plants. In M. Siddiqui, M. Al-Whaibi, & F. Mohammad (Eds.), *Nanotechnology and Plant Sciences.* Cham: Springer. https://doi.org/10.1007/978-3-319-14502-0_1.

Rizwan, M., Ali, S., Ali, B., Adrees, M., Arshad, M., Hussain, A., Zia ur Rehman, M., & Waris, A. A. (2019). Zinc and iron oxide nanoparticles improved the plant growth and reduced the oxidative stress and cadmium concentration in wheat. *Chemosphere, 214,* 269–277. https://doi.org/10.1016/j.chemosphere.2018.09.120

Rout, G. R., & Sahoo, S. (2015). Role of iron in plant growth and metabolism. *Reviews in Agricultural Science, 3,* 1–24.

Sachdev, S., Ansari, S. A., Ansari, M. I., Fujita, M., & Hasanuzzaman, M. (2021). Abiotic stress and reactive oxygen species: Generation, signaling, and defense mechanisms. *Antioxidants, 10*(2), 277. https://doi.org/10.3390/antiox10020277

Sadeghi, A., Ataabadi, M., & Abolhasani, M. H. (2021). Chromium removal from a contaminated soil using nano zero-valent iron and magnetite affected by temperature and moisture. *Soil and Sediment Contamination: An International Journal, 30*(5), 610–621.

Salam, A., Khan, A. R., Liu, L., Yang, S., Azhar, W., Ulhassan, Z., Zeeshan, M., Wu, J., Fan, X., & Gan, Y. (2022). Seed priming with zinc oxide nanoparticles downplayed ultrastructural damage and improved photosynthetic apparatus in maize under cobalt stress. *Journal of Hazardous Materials, 423,* 127021. https://doi.org/10.1016/j.jhazmat.2021.127021

Salama, H. M. (2012). Effects of silver nanoparticles in some crop plants, common bean (*Phaseolus vulgaris* L.) and corn (*Zea mays* L.). *International Research Journal of Biotechnology, 3,* 190–197.

Saleh, A. M., Hassan, Y. M., Selim, S., & AbdElgawad, H. (2019). NiO-nanoparticles induce reduced phytotoxic hazards in wheat (*Triticum aestivum* L.) grown under future climate CO2. *Chemosphere, 220,* 1047–1057.

Sanzari, I., Leone, A., & Ambrosone, A. (2019). Nanotechnology in plant science: To make a long story short. *Frontiers in Bioengineering and Biotechnology, 7*. https://doi.org/10.3389/fbioe.2019.00120, 120-120.

Schwab, F., Zhai, G., Kern, M., Turner, A., Schnoor, J. L., & Wiesner, M. R. (2015). Barriers, pathways and processes for uptake, translocation and accumulation of nanomaterials in plants — critical review. *Nanotoxicology, 10*(3), 257—278. https://doi.org/10.3109/17435390.2015.1048326

Sedghi, M., Hadi, M., & Toluie, S. G. (2013). Effect of nano zinc oxide on the germination parameters of soybean seeds under drought stress. *Series of Biology, 16*(2), 73. Annales of West University of Timisoara.

Seleiman, M. F., Al-Suhaibani, N., Ali, N., Akmal, M., Alotaibi, M., Refay, Y., Dindaroglu, T., Abdul-Wajid, H. H., & Battaglia, M. L. (2021). Drought stress impacts on plants and different approaches to alleviate its adverse effects. *Plants, 10*(2), 259. https://doi.org/10.3390/plants10020259

Semida, W. M., Abdelkhalik, A., Mohamed, G. F., Abd El-Mageed, T. A., Abd El-Mageed, S. A., Rady, M. M., & Ali, E. F. (2021). Foliar application of zinc oxide nanoparticles promotes drought stress tolerance in eggplant (solanum melongena L.). *Plants, 10*(2), 421. https://doi.org/10.3390/plants10020421

Shabala, S., & Cuin, T. A. (2008). Potassium transport and plant salt tolerance. *Physiologia Plantarum, 133*(4), 651—669. https://doi.org/10.1111/j.1399-3054.2007.01008.x

Shahzad, B., Tanveer, M., Che, Z., Rehman, A., Cheema, S. A., Sharma, A., Song, H., Rehman, S.u., & Zhaorong, D. (2018). Role of 24-epibrassinolide (EBL) in mediating heavy metal and pesticide induced oxidative stress in plants: A review. *Ecotoxicology and Environmental Safety, 147*, 935—944. https://doi.org/10.1016/j.ecoenv.2017.09.066

Shang, H., Guo, H., Ma, C., Li, C., Chefetz, B., Polubesova, T., & Xing, B. (2019). Maize (*Zea mays* L.) root exudates modify the surface chemistry of CuO nanoparticles: Altered aggregation, dissolution and toxicity. *Science of the Total Environment, 690*, 502—510. https://doi.org/10.1016/j.scitotenv.2019.07.017

Sharma, P., Jha, A. B., Dubey, R. S., & Pessarakli, M. (2012). Reactive oxygen species, oxidative damage, and antioxidative defense mechanism in plants under stressful conditions. *Journal of Botany, 2012*, 1—26. https://doi.org/10.1155/2012/217037

Shen, R., Zhang, L., Chen, X., Jaroniec, M., Li, N., & Li, X. (2020). Integrating 2D/2D $CdS/\alpha\text{-}Fe_2O_3$ ultrathin bilayer Z-scheme heterojunction with metallic β-NiS nanosheet-based ohmic-junction for efficient photocatalytic H_2 evolution. *Applied Catalysis B: Environmental, 266*, 118619. https://doi.org/10.1016/j.apcatb.2020.118619

Shukla, P. K., Misra, P., & Kole, C. (2016). Uptake, translocation, accumulation, transformation, and generational transmission of nanoparticles in plants. In *Plant nanotechnology* (pp. 183—218). Springer International Publishing.

Shukla, P. K., Shukla, S., Rajoriya, P., & Misra, P. (2018). Enhancing crop productivity in saline environment using nanobiotechnology. *Salinity Responses and Tolerance in Plants, 2*, 289—305. Springer International Publishing.

Siddiqui, M. H., Al-Whaibi, M. H., Firoz, M., & Al-Khaishany, M. Y. (2015). Role of nanoparticles in plants. *Nanotechnology and Plant Sciences*, 19—35.

Siddiqui, H., Singh, P., Arif, Y., Sami, F., Naaz, R., & Hayat, S. (2022). Role of micronutrients in providing abiotic stress tolerance. In *Microbial biofertilizers and micronutrient availability* (pp. 115—136). Cham: Springer.

Singh, A., Vihinen, J., Erkka, F., Leo, H., Mari, H., & Erkki, L. (2016). Pulsed laser ablation-induced green synthesis of TiO2 nanoparticles and application of novel small angle X-ray scattering technique for nanoparticle size and size distribution analysis. *Nanoscale Research Letters, 11*(1), 1—9.

Su, Y., Ashworth, V., Kim, C., Adeleye, A. S., Rolshausen, P., Roper, C., White, J., & Jassby, D. (2019). Delivery, uptake, fate, and transport of engineered nanoparticles in plants: A critical review and data analysis. *Environmental Science: Nano, 6*(8), 2311–2331. https://doi.org/10.1039/c9en00461k

Tanveer, M., & Shabala, S. (2018). Targeting redox regulatory mechanisms for salinity stress tolerance in crops. In *Salinity responses and tolerance in plants* (pp. 213–234). Springer International Publishing.

Tanveer, M., & Shah, A. N. (2017). An insight into salt stress tolerance mechanisms of Chenopodium album. *Environmental Science and Pollution Research, 24*(19), 16531–16535.

Tanveer, M., Shahzad, B., & Ashraf, U. (2020). Nanoparticle application and abiotic-stress tolerance in plants. In *Plant life under changing environment* (pp. 627–641). Academic Press.

Tariq, M., Choudhary, S., Singh, H., Siddiqui, M. A., Kumar, H., Amir, A., & Kapoor, N. (2021). Role of nanoparticles in abiotic stress. *Technology in Agriculture*, 323.

Torabian, S., Shakiba, M. R., Dabbagh Mohammadi Nasab, A., & Toorchi, M. (2018). Leaf gas exchange and grain yield of common bean exposed to spermidine under water stress. *Photosynthetica, 56*(4), 1387–1397. https://doi.org/10.1007/s11099-018-0834-4

Tripathi, D. K., Shweta, S., Singh, S., Pandey, R., Singh, V. P., Sharma, N. C., Prasad, S. M., Dubey, N. K., & Chauhan, D. K. (2017). An overview on manufactured nanoparticles in plants: Uptake, translocation, accumulation and phytotoxicity. *Plant Physiology and Biochemistry, 110*, 2–12. https://doi.org/10.1016/j.plaphy.2016.07.030

Tripathi, D. K., Singh, V. P., Prasad, S. M., Chauhan, D. K., & Dubey, N. K. (2015). Silicon nanoparticles (SiNp) alleviate chromium (VI) phytotoxicity in Pisum sativum (L.) seedlings. *Plant Physiology and Biochemistry, 96*, 189–198.

Tripathi, D. K., Singh, S., Singh, V. P., Prasad, S. M., Dubey, N. K., & Chauhan, D. K. (2017). Silicon nanoparticles more effectively alleviated UV-B stress than silicon in wheat (*Triticum aestivum*) seedlings. *Plant Physiology and Biochemistry, 110*, 70–81. https://doi.org/10.1016/j.plaphy.2016.06.026

Uçarlı, C. (2020). Effects of salinity on seed germination and early seedling stage. *Abiotic Stress in Plants*, 211.

Usman, M., Farooq, M., Wakeel, A., Nawaz, A., Cheema, S. A., Rehman, H. U., Ashraf, I., & Sanaullah, M. (2020). Nanotechnology in agriculture: Current status, challenges and future opportunities. *Science of the Total Environment, 721*, 137778. https://doi.org/10.1016/j.scitotenv.2020.137778

Van Nguyen, D., Nguyen, H. M., Le, N. T., Nguyen, K. H., Nguyen, H. T., Le, H. M., Nguyen, A. T., Dinh, N. T. T., Hoang, S. A., & Van Ha, C. (2021). Copper nanoparticle application enhances plant growth and grain yield in maize under drought stress conditions. *J. Plant Growth Regul., 41*, 364–375. https://doi.org/10.1007/s00344-021-10301-w

Verma, K. K., Song, X. P., Joshi, A., Tian, D. D., Rajput, V. D., Singh, M., Arora, J., Minkina, T., & Li, Y. R. (2022). Recent tTriticum aestivumrends in nano-fertilizers for sustainable agriculture under climate change for global food security. *Nanomaterials, 12*(1), 173.

Wu, H., Shabala, L., Liu, X., Azzarello, E., Zhou, M., Pandolfi, C., Chen, Z.-H., Bose, J., Mancuso, S., & Shabala, S. (2015). Linking salinity stress tolerance with tissue-specific Na(+) sequestration in wheat roots. *Frontiers in Plant Science, 6*. https://doi.org/10.3389/fpls.2015.00071, 71-71.

Xin, X., Zhao, F., Judy, J. D., & He, Z. (2022). Copper stress alleviation in corn (Zea mays L.): Comparative efficiency of carbon nanotubes and carbon nanoparticles. *NanoImpact, 25*, 100381. https://doi.org/10.1016/j.impact.2022.100381

Yadav, T., Kumar, A., Yadav, R. K., Yadav, G., Kumar, R., & Kushwaha, M. (2020). Salicylic acid and thiourea mitigate the salinity and drought stress on physiological traits governing yield in pearl millet-wheat. *Saudi Journal of Biological Sciences, 27*(8), 2010–2017. https://doi.org/10.1016/j.sjbs.2020.06.030

Yang, F., Hong, F., You, W., Liu, C., Gao, F., Wu, C., & Yang, P. (2006). Influences of nano-anatase TiO_2 on the nitrogen metabolism of growing spinach. *Biological Trace Element Research, 110*, 179–190. https://doi.org/10.1385/bter:110:2:179

Yang, Y., Xu, S., An, L., & Chen, N. (2007). NADPH oxidase-dependent hydrogen peroxide production, induced by salinity stress, may be involved in the regulation of total calcium in roots of wheat. *Journal of Plant Physiology, 164*(11), 1429–1435. https://doi.org/10.1016/j.jplph.2006.08.009

Yin, L., Colman, B. P., McGill, B. M., Wright, J. P., & Bernhardt, E. S. (2012). Effects of silver nanoparticle exposure on germination and early growth of eleven wetland plants. *PloS One,* 7(10). https://doi.org/10.1371/journal.pone.0047674. e47674-e47674.

Zahedi, S. M., Moharrami, F., Sarikhani, S., & Padervand, M. (2020a). Selenium and silica nanostructure-based recovery of strawberry plants subjected to drought stress. *Scientific Reports, 10*(1). https://doi.org/10.1038/s41598-020-74273-9, 17672-17672.

Zahedi, S. M., Moharrami, F., Sarikhani, S., & Padervand, M. (2020b). Selenium and silica nanostructure-based recovery of strawberry plants subjected to drought stress. *Scientific Reports, 10*(1), 1–18.

Zahra, Z., Arshad, M., Rafique, R., Mahmood, A., Habib, A., Qazi, I. A., & Khan, S. A. (2015). Metallic nanoparticle (TiO_2 and Fe_3O_4) application modifies rhizosphere phosphorus availability and uptake by Lactuca sativa. *Journal of Agricultural and Food Chemistry, 63*, 6876–6882.

Zaidi Mgh, S. A. (2014). Impact of gold nanoparticles on physiological and biochemical characteristics of Brassica juncea. *Journal of Plant Biochemistry & Physiology, 02*(02). https://doi.org/10.4172/2329-9029.1000133

Zhang, L., Zhu, Y., Zhang, J., Zeng, G., Dong, H., Cao, W., Fang, W., Cheng, Y., Wang, Y., & Ning, Q. (2019). Impacts of iron oxide nanoparticles on organic matter degradation and microbial enzyme activities during agricultural waste composting. *Waste Management, 95*, 289–297. https://doi.org/10.1016/j.wasman.2019.06.025

CHAPTER SEVENTEEN

Insight on abiotic stress management in plants by improving plant nutritional status

Amir Hossein Vafaei[a], Mohammad Reza Gohari Kia[b], Vahideh Gougerdchi[b], Zahra Dehghanian[c], Behnam Asgari Lajayer[d], Tariq Aftab[e], Tess Astatkie[f]

[a]Department of Microbial Biotechnology, Faculty of Basic Sciences, Higher Education Institute of Mizan, Tabriz, Iran
[b]Department of Plant Breeding and Biotechnology, Faculty of Agriculture, University of Tabriz, Tabriz, Iran
[c]Department of Biotechnology, Faculty of Agriculture, Azarbaijan Shahid Madani University, Tabriz, Iran
[d]Department of Soil Science, Faculty of Agriculture, University of Tabriz, Tabriz, Iran
[e]Department of Botany, Aligarh Muslim University, Aligarh, Uttar Pradesh, India
[f]Faculty of Agriculture, Dalhousie University, Truro, Canada

1. Introduction

The changing climate had already significantly diminished ecological remediation, which is making crops more vulnerable to biotic sources of stress throughout ontogeny (Bulgari et al., 2019). This reduces their struggle for survival, growth, and economic production (Sachdev & Singh, 2018a). Heat shock, cold shock, freezing, water shortage, waterlogging, salt, unbalanced nutrition, toxic metals, and xenobiotic pressure are all abiotic stressors that contribute to 50% of output loss. Temperature extremes, fluctuations, excessive UVA and UVB ray's irradiation, light intensity fluctuations, powerful floods, precipitation patterns (shortfall or abundance), metal ions release and concentration, as well as other xenobiotic contaminants (pesticides, fertilizers, hydrocarbons) are all contributing environmental factors. Abiotic stressors frequently occur in combination (Kumar et al., 2017) in a natural environment. Plants are affected by their linked pathways, which have distinct and compounding effects on their cell membrane, metabolism, and biological performances (Bulgari et al., 2019). Under various climatic circumstances, reactive oxygen species (ROS) such as superoxide radical (O_2^-), hydroxyl radical (OH), hydrogen peroxide (H_2O_2), peroxy radical (ROO), singlet oxygen ($^1O_2$), and alkoxyl radicals (RO) have been shaped

Sustainable Plant Nutrition
ISBN: 978-0-443-18675-2
https://doi.org/10.1016/B978-0-443-18675-2.00007-9

at cold temperatures in plant tissues. Aggressive climate conditions, on the other hand, cause excessive production of ROS (Rane et al., 2021). ROS induce plant mortality by oxidizing biomolecules (lipids, carbohydrates, proteins, enzymes, and DNA) and causing damage to molecular and cellular components (Bhuyan et al., 2020). Plants carefully regulate ROS generation to prevent harm by recruiting antioxidants, both enzymatic and non-enzymatic. Superoxide dismutase (SOD), catalase (CAT), ascorbate peroxidase (APX), glutathione reductase (GR), peroxidase (POX), and others are all components of the antioxidant enzymatic structure. Non-enzymatic antioxidant properties like vitamins, stilbenes, flavonoids, and carotenoids quench excess ROS, acting as a protectant against osmotic damage (Hasanuzzaman et al., 2020). The unrestrained propagation of reactive species derived from oxygen (O_2) is harmful to plant health. In contrast, controlled ROS production contributes to plant development and growth, as well as redox signifying (Florez-Sarasa et al., 2020). Cell-to-cell communication is facilitated by fine-tuned ROS production that induces nicotinamide adenine dinucleotide phosphate (NADPH) oxidase, defined as respiratory burst oxidase homolog (RBOH), and responds to stress by modifying the structure of proteins and amplifying the involved defense and security genes (Hasanuzzaman et al., 2018). The presence of abiotic stressors, either separately or concurrently, that causes an oversupply of ROS in plant tissue is a serious barrier to optimizing plant growth and production. Investigating the molecular procedures of ROS signal transduction pathways is crucial for mitigating stress or promoting signaling in economically important plants while monitoring and predicting climatic conditions, maintaining sensitivity, and maintaining economic efficiency. The current study investigates the mechanisms of antioxidant defense that regulate ROS generation and dissipate oxidative stress, as well as the wellness of plants under anaerobic environment challenges. The review examines ROS's dual roles as a stress response molecule that aids plant's adaptation and development. Furthermore, the importance of their analysis and description, which includes systems biology, molecular priming, plant simulation, and phenomics in improving plant resistance to oxidative stress is discussed.

2. Global warming causes abiotic stress and ROS production

On a worldwide scale, the preponderance of abiotic stress and its effects on crops has increased as a result of climate change. According to

Raza (2019), one or more types of stress affect 96.5% of the global agricultural area (Raza et al., 2019). Plants' water requirements will rise by 4%−4.5% for every 1°C temperature increase (Ye et al., 2015). In drought-stricken places, this makes plant cultivation more challenging. Temperature rises also cause chemical pollutants to bio-transform further into dangerous or biomaterial shapes, exacerbating ecological irritability and disturbing plant metabolism (Noyes et al., 2009). Temperature rises combined with precipitation decreases increase the evaporation of xenobiotics as environmental contaminants, contributing to increased pollution of the air. On the other hand, excessive precipitation, causes soil pollution, groundwater flow pollutants, and salt content by increasing air pollutant deposits onto land or speeding up the dumping of soil quality or contaminants into underground (Zheng et al., 2020). Moderate temperature and rainfall changes, nutritional imbalances, waterlogging, and so forth are all common occurrences under normal circumstances, which may temporarily suppress the growth of plants. In contrast, major weather patterns and oscillations extend and magnify the severity and duration of stressors, pushing plant's performance beyond recovery (Saijo & Loo, 2020). Climate change has been designed to generate unfavorable conditions from which plants cannot escape, as well as to subject them to a variety of debilitating abiotic stressors (Table 17.1). Enhancing plant growth and productivity to nourish the world's current population is not the only goal; meeting the nutritional needs of future generations is also critically important. As a result, it is critical to assess the extent to which certain persistent abiotic stressors impinge on plants. Combining this data with simulation models might aid in developing long-term crop protection plans in response to predicted changes in climatic circumstances.

2.1 Stress due to temperature

Intense climate change (10−15°C deviation), a step up or a step down from the ideal environment, produces suffocating heat or harrowing stress or inhibits photosynthetic plant structure, propagation, and yield (Bita & Gerats, 2013). Extreme heat creates morphometric, cellular, and energy metabolism effects on plants that reduce the performance of photosynthesizing and breathing equipment, decrease enzyme reactions, increase heat shock protein (HSP) alternative splicing, raise calcium (Ca^{2+}) emergence, and amplify the production of ROS (Bita & Gerats, 2013). Heat stress impairs cell development, which reduces leaf area (Hussain et al., 2018). High-temperature diagnosis of hyacinth bean (*Lablab purpureus* L.) reduces cell membrane permeability, lipid peroxidation, and ROS production; it inhibits plant

Table 17.1 List of genes used to generate drought-tolerant transgenic plants.

Goal gene	Gene's source	Goal plant	Assessment	Changes in functionality	References
AtABF3	*Arabidopsis thaliana*	*Oryza sativa* L.	Greenhouse	There are no obvious growth abnormalities, and the plant is more drought tolerant.	(Oh et al., 2005)
SNAC1	Rice IRAT109	*Oryza sativa* L.	Greenhouse, field	There are no growth anomalies, and the plant is drought tolerant.	(Hu et al., 2006)
OsNAC6	*Oryza sativa* L.	*Oryza sativa* L.	Greenhouse	Growth was slow, the reproductive rate was low, tolerance to water loss was higher, and blast disease resistance was improved.	(Nakashima et al., 2007)
DREB1A	*Arabidopsis thaliana*	*Triticum aestivum*	Greenhouse	Symptoms of a delayed drought.	(Pellegrineschi et al., 2004)
DREB1A	*Arabidopsis thaliana*	*Arachis hypogaea* L.	Greenhouse	Transpiration performance is 40% higher than that in untransformed controls.	(Bhatnagar-Mathur et al., 2007)
OsDREB1G	Oryza sativa L.	*Oryza sativa* L.	Greenhouse	Increased resistance to drought stress.	(Chen et al., 2008)
OsDREB1F	*Oryza sativa* L.	*Oryza sativa* L.	Greenhouse	Increased resistance to salinity, drought, and low temperatures.	(Wang et al., 2008)
GhDREB	*Gossypium hirsutum*	*Triticum aestivum* L.	Greenhouse	Increased resistance to drought, salinity, and chilling stresses, as well as enhanced aggregation of solvable sugar and chlorophyllin in leaves under stress circumstances.	(Gao et al., 2009)

HhDREB2	*Halimodendron halodendron*	*Arabidopsis thaliana*	Greenhouse	Increased resistance to salinity and drought stress.	(Ma et al., 2015)
AtDREB2A-CA	*Arabidopsis thaliana*	*Gossypium hirsutum* L.	Greenhouse	Enhanced shoot growth and root trait morphometrics under water-stressed conditions.	(LiseideSá et al., 2017)
OsNHXI	*Oryza sativa*	*Zea mays* L.	Greenhouse, field	Increased biomass and grain yield at the germination stage, and the coleoptiles and radicles are longer.	(Chen, Chen, et al., 2007, Chen, Wang, et al., 2007)
GmDREBI	*Glycine* max L.	*Triticum aestivum* L.	Field	At the seedling stage, the root length and number of tillers per plant are increased; oxidative and osmotic stress-related proteins are upregulated.	(Jiang et al., 2014)
AhDREBI	*Atriplex hortensis*	*Populstomentosa Populus*	Field	Increased proline and chlorophyll levels; increased POX and SOD activity.	(Lu et al., 2014)
EcNAC1	*Eleusine coracana* L.	*Nicotiana tabacum*	Greenhouse	Decreased ROS harm and increased resistance to oxidative stress.	(Ramegowda et al., 2012)
TaMYB2A		*Arabidopsis thaliana*	Greenhouse	Reduced water loss and improved cell membrane stability, lowered osmotic potential and increased photosynthetic potential.	(Mao et al., 2011)
DnWRKY11	*Dendrobium nobile*	*Nicotiana tabacum*	Greenhouse	Increased germination rate, increased root length, increased CAT, POX, and SOD activity, and decreased malondialdehyde amount.	(Xu et al., 2014)

development and efficiency, and reduces the number of leaves, the amount of chlorophyll, and the carotenoid in the leaves and enhances a disparity among H_2O_2 and O_2 production for quenching (Balal et al., 2016). Heat stress affects cucumber plants while boosting the production of ROS, lipid peroxidation, intercellular CO_2 content, and non-photochemical dehydrating as well as the development, production, total chlorophyll, photosynthetic activity, stomatal closure, evapotranspiration, anti-oxidants, and stomatal ratio (NPQ). The creation of ozone (O_3) in the troposphere increases as the temperature rises, causing plant oxidative stress (Rai et al., 2018).

Minimal temperature events, known as chilling stress, enhance O_2 solubility and accumulation, as well as leakage of electrons from the photosynthetic electron transport chain (ETC) and respiratory ETC decrease. both of which increase ROS generation in Cells from plants (Coates et al., 2016), influences a fluency of both the cell wall and enzymatic activity (Jalil et al., 2017). Cucumber seedlings have been found to have increased electrolyte leakage, and also reduced levels of chlorophyll and water in the tissues while exposed to cold exposure (Zhang et al., 2012). Under low-temperature stress, tomato leaves had RBOH1 expression, H_2O_2 and O_2 buildup, increased malondialdehyde (MDA) content, as well as decreased net photosynthetic rate and chlorophyll fluorescence (Liu et al., 2019).

2.2 Water stress

Changes in weather patterns have had a profound influence on rainfall patterns during the last several decades, resulting in irregular precipitation of varying magnitudes and seasonal changes (Feng et al., 2013).

2.2.1 Drought

Water scarcity is possibly the most significant ecological stressor influencing agricultural production, accounting for 70% loss of the global yield (Shiferaw et al., 2014). Daryanto et al. (2016) examined data from 1980 to 2015 around the world and discovered that droughts caused yield decreases of up to 21% and 40% in wheat and rice, respectively. With shifting climatic conditions and scarcity of water supplies, agricultural plants that can withstand drought without reducing production are needed. Furthermore, due to lack of water, many fields have become barren. Plants that can endure drought conditions would improve food security of the world's increasing population. Drought tolerance in plants requires knowledge of the physiological and metabolic installations of drought reaction, and genetic regulation connections is important for plant drought tolerance. Drought tolerance in plants requires knowledge of the biochemical and physiological bases of drought resistance; and genetic regulation systems are crucial for plant drought resistance. They are crucial for several notable investigations that have been performed to determine the primary regulators of water deficit response at various steps. Drought induced transcription factors include the element that responds to dehydration binding proteins (DREBs), abscisic acid response element binding proteins (AREBs)/abscisic acid responsive element binding factors (ABFs), nuclear factor Y.B subunits (NF.YB), and tryptophan–arginine–lysine–tyrosine (WRKY) (Bhaskara et al., 2015). In the lab, drought responsive genes have been overexpressed or downregulated successfully. Field studies are needed to ascertain the findings. Drought stress causes the phytohormone abscisic acid (ABA) to be synthesized or transported, which is a crucial molecule in signaling processes while there is a drought (Fang & Xiong, 2015). The perspective of ABA absorption is initially mediated by a pyrabactin resistance (PYR)/pyrabactin1 like (PYL)/regulatory component of abscisic acid response (RCARb), protein phosphatase 2C (PP2C), and sucrose nonfermenting related protein kinase 2 (SnRK2) complex, that causes the transcription factors NF.YA and SN to be expressed (Roychoudhury & Paul, 2012). These proteins control the opening and closing of stomata to limit transpirational water loss. Drought stress is detected by a regulation cycle that incorporates calcineurin B like protein interacting protein kinase (CIPK) and, Ca^{2+} dependent protein kinase (CDPK) where it induces DREBs and, AREB that connect toward the dehydration and abscisic acid responsive element cis elements of downstream genes, producing transcription factors such as late embryogenesis abundant protein (LEA), heat shock protein (HS) (Yang et al., 2010). Drought sensitive plants with these transcription factors overexpressed have increased their tolerance to water scarcity. Furthermore, several crops with constitutively expressed drought.responsive transcriptional regulators showed developmental delays (Suo et al., 2012). Stress inducible promoters, such as HVA22P, have been used to induce the expression of these transgenes in transgenic plants to mitigate this unfavorable effect (Xiao et al., 2009). When water deficit is prolonged, these genes are continuously expressed in transgenic plants, resulting in growth abnormalities. To overcome this barrier, researchers created stress.inducible tissue.specific promoters such as Responsive to Dehydration 29A (RD29A) to represent these transgenes (Ito et al., 2006). When rice plants are under abiotic stress, the RD29A promoter is found in roots; however, the little issue it creates with root growth might prevent it from being used. Kudo et al. (2017)

used transgenic Arabidopsis plants to stack two transcription factors: DREB1A to enhance drought resistance, and Phytochrome Interacting Factor Like 1 (OsPIL1) to aid the growth of the plant. By modulating cell wall related gene expression, OsPIL1 promotes cell elongation while avoiding the detrimental consequences of increased expression of the DREB1A gene. Each of these separate strategies can be combined with gene stacking to ensure the effective use of pressure tissue specific promoters to confer drought resistance while removing growth retarding effects. The technique will be more successful and acceptable if such genetic mutations and promoters are extracted from the same plant and expressed in the same plant. The above regulatory network governs drought stress tolerance via post translational modifications to drought.responsive proteins like phosphorylation, farnesylation, sumoylation, and poly (ADP.ribosyl) ation (PAR) (Kim et al., 2017). Under field settings, the use of AtHPR1 promoter to conditionally and selectively downregulate the farnesyl transferase gene throughout canola tends to result in drought stress yield safety (Wang et al., 2009). In transgenic plants, drought tolerance was increased by overexpression of protein kinases, including CDPKs and CIPKs (Wang et al., 2018). LEAs are hydrophilic proteins that accrue in embryos through seed evaporation. They are considered an important element in interactive dehydration reactions through conditional molecules of water, trying to stabilize proteins or membranes in the same way that HSPs do. There is an increase in ABA sensitivity and osmotic tolerance when LEA genes are expressed in transgenic plants (Banerjee & Roychoudhury, 2016). Glycine betaine, Proline, polyamines like putrescine, spermidine, spermine, and sugars like trehalose, accumulate to reduce dehydration and prevent cell structures from osmolarity destruction (Montilla-Bascón et al., 2017). Drought resistance and enhanced production have been provided by plants, or bacterial cold shock proteins are overexpressed in essential ingredient crops like wheat, rice, and maize in field cold shock proteins from plants or microorganisms are overexpressed in circ umstances(Yu et al., 2017). Cellular energy homeostasis variations during stress situations is essential for improved plant production and production consistency. Poly (ADP-ribosyl) ation is a one of a kind post translational protein alteration (induced mostly by the PARP enzyme) that happens in crops in an oxidative stress response. DNA synthesis and restoration, transcription, and cell cycle action have all been related to PAR (d'amours et al., 1999). PARP suppression alters photosynthesis and improves plant stress resistance (Schulz et al., 2012). Micro RNAs and noncoding RNAs have been recognized in response to drought stress (Qin et al., 2017). Drought alters the expression of a number of miRNAs (Banerjee et al., 2016). By using specialized promoters, these miRNAs could be used to generate transgenic crops. Drought and salt tolerance were improved in transgenic creeping bentgrass (*Agrostis stolonifera*) after rice Osa-miR319a was overexpressed. The wax content of the leaf and its ability to retain water were higher in these plants, while sodium absorption was lower (Zhou et al., 2013). Chickpea and tomato plants with high levels of miRNA408 and miR169 exhibited improved drought tolerance (Hajyzadeh et al., 2015).

Enhanced sequencing ways of achieving genomic and transcriptome statistics throughout drought resistance have become accessible, that could allow the discovery of the regulatory networks that are active throughout drought (Bai et al., 2017). During a drought, regulatory networks are engaged. These regulatory networks' components can be engineered to aid in the development of water stressed plants. Because some of the previous studies have been conducted in greenhouses or on drought-tolerant crops, they need to be evaluated for their maximum potential in innate field circumstances in which other external factors precede the drought condition. In drought conditions, it is really difficult to evaluate the success of transgenic crops, because drought varies in severity and length. Drought GardTM was marketed by Monsanto as a drought-tolerant transgenic maize line that contains the bacterial cspB gene (Castiglioni et al., 2008). When there is different drought circumstances and varying temperature and soil properties, a gene's credibility could not be properly tested. The transgenic line was drought tolerant but could withstand high water shortages (Gurian-sherman, 2012). Performance Plants, Inc., a subsidiary of Bayer Crop Science, was also adopted. In 2009, Yield Protection Technology was used to generate drought.resistant and high-yielding cotton. After 3 years of repeated field experiments in the normal ground settings, such cotton plantations demonstrated considerable produce benefits during water stress without any negative impact on growth under ideal circumstances (Suprasanna & Ghag, 2019).

2.2.2 Waterlogging and flooding

Soil flooding or waterlogging is caused by an excessive buildup as a result of the presence of water in the soil, high precipitation over time, inadequate drainage, or other factors (Fukao et al., 2019). Waterlogging has had a negative impact on approximately 10% of the world's largest land area surface (Hossain & Uddin, 2011). Waterlogging engulfs root system of plants and is distinguished by insufficient light, poor gas share, hypoxia, and anoxia (Fukao et al., 2019). It inhibits aerobic conditioning, which includes root breathing through soil, by reducing O_2 dispersion by 10,000 points of time when contrasted to air (Yamauchi et al., 2019). ROS is produced when the anoxic situation inhibits the ETC of chloroplasts and mitochondria (Sasidharan et al., 2018). Waterlogging enhances lipid peroxidation, ROS accumulation, and methylglyoxal concentrations in sesame plants, all of which cause oxidative stress (Anee et al., 2019).

Light reaches to the immersed parts of the plant easily in submerged water, causing photorespiration and the production of paroxysmal H_2O_2 (Sasidharan et al., 2018). Flooding already depletes soil nutrients, causes salt buildup, and raises metal ions obtained by adjusting to pH changes. These negative alterations eventually cause nutritional deficiencies as well as other stressors (salinity, heavy metals) in plants (Steffens, 2014).

2.3 Salinity and salt stress

Salinity arises due to several reasons including improper agricultural land use, a lack of rain, excessive evaporation, and a lack of drainage. Soils with salt concentrations that hinder plant development ($EC_e > 4$ ds mL^{-1}) and soils without salt concentrations that affect soil structure (ESP 15) are referred to as saline soil. Salt stress is caused by large accumulations of these salts, which have an impact on plant development and germination, and also seed size and quality. According to the World Irrigation Organization (WIO), irrigation is now used in 17% of the world's dry and semi.arid areas, and owing to improper management, around 20% of these irrigated areas are unproductive, according to the WIO, a United Nations organization that promotes sustainable development (Sachdev & Singh, 2018a). Irrigation-necessary fields cover 2% of the surface area, and 74% of these lands (about 12 1000 hectares) are vulnerable to the detrimental impacts of salt (Ye et al., 2015). Soil salinity affects plants by disrupting ground-water movement from roots owing to excessive concentrations of salt and creating toxicity due to buildup in the plant (Pereira, 2016). Plants that struggle with salt attempt to maintain their life cycles by exhibiting morphological, biochemical, and physiological responses. In the world, salinity has a deleterious impact on more than 800 million hectares of agricultural land. This ratio accounts for more than 6% of all agricultural land on the globe (Noyes et al., 2009). There are two kinds of salinity levels: primary and secondary. Primary salinity is caused by natural factors such as seas and rock weathering, whereas secondary salinity is influenced by human factors such as excessive irrigation in farmlands and damage to farmland structure (Pereira, 2016). This problem is becoming more pronounced by the day. It is anticipated that by 2050, half of all planted agricultural fields will be salt.stressed (Jarsjö et al., 2020). The major factors influencing agricultural land salinity are NaCl and Na_2SO_4 salts (Zheng et al., 2020). It is easier to determine if plants are salt resistant at non.lethal salt concentrations at germination and early seedling phases, as these are the phases where the most harm occurs (Saijo & Loo, 2020). However, along with well known ideas such as drainage and irrigation resource management, cultural practices and agricultural land development efforts should be addressed while attempting to regulate salinity (Esim & Atici, 2014). Fertilization, planting procedures, irrigation treatment, and soil leveling are all examples of cultural practices. Farmland enhancements include drainage, land leveling, and breeding irrigation (Esim & Atici, 2014). The absence of drainage is the most critical reason for salinity, and millions of acres of rich agricultural land are destroyed as a result.

2.3.1 Effect of salinity on plants

Stress is exacerbated by salinity, which disrupts ionic and osmotic equilibrium in plants. Osmotic stress, generated by increasing quantity of salt in the soil, reduces the amount of water that plants utilize, resulting in physiological dryness. Ionic stress develops in the plant as a result of these circumstances, resulting in a degradation of the plant's ion balance. Ions of Cl and Na, which rise in the medium due to ionic stress, compete with important nutrients such as K^+, Ca^{2+}, and Mg^{2+}, resulting in a nutritional deficit in the plant. While osmotic and ionic stressors are the primary effects of salinity, deterioration in structure and the production of hazardous components are the subsequent effects (Megha et al., 2018).

2.3.2 Salt stress's secondary effect

Complications in getting K^+ into cells, which is effective in closing stomata, decreased photosynthetic activity, ROS production, and programmed cell death (Ansari & Lin, 2010) that are the main secondary causes produced by NaCl. Because K is one of the most important components for plants, Na^+ and K^+ ions end up competing for cell entry. Na^+ contributes to stress by preventing K^+ entry into the cell. Secondary metabolic effects are caused by ion and hyperosmotic stress in plants, and the plant must reduce these stressors in order to continue its growth (Hasanuzzaman et al., 2020).

2.3.3 Salinity tolerance

Plants can withstand salt stress by avoiding or sequestering and storing salts in vacuoles, reducing osmotic pressure through the production of osmolytes like proline, glycine betaine or trehalose, and the production of antioxidants or enzymes to battle ROS (Das & Roychoudhury, 2014). Through breeding or genetic engineering, to increase salt tolerance in agricultural crops, numerous genes and gene families are used. A regular plant breeding is used to create salt stress lines; however, its application is restricted due to lack of genetic diversity. To acquire salt tolerance, genetic engineering has been proven to be useful in the transfer of genes from salt-tolerance plants (halophytes and non.halophytes). When plants are exposed to salt tolerance, an indicating cascade detects transcription factors like DREBs, MYBs, bZIPs, NACs, and WRKYs (Xu et al., 2014). All such genes collaborate to affect a metabolic transition that accumulates osmoprotectants as well as other cellular metabolic products, reducing stress.induced damage. In plants, toleration of salt has been achieved through the interpretation of transporter proteins such as the Na^+/H^+ antiporter, Na^+/K^+ transporter, and vacuolar H^+ pyrophosphatase (Roychoudhury & Chakraborty, 2013). Salinity stress tolerance in plants was demonstrated in transgenic studies by gene mutations due to a rise in the concentration of water.soluble osmolytes. Rice's resistance to high salinity was increased by OsTPS1 gene overexpression (encoding the enzyme trehalose.6.phosphate synthase) (Liu et al., 2011). It has already been proven that the expression levels known to accumulate glycine betaine, proline, and myoinositolide lead to salt sensitivity (Chen, Yang, et al., 2013a, Chen, Chen, et al., 2013b). Aquaporins are membrane proteins that are known to preserve water cycle in cell lines and even to play an important role in abiotic and biotic stress civility (Afzal et al., 2016). PeTIP4; 1.1, a bamboo aquaporin, and JcPIP2; 7 and JcTIP1; 3, two Jatropha aquaporins, conferred salt tolerance in Arabidopsis (Sun, Li, et al., 2017a, Sun,-Deng, et al., 2017b). When overexpressed in transgenic plants, other aquaporins including MzPIP2; 1, TaNIP, and MusaPIP1; two conferred salt tolerance as well (Wang et al., 2015). Furthermore, antioxidative enzymes such as CAT, glutathione synthase, SOD, and APXs were found to decrease salt induced destruction by producing Free radicals, thereby imparting salinity plant tolerance (Liao et al., 2016). Plant DNA and RNA helicases have been found to be salt-stress-regulated, and transgenic plants trying to express them have demonstrated salt tolerance. Salt tolerance is conferred by these helicases, which maintain osmotic equilibrium, boost photosynthetic activity, and boost ROS scavenging task (Tuteja et al., 2014). Pea

DNA helicase 45 increased expression enhanced salt tolerance in transgenic rice, tobacco, and chili (Shivakumara et al., 2017). Salinity tolerance was provided in cotton plants through increased expression of AvDH1, a DEAD-Box helicase gene from *Apocynum venetum* (Chen et al., 2016). Overexpression of mRNA might be exploited to produce salt tolerance in transgenic plants with an understanding of short RNAs that regulate stress.responsive conditions (Zhang & Wang, 2016). Moreover, research is required to investigate new mRNA regulatory under salt stress and to communicate these distinct mRNAs under powerful tissue specific inactivating promoters. In conclusion, the majority of the salt-tolerant transgenic lines that have been developed have been restricted to the greenhouse level, and they need to be tested in field conditions. Also, further study to create appropriate ways of producing salt-tolerant lines that can thrive in organic salt and non-saline ground situations that do not jeopardize development or produce research is required.

2.4 Nutrient deficiency

Because of the constantly changing characteristics of the global environment, ensuring optimal plant growth and development has become a serious matter. The paucity of critical plant nutrients in soil has a negative impact on their physiological processes, including ETC, water relations, and gaseous exchange, all of these factors contribute to the formation of ROS and the development of oxidative stresses (Hajiboland, 2012). Plasma membrane bound NADPH oxidase is a major source of production of ROS in plant tissue (Gupta et al., 2017). Plant minerals like zinc (Zn^{2+}) and potassium (K^{+}) influence NADPH oxidase activity, thus their shortage increases enzymatic activity, which acts as a catalyst for the production of O_2 or H_2O_2 (Hajiboland, 2012). Plants that are deprived of nutrients produce ROS as a result of the ethylene signaling pathways. The scarcity of K^{+} promotes ethylene metabolites, which then in turns increases upriver ROS production (Jung et al., 2009). Nutrient minerals like nitrogen (N), magnesium (Mg), copper (Cu), manganese (Mn), and zinc (Zn) are required by a variety of enzymes (Cu/Zn-SOD, Mn-SOD, etc.) and antioxidants associated with energy metabolism and the scrounging of ROS (Hajiboland, 2012). The ability of plants to scavenge ROS is harmed by nutritional deprivation, which leads to ROS generation indirectly (Hajiboland, 2012). The Heber-Weiss reaction, for example, raises the quantity of OH by reducing enzymes' capability to remove H_2O_2 and O_2 interior plant tissue

(Hajiboland, 2012). Further deprivation of minerals such as Mg, a key component of chlorophyll, decrease photosynthesis process, resulting in the production of free radicals (ROS) (Guo et al., 2016).

2.5 Heavy metal stress

Toxicity is caused by the accumulation of non.essential metals in crops through the use of production of ROS. On the other hand, the unrestricted intake of vital nutrients causes ROS generation (Saleem et al., 2020). Because of their large concentrations in soil, Iron (Fe), chromium (Cr), and Cu have become essential redox-active metals that induce oxidative stress in crops (Schutzendubel & Polle, 2002). ROS generation is triggered by heavy metal stress, which is mediated through the ETC of chloroplasts, mitochondria, apoplasts, and peroxisomes (Trinh et al., 2014). Cadmium (Cd) is a toxic non-essential metal that can be found in plants. Because of their deficient actions, Cd ions replace Cu or Fe ions in antioxidant metalloenzymes, implicitly boosting ROS production, impacting breathing ETC, and interfering with the cell redox system (Shin et al., 2005). Despite being a necessary micronutrient, excessive Fe buildup in plants triggers the generation of ROS through a series of processes (Becana et al., 1998) and damages lipid membrane and chlorophyll (Hajiboland, 2012). H_2O_2 and O_2 are produced when the reduced form of Fe oxidizes. H_2O_2 then oxidizes the reduced Fe molecules, producing an extremely poisonous OH (Hajiboland, 2012). The Fenton-type reaction catalyzes the auto-oxidation of redox-active metals such as Cu and Fe, due to the production of ROS (Schutzendubel & Polle, 2002). Xenobiotic chemicals, like pesticides, are similar to heavy metals in that they cause ROS generation and oxidative stress (Sharma et al., 2019). Only 1% of pesticides administered reach their intended target, with the rest accumulating in soil and non-target live species (Sachdev & Singh, 2018b). Pesticides slow plant development, reduce photosynthetic efficiency, cause molecular changes, boost ROS generation, and affect antioxidant status (Sachdev & Singh, 2018b). In tomato leaves treated with thiram, chlorophyll breakdown was seen along with an increase in H_2O_2 and MDA levels (Sachdev & Singh, 2018b). Another study showed the reduction of chlorophyll content of *B. juncea* seedlings by imidacloprid. The increased expression of the gene CHLASE, which encodes the chlorophyllase enzyme that catalyzes chlorophyll breakdown, is to blame for the reduction in chlorophyll (Sharma et al., 2019). Furthermore, pesticides increase the amount of RBOH transcripts and the buildup of ROS.

2.5.1 Plants' defense mechanisms against heavy metal stress

To avoid or withstand heavy metal (HM) poisoning, plants have a complex and interconnected network of defensive systems. Metal-resistant plants have physical barriers as their initial line of protection. When plants are subjected to HM stress, several morphological features like cuticle thickness can be formed by physiologically active body tissue, including trichomes and cell walls, as well as mycorrhizal symbiosis (Florez.Sarasa et al., 2020). For example, trichomes can either store HM for detoxification or produce a variety of secondary metabolites to counteract metal toxicity (Singh et al., 2006). Plants, on the other hand, initiate various cell defense systems to inactivate and mitigate the impact of HMs before they absorb tissues and conquer physical and biological obstacles. The fundamental method for tolerating or neutralizing metal toxicity is biosynthesis of several cellular macromolecules. It also contains plenty of low-molecular-weight protein metallochaperones or chelators like nicotianamine, spermine, putrescine, glutathione, mugineic acids, organic acids, metallothioneins, and phytochelatins; and cell membrane secretions like flavonoid and factors affecting heat shock proteins, protons, and specific amino acids like ethylene and proline (Saijo & Loo, 2020). When the aforementioned methods fail to prevent metal poisoning, plants' cell-cycle systems become out of balance, resulting in increased ROS induction (Hameed et al., 2011). Plant species have developed an antioxidative defense structure that incorporates enzymatic antioxidants like CAT, SOD, GR, and APX, which serve as free radical scavengers (Zheng et al., 2020). Multifunctional biological compounds involved in cellular metal detoxification may have antiradical, chelating, or antioxidant properties. Plant species' metal tolerance (Kumari et al., 2015) and development stage may influence the use and overexpression of these processes and biomolecules.

3. Conclusions

In the next 50–100 years, the earth's temperature is anticipated to rise by 3–5°C. Because of the constant rise in temperature and inconsistent rainfall, floods and droughts are always a possibility. Excessive fertilizers, improper irrigation, and the exploitation of metal resources are all examples of anthropogenic activities that can cause salt stress to a great extent. Plants will most likely be exposed to both biotic and abiotic stressors more often under these conditions. Plant breeders have a responsibility to produce stress-tolerant cultivars in order to assure food security and agriculturalist safety. Molecular studies have to be done at the genetic level to develop mechanisms in plants in order to protect them from various kinds of stress circumstances. So until response mechanisms against biotic and abiotic stressors are not discovered, plants will be continually subjected to such stresses, posing a significant danger to global agriculture.

References

d'amours, D., Desnoyers, S., d'silva, I., & Poirier, G. G. (1999). Poly (ADP.ribosyl) ation reactions in the regulation of nuclear functions. *Biochemical Journal, 342*(2), 249–268.

Afzal, Z., Howton, T., Sun, Y., & Mukhtar, M. S. (2016). The roles of aquaporins in plant stress responses. *Journal of Developmental Biology, 4*(1), 9.

Anee, T. I., Nahar, K., Rahman, A., Mahmud, J. A., Bhuiyan, T. F., Alam, M. U., Fujita, M., & Hasanuzzaman, M. (2019). Oxidative damage and antioxidant defense in Sesamum indicum after different waterlogging durations. *Plants, 8*(7), 196.

Ansari, M. I., & Lin, T. P. (2010). Molecular analysis of dehydration in plants. *International Research Journal of Plant Science, 1*(2), 21–25.

Bai, Z. y., Wang, T., Wu, Y. h., Wang, K., Liang, Q. y., Pan, Y. z., Jiang, B. B., Zhang, L., Liu, G. L., Jia, Y., & Liu, Q. L. (2017). Whole.transcriptome sequence analysis of differentially expressed genes in *Phormium tenax* under drought stress. *Scientific Reports,* 7(1), 1–9.

Balal, R. M., Shahid, M. A., Javaid, M. M., Iqbal, Z., Anjum, M. A., GarciaSanchez, F., & Mattson, N. S. (2016). The role of selenium in amelioration of heat.induced oxidative damage in cucumber under high temperature stress. *Acta Physiologiae Plantarum, 38*(6), 1–14.

Banerjee, A., & Roychoudhury, A. (2016). Group II late embryogenesis abundant (LEA) proteins: Structural and functional aspects in plant abiotic stress. *Plant Growth Regulation, 79*(1), 1–17.

Banerjee, A., Roychoudhury, A., & Krishnamoorthi, S. (2016). Emerging techniques to decipher microRNAs (miRNAs) and their regulatory role in conferring abiotic stress tolerance of plants. *Plant Biotechnology Reports, 10*(4), 185–205.

Becana, M., Moran, J., & IturbeOrmaetxe, I. (1998). Iron.dependent oxygen free radical generation in plants subjected to environmental stress: Toxicity and antioxidant protection. *Plant and Soil, 201*(1), 137–147.

Bhaskara, G. B., Yang, T. H., & Verslues, P. E. (2015). Dynamic proline metabolism: Importance and regulation in water limited environments. *Frontiers in Plant Science, 6*, 484.

Bhatnagar-Mathur, P., Devi, M. J., Reddy, D. S., Lavanya, M., Vadez, V., Serraj, R., Yamaguchi-Shinozaki, K., & Sharma, K. K. (2007). Stress.inducible expression of at DREB1A in transgenic peanut (*Arachis hypogaea* L.) increases transpiration efficiency under water.limiting conditions. *Plant Cell Reports, 26*(12), 2071–2082.

Bhuyan, M. B., Hasanuzzaman, M., Parvin, K., Mohsin, S. M., Al Mahmud, J., Nahar, K., & Fujita, M. (2020). Nitric oxide and hydrogen sulfide: Two intimate collaborators regulating plant defense against abiotic stress. *Plant Growth Regulation, 90*(3), 409–424.

Bita, C., & Gerats, T. (2013). Plant tolerance to high temperature in a changing environment: Scientific fundamentals and production of heat stress.tolerant crops. *Frontiers in Plant Science, 4*, 273.

Bulgari, R., Franzoni, G., & Ferrante, A. (2019). Biostimulants application in horticultural crops under abiotic stress conditions. *Agronomy, 9*(6), 306.

Castiglioni, P., Warner, D., Bensen, R. J., Anstrom, D. C., Harrison, J., Stoecker, M., Abad, M., Kumar, G., Salvador, S., D'Ordine, R., & Navarro, S. (2008). Bacterial

RNA chaperones confer abiotic stress tolerance in plants and improved grain yield in maize under water.limited conditions. *Plant Physiology, 147*(2), 446–455.
Chen, Y., Chen, Z., Kang, J., Kang, D., Gu, H., & Qin, G. (2013). AtMYB14 regulates cold tolerance in Arabidopsis. *Plant Molecular Biology Reporter, 31*(1), 87–97.
Chen, M., Chen, Q., Niu, X., Zhang, R., Lin, H., Xu, C., Wang, X., Wang, G., & Chen, J. (2007). Expression of OsNHX1 gene in maize confers salt tolerance and promotes plant growth in the field. *Plant Soil and Environment, 53*(11), 490.
Chen, J. Q., Meng, X. P., Zhang, Y., Xia, M., & Wang, X. P. (2008). Over.expression of OsDREB genes lead to enhanced drought tolerance in rice. *Biotechnology Letters, 30*(12), 2191–2198.
Chen, M., Wang, Q. Y., Cheng, X. G., Xu, Z. S., Li, L. C., Ye, X. G., Xia, L. Q., & Ma, Y. Z. (2007). GmDREB2, a soybean DRE.binding transcription factor, conferred drought and high.salt tolerance in transgenic plants. *Biochemical and Biophysical Research Communications, 353*(2), 299–305.
Chen, J., Wan, S., Liu, H., Fan, S., Zhang, Y., Wang, W., Xia, M., Yuan, R., Deng, F., & Shen, F. (2016). Overexpression of an Apocynum venetum DEAD.box helicase gene (AvDH1) in cotton confers salinity tolerance and increases yield in a saline field. *Frontiers in Plant Science, 6*, 1227.
Chen, J. B., Yang, J. W., Zhang, Z. Y., Feng, X. F., & Wang, S. M. (2013). Two P5CS genes from common bean exhibiting different tolerance to salt stress in transgenic Arabidopsis. *Journal of Genetics, 92*(3), 461–469.
Coates, J., Mar, K. A., Ojha, N., & Butler, T. M. (2016). The influence of temperature on ozone production under varying NO x conditions—a modelling study. *Atmospheric Chemistry and Physics, 16*(18), 11601–11615.
Daryanto, S., Wang, L., & Jacinthe, P. A. (2016). Global synthesis of drought effects on maize and wheat production. *PloS One, 11*(5), e0156362.
Das, K., & Roychoudhury, A. (2014). Reactive oxygen species (ROS) and response of antioxidants as ROS.scavengers during environmental stress in plants. *Frontiers in Environmental Science, 2*, 53.
Esim, N., & Atici, O. (2014). Nitric oxide improves chilling tolerance of maize by affecting apoplastic antioxidative enzymes in leaves. *Plant Growth Regulation, 72*(1), 29–38.
Fang, Y., & Xiong, L. (2015). General mechanisms of drought response and their application in drought resistance improvement in plants. *Cellular and Molecular Life Sciences, 72*(4), 673–689.
Feng, X., Porporato, A., & RodriguezIturbe, I. (2013). Changes in rainfall seasonality in the tropics. *Nature Climate Change, 3*(9), 811–815.
FlorezSarasa, I., Fernie, A. R., & Gupta, K. J. (2020). Does the alternative respiratory pathway offer protection against the adverse effects resulting from climate change? *Journal of Experimental Botany, 71*(2), 465–469.
Fukao, T., BarreraFigueroa, B. E., Juntawong, P., & PeñaCastro, J. M. (2019). Submergence and waterlogging stress in plants: A review highlighting research opportunities and understudied aspects. *Frontiers in Plant Science, 10*, 340.
Gao, S. Q., Chen, M., Xia, L. Q., Xiu, H. J., Xu, Z. S., Li, L. C., Zhao, C. P., Cheng, X. G., & Ma, Y. Z. (2009). A cotton (Gossypium hirsutum) DRE.binding transcription factor gene, GhDREB, confers enhanced tolerance to drought, high salt, and freezing stresses in transgenic wheat. *Plant Cell Reports, 28*(2), 301–311.
Guo, W., Nazim, H., Liang, Z., & Yang, D. (2016). Magnesium deficiency in plants: An urgent problem. *The Crop Journal, 4*(2), 83–91.
Gupta, D., Pena, L. B., Romero-Puertas, M. C., Hernández, A., Inouhe, M., & Sandalio, L. M. (2017). NADPH oxidases differentially regulate ROS metabolism and nutrient uptake under cadmium toxicity. *Plant, Cell & Environment, 40*(4), 509–526.

Guriansherman, D. (2012). *High and dry: Why genetic engineering is not solving agriculture's drought problem in a thirsty world. High and dry: Why genetic engineering is not solving agriculture's drought problem in a thirsty world.*
Hajiboland, R. (2012). *Effect of micronutrient deficiencies on plants stress responses, Abiotic stress responses in plants* (pp. 283–329). Springer.
Hajyzadeh, M., Turktas, M., Khawar, K. M., & Unver, T. (2015). miR408 overexpression causes increased drought tolerance in chickpea. *Gene, 555*(2), 186–193.
Hameed, A., Bibi, N., Akhter, J., & Iqbal, N. (2011). Differential changes in antioxidants, proteases, and lipid peroxidation in flag leaves of wheat genotypes under different levels of water deficit conditions. *Plant Physiology and Biochemistry, 49*(2), 178–185.
Hasanuzzaman, M., Bhuyan, M., Zulfiqar, F., Raza, A., Mohsin, S. M., Mahmud, J. A., Fujita, M., & Fotopoulos, V. (2020). Reactive oxygen species and antioxidant defense in plants under abiotic stress: Revisiting the crucial role of a universal defense regulator. *Antioxidants, 9*(8), 681.
Hasanuzzaman, M., Fujita, M., Nahar, K., & Biswas, J. K. (2018). *Advances in rice research for abiotic stress tolerance.* Woodhead publishing.
Hossain, M. A., & Uddin, S. N. (2011). Mechanisms of waterlogging tolerance in wheat: Morphological and metabolic adaptations under hypoxia or anoxia. *Australian Journal of Crop Science, 5*(9), 1094–1101.
Hu, H., Dai, M., Yao, J., Xiao, B., Li, X., Zhang, Q., & Xiong, L. (2006). Overexpressing a NAM, ATAF, and CUC (NAC) transcription factor enhances drought resistance and salt tolerance in rice. *Proceedings of the National Academy of Sciences, 103*(35), 12987–12992.
Hussain, H. A., Hussain, S., Khaliq, A., Ashraf, U., Anjum, S. A., Men, S., & Wang, L. (2018). Chilling and drought stresses in crop plants: Implications, cross talk, and potential management opportunities. *Frontiers in Plant Science, 9*, 393.
Ito, Y., Katsura, K., Maruyama, K., Taji, T., Kobayashi, M., Seki, M., Shinozaki, K., & amaguchi-Shinozaki, K. (2006). Functional analysis of rice DREB1/CBF.type transcription factors involved in cold.responsive gene expression in transgenic rice. *Plant and Cell Physiology, 47*(1), 141–153.
Jalil, S. U., Ahmad, I., & Ansari, M. I. (2017). Functional loss of GABA transaminase (GABA.T) expressed early leaf senescence under various stress conditions in Arabidopsis thaliana. *Current Plant Biology, 9*, 11–22.
Jarsjö, J., AnderssonSköld, Y., Fröberg, M., Pietroń, J., Borgström, R., Löv, Å., & Kleja, D. B. (2020). Projecting impacts of climate change on metal mobilization at contaminated sites: Controls by the groundwater level. *Science of the Total Environment, 712*, 135560.
Jiang, Q., Hu, Z., Zhang, H., & Ma, Y. (2014). Overexpression of GmDREB1 improves salt tolerance in transgenic wheat and leaf protein response to high salinity. *The Crop Journal, 2*(2.3), 120–131.
Jung, J. Y., Shin, R., & Schachtman, D. P. (2009). Ethylene mediates response and tolerance to potassium deprivation in Arabidopsis. *The Plant Cell, 21*(2), 607–621.
Kim, J. Y., Song, J. T., & Seo, H. S. (2017). Post-translational modifications of Arabidopsis E3 SUMO ligase at SIZ 1 are controlled by environmental conditions. *FEBS Open Bio, 7*(10), 1622–1634.
Kudo, M., Kidokoro, S., Yoshida, T., Mizoi, J., Todaka, D., Fernie, A. R., Shinozaki, K., & Yamaguchi-Shinozaki, K. (2017). Double overexpression of DREB and PIF transcription factors improves drought stress tolerance and cell elongation in transgenic plants. *Plant Biotechnology Journal, 15*(4), 458–471.
Kumari, S., Agrawal, M., & Singh, A. (2015). Effects of ambient and elevated CO_2 and ozone on physiological characteristics, antioxidative defense system and metabolites of potato in relation to ozone flux. *Environmental and Experimental Botany, 109*, 276–287.

Kumar, V., Khare, T., Sharma, M., & Wani, S. H. (2017). *ROS.induced signaling and gene expression in crops under salinity stress, reactive oxygen species and antioxidant systems in plants: Role and regulation under abiotic stress* (pp. 159–184). Springer.

Liao, Y. D., Lin, K. H., Chen, C. C., & Chiang, C. M. (2016). Oryza sativa protein phosphatase 1a (OsPP1a) involved in salt stress tolerance in transgenic rice. *Molecular Breeding, 36*(3), 22.

LiseideSá, M. E., ARRAES, F., Brito, G. G., Beneventi, M. A., LourençoTessutti, I., Basso, A. M., Amorim, R. M., Silva, M. C., Faheem, M., Oliveira, N. G., & Mizoi, J. (2017). *AtDREB2A.CA influences root architecture and increases drought tolerance in transgenic cotton.* Embrapa Recursos Genéticos e Biotecnologia.Artigo em periódico indexado (ALICE).

Liu, X., Hong, L., Li, X. Y., Yao, Y., Hu, B., & Li, L. (2011). Improved drought and salt tolerance in transgenic Arabidopsis overexpressing a NAC transcriptional factor from *Arachis hypogaea. Bioscience, Biotechnology, and Biochemistry, 75*(3), 443–450.

Liu, T., Xu, J., Li, J., & Hu, X. (2019). NO is involved in JA.and H2O2.mediated ALA.induced oxidative stress tolerance at low temperatures in tomato. *Environmental and Experimental Botany, 161*, 334–343.

Lu, N., Wei, B., Sun, Y., Liu, X., Chen, S., Zhang, W., Zhang, Y., & Li, Y. (2014). Field supervisory test of DREB.transgenic populus: Salt tolerance, long.term gene stability and horizontal gene transfer. *Forests, 5*(5), 1106–1121.

Mao, X., Jia, D., Li, A., Zhang, H., Tian, S., Zhang, X., Jia, J., & Jing, R. (2011). Transgenic expression of TaMYB2A confers enhanced tolerance to multiple abiotic stresses in Arabidopsis. *Functional & Integrative Genomics, 11*(3), 445–465.

Ma, J. T., Yin, C. C., Zhou, M. L., Wang, Z. L., & Wu, Y. M. (2015). A novel DREB transcription factor from Halimodendron halodendron leads to enhance drought and salt tolerance in Arabidopsis. *Biologia Plantarum, 59*(1), 74–82.

Megha, S., Basu, U., & Kav, N. N. (2018). Regulation of low temperature stress in plants by microRNAs. *Plant, Cell & Environment, 41*(1), 1–15.

Montilla-Bascón, G., Rubiales, D., Hebelstrup, K. H., Mandon, J., Harren, F. J., Cristescu, S. M., Mur, L. A., & Prats, E. (2017). Reduced nitric oxide levels during drought stress promote drought tolerance in barley and is associated with elevated polyamine biosynthesis. *Scientific Reports, 7*(1), 1–15.

Nakashima, K., Tran, L. S. P., Van Nguyen, D., Fujita, M., Maruyama, K., Todaka, D., Ito, Y., Hayashi, N., Shinozaki, K., & Yamaguchi-Shinozaki, K. (2007). Functional analysis of a NAC-type transcription factor OsNAC6 involved in abiotic and biotic stress-responsive gene expression in rice. *The Plant Journal, 51*(4), 617–630.

Noyes, P. D., McElwee, M. K., Miller, H. D., Clark, B. W., Van Tiem, L. A., Walcott, K. C., & Erwin, K. N. (2009). The toxicology of climate change: Environmental contaminants in a warming world. *Environment International, 35*(6), 971–986.

Oh, S. J., Song, S. I., Kim, Y. S., Jang, H. J., Kim, S. Y., Kim, M., Kim, Y. K., Nahm, B. H., & Kim, J. K (2005). Arabidopsis CBF3/DREB1A and ABF3 in transgenic rice increased tolerance to abiotic stress without stunting growth. *Plant Physiology, 138*(1), 341–351.

Pellegrineschi, A., Reynolds, M., Pacheco, M., Brito, R. M., Almeraya, R., YamaguchiShinozaki, K., & Hoisington, D. (2004). Stress.induced expression in wheat of the Arabidopsis thaliana DREB1A gene delays water stress symptoms under greenhouse conditions. *Genome, 47*(3), 493–500.

Pereira, A. (2016). Plant abiotic stress challenges from the changing environment. *Frontiers in Plant Science, 7*, 1123.

Qin, T., Zhao, H., Cui, P., Albesher, N., & Xiong, L. (2017). A nucleus.localized long non.-coding RNA enhances drought and salt stress tolerance. *Plant Physiology, 175*(3), 1321–1336.

Rai, K. K., Rai, N., & Rai, S. P. (2018). Salicylic acid and nitric oxide alleviate high temperature induced oxidative damage in Lablab purpureus L plants by regulating bio.physical processes and DNA methylation. *Plant Physiology and Biochemistry, 128*, 72–88.

Ramegowda, V., SenthilKumar, M., Nataraja, K. N., Reddy, M. K., Mysore, K. S., & Udayakumar, M. (2012). Expression of a finger millet transcription factor, EcNAC1, in tobacco confers abiotic stress.tolerance. *PloS One, 7*(7), e40397.

Rane, J., Singh, A. K., Kumar, M., Boraiah, K. M., Meena, K. K., Pradhan, A., & Prasad, P. (2021). The adaptation and tolerance of major cereals and legumes to important abiotic stresses. *International Journal of Molecular Sciences, 22*(23), 12970.

Raza, A., Razzaq, A., Mehmood, S. S., Zou, X., Zhang, X., Lv, Y., & Xu, J. (2019). Impact of climate change on crops adaptation and strategies to tackle its outcome: A review. *Plants, 8*(2), 34.

Roychoudhury, A., & Chakraborty, M. (2013). Biochemical and molecular basis of varietal difference in plant salt tolerance. *Annual Research & Review in Biology, 422*, 454.

Roychoudhury, A., & Paul, A. (2012). Abscisic acid.inducible genes during salinity and drought stress. *Advances in Medicine and Biology, 51*, 1–78.

Sachdev, S., & Singh, R. (2018a). Root colonization: Imperative mechanism for efficient plant protection and growth. *MOJ Ecology Environmental Science, 3*(4), 240–242.

Sachdev, S., & Singh, R. P. (2018b). Isolation, characterisation and screening of native microbial isolates for biocontrol of fungal pathogens of tomato. *Climate Change and Environmental Sustainability, 6*(1), 46–58.

Saijo, Y., & Loo, E. P.i. (2020). Plant immunity in signal integration between biotic and abiotic stress responses. *New Phytologist, 225*(1), 87–104.

Saleem, M. H., Fahad, S., Khan, S. U., Din, M., Ullah, A., Sabagh, A. E., Hossain, A., Llanes, A., & Liu, L. (2020). Copper.induced oxidative stress, initiation of antioxidants and phytoremediation potential of flax (Linum usitatissimum L.) seedlings grown under the mixing of two different soils of China. *Environmental Science and Pollution Research, 27*(5), 5211–5221.

Sasidharan, R., Hartman, S., Liu, Z., Martopawiro, S., Sajeev, N., van Veen, H., Yeung, E., & Voesenek, L. A. (2018). Signal dynamics and interactions during flooding stress. *Plant Physiology, 176*(2), 1106–1117.

Schulz, P., Neukermans, J., Van Der Kelen, K., Mühlenbock, P., Van Breusegem, F., Noctor, G., Teige, M., Metzlaff, M., & Hannah, M. A. (2012). Chemical PARP inhibition enhances growth of Arabidopsis and reduces anthocyanin accumulation and the activation of stress protective mechanisms. *PloS One, 7*(5), e37287.

Schutzendubel, A., & Polle, A. (2002). Plant responses to abiotic stresses: Heavy metal-induced oxidative stress and protection by mycorrhization. *Journal of Experimental Botany, 53*(372), 1351–1365.

Sharma, A., Yuan, H., Kumar, V., Ramakrishnan, M., Kohli, S. K., Kaur, R., Thukral, A. K., Bhardwaj, R., & Zheng, B. (2019). Castasterone attenuates insecticide induced phytotoxicity in mustard. *Ecotoxicology and Environmental Safety, 179*, 50–61.

Shiferaw, B., Tesfaye, K., Kassie, M., Abate, T., Prasanna, B., & Menkir, A. (2014). Managing vulnerability to drought and enhancing livelihood resilience in sub.Saharan Africa: Technological, institutional and policy options. *Weather and Climate Extremes, 3*, 67–79.

Shin, R., Berg, R. H., & Schachtman, D. P. (2005). Reactive oxygen species and root hairs in Arabidopsis root response to nitrogen, phosphorus and potassium deficiency. *Plant and Cell Physiology, 46*(8), 1350–1357.

Shivakumara, T. N., Sreevathsa, R., Dash, P. K., Sheshshayee, M., Papolu, P. K., Rao, U., Tuteja, N., & UdayaKumar, M. (2017). Overexpression of Pea DNA Helicase 45 (PDH45) imparts tolerance to multiple abiotic stresses in chili (Capsicum annuum L.). *Scientific Reports, 7*(1), 1–12.

Singh, S., Eapen, S., & D'souza, S. (2006). Cadmium accumulation and its influence on lipid peroxidation and antioxidative system in an aquatic plant, Bacopa monnieri L. *Chemosphere, 62*(2), 233–246.

Steffens, B. (2014). The role of ethylene and ROS in salinity, heavy metal, and flooding responses in rice. *Frontiers in Plant Science, 5*, 685.

Sun, X., Deng, Y., Liang, L., Jia, X., Xiao, Z., & Su, J. (2017). Overexpression of a PIP1 gene from Salicornia bigelovii in tobacco plants improves their drought tolerance. *Journal of the American Society for Horticultural Science, 142*(4), 235–245.

Sun, H., Li, L., Lou, Y., Zhao, H., Yang, Y., Wang, S., & Gao, Z. (2017). The bamboo aquaporin gene PeTIP4; 1–1 confers drought and salinity tolerance in transgenic Arabidopsis. *Plant Cell Reports, 36*(4), 597–609.

Suo, H., Ma, Q., Ye, K., Yang, C., Tang, Y., Hao, J., Zhang, Z. J., et al. (2012). Overexpression of AtDREB1A causes a severe dwarf phenotype by decreasing endogenous gibberellin levels in soybean [*Glycine max* (L.) Merr.]. *PLoS One,* 7, e45568. https://doi.org/10.1371/journal.pone.0045568

Suprasanna, P., & Ghag, S. (2019). Plant tolerance to environmental stress: Translating research from lab to land. *Molecular Plant Abiotic Stress: Biology and Biotechnology, 1*, 28.

Trinh, N. N., Huang, T. L., Chi, W. C., Fu, S. F., Chen, C. C., & Huang, H. J. (2014). Chromium stress response effect on signal transduction and expression of signaling genes in rice. *Physiologia Plantarum, 150*(2), 205–224.

Tuteja, N., Banu, M. S. A., Huda, K. M. K., Gill, S. S., Jain, P., Pham, X. H., & Tuteja, R. (2014). Pea p68, a DEAD.box helicase, provides salinity stress tolerance in transgenic tobacco by reducing oxidative stress and improving photosynthesis machinery. *PloS One, 9*(5), e98287.

Wang, Y., Beaith, M., Chalifoux, M., Ying, J., Uchacz, T., Sarvas, C., Griffiths, R., Kuzma, M., Wan, G., & Huang, Y. (2009). Shoot.specific down.regulation of protein farnesyltransferase (α.subunit) for yield protection against drought in canola. *Molecular Plant, 2*(1), 191–200.

Wang, Q., Guan, Y., Wu, Y., Chen, H., Chen, F., & Chu, C. (2008). Overexpression of a rice OsDREB1F gene increases salt, drought, and low temperature tolerance in both Arabidopsis and rice. *Plant Molecular Biology, 67*(6), 589–602.

Wang, Y., Li, T., John, S. J., Chen, M., Chang, J., Yang, G., & He, G. (2018). A CBL.interacting protein kinase TaCIPK27 confers drought tolerance and exogenous ABA sensitivity in transgenic Arabidopsis. *Plant Physiology and Biochemistry, 123*, 103–113.

Wang, L., Li, Q., Lei, Q., Feng, C., Gao, Y., Zheng, X., Zhao, Y., Wang, Z., & Kong, G. (2015). MzPIP2; 1: An aquaporin involved in radial water movement in both water uptake and transportation, altered the drought and salt tolerance of transgenic Arabidopsis. *PLoS One, 10*(11), e0142446.

Xiao, B. Z., Chen, X., Xiang, C. B., Tang, N., Zhang, Q. F., & Xiong, L. Z. (2009). Evaluation of seven function.known candidate genes for their effects on improving drought resistance of transgenic rice under field conditions. *Molecular Plant, 2*(1), 73–83.

Xu, X. B., Pan, Y. Y., Wang, C. L., Ying, Q. C., Song, H. M., & Wang, H. Z. (2014). Overexpression of DnWRKY11 enhanced salt and drought stress tolerance of transgenic tobacco. *Biologia, 69*(8), 994–1000.

Yamauchi, T., Abe, F., Tsutsumi, N., & Nakazono, M. (2019). Root cortex provides a venue for gas.space formation and is essential for plant adaptation to waterlogging. *Frontiers in Plant Science, 10*, 259.

Yang, S., Vanderbeld, B., Wan, J., & Huang, Y. (2010). Narrowing down the targets: Towards successful genetic engineering of drought.tolerant crops. *Molecular Plant, 3*(3), 469–490.

Ye, Q., Yang, X., Dai, S., Chen, G., Li, Y., & Zhang, C. (2015). Effects of climate change on suitable rice cropping areas, cropping systems and crop water requirements in southern China. *Agricultural Water Management, 159*, 35–44.

Yu, T. F., Xu, Z. S., Guo, J. K., Wang, Y. X., Abernathy, B., Fu, J. D., Chen, X., Zhou, Y. B., Chen, M., Ye, X. G., & Ma, Y. Z. (2017). Improved drought tolerance in wheat plants overexpressing a synthetic bacterial cold shock protein gene SeCspA. *Scientific Reports,* 7(1), 1–14.

Zhang, Y., Jiang, W., Yu, H., & Yang, X. (2012). Exogenous abscisic acid alleviates low temperature.induced oxidative damage inseedlings of Cucumis sativus. L. *Transactions of the Chinese Society of Agricultural Engineering, 28*(1), 221–228.

Zhang, B., & Wang, Q. (2016). MicroRNA, a new target for engineering new crop cultivars. *Bioengineered,* 7(1), 7–10.

Zheng, W., Wang, S., Tan, K., & Lei, Y. (2020). Nitrate accumulation and leaching potential is controlled by land.use and extreme precipitation in a headwater catchment in the North China Plain. *Science of the Total Environment,* 707, 136168.

Zhou, M., Li, D., Li, Z., Hu, Q., Yang, C., Zhu, L., & Luo, H. (2013). Constitutive expression of a miR319 gene alters plant development and enhances salt and drought tolerance in transgenic creeping bentgrass. *Plant Physiology, 161*(3), 1375–1391.

Index

Note: 'Page numbers followed by "*f*" indicate figures and "*t*" indicate tables.'

Q

R

S

Kolibri 360 GmbH
Ferdinand-Jühlke-Straße 7
99095 Erfurt, Deutschland
produktsicherheit@kolibri360.de